# HPLC AND UHPLC FOR PRACTICING SCIENTISTS

# HPLC AND UHPLC FOR PRACTICING SCIENTISTS

Second Edition

Michael W. Dong

WILEY

This edition first published 2019


*Registered Office*
John Wiley & Sons, Inc., 111 River Street, Hoboken, NJ 07030, USA

*Editorial Office*
111 River Street, Hoboken, NJ 07030, USA

For details of our global editorial offices, customer services, and more information about Wiley products visit us at www.wiley.com.

Wiley also publishes its books in a variety of electronic formats and by print-on-demand. Some content that appears in standard print versions of this book may not be available in other formats.



*Library of Congress Cataloging-in-Publication Data*

Names: Dong, M. W., author.
Title: HPLC and UHPLC for practicing scientists, second edition / Michael W. Dong.
Other titles: Modern HPLC for practicing scientists
Description: Second edition. | Hoboken, NJ : Wiley, 2019. | Originally published: Modern HPLC for practicing scientists. 2006. | Includes bibliographical references and index. |
Identifiers: LCCN 2019002892 (print) | LCCN 2019003814 (ebook) | ISBN 9781119313793 (Adobe PDF) | ISBN 9781119313779 (ePub) | ISBN 9781119313762 (paperback)
Subjects: LCSH: High performance liquid chromatography. | Drugs–Analysis. | BISAC: SCIENCE / Chemistry / Analytic.
Classification: LCC RS189.5.H54 (ebook) | LCC RS189.5.H54 D66 2019 (print) | DDC 615.1/901–dc23
LC record available at https://lccn.loc.gov/2019002892

Cover Design: Wiley
Cover Images: Courtesy of Michael W. Dong; Background: © Wasant/Shutterstock

Set in 10/12pt TimesTenLTStd by SPi Global, Chennai, India

MIX
Paper from responsible sources
FSC www.fsc.org FSC® C013604

# CONTENTS

# AUTHOR'S BIOGRAPHY

Dr. Michael W. Dong is a principal consultant in MWD Consulting focusing on consulting and training services on HPLC, pharmaceutical analysis, and drug quality. He was formerly Senior Scientist in Analytical Chemistry and Quality Control at Genentech, Research Director at Synomics Pharma, Research Fellow at Purdue Pharma, Senior Staff Scientist at Applied Biosystems/Perkin-Elmer, section head at Celanese Research Company, and postdoctoral research fellow at Naylor-Dana Institute for Disease Prevention. He holds a PhD in Analytical Chemistry from the City University of New York and a certificate in Biotechnology from University of California Santa Cruz. He has 120+ publications including a bestselling book on chromatography (Modern HPLC for Practicing Scientists, Wiley). He is an advisory board member of LCGC magazine, American Pharmaceutical Review, Chinese American Chromatography Association, and Connecticut Separation Science Council. He has been a columnist of "Perspectives in Modern HPLC" for LCGC North America since 2013. Michael was born in Shanghai and raised in Hong Kong. He is multilingual, a former Eagle Scout, and a Toastmaster.

# BIOGRAPHIES OF CONTRIBUTORS

**Christine Gu**

Dr. Christine Gu is currently a Senior Scientist III in the Department of Process Sciences at AbbVie Stemcentrx LLC. She oversees the analytical and DMPK support of Small-Molecule Discovery and Development by using cutting-edge HPLC and mass spectrometric technologies. Before joining AbbVie, she worked as a Scientist in the Department of Small Molecule Pharmaceutical Science at Genentech and as a Senior Application Scientist at Thermo Fisher Scientific. She holds a PhD Degree in Toxicology at the University of California at Riverside.

**Jennifer Rea**

Dr. Jennifer Rea is a Senior Scientist in the Protein Analytical Chemistry Department at Genentech, a Roche company, and has been working in analytical chemistry since 2009. Dr. Rea leads several teams at Genentech, supporting clinical projects in both early-stage and late-stage clinical development. Dr. Rea also leads a global chromatography expert team at Roche that develops and validates analytical methods for global QC and R&D use. Jennifer received her BS from the University of California, Berkeley, in 2003 and PhD from Northwestern University in 2009. She has authored 15 publications in the fields of bioanalytical chemistry, biotherapeutics, and gene delivery. Dr. Rea's research interests include developing novel chromatography techniques for protein therapeutics and working with global partners on analytical strategies for next-generation biotherapeutics.

**Taylor Zhang**

Dr. Taylor Zhang is a director at Bioanalytical Department at Juno Therapeutics, a Celgene company, leading a group that is responsible for the bioanalytical assays development for the CAR-T therapies. He was formerly a Principal Scientist and group leader in Genentech and has been in Protein Analytical Chemistry Department since 2006, supporting multiple commercial and clinical stages of biologic process development. He received his PhD in Analytical Chemistry from Iowa State University in 2000. He has authored 30+ publications in the field of bioanalytical chemistry. Dr. Zhang's research interests involve utilizing different bioanalytical techniques for cellular therapeutics and protein therapeutics and developing methods for their quality control, process, and product characterization. He is an advisory board member of LCGC magazine.

# PREFACE

It has been 12 years since the publication of the first edition of "Modern HPLC for Practicing Scientists," and much has changed in HPLC, the world, and my personal life. I wrote the first edition in just 10 months when I worked at Purdue Pharma as a research fellow. I left Purdue and worked at Synomics Pharma at Massachusetts when the book was published in 2006. I relocated shortly afterward to California to work for Genentech as a Senior Scientist for the next eight years. After moving back to Connecticut in 2015, I was contacted by Bob Esposito of Wiley about updating the first edition.

I was hesitant at first to take on this big time commitment, though the good reasons for this project eventually prevailed. First, HPLC has progressed much, and most current HPLC books are in dire need for updates. Second, the format of the first edition paperback book with many figures and tables was well-received with 8000 copies sold; so the task for a second edition seemed somewhat less formidable. Third, the timing was right with my new role as a self-employed consultant with more time flexibility and capacity.

I signed the contract with Wiley in April of 2016 and started writing that July. I made steady progress in the next 18 months, completing a rough chapter draft every one or two months and allowing time for national meetings, personal trips, and clients' visits. I took on yoga and swimming so having a writing project provided a good balance between aerobic and intellectual activities. I recruited contributors for two new chapters on LC/MS and biopharmaceutical analysis. By January of 2018, I had reasonable drafts for all my 11 chapters.

Next came the review process. It has not been easy finding good and responsive reviewers. While I had 20+ friends and colleagues whom I enlisted in the past for my various publications, it would be an imposition to ask them again for this projects. Luckily, two remedies came for the rescue.

First was a software called "Grammarly," which flagged most of my grammatical issues, reducing the editing time for my reviewers. Second, I decided to solicit volunteers from my LinkedIn network (a social media website for professionals), an unconventional move that reaped huge dividends. The review process lasted for four months and significantly improved the quality and clarity of each chapter.

## WHAT'S NEW IN THE SECOND EDITION?

The title of the second edition was changed to "HPLC and UHPLC for Practicing Scientists" to reflect the eminence of UHPLC as the new equipment platform. The scope of the second edition remained an all-inclusive but concise overview of HPLC presented at an intermediate level. The format was similar to the first edition with an abundance of figures and tables, supported by discussion, case studies, and key references.

The number of chapters increased from 11 to 13, with three new chapters on UHPLC, LC/MS, and biopharmaceutical analysis. Five chapters (1, 2, 5, 6, and 13) were updates while the remaining five were substantial rewrites. The page number increased from 300 to ~400, reflecting an expanded scope with longer discussions. A quiz section was added to each chapter as a teaching aid as a textbook.

## MY DECISION TO BE A SOLE AUTHOR

I pondered much about the decision to write the second edition as a sole author. Edited books with multiple authors have many pros and cons such as lack of consistency, style, and a potential loss of editorial control of the content.

I started my career in HPLC as a graduate student in chromatography and continued in the chemical industry as a research chemist, an applications scientist and a marketing specialist for an instrument company, and finally as an analytical and quality control chemist/manager for two pharmaceutical companies and a contract research organization. HPLC has always been my career anchor. I taught short courses in HPLC at national meetings for 20 years and published 120+ articles. These hands-on experiences in instrumental hardware, data systems, marketing, pharmaceutical analysis, methods development, quality control, and regulatory filings have given me a wide perspective on HPLC as a subject matter expert. As a sole author, I believe that I am in a better position to deliver a concise and comprehensive overview of modern HPLC for practicing scientists.

*Norwalk, Connecticut*

Michael W. Dong
October 2018

# FOREWORD

The analytical high-performance liquid chromatography (HPLC) market represents about 9% of the overall demand for laboratory analytical instrumentation. The technology continues to evolve with expanding applications not only from life science and industrial sectors, but also from applied markets such as environmental, food, and clinical/diagnostics.

HPLC was transformed by the introduction of ultra-high-pressure liquid chromatography (UHPLC) systems, made popular by Waters Corporation in 2004 with its ACQUITY UPLC system. At the time of the introduction, HPLC was somewhat of a commodity instrument with more than a dozen HPLC instrument manufacturers. The market was maturing with some vendor consolidation, yet continuing to increase 4–6% annually. Users looked to increase lab productivity and invested in automated solutions.

UHPLC technology revived the HPLC market and challenged chromatographers to rethink chromatography. Additionally, acetonitrile prices were increasing, the global economy was faltering in the financial crisis of the late 2000s, and as a consequence, UHPLC technology became increasingly favorable.

UHPLC had become mainstream, forcing instrument manufacturers to make business decisions to compete at this higher level of performance. As more users migrated to UHPLC, there was a bit of a shakeout from instrument vendors. Users preferred mainstream HPLC instrument companies (Agilent, Shimadzu, Thermo Fisher Scientific, and Waters), which collectively represent more than 85% of the total market for HPLC instruments today.

Many other factors shaped the market for HPLC. Column chemistries, smaller particle sizes, sample prep, automation, and software had played key roles in re-energizing the market. Chromatographers had a shiny new tool in the lab toolbox, and scientists were finding new ways to tap into its potentials. Manufactures introduced new products to keep up with the speed and efficiency of UHPLC technology that ultimately increased overall laboratory productivity.

Overall, the market for HPLC is forecasted to be robust with an estimated installed base of about 300 000 active units. The drivers for today's market continue to be fueled by pharmaceutical and biotechnology R&D. Innovation has been centered on the use of LC/MS instrumentation, particularly high-resolution mass spectrometry. Proteomics and metabolomics/lipidomics have been the growth engine of the $4.7 billion analytical HPLC market (2017). The use of HPLC in clinical diagnostics is another area that is becoming more widely adopted with Vitamin D tests, newborn screening, and pain management fueling growth. Other key applications buoying the applied market include environmental testing for pesticides, cannabis potency, and food adulteration.

While North America and Europe account for the lion's share of the market, the expansion in Asia, including China, India, and South Korea, as well as other Southeast Asian countries remains strong. Currently, the world economic situation is somewhat in flux due to talks of trade wars and tariffs. Such issues are expected to have a minimal impact on the overall market, mainly due to a small level of cautious spending from chemicals and other industrial sectors. However, overall, the near-term future for analytical HPLC instruments is strong and projected to increase 6–8%.

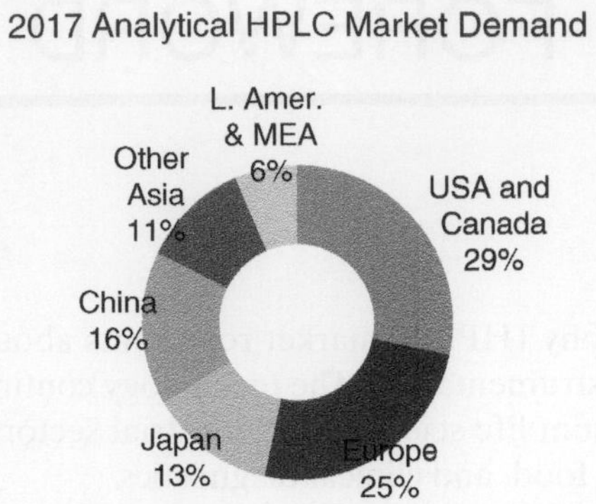


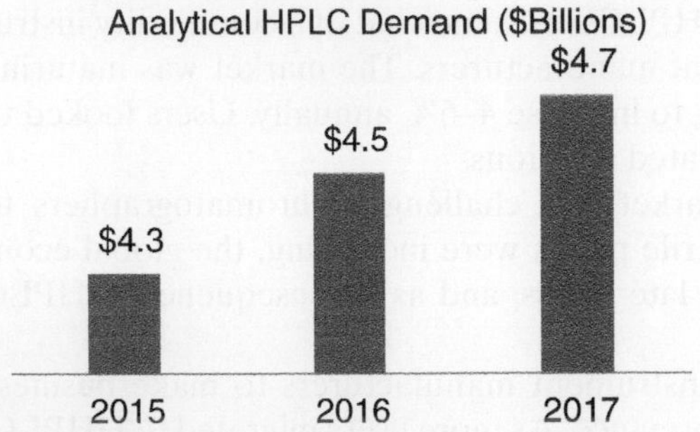


*Glenn Cudiamat* is a market research expert and has been covering chromatography and other the analytical instrumentation since the 1990s. Market size and growth estimates were adopted from ***TDA's Industry Data***, a database of technology market profiles and benchmarks, as well as the ***2018 Instrument Industry Outlook*** report from Top-Down Analytics (TDA).

August 2018

# ACKNOWLEDGMENTS

The author offers sincere appreciation to the 50+ reviewers listed below.

| | Chapter | Reviewer |
|---|---|---|
| 1 | Introduction | James Stuart of University of Connecticut, Diane Diehl of Waters |
| 2 | Basic Terms and Concepts | J. Stuart, David Locke of City University of New York, Greg Slack of PharmAssist, R. Ornaf of Vertex, Brian Holder of Merck |
| 3 | HPLC Columns and Trends | Thomas Waeghe of MacMod, Ron Majors of ChromPrep, Tom Walter of Waters, D. Locke, Alan McKeown of ACT, M. Farooq Wahab of University of Texas Arlington |
| 4 | HPLC Instrumentation and Trends | Wilhad Reuter of PerkinElmer, Yi He of John Jay College, Michael Heidorn of Thermo Fisher Scientific, Yves Ley of Nestle Skin Health, J. Stuart, Christian Skonberg of Lundbeck, Anthony Provatas of University of Connecticut. David Schiessel of Orange County Water District |
| 5 | UHPLC | James Jorgenson of University of North Carolina at Chapter Hill, J. Stuart, Pankaj Aggarwal of Pfizer, Naijun Wu of Celgene, Chris Vanselow of Thermo Fisher Scientific, Szabolcs Fekete of University of Geneva, Davy Guillarme of University of Geneva, Justin Shearer of GSK |
| 6 | LC/MS | Perry Wang of US FDA, Calin Znamirovschi of Sancilio & Co., Tom Buchanan of Thermo Fisher Scientific, Hardik Shah of Massachusetts General Hospital |
| 7 | HPLC/UHPLC Operation | Matt Mullaney |
| 8 | Maintenance and Troubleshooting | M. Mullaney, Dwight Stoll of Gustavus Adolphus College, Mangesh Chatre, Giacomo Chiti of Manetti & Roberts |
| 9 | Pharmaceutical Analysis | David Van Meter of Eurofins EAG Laboratories, Oscar Liu of BeyondSpring, G. Chiti, Harika Vemula of Merck KGaA, Shashank Gorityala of Covance, Dev Kant Shandilya of Eisai Pharmaceuticals, M. Mullaney, |

| | Chapter | Reviewer |
|---|---|---|
| 10 | HPLC Method Development | A. McKeown of ACT, Dawen Kou of Genentech, Tao Chen of Genentech, Imad A. Haidar Ahmad of Merck, B. Holder, Richard Goodin of OMI Industries |
| 11 | System Qualifications and Method Validation | Kim Huynh of Pharmalytik, Melody Jones of Recro Pharma, Kate Evans of Longboard Consulting, Arindam Roy of Merck, G. Chiti, Patrick Kenny of Thermo Fisher Scientific, Alice Krumenaker of TW Metals LLC |
| 12 | Biopharmaceutical Analysis | Reed Harris of Genentech, J. Shearer, Phil Humphrey of GSK, Ira Krull of Northeastern University |
| 13 | HPLC Applications | M. Mullaney, G. Slack of PharmAssist, Delia A. Serna Guerrero of Analitek, Shilpi Chopra or Merial, Cassy Chong of Synthomer |

I am particularly in-debt to the reviewers of multiple chapters: Prof. David Locke (my mentor), Prof. James Stuart, Dr. Greg Slack, Dr. Alan McKeown, Dr. Justin Shearer, Matt Mullaney, Giacomo Chiti, and Brian Holder.

I also thank my contributors Christine Gu of Abbvie, Jennifer Rea of Genentech, and Taylor Zhang of Juno Therapeutics for the delivery of two fine chapters on LC/MS and biopharmaceutical analysis and Glenn Cudiamat of Top-Down Analytics for a brief overview of the HPLC market in the Preface.

*Norwalk, Connecticut*

Michael W. Dong
October 2018

# 1

# INTRODUCTION

## 1.1 INTRODUCTION

### 1.1.1 Scope

High-performance liquid chromatography (HPLC) is a versatile analytical (separation) technique widely used for the analysis of pharmaceuticals, biomolecules, polymers, and many organic and ionic compounds. There is no shortage of excellent books on chromatography [1–3] and on HPLC [4–11], though many are outdated and others tend to focus more on academic theories or specialized topics. This book strives to be a concise and an all-inclusive text that "capsulizes" the essence of HPLC fundamentals, applications, and developments. It describes the fundamental theories and terminologies for the novice and reviews relevant concepts, best practices, and modern trends for the experienced practitioner. While broad in scope, this book focuses on reversed-phase HPLC (the most common separation mode) and pharmaceutical applications (the most significant user segment). Information is presented straightforwardly and is illustrated with an abundance of figures, chromatograms, tables, and case studies, supported by selected key references or web resources.

Most importantly, this book was written as an updated reference guide for busy laboratory analysts and researchers. Topics covered include HPLC operation, method development, maintenance/troubleshooting, and the regulatory aspects of pharmaceutical analysis. This book can serve as a supplementary text for students pursuing a career in analytical chemistry and pharmaceutical science. A reader with a science degree and a basic understanding of chemistry is assumed. This second edition continues the same theme as the first edition [4] with updates on all chapters plus three new chapters on ultra-high-pressure liquid chromatography (UHPLC), liquid chromatography–mass spectrometry (LC/MS), and the analysis of recombinant biologics (biopharmaceuticals). A quiz section at the end of each chapter serves as a teaching/evaluation aid.

This book offers the following benefits:

- A broad scope overview of fundamental principles, instrumentation, columns, and applications.

*HPLC and UHPLC for Practicing Scientists*, Second Edition. Michael W. Dong.

- A concise review of concepts and trends of modern HPLC.
- An update of best practices in HPLC operation, method development, maintenance, troubleshooting, and regulatory aspects in analytical testing.
- New standalone overview chapters on UHPLC, LC/MS, and analysis of recombinant biologics.

### 1.1.2 What Is HPLC?

Liquid chromatography (LC) is a physical separation technique conducted between two phases – a solid phase and a liquid phase. A sample is separated into its constituent components (or analytes) by distributing (via partitioning, adsorption, or other interactions) between the mobile phase (a flowing liquid) and a solid stationary phase (sorbents packed inside a column). For example, the flowing liquid can be an organic solvent such as hexane and the stationary phase can be the porous silica particles packed into a column. HPLC is a modern form of LC that uses small-particle columns through which the mobile phase is pumped at high pressure.

Figure 1.1a is a schematic of the chromatographic process, where a mixture of components A and B are separated into two distinct bands as they migrate down the column filled with packing (stationary phase). Figure 1.1b is a microscopic representation of the dynamic partitioning process of the analytes between the flowing liquid and the stationary phase attached to a spherical packing particle. Note that the movement of component B is retarded in the column because each B molecule has a stronger affinity for the stationary phase than the A

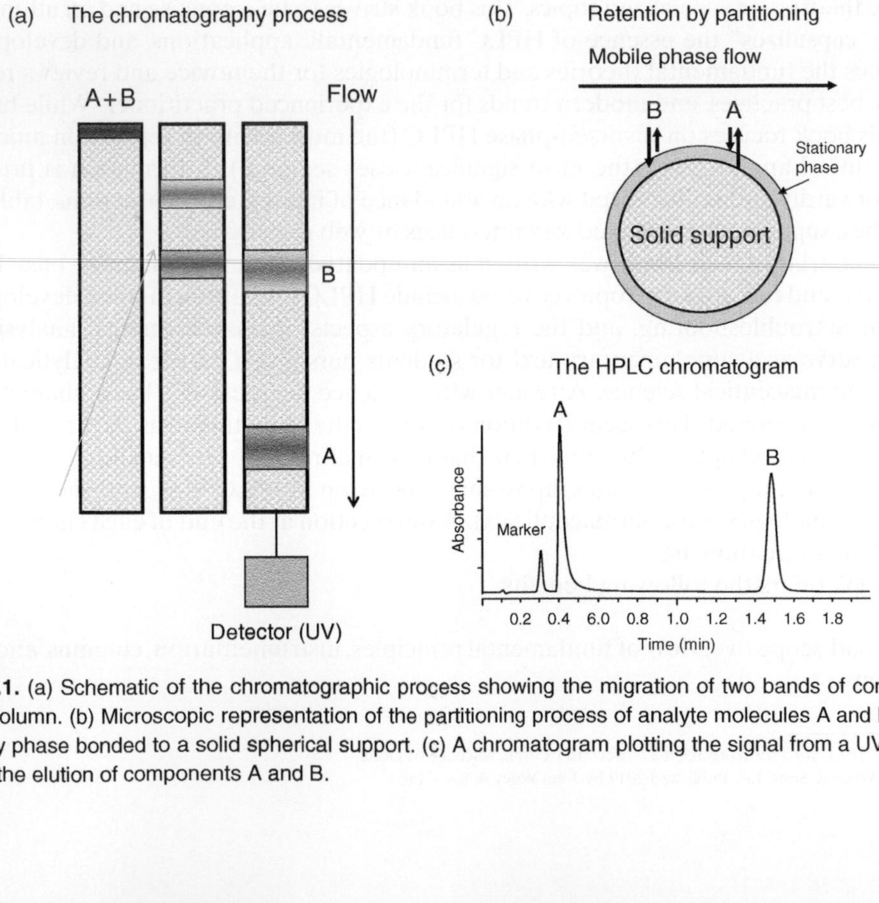


**Figure 1.1.** (a) Schematic of the chromatographic process showing the migration of two bands of components down a column. (b) Microscopic representation of the partitioning process of analyte molecules A and B into the stationary phase bonded to a solid spherical support. (c) A chromatogram plotting the signal from a UV detector displays the elution of components A and B.

molecule. An inline detector monitors the concentration of each separated component band in the effluent and generates a signal trace called the "chromatogram," shown in Figure 1.1c.

### 1.1.3 A Brief History

The term *chromatography* meaning "color writing" was first used by Mikhail Tsvet, a Russian botanist who separated plant pigments on chalk ($CaCO_3$) packed in glass columns in 1903. Since the 1930s, chemists have used gravity-fed silica columns to purify organic materials and ion-exchange resin columns to separate ionic compounds and radionuclides. The invention of gas chromatography (GC) by the British biochemists A. J. P. Martin and R. L. M. Synge in 1952 and its successful applications provided the theoretical foundation and the incentive for the development of LC. In the late 1960s, LC turned "highperformance" with the use of small-particle columns that required high-pressure pumps. The first generation of HPLCs was developed by researchers in the 1960s, including Joseph Huber in Europe and Csaba Horváth and Jack Kirkland in the United States. Commercial development of inline detectors and reliable injectors allowed HPLC to become a sensitive and quantitative technique leading to an explosive growth of applications [4, 5]. In the 1980s, the versatility and precision of HPLC rendered it virtually indispensable in pharmaceutical and many diverse industries. The annual worldwide sales of HPLC systems and accessories were about four billion US$ in 2016 [12]. (http://www.marketsandmarkets.com/PressReleases/chromatography-instrumentation.asp) Today, HPLC continues to evolve rapidly toward higher speed, efficiency, and sensitivity, driven by the emerging needs of life sciences and pharmaceutical applications. Figure 1.2a depicts the classical technique of LC with a glass column packed with coarse adsorbents and gravity fed with solvents. Fractions of the eluent containing separated components are collected manually and subsequently analyzed by spectrometry. This low-pressure LC is contrasted with the latest computer-controlled

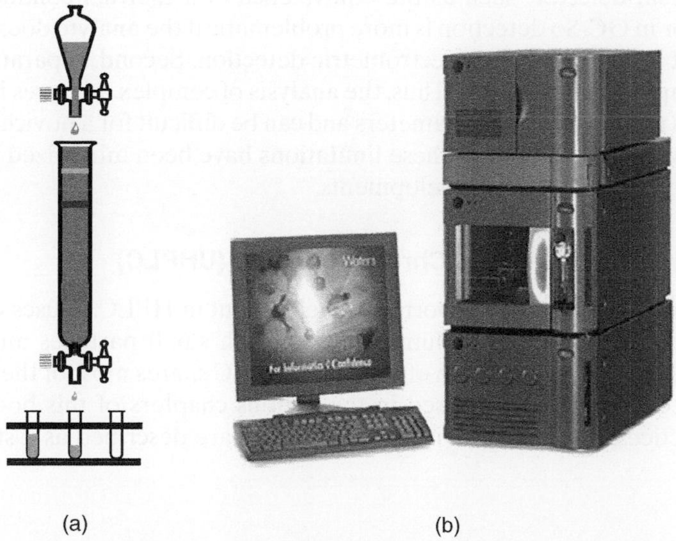

(a) (b)

**Figure 1.2.** (a) The traditional technique of low-pressure liquid chromatography using a glass column and gravity-fed solvent with manual fraction collection. (b) A modern automated UHPLC instrument (Waters Acquity UPLC system) capable of very high efficiency and pressure up to 15 000 psi.

**Table 1.1. Advantages and Limitations of HPLC**

Advantages

- Applicable to diverse analyte types
- Precise and highly reproducible quantitative analysis
- HPLC coupled with mass spectrometry (HPLC/MS)
- High separation power with sensitive detection

Perceived limitations

- Lack of an ideal universal detector
- Less separation efficiency than capillary gas chromatography (GC)
- Still arduous for regulatory or quality control (QC) testing

UHPLC, depicted in Figure 1.2b, operated at very high pressures, and capable of exceptional high separation efficiency.

### 1.1.4 Advantages and Limitations

Table 1.1 highlights the advantages and limitations of HPLC. HPLC is a premier separation technique capable of multicomponent analysis of complex mixtures. Few analytical techniques can match its versatility and precision of <0.1–0.5% relative standard deviation (RSD). HPLC can be highly automated, using sophisticated autosamplers and data systems for unattended analysis and report generation. A host of highly sensitive and specific detectors extend detection limits to nanogram, picogram, and even femtogram levels. As a preparative technique, it provides quantitative recovery of many labile components in milligram to kilogram quantities. Most importantly, HPLC is amenable to 60–80% of all existing compounds, as compared with about 15% for gas chromatography (GC) [3, 4].

Historically, HPLC is known to have several disadvantages or perceived limitations. First, there is no universal detector, such as the equivalence of a thermal conductivity or flame ionization detector in GC. So detection is more problematic if the analyte does not absorb UV radiation or is not ionized for mass spectrometric detection. Second, separation efficiency is less than that of capillary column GC. Thus, the analysis of complex mixtures is more difficult. Finally, HPLC has many operating parameters and can be difficult for a novice to develop new methods. As shown in later chapters, these limitations have been minimized mainly through the recent instrument and column developments.

### 1.1.5 Ultra-High-Pressure Liquid Chromatography (UHPLC)

UHPLC is the latest and the most important development in HPLC. It uses equipment with very high pressures together with columns packed with small particles and is capable of facilitating faster separations with high efficiency. UHPLC shares most of theories and applications with HPLC, which are discussed in the various chapters of this book. The history, benefits, best practices, and potential issues of UHPLC are described as a standalone topic in Chapter 5.

## 1.2 PRIMARY MODES OF HPLC

In this section, the four primary separation modes of HPLC are introduced and illustrated with application examples, labeled with the pertinent parameters: column (stationary phase), mobile phase, flow rate, detector, and sample information. These terminologies will be elaborated in later chapters.

### 1.2.1 Normal-Phase Chromatography (NPC)

Also known as liquid–solid chromatography or adsorption chromatography, normal-phase chromatography (NPC) is the traditional separation mode based on adsorption/desorption of the analyte onto a polar stationary phase (typically silica or alumina) [3–5]. Figure 1.3a shows a schematic diagram of a porous silica particle with silanol groups (Si-OH) residing at the surface and inside its pores. Polar analytes migrate slowly through the column due to strong interactions with the silanol groups. Figure 1.4 shows a chromatogram of four vitamin E isomers in a palm olein sample using a nonpolar mobile phase of hexane modified with ethanol. It is believed that a surface layer of water reduces the activity of the silanol groups and allows for more symmetrical peaks [3]. NPC is particularly useful for the separation of nonpolar compounds and isomers, as well as for the fractionation of complex samples by functional groups or sample clean-up. One significant disadvantage of this mode is the easy contamination of the polar surfaces by highly retained sample components rendering it a less reproducible technique. This problem is reduced by bonding polar functional groups such as amino- or cyano-moiety to the silanol groups. Today, NPC is primarily used in chiral separations and preparative applications.

### 1.2.2 Reversed-Phase Chromatography (RPC)

The separation is based on analytes' partition coefficients between a polar mobile phase and a hydrophobic (nonpolar) stationary phase. The earliest stationary phases were solid particles coated with nonpolar liquids. These were quickly replaced by permanently covalently bonded hydrophobic groups, such as octadecyl (C18) bonded groups on a silica support. A simplified schematic view of reversed-phase chromatography (RPC) is shown in Figure 1.3b, where

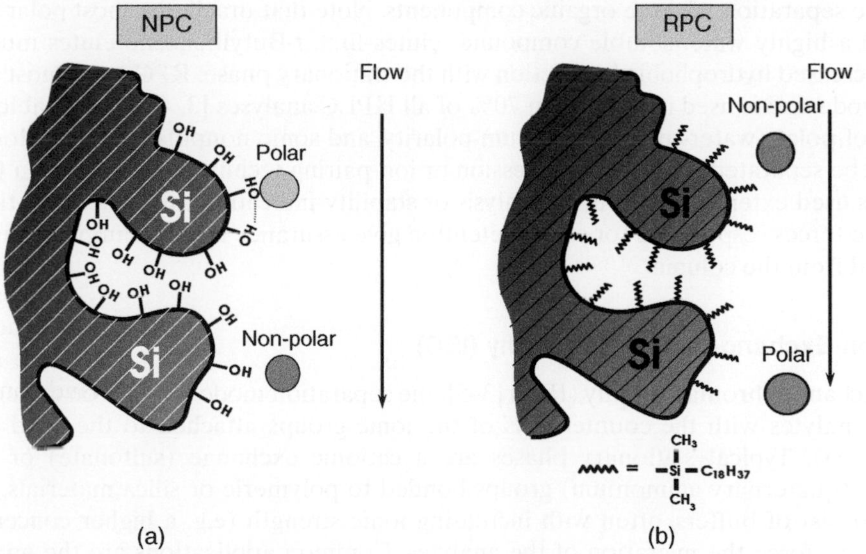


**Figure 1.3.** Schematic diagrams depicting separation modes of (a) normal-phase chromatography (NPC) and (b) reversed-phase chromatography (RPC).

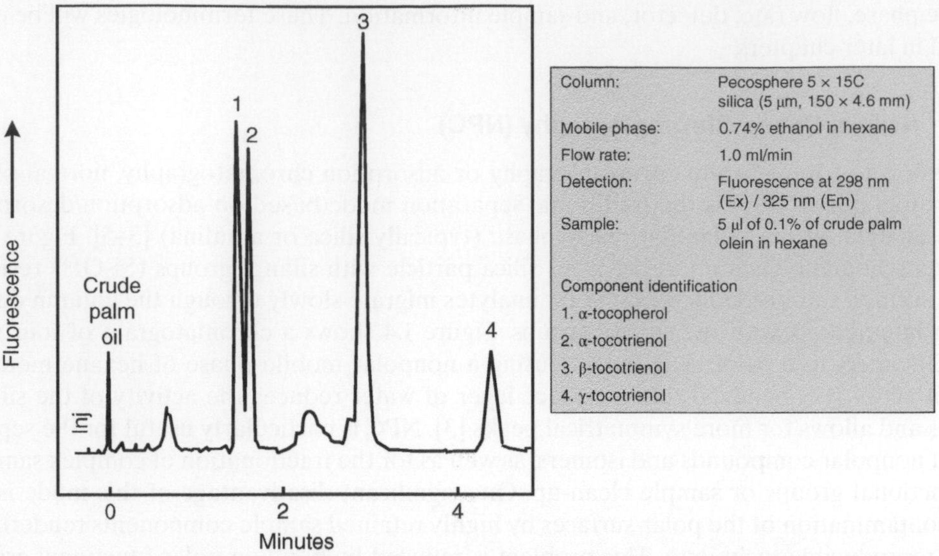


**Figure 1.4.** A normal-phase HPLC chromatogram of a palm olein sample showing the separation of various isomers of vitamin E. Source: Courtesy of PerkinElmer.

polar analytes elute first while nonpolar analytes elute later by interacting more strongly with the hydrophobic C18 groups that form a "liquid-like" layer around the solid silica support. This elution order of "polar first and nonpolar last" is the reverse order of that observed in NPC, and thus the term "reversed-phase chromatography." RPC typically uses a mixture of methanol or acetonitrile with water. The mechanism of separation is mainly attributed to hydrophobic or "solvophobic" interaction [13, 14]. The term "solvophobic interaction" refers to the relatively strong cohesive forces between the polar solvent molecules themselves and with the hydrated analytes and their interaction with the nonpolar stationary phase. Figure 1.5 shows the separation of three organic components. Note that uracil, the most polar component and a highly water-soluble compound, elutes first. *t*-Butylbenzene elutes much later due to increased hydrophobic interaction with the stationary phase. RPC is the most popular HPLC mode and is used in more than 70% of all HPLC analyses [3, 4]. It is suitable for the analysis of polar (water-soluble), medium-polarity, and some nonpolar analytes. Ionic analytes can be separated using ion-suppression or ion-pairing techniques discussed in Chapter 2. RPC is used extensively in purity analysis or stability-indicating assays because the weak dispersive forces responsible for solute retention give assurance that all sample components are eluted from the column.

### 1.2.3 Ion-Exchange Chromatography (IEC)

In ion-exchange chromatography (IEC) [3–5], the separation mode is based on the exchange of ionic analytes with the counter ions of the ionic groups attached to the solid support (Figure 1.6a). Typical stationary phases are a cationic exchange (sulfonate) or anionic exchange (quaternary ammonium) groups bonded to polymeric or silica materials. Mobile phases consist of buffers, often with increasing ionic strength (e.g. a higher concentration of NaCl), to force the migration of the analytes. Common applications are the analysis of ions and biological components such as amino acids, proteins/peptides, and polynucleotides.

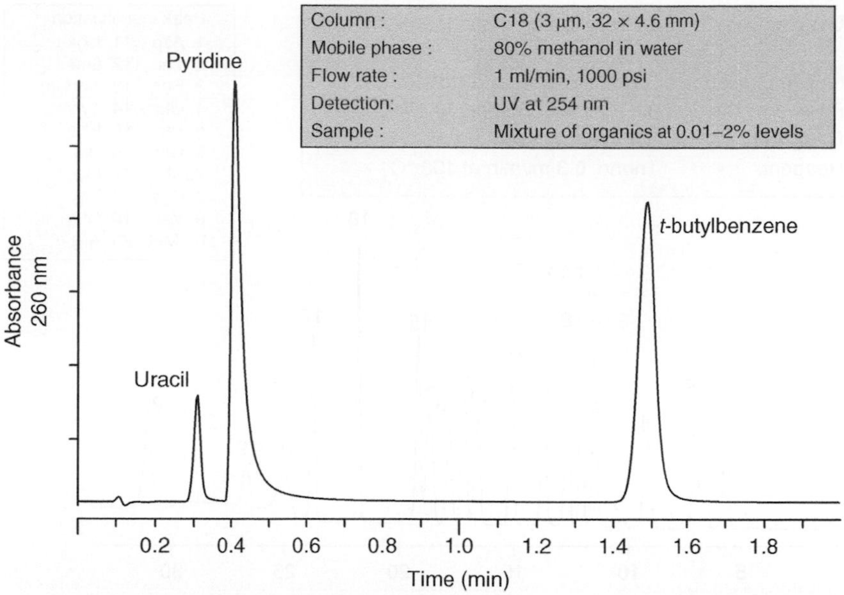


**Figure 1.5.** A reversed-phase HPLC chromatogram of three organic components eluting in the order of "polar first and nonpolar last." The basic pyridine peak is tailing due to a secondary interaction of the nitrogen lone pair with residual silanol groups of the silica-based bonded phase. Source: Ahuja and Dong 2005 [9]. Copyright 2005. Reprinted with permission of Elsevier.

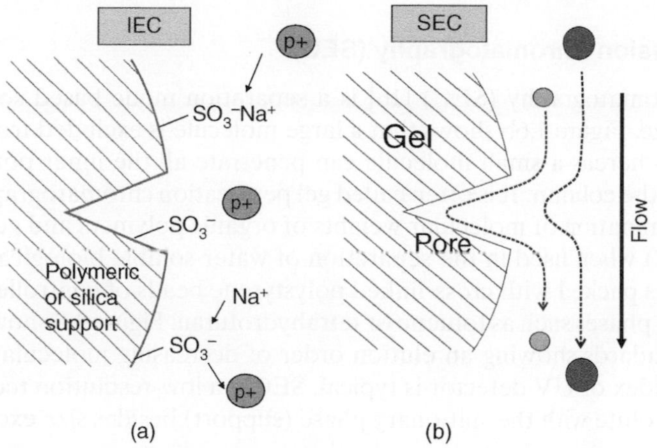


**Figure 1.6.** (a) Schematic diagrams depicting separation modes of (a) ion-exchange chromatography (IEC), showing the exchange of analyte ion p+ with the sodium counter ions of the bonded sulfonate groups. (b) Size-exclusion chromatography (SEC), showing the faster migration of large molecules.

Figure 1.7 shows the separation of amino acids on a sulfonated polymer column and a mobile phase of increasing sodium ion concentration and increasing pH. Since amino acids do not absorb strongly in the UV or visible region, a postcolumn reaction technique is used to form colored derivatives to allow sensitive detection at 550 nm. Ion chromatography [15] is a segment of IEC pertaining to the analysis of cations or anions using a high-performance ion-exchange column, usually with a specialized suppressed conductivity detector.

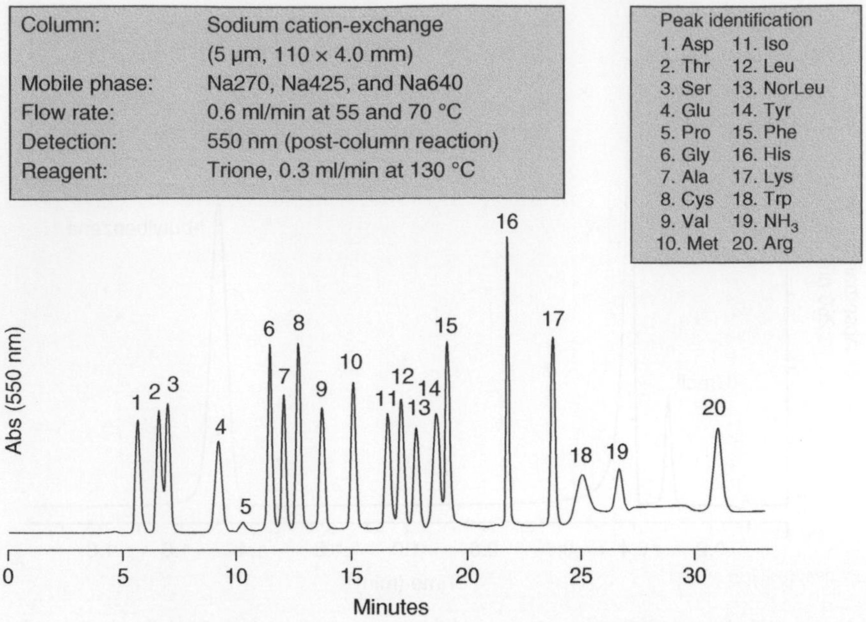


**Figure 1.7.** An IEC chromatogram of essential amino acids using a cationic sulfonate column and detection with postcolumn reaction. Note that Na270, Na425, and Na640 are prepackaged eluents containing sodium ions (NaCl) and buffered at pH of 2.70, 4.25, and 6.40, respectively. Trione is a derivatization reagent similar to ninhydrin used for postcolumn derivatization. Source: Courtesy of Pickering Laboratories.

### 1.2.4 Size-Exclusion Chromatography (SEC)

Size-exclusion chromatography (SEC) [16] is a separation mode based solely on the analyte's molecular size. Figure 1.6b shows that a large molecule is excluded from the pores and migrates quickly, whereas a small molecule can penetrate all the inner pores and migrates more slowly down the column. It is often called gel permeation chromatography (GPC) when used for the determination of molecular weights of organic polymers and gel-filtration chromatography (GFC) when used in the separation of water-soluble biological compounds. In GPC, the column is packed with cross-linked polystyrene beads of controlled pore sizes and eluted with mobile phases such as toluene or tetrahydrofuran. Figure 1.8 shows the separation of polystyrene standards showing an elution order of decreasing molecular size. Detection with a refractive index or UV detector is typical. SEC is a low-resolution technique in which interaction of the solute with the stationary phase (support) besides size exclusion should be avoided.

### 1.2.5 Other Separation Modes

Besides the four primary HPLC separation modes, several other modes or related techniques are noted below.

- *Affinity chromatography* [5]: Based on a receptor/ligand interaction in which immobilized ligands (enzymes, antigens, or hormones) on solid supports are used to isolate selected components from a mixture. The retained components can later be released in as a purified fraction.

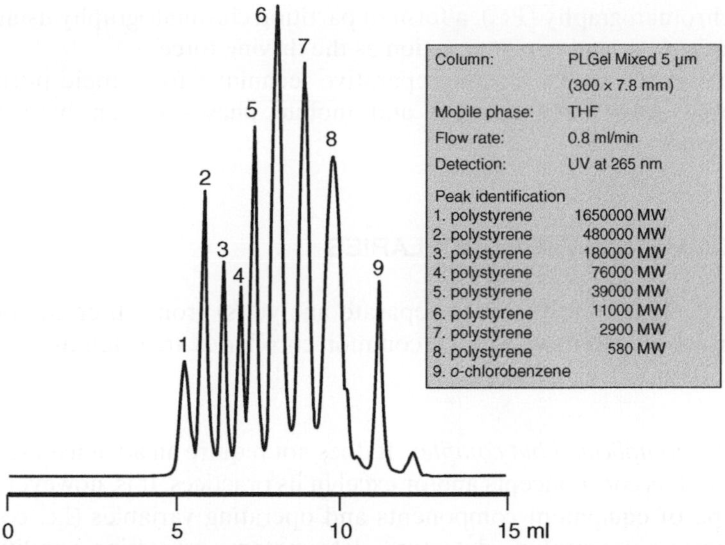


**Figure 1.8.** A GPC chromatogram of polystyrene standards on a mixed-bed polystyrene column. Source: Courtesy of Polymer Laboratories (Agilent Technologies).

- *Chiral chromatography* [17]: For the separation of enantiomers using a chiral-specific stationary phase. Both NPC and RPC chiral columns are available.
- *Hydrophilic interaction chromatography (HILIC)*: This is somewhat similar to NPC using a polar stationary phase such as silica or ion-exchange materials but eluted with polar mobile phases of organic solvents and aqueous buffers. It is commonly used to separate polar analytes, ions, hydrophilic peptides, glycans, and polar metabolites.
- *Hydrophobic interaction chromatography (HIC)*: Analogous to RPC using the same type of stationary phases except that it uses mobile phases of low organic solvent content and high salt concentrations. It is commonly used for the separation of proteins that are readily denatured by RPC mobile phases. HIC is particularly useful in the analysis of antibody–drug conjugates (ADCs).
- *Electrochromatography*: Uses capillary electrophoresis [18] (CE) equipment with a packed capillary column. The mobile phase is driven by the electromotive force from a high-voltage source as opposed to a mechanical pump. It is capable of very high efficiencies.
- *Supercritical fluid chromatography (SFC)* [19]: Uses HPLC-type packed columns and a mobile phase of pressurized supercritical fluids (i.e. carbon dioxide modified with a polar organic solvent). It is useful for nonpolar analytes and preparative applications where purified materials can be recovered quickly by evaporating the carbon dioxide. HPLC pumps with cryogenic cooling are often used with UV and MS detectors. There was a recent resurgence of interest in SFC brought forth by newer instruments of improved precision, sensitivity, and reliability. SFC is becoming the preferred technique in chiral separation and high-throughput purification in drug discovery research.
- *Other forms of low-pressure liquid chromatography*:
  - — Thin-layer chromatography (TLC) [20] uses glass plates coated with adsorbents and capillary action as the driving force. Useful for sample screening and semiquantitative analysis.

— Paper chromatography (PC), a form of partition chromatography using paper as the stationary phase and capillary action as the driving force.
— Flash chromatography, a semipreparative technique for sample purification using disposable glass NPC columns and mobile phases driven by gas-pressure or low-pressure pumps.

## 1.3 SOME COMMON-SENSE COROLLARIES

The goal of most HPLC analysis is to separate analyte(s) from other components in the sample for accurate quantitation. Several common corollaries are often overlooked by practitioners:

1. *HPLC is not complicated but complex*: It does not require an advanced scientific degree to understand its core concepts and/or excel in its practices. It is, however, complex due to the scope of equipment components and operating variables (i.e. column, mobile phases, pump, autosamplers, detectors, data systems, operating conditions, samples, standards, diluent) working in tandem to generate robust and accurate results.
2. *The sample must be soluble: "If it's not in solution, it cannot be analyzed by HPLC."* Solubility issues often complicate assays of low-solubility analytes or components, which are difficult to be extracted from sample matrices. Low recoveries usually stem from poor sample preparation steps rather than the HPLC analysis itself.
3. *For separation to occur, analytes must be retained and have differential migration in the column*: Separation cannot happen without retention and sufficient differential interaction with the stationary phase. For quantitative analysis with UV detection, analytes must have different retention on the column vs. other components. Baseline resolution is usually not needed for MS detection since specific signals can be customized for each ion according to its molecular weight.
4. *The mobile phase controls the separation*: Whereas the stationary phase provides a media for analyte interaction, the mobile phase controls the overall separation. In HPLC method development, efforts focus on finding a set of mobile phase conditions with the appropriate stationary phase for separating the analyte(s) from other components. Exceptions to this rule are size exclusion, chiral, and affinity chromatography where the mobile phase plays a minor role.
5. *All C18-bonded phase columns are not the same and cannot be interchanged for critical assays*: There are hundreds of C18 columns on the market. They vary tremendously in their retention and silanol characteristics [9]. For critical assays of complex samples, C18 columns are not interchangeable and the exact column from the specific manufacturer should be used. For more straightforward potency assays of the main component, a similar C18 bonded column of identical dimension can often be substituted.
6. *The final analyte solution should be prepared in the mobile phase A*: The final analyte solution, if possible, should be dissolved in the mobile phase or a solvent of "weaker" strength (mobile phase A by convention) or the starting mobile phase in a gradient analysis. Many anomalies such as splitting or fronting peaks are caused by injecting samples dissolved in diluents stronger than the starting mobile phase. If a stronger diluent must be used to dissolve the sample, a smaller injection volume (i.e. 2–5 μl) should be considered to minimize these problems.

7. *There are no perfect methods, and every analytical method has its caveats, limitations, or pitfalls*: An experienced scientist can identify these potential pitfalls and find conditions to minimize any issues.

An article on common-sense corollaries in HPLC was published elsewhere with more extensive discussions [21].

## 1.4 HOW TO GET MORE INFORMATION

The reader is encouraged to obtain more information from the following sources:

- Courses sponsored by training institutions [22], manufacturers, or national meetings (American Chemical Society, Pittsburgh Conference, Eastern Analytical Symposium).
- Computer-based training programs [23].
- Useful books [4–9] and websites [24–26] of universities and other government or compendia agencies, such as the U.S. Food and Drug Administration (FDA), U.S. Environmental Protection Agency (EPA), International Conference on Harmonization (ICH), the United States Pharmacopoeia, (USP), Association of Official Analytical Chemist International (AOAC), and American Society of Testing and Materials (ASTM).
- Research and review articles published in journals [27–30] such as the *Journal of Chromatography*, *Journal of Chromatographic Science*, *Journal of Liquid Chromatography*, *LCGC Magazine*, *Analytical Chemistry*, *American Pharmaceutical Review*, *Journal of Separation Science*, and *American Laboratories*.

## 1.5 SUMMARY

This introductory chapter describes the scope of the book and gives a summary of the history, advantages, limitations, and common-sense axioms of HPLC. Primary separation modes are discussed and illustrated with examples. Information resources on HPLC are also listed.

## 1.6 QUIZZES

**1.** The most popular chromatographic mode is
(a) IEC
(b) RPC
(c) SEC
(d) NPC

**2.** Which molecule elutes the latest in SEC?
(a) large molecule
(b) polar
(c) nonpolar
(d) small molecule

**3.** Which is NOT a major advantage of HPLC?
(a) amenable to diverse sample types
(b) HPLC/MS

(c) high sensitivity
(d) simplicity for regulatory testing

4. NPC is NOT useful for
   (a) protein separation
   (b) SFC
   (c) chiral separation
   (d) purification

5. Analysis of amino acid is likely to use
   (a) NPC
   (b) IEC
   (c) SEC
   (d) RPC

### 1.6.1 Bonus Quiz

In your own words, describe the reasons why HPLC is the most popular analytical technique for quantitative analysis of complex samples.

## 1.7 REFERENCES


1. Poole, C. (2003). *The Essence of Chromatography*. Amsterdam, the Netherlands: Elsevier. *(A comprehensive survey of the practice of chromatography including theory and quantitative principles. It covers HPLC, GC, TLC, SFC, and CE.)*.
2. Miller, J. (2004). *Chromatography: Concepts and Contrasts*, 2e. Hoboken, NJ: Wiley. *(A textbook on all branches of chromatography, including HPLC, GC, and CE, as well as industrial practices in regulated industries.)*.
3. Vitha, M.F. (2016). *Chromatography: Principles and Instrumentation*. Hoboken, NJ: Wiley. *(An updated textbook on LC and GC for students.)*.
4. Dong, M.W. (2006). *Modern HPLC for Practicing Scientists*. Hoboken, NJ: Wiley. *(The first edition to this current book.)*.
5. Snyder, L.R. and Kirkland, J.J. (2010). *Introduction to Modern Liquid Chromatography*, 3e. Hoboken, NJ: Wiley. *(This is the third edition of the classic book on HPLC fundamentals and applications.)*.
6. Meyer, V.R. (2010). *Practical High-Performance Liquid Chromatography*, 5e. Chichester, United Kingdom: Wiley. *(This updated text provides a systematic treatment of HPLC with appeals to students and laboratory professionals.)*.
7. Corradini, D. (ed.) (2011). *Handbook of HPLC*, 2e. Boca Raton, FL: CRC Press. *(This 700-page handbook covers many current topics on fundamentals and instrumentation of HPLC and related techniques.)*.
8. Kazakevich, Y.V. and LoBrutto, R. (eds.) (2007). *HPLC for Pharmaceutical Scientists*. Hoboken, NJ: Wiley. *(This 1000-page book is a comprehensive text for HPLC as practiced by pharmaceutical scientists.)*.
9. Ahuja, S. and Dong, M.W. (eds.) (2005). *Handbook of Pharmaceutical Analysis by HPLC*. Amsterdam, the Netherlands: Elsevier. *(A reference guide on the practice of HPLC in pharmaceutical analysis.)*.
10. Snyder, L.R., Kirkland, J.J., and Glajch, J.L. (1997). *Practical HPLC Method Development*, 2e. New York: Wiley-Interscience. *(A comprehensive though a somewhat outdated book on HPLC method development.)*.

11. Neue, U.D. (1997). *HPLC Columns: Theory, Technology, and Practice.* New York, NY: Wiley-VCH. *(This book focuses on column technologies, including theory, design, packing, chemistry, modes, method development, and maintenance.).*
12. Arnaud, C.H. (2016). *Chem. Eng. News* 94 (24): 29–35.
13. Melander, W.R. and Horvath, C.(ed. C. Horvath). *High-Performance Liquid Chromatography: Advances and Perspectives*, vol. 2, 1980, 113. New York, NY: Academic Press.
14. Carr, P.W., Martire, D.E., and Snyder, L.R. (1993). *J. Chromatogr. A* 656: 1.
15. Fritz, J.S. and Gjorde, D.T. (2009). *Ion Chromatography*, 4e. Weinheim, Germany: Wiley-VCH.
16. Striegel, A., Yau, W.W., Kirkland, J.J., and Bly, D.D. (2009). *Modern Size-Exclusion Chromatography: Practice of Gel Permeation and Gel Filtration Chromatography*, 2e. Hoboken, NJ: Wiley.
17. Subramania, G. (2006). *Chiral Separation Techniques: A Practical Approach*, 3e. Weinheim, Germany: Wiley-VCH.
18. Weinberger, R. (2000). *Practical Capillary Electrophoresis*, 2e. New York, NY: Academic Press.
19. Webster, G.K. (ed.) (2014). *Supercritical Fluid Chromatography: Advances and Applications in Pharmaceutical Analysis*. Singapore: Pan Stanford.
20. Bobbitt, J.M. (2013). *Thin Layer Chromatography*. Whitefish, MT: Literary Licensing.
21. Dong, M.W. (2018). *LCGC North Am.* 36 (8): 506.
22. LC Resources, training courses, Lafayette California, http://www.lcresources.com/training.html.
23. Academy Savant. Introduction to HPLC, CLC-10 (Computer-based Instruction), Academy Savant, Fullerton, California http://www.academysavant.com/clc-10.htm (accessed 06 January 2019).
24. Forumsci.co.il http://www.forumsci.co.il/HPLC/ (accessed 06 January 2019). *(by Dr. Shulamit Levin of Hebrew University with many training resources and useful links on HPLC and LC/*MS).
25. http://www.separationsnow.com (accessed 06 January 2019). (a free-access web portal on all separation sciences, sponsored by Wiley, Hoboken, NJ).
26. CHROMacademy http://www.chromacademy.com/ (accessed 06 January 2019). an e-learning website for analytical scientists (associated with LCGC Magazine).
27. Dong, M.W. (2013). *LCGC North Am.* 31 (6): 472.
28. Guillarme, D. and Dong, M.W. (2013). *Amer. Pharm. Rev.* 16 (4): 36.
29. Dong, M.W. (2007). *LCGC North Am.* 25: 89.
30. Guillarme, D. and Dong, M.W. (eds.) (2014). UHPLC: where we are ten years after its commercial introduction. *Trends Anal. Chem.* 63: 1–188. (Special issue).

# 2

# BASIC TERMS AND CONCEPTS

## 2.1 SCOPE

The objective of this chapter is to provide the reader with a concise overview of high-performance liquid chromatography (HPLC) terms and concepts. Both basic and some advanced theories are covered. The reader is referred to other HPLC textbooks [1–7], training courses [8, 9], journals, and Internet resources for a detailed treatment of HPLC theory. Terminologies and concepts important to ultra-high-pressure liquid chromatography (UHPLC) are discussed separately in Chapter 5, though chromatographic fundamentals are identical for both HPLC and UHPLC. This chapter covers the following:

- Basic terminologies and concepts of retention, selectivity, efficiency, resolution, and peak tailing
- Choice of mobile phases and parameters (solvent strength, pH, ion-pairing reagent, flow rate, and temperature), and the linear solvent strength theory in reversed-phase chromatography (RPC)
- The resolution equation (effects of efficiency, retention, and selectivity)
- The Van Deemter equation (effects of particle size and flow rate on plate height)
- Concepts in gradient analysis (peak capacity, effects of flow rate, gradient time) and orthogonal separations

The focus of this book is on RPC. However, the same concepts are often applicable to other modes of HPLC. The International Union of Pure and Applied Chemistry (IUPAC) [10] nomenclature is used. The term "sample component" is used interchangeably with "analyte" and "solute" in the context of this book. As mentioned in Chapter 1, the most common stationary phase is a hydrophobic C18-bonded phase on a porous silica support used with a mixed organic and aqueous mobile phase. The term "sorbent" refers to the bonded phase, whereas "support" refers to the unbonded silica material.

*HPLC and UHPLC for Practicing Scientists*, Second Edition. Michael W. Dong.

## 2.2 BASIC TERMS AND CONCEPTS

### 2.2.1 Retention Time ($t_R$), Void Time ($t_M$), Peak Height ($h$), and Peak Width ($w_b$)

Figure 2.1 shows a chromatogram with a single sample component. The time between the sample injection and the peak maximum is called the retention time ($t_R$). The retention time of an unretained component (often the first baseline disturbance caused by the sample diluent) is called the void time ($t_M$). $t_M$ is the total time spent by any unretained component in the mobile phase within the column. The adjusted retention time, $t_R'$ is equal to ($t_R - t_M$), that is, the time the solute resides in the stationary phase. Thus, $t_R = t_R' + t_M$ or the retention time is the total time the solute spends in the stationary phase ($t_R'$) and the mobile phase ($t_M$). The solute peak has both a peak width and a peak height ($h$). The peak width is usually measured at the base ($w_b$) or the peak's half-height ($w_{1/2}$).

Figure 2.2 shows how $w_b$ and $w_{1/2}$ are measured. Two tangent lines are drawn from the inflection points of the ascending and descending boundaries of the peak to the baseline. The distance between the two points at which the two tangents intercept with the baseline is $w_b$. Note that the peak area is roughly equal to $\frac{1}{2} \times (w_b \cdot h)$ [1, 6]. For Gaussian peaks, $w_b$ is approximately equal to four times the standard deviation ($4\sigma$), which brackets 95% of the total peak area. The width at half height ($w_{1/2}$) is easier to measure and is often used to calculate column efficiency ($N$).

The height or the area of a peak is proportional to the amount of analyte component present in the sample. Peak area is more commonly used to perform quantitative calculations.

### 2.2.2 Retention Volume ($V_R$), Void Volume ($V_M$), and Peak Volume

The retention volume ($V_R$) is the volume of mobile phase needed to elute the analyte at a given flow rate ($F$).

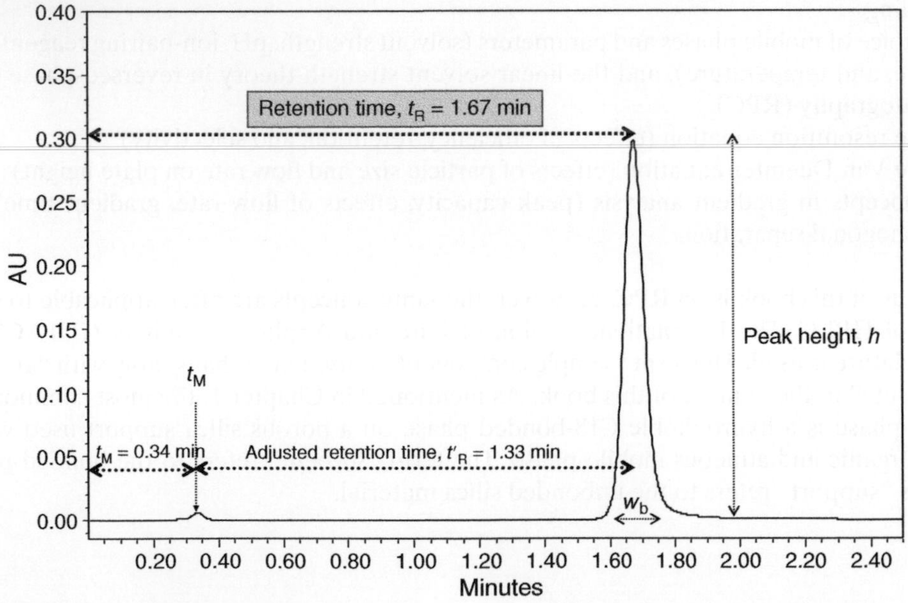


**Figure 2.1.** A single-component chromatogram showing a retention time ($t_R$), void time ($t_M$), peak width ($w_b$), and peak height ($h$).

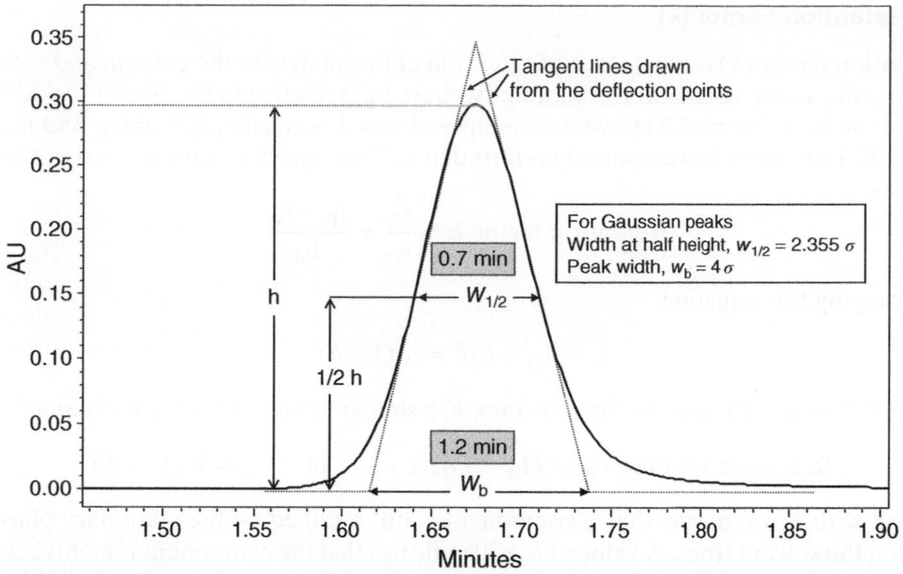


**Figure 2.2.** Diagram illustrating how peak width ($w_b$) and peak width at half height ($w_{1/2}$) are measured.

Here,

$$\text{Retention volume, } V_R = t_R F \tag{2.1}$$

Similarly,

$$\text{Void volume, } V_M = t_M F \tag{2.2}$$

The void volume ($V_M$) is the total volume of the mobile phase contained in the column (also called the liquid holdup volume). It is the volume of the empty column ($V_c$) minus the volume of the solid packing. $V_M$ is the sum of the intraparticular volume ($V_0$) and the interstitial volumes ($V_e$) inside the pores of the solid support. For most columns, the void volume can be estimated by:

$$V_M = 0.65\,V_c = 0.65\,\pi r^2 L \tag{2.3}$$

where $r$ is the inner radius of the column and $L$ is the length of the column.

$V_M$ can also be calculated from $t_M$ in the chromatogram since $V_M = t_M F$ (Eq. (2.2)). Note that $V_M$ is proportional to $r^2$, which also dictates the operating flow rate through the column. Note that $V_0$ does not include the interstitial pore volume and is equal to $V_M$ only for columns packed with nonporous particles. The peak volume, also called the peak bandwidth, is the volume of the mobile phase containing the eluted peak:

$$\text{Peak volume} = w_b F \tag{2.4}$$

Peak volume is proportional to $V_M$, and therefore smaller columns produce smaller peak volumes (see Section 2.2.6 and Eq. (2.14)). The understanding of $V_M$ and peak volume becomes vital in the practice of UHPLC, which uses smaller columns producing smaller peak volumes.

### 2.2.3 Retention Factor (*k*)

The retention factor ($k$) is the degree of retention of the analyte in the column and is defined as the time the solute resides in the stationary phase ($t_R'$) relative to the time it resides in the mobile phase ($t_M$). Figure 2.3 shows an example of how $k$ is calculated from $t_R$ and $t_M$. Note that $k$, an IUPAC term, was commonly referred to as $k'$ or capacity factor in most of the older references.

$$\text{Retention factor, } k = \frac{t'_R}{t_M} = \frac{t_R - t_M}{t_M} \tag{2.5}$$

Rearranging this equation,

$$t_R = t_M + t_M k = t_M(1 + k) \tag{2.6}$$

By multiplying both sides by the flow rate, $F$, a similar equation for $V_R$ is obtained:

$$\text{Retention volume, } V_R = Ft_R = Ft_M(1 + k) \quad \text{or} \quad V_R = V_M(1 + k) \tag{2.7}$$

A peak with $k = 0$ means that a component is not retained by the stationary phase and elutes with the solvent front. A value of $k > 20$ indicates that the component is highly retained. The desired $k$ values for most analyses are between 1 and 20. Analytes eluting with $k > 20$ are difficult to detect due to an excessive broadening of the peak. Figure 2.3 shows an example of how $k$ is calculated from $t_R$ and $t_M$.

Chromatography is a thermodynamically based method of separation, where each analyte component in a sample is distributed between the mobile phase and the stationary phase [1, 7, 11].

$$X_m \leftrightarrow X \text{ Partition coefficient, } K = \frac{[X_s]}{[X_m]} \tag{2.8}$$

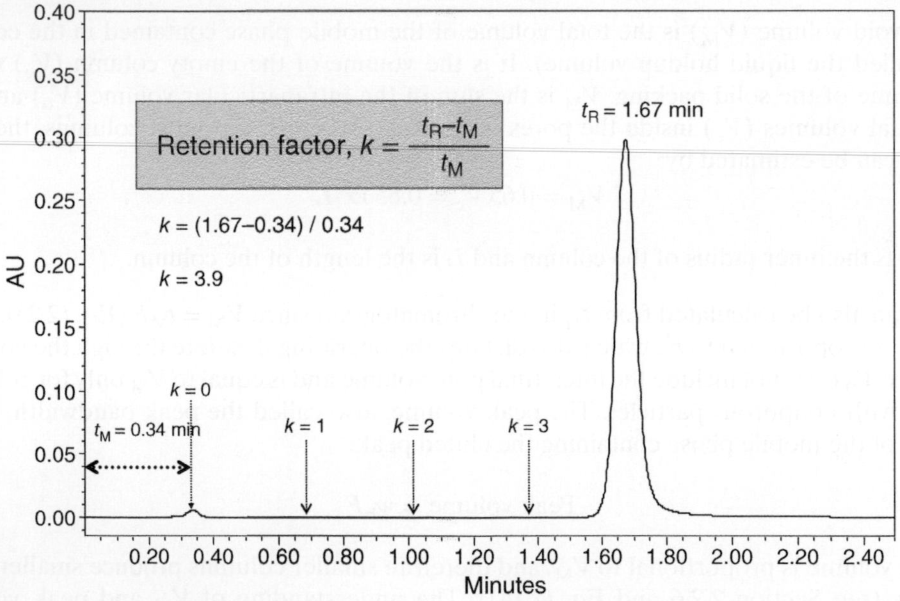


**Figure 2.3.** A chromatogram showing the calculation of retention factor ($k$), which is equal to $t_R/t_M$. $k$ is an important parameter defining the retention of the analyte. Desirable $k$ values for isocratic analyses are 1–20.

where $[X_m]$ and $[X_s]$ are the equilibrium concentrations of analyte $X$ in the mobile phase and stationary phase, respectively. The distribution of analyte $X$ is governed by the partition coefficient, $K$, which is a fundamental physical property of the analyte that is distinctive of the solute and solvent pair and is constant at a given temperature. The retention factor $k$ can also be described by the ratio of the total number of moles of analytes present in each phase at equilibrium [7].

$$\text{Retention factor}, k = \frac{\text{Moles of } X \text{ in stationary phase}}{\text{Moles of } X \text{ in mobile phase}} = \frac{[X_s]}{[X_m]}\frac{V_s}{V_M} = K\frac{V_s}{V_M} \tag{2.9}$$

where $V_s$ is the volume of the stationary phase, and $V_m$ is the volume of the mobile phase in the column void volume. The retention factor of an individual analyte component is primarily controlled by the strength of the mobile phase, the nature of the stationary phase, the column temperature, and the physicochemical properties of the analyte.

### 2.2.4 Separation Factor ($\alpha$)

The separation factor or selectivity ($\alpha$) [1, 11] is a measure of the relative retention ($k_2/k_1$) of two sample components as shown in Figure 2.4. Selectivity must be >1.0 for physical peak separation to occur. Selectivity is a measure of differential migration and is dependent on many factors that affect $K$, such as the nature of the stationary phase, the mobile phase composition, and properties of the solutes. Changing $\alpha$ is the most effective way to separate two closely eluting solutes. Experienced chromatographers can skillfully exploit the selectivity effects during method development to increase the separation of unresolved analytes in the sample [2].

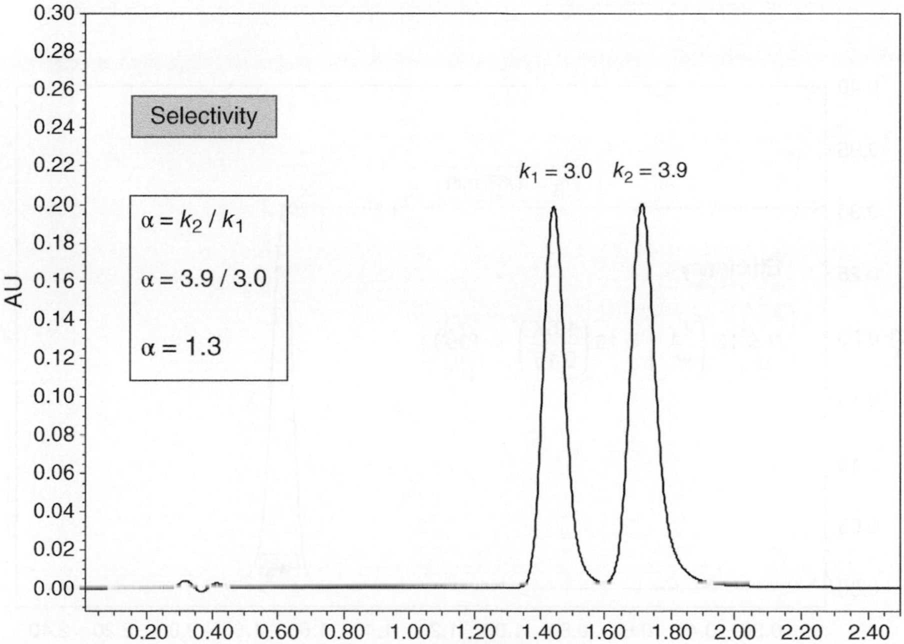


**Figure 2.4.** A chromatogram of two peaks with a selectivity factor ($\alpha$) of 1.3.

### 2.2.5 Column Efficiency and Plate Number (*N*)

An efficient column produces sharp peaks and separates many sample components in a given time. As seen in most chromatograms, peaks tend to be Gaussian and broaden with retention time, and $w_b$ becomes wider with $t_R$. This band broadening inside the column is fundamental to the chromatographic process under isocratic conditions [1, 6, 11]. The number of theoretical plates or plate number ($N$) is a measure of the efficiency of the column. $N$ is defined as the square of the ratio of the retention time divided by the standard deviation of the peak ($\sigma$). For a Gaussian peak, $w_b$ is equal to $4\sigma$,

$$\text{Number of theoretical plates, } N = \left(\frac{t_R}{\sigma}\right)^2 = \left(\frac{4t_R}{w_b}\right)^2 = 16\left(\frac{t_R}{w_b}\right)^2 \quad (2.10)$$

Figure 2.5 shows an example of how $N$ is calculated using the equation above. Since it is more difficult to measure $\sigma$ or $w_b$, the width at half height ($w_{1/2}$) is often used to calculate $N$ as described in the United States Pharmacopoeia (USP) [12]. Note that for a Gaussian peak, $w_{1/2}$ is equal to $2.355\sigma$ (Figure 2.2) [6]. Also, note that $N$ can only be measured under isocratic conditions.

$$N = \left(\frac{t_R}{\sigma}\right)^2 = \left(\frac{2.355t_R}{w_{1/2}}\right)^2 = 5.546\left(\frac{t_R}{w_{1/2}}\right)^2 \quad (2.11)$$

### 2.2.6 Peak Volume

Peak volume is the volume of the mobile phase, or eluent, containing the eluting peak. Peak volume is proportional to $k$ and $V_M$. The relationship between peak volume and these factors

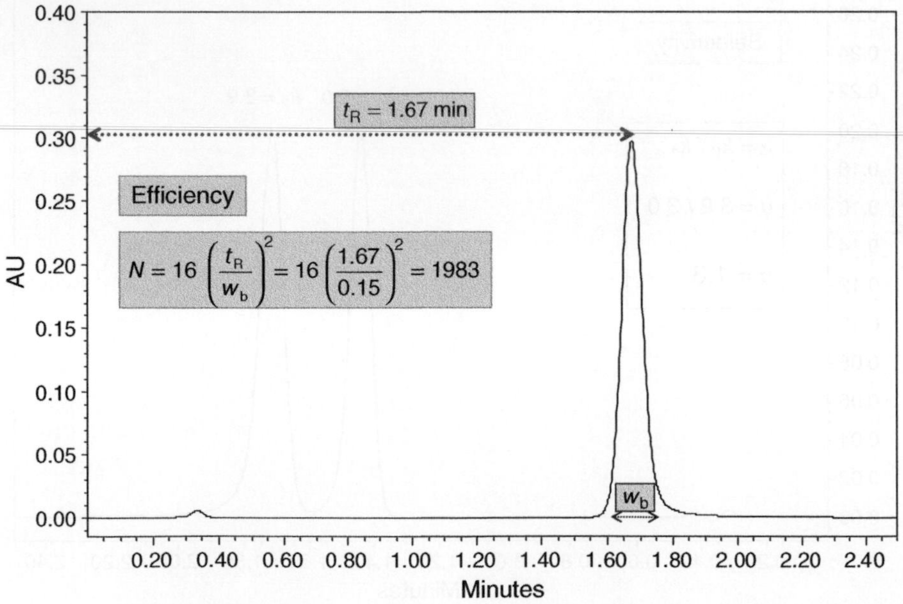


**Figure 2.5.** A chromatogram showing a peak from a column with $N$ = 1983.

can be derived by rearranging Eq. (2.10):

$$N = 16\left(\frac{t_R}{w_b}\right)^2 \qquad \frac{\sqrt{N}}{4} = \left(\frac{t_R}{w_b}\right) \tag{2.12}$$

Then, multiplying both numerator and denominator by $F$ leads to Eq. (2.13) and Eq. (2.14),

$$\frac{\sqrt{N}}{4} = \left(\frac{F}{F}\frac{t_R}{w_b}\right) = \left(\frac{V_R}{Fw_b}\right) = \left(\frac{V_R}{\text{Peak Volume}}\right) \tag{2.13}$$

$$\text{Peak Volume} = \frac{4V_R}{\sqrt{N}} = \frac{4V_M(1+k)}{\sqrt{N}} \tag{2.14}$$

Thus for a given column, peak volume is roughly proportional to $k$ and $V_M$ under isocratic conditions. The smaller peak volumes from small-diameter columns are more affected by the deleterious effect of system dispersion by the instrument (extra-column band broadening), as discussed later in Section 4.10.

### 2.2.7 Height Equivalent to a Theoretical Plate or Plate Height (HETP or *H*)

The concept of a "plate" was adapted from the industrial distillation process in the fractionation of crude oil using a distillation column consisting of individual plates where the condensed liquid is equilibrated with the rising vapor. A longer distillation column has more "plates" or separation power and therefore can separate a complex mixture such as crude oil into more fractions of distillates such as aviation fuels, gasoline, kerosene, heating oils, and tars. Although there are no plates inside the HPLC column, the same concept of plate number ($N$) or plate height ($H$) can be applied. The height equivalent to a theoretical plate (HETP or $H$) is equal to the length of the column ($L$) divided by the plate number ($N$): [6, 7]

$$\text{HETP}, H = L/N \tag{2.15}$$

In HPLC, the main factor controlling $H$ is the diameter of the particles ($d_p$) used as the packing material. For a well-packed column, $H$ is roughly equal to $2d_p$. A typical 150-mm-long column packed with 5-μm material should have $N = L/H = 150\,000\,\mu\text{m}/(2 \times 5\,\mu\text{m})$, or about 15 000 plates. Similarly, a 150-mm column packed with 3-μm material should have $N = L/H = 150\,000\,\mu\text{m}/6\,\mu\text{m}$, or about 25 000 plates. Thus, columns packed with smaller particles are more efficient and have a higher plate number per unit length.

### 2.2.8 Resolution ($R_s$)

The goal of most HPLC analyses is the separation of analytes in the sample from each other, usually for quantitative measurements. The degree of separation of two adjacent peaks is known as resolution ($R_s$). $R_s$ is defined as the difference in retention times of the two peaks divided by the average peak width (Figure 2.6). Since peak widths of adjacent peaks tend to be similar, the average peak width is approximated by one of the $w_b$.

$$\text{Resolution}, R_s = \frac{t_{R2} - t_{R1}}{\left(\frac{w_{b1} + w_{b2}}{2}\right)} = \frac{\Delta t_R}{w_b} \tag{2.16}$$

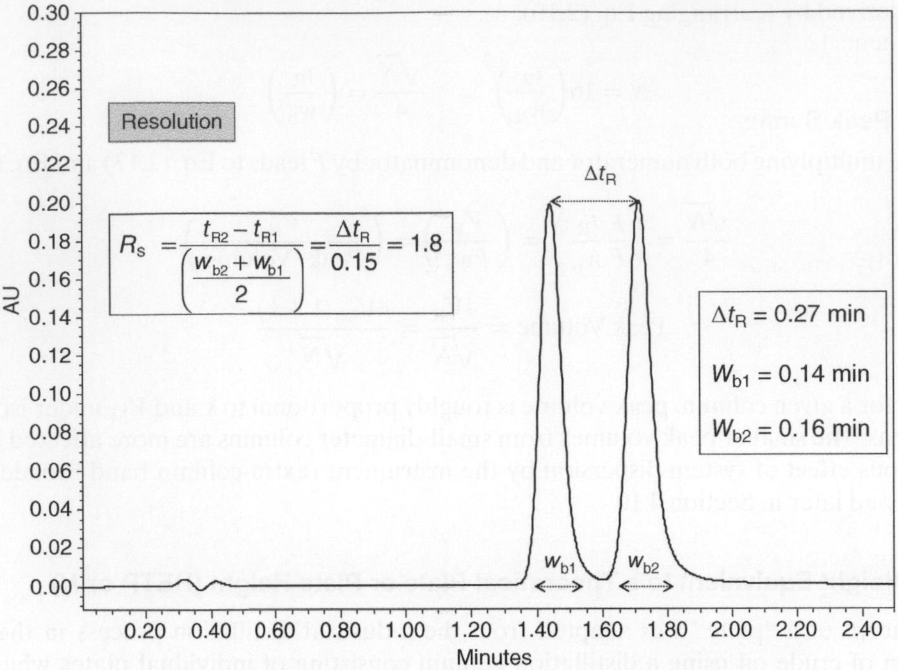


**Figure 2.6.** A chromatogram of two peaks with a resolution ($R_s$) of 1.8.

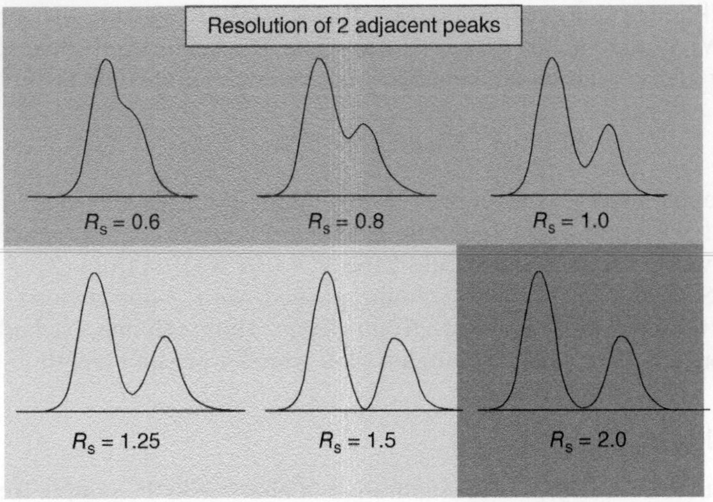


**Figure 2.7.** Chromatographic profiles two closely eluting peaks at various resolution values from 0.6 to 2.0. Source: Reproduced with permission of Academy Savant.

Figure 2.7 shows a graphical representation of resolution for two peaks with $R_s$ ranging from 0.6 to 2.0. Note that $R_s = 0$ indicates complete co-elution or no separation. $R_s = 0.6$ indicates that a shoulder is discernible or slight partial separation has been achieved. $R_s = 1$ indicates a partial separation, allowing for a rough quantitation. $R_s = 1.5$ indicates that a baseline separation of the two components has been achieved. The goal of most HPLC methods is to achieve baseline separation, $R_s = 1.5$ for all key analytes for accurate quantitation of these components [2, 3].

### 2.2.9 Peak Symmetry: Asymmetry Factor ($A_s$) and Tailing Factor ($T_f$)

Ideally, chromatographic peaks should be Gaussian, with perfect symmetry. In reality, most peaks are either slightly fronting or tailing (see Figure 2.8). The asymmetry factor ($A_s$) is used to measure the degree of peak symmetry and is defined as the peak width at 10% of the peak height ($W_{0.1}$) as shown in Figure 2.8.

$$\text{Asymmetry factor}, A_s = B/A \tag{2.17}$$

Tailing factor ($T_f$) is a similar term for peak symmetry as defined by the USP [12]. $T_f$ is calculated using the peak width at 5% peak height ($W_{0.05}$) as shown in Figure 2.8:

$$\text{Tailing factor } T_f = W_{0.05}/2f \tag{2.18}$$

Note that $T_f$ is used here instead of $T$ in the USP because $T$ often stands for temperature. Tailing factors of the main component(s) are required calculations in most regulatory pharmaceutical methods. $T_f = 1.0$ indicates a perfectly symmetrical peak. $T_f > 2$ indicates a tailing peak that is less acceptable due to difficulty in integrating the peak area precisely. With a $T_f$ between 0.5 and 2.0, the values of $A_s$ and $T_f$ are similar. For severely tailing peaks, $A_s$ tends to be somewhat larger than $T_f$.

Peak tailing is typically caused by adsorption, strong interaction with the stationary phase, or extra-column band broadening of the solute. Peak fronting is caused by column overloading or occasional chemical reaction of the analyte (e.g. isomerization) during chromatography [1]. For instance, many basic analytes (e.g. amines) display some peak tailing due to their polar interaction with acidic residual silanol groups in silica-based stationary phases [6]. Figure 2.9

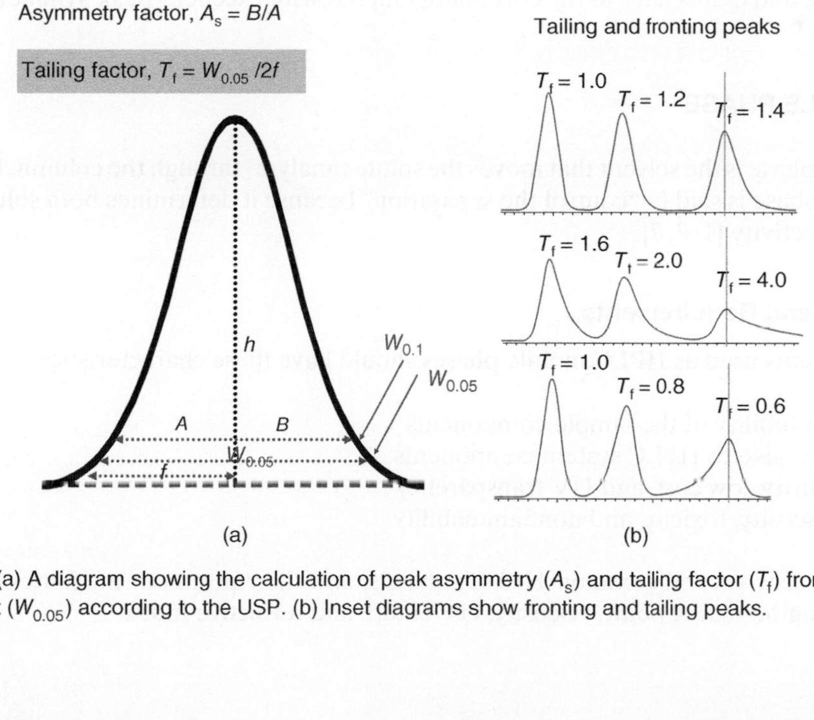


**Figure 2.8.** (a) A diagram showing the calculation of peak asymmetry ($A_s$) and tailing factor ($T_f$) from peak width at 5% height ($W_{0.05}$) according to the USP. (b) Inset diagrams show fronting and tailing peaks.

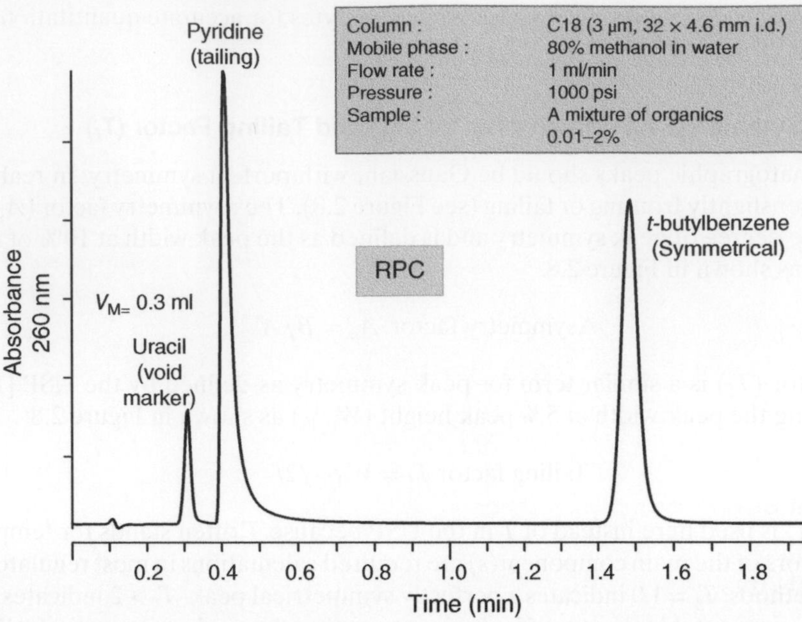


**Figure 2.9.** An HPLC chromatogram of three components using RPC conditions shown in the inset. Note that the basic pyridine peak is tailing although the *t*-butylbenzene peak (neutral) is symmetrical. Source: Ahuja and Dong 2005 [3]. Copyright 2005. Reprint with permission of Elsevier.

shows an RPC chromatogram with three components. Uracil is very water soluble and elutes with the solvent front and $k = 0$. For that purpose, uracil is often used as a void volume marker for the measurement of $V_M$ in RPC. Pyridine is a base and exhibits considerable peak tailing due to a secondary polar interaction with residual silanols in addition to the primary hydrophobic interaction by the C18-bonded phase. *t*-Butylbenzene is a neutral and hydrophobic molecule and elutes later in the chromatogram but with excellent peak symmetry.

## 2.3 MOBILE PHASE

The mobile phase is the solvent that moves the solute (analyte) through the column. In HPLC, the mobile phase is said to "control the separation" because it determines both solute retention and selectivity [1–3, 7].

### 2.3.1 General Requirements

Ideally, solvents used as HPLC mobile phases should have these characteristics:

- Good solubility of the sample components
- Noncorrosive to HPLC system components
- High purity, low cost, and UV transparency
- Low viscosity, toxicity, and nonflammability

Table 2.1 lists several common HPLC solvents, and their pertinent attributes including solvent strength, boiling point, viscosity, UV cutoff, and refractive index.

**Table 2.1. HPLC Solvents and their Pertinent Attributes**

| Solvent | Solvent strength ($E°$) | bp (°C) | Viscosity (cP) at 20 °C | UV cut off (nm) | Refractive Index |
|---|---|---|---|---|---|
| *n*-Hexane | 0.01 | 69 | 0.31 | 190 | 1.37 |
| Toluene | 0.29 | 111 | 0.59 | 285 | 1.49 |
| Methylene chloride | 0.42 | 40 | 0.44 | 233 | 1.42 |
| Tetrahydrofuran (THF) | 0.45 | 66 | 0.55 | 212 | 1.41 |
| Acetonitrile (ACN) | 0.55–0.65 | 82 | 0.37 | 190 | 1.34 |
| 2-Propanol | 0.82 | 82 | 2.30 | 205 | 1.38 |
| Methanol (MeOH) | 0.95 | 65 | 0.54 | 205 | 1.33 |
| Water | Large | 100 | 1.00 | <190 | 1.33 |

$E°$ (solvent elution strength as defined by Hildebrand on alumina sorbent).
Source: Data extracted from Ref. [2] and other sources.

### 2.3.2 Solvent Strength and Selectivity

Solvent strength refers to the ability of a solvent to elute solutes from a column [1, 2, 7, 11]. Solvent strengths under normal phase conditions are often characterized by the Hildebrand's elution strength scale ($E°$), which are listed in Table 2.1. Solvent strength is related to its polarity. Nonpolar hexane is a weak solvent in normal-phase chromatography (NPC), whereas water is a strong solvent. The opposite is true in RPC since the stationary phase is hydrophobic. Here water is a weak solvent, and organic solvents are strong and in a somewhat reversed order of the Hildebrand scale (e.g. THF > ACN > MeOH ≫ water). Water is a weak solvent because it is a poor solvent for hydrophobic components, which preferentially partition into the hydrophobic stationary phase in RPC. ACN, an aprotic solvent, is the preferred RPC solvent because of its low viscosity (0.37 cP) leading to higher column efficiency, good UV transparency (to 190 nm), and strong elution strength. It acts as a proton acceptor and is capable of $\pi$–$\pi$ interaction with the solute. MeOH is the second choice, which yields different selectivity shown by ACN due to its ability to hydrogen bond with analyte molecules. It acts as both a proton acceptor and a donor. MeOH has end absorbance below 205 nm and has substantially higher viscosity when mixed with water. THF has strong solubilizing power but is highly toxic and has a safety issue because of peroxide formation. Methyl *tert*-butyl ether (MTBE) can be a substitute for THF though it has limited miscibility with water; it does not form peroxides.

Figure 2.10 shows a series of six chromatograms to illustrate the effect of solvent strength in RPC. Here, the two components (nitrobenzene and propylparaben) are eluted with an aqueous/ACN mobile phases of decreasing solvent strength (i.e. decreasing concentration of acetonitrile [ACN]). At 100% ACN, both components are not retained by the column and elute with a $k$ close to zero. At 60% ACN and 40% water, the peaks are slightly retained ($k$ close to 1) and are partially separated. The two components merge back together at 40% ACN (this is somewhat unusual). At 30% ACN, the two components are well separated, though propylparaben now elutes after nitrobenzene. At 20% ACN, propylparaben is highly retained with a $k$ of 31. Note that with identical injection amounts, the peak heights of the analytes are significantly lower at low solvent strengths due to the wider peak widths and larger peak volumes.

Table 2.2 summarizes the $t_R$, $k$, and $\alpha$ of both analytes (nitrobenzene and propylparaben) as a function of the percent of ACN or MeOH content in the mobile phase. The following observations can be made:

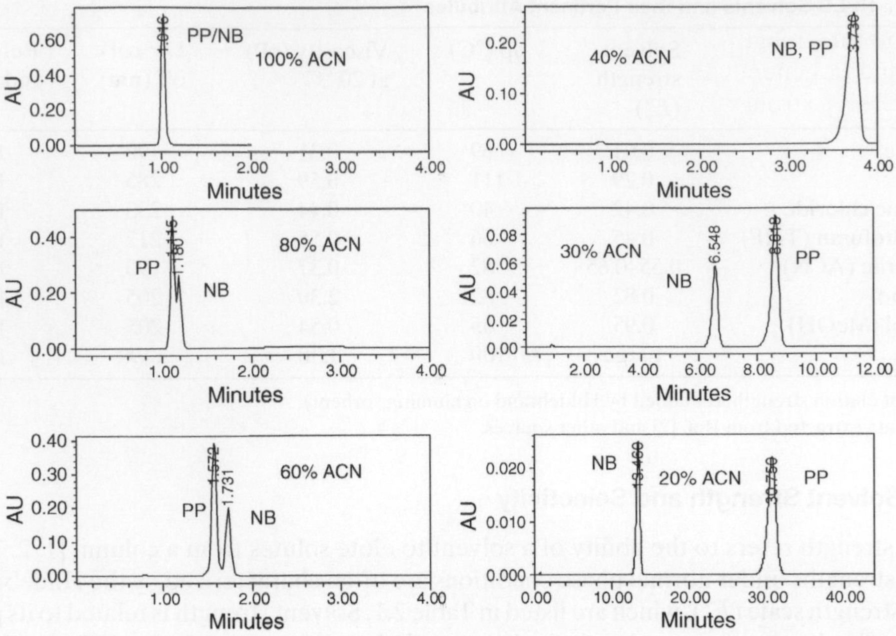


**Figure 2.10.** Six RPC chromatograms illustrating the effect of mobile phase solvent strength on solute retention and resolution. LC conditions were: column – Waters symmetry C18, 3 μm, 75 × 4.6 mm, 1 ml/min, 40 °C, detection at 258 nm. The mobile phase is a mixture of acetonitrile (ACN) and water. Solutes were nitrobenzene (NB) and propylparaben (PP).

**Table 2.2. Retention Data for Nitrobenzene (NB) and Propylparaben (PP) in Mobile Phases Containing Acetonitrile or Methanol**

| % ACN | $t_R$(NB) (min) | $k$(NB) | $t_R$(PP) (min) | $k$(PP) | $\alpha$ (PP/NB) |
|---|---|---|---|---|---|
| 100 | 1.02 | 0.28 | 1.02 | 0.28 | 1.00 |
| 90 | 1.04 | 0.30 | 1.04 | 0.30 | 1.00 |
| 80 | 1.18 | 0.48 | 1.12 | 0.39 | 0.83 |
| 70 | 1.38 | 0.73 | 1.27 | 0.59 | 0.81 |
| 60 | 1.73 | 1.16 | 1.57 | 0.96 | 0.83 |
| 50 | 2.37 | 1.96 | 2.29 | 1.86 | 0.95 |
| 40 | 3.73 | 3.66 | 3.73 | 3.66 | 1.00 |
| 30 | 6.55 | 7.19 | 8.62 | 9.78 | 1.36 |
| 25 | 9.25 | 10.56 | 15.35 | 18.19 | 1.72 |
| 20 | 13.46 | 15.83 | 30.75 | 37.44 | 2.37 |
| **% MeOH** | **$t_R$(NB), min** | **$k$(NB)** | **$t_R$(PP) (min)** | **$k$(PP)** | **$\alpha$ (PP/NB)** |
| 100 | 1.02 | 0.28 | 1.02 | 0.28 | 1.00 |
| 90 | 1.08 | 0.35 | 1.08 | 0.35 | 1.00 |
| 80 | 1.25 | 0.56 | 1.25 | 0.56 | 1.00 |
| 70 | 1.5 | 0.88 | 1.68 | 1.10 | 1.26 |
| 60 | 2.02 | 1.53 | 2.73 | 2.41 | 1.58 |
| 50 | 3.05 | 2.81 | 5.65 | 6.06 | 2.16 |
| 40 | 5.07 | 5.34 | 14.36 | 16.95 | 3.18 |
| 30 | 8.91 | 10.14 | 41 | 50.25 | 4.96 |
| 25 | 11.78 | 13.73 | 74 | 91.50 | 6.67 |
| 20 | 15.4 | 18.25 | 130.75 | 162.44 | 8.90 |

Column: Waters symmetry C18, 3-μm, 75 × 4.6 mm, 1 ml/min, 40 °C.
Solute: NB = Nitrobenzene, PP = propylparaben.

- Both $t_R$ and $k$ increase exponentially with decreasing percentage of organic solvents (solvent strength) in the mobile phase.
- $\alpha$ and $R_s$ generally increase at lower solvent strength.
- ACN is a stronger solvent than MeOH and has different selectivity.

Figure 2.11 is a plot of the retention time and log $k$ of nitrobenzene versus percent ACN (%B). Note that log $k$ is inversely proportional to solvent strength or percent of organic solvent. This relationship is termed "the Linear Solvent Strength Theory." [2, 13]

Figure 2.12 compares the chromatograms of the two solutes with 60% and 20% of ACN and MeOH, respectively. Notice the considerable differences in retention time and selectivity (elution order at 60% organic solvent) of the two solutes. Selectivity is a function of dipole moment, hydrogen bonding, and dispersive characteristics of the stationary phase, the solutes, and mobile phase components, among other factors. Chapter 10 discusses how these selectivity effects can be manipulated to great advantage during HPLC method development.

### 2.3.3 pH Modifiers and Buffers

The pH of the aqueous component in the mobile phase can have a dramatic effect on the retention of ionizable (acidic or basic) analytes. In RPC, the ionized form of the solute does not partition well into the hydrophobic stationary phase and has a significantly lower $k$ than the neutral or nonionized form.

Buffers are required to control the pH of the mobile phase for critical assays. Table 2.3 summarizes the common additives and buffers for HPLC and their respective p$K_a$ and UV cutoffs. Buffers of ammonium salts of volatile acids are used for the development of mass spectrometer (MS)-compatible HPLC methods. Note that buffers are only effective within ±1.0 to ±1.5 pH units from their p$K_a$ (Henderson–Hasselbalch equation). Although 50 mM

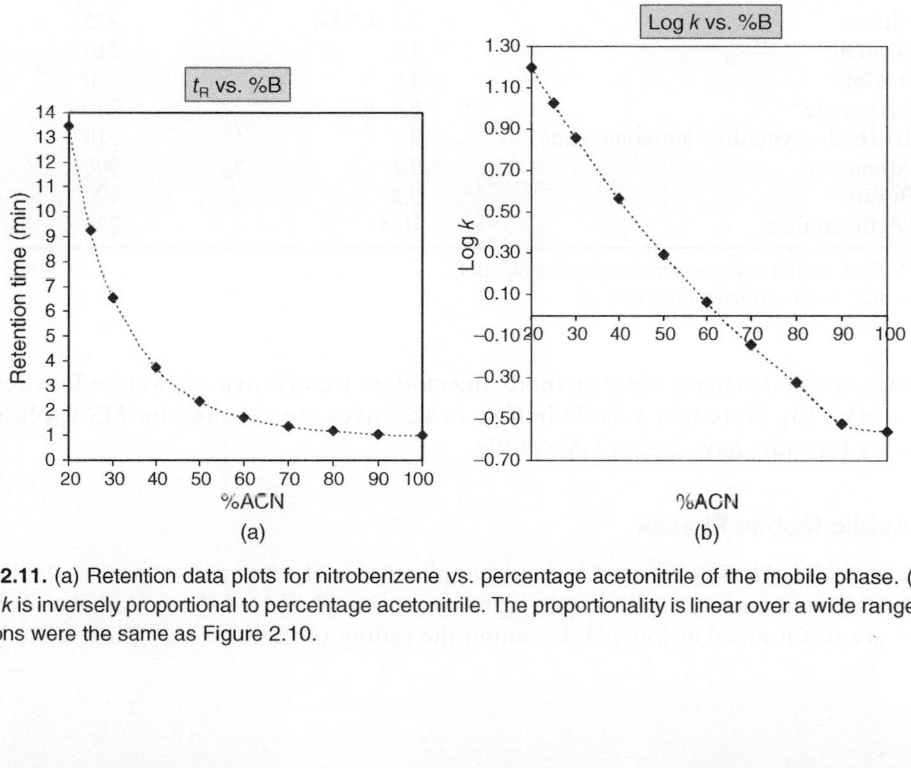


**Figure 2.11.** (a) Retention data plots for nitrobenzene vs. percentage acetonitrile of the mobile phase. (b) Note that log $k$ is inversely proportional to percentage acetonitrile. The proportionality is linear over a wide range. HPLC conditions were the same as Figure 2.10.

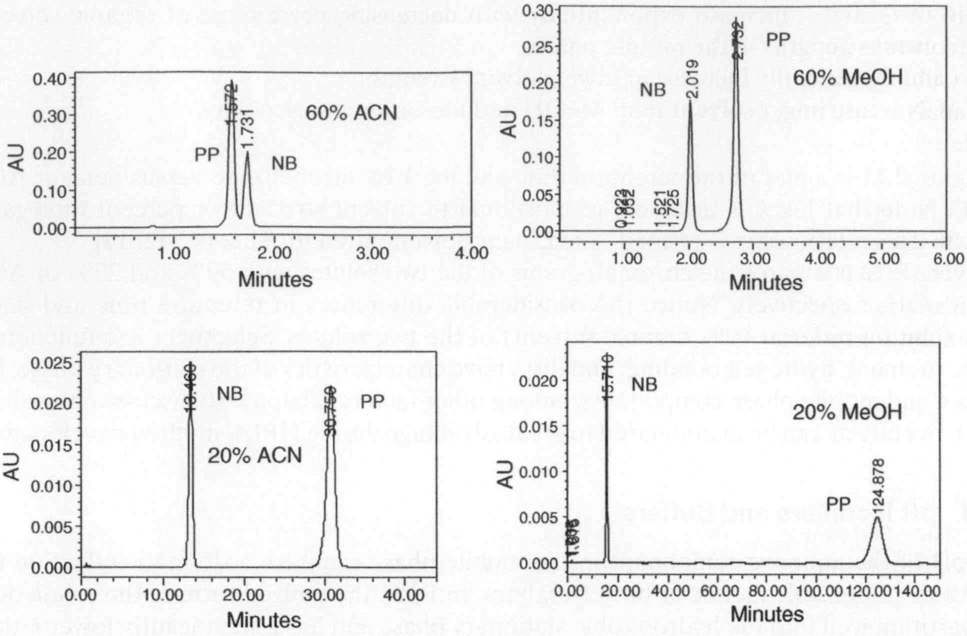


**Figure 2.12.** Four RPC chromatograms illustrating the effect of mobile phase strength and selectivity of acetonitrile (ACN) and methanol (MeOH). See Figure 2.10 for LC conditions.

**Table 2.3. Common HPLC Buffers and their Respective p$K_a$ and UV Cutoff**

| Buffer | p$K_a$ | UV cutoff (nm) |
|---|---|---|
| Trifluoroacetic acid[a] | 0.3 | 210 |
| Phosphate | 2.1, 7.2, 12.3 | 190 |
| Citrate | 3.1, 4.7, 5.4 | 225 |
| Formate[a] | 3.8 | 210 |
| Acetate[a] | 4.8 | 210 |
| Carbonate[a] | 6.4, 10.3 | 200 |
| Tris(hydroxymethyl) aminomethane | 8.3 | 210 |
| Ammonia[a] | 9.2 | 200 |
| Borate | 9.2 | 190 |
| Diethylamine | 10.5 | 235 |

[a]Volatile buffer systems, which are MS-compatible.
Source: Data extracted from Ref. [2].

buffers are specified in many older methods, the modern trend is to use lower buffer strengths (e.g. 5–20 mM) [3]. Note that volatile buffers or additives are required for MS applications though they typically have higher UV cutoffs.

### 2.3.4 Acidic Mobile Phases

In RPC, an acidic pH of 2–4 is used for many applications. The low pH suppresses the ionization of weakly acidic analytes, leading to higher retention [2, 3, 11, 14]. Surface silanols in the packings are not ionized at low pH, lessening the tailing of basic solutes. Most silica-based bonded phases are not stable below pH 2 due to acid-catalyzed hydrolytic cleavage of the bonded groups [2, 3]. Common acids used for mobile phase preparations are phosphoric acid (PA), trifluoroacetic acid (TFA), formic acid (FA), and acetic acid. However, water-soluble basic analytes are ionized at low pH and might not be retained in RPC. Note that very simple mobile phases such as 0.1% TFA (pH at 2.1) or 0.1% FA (pH 2.8) are easily prepared for HPLC and LC/MS analysis by pipetting 1 ml of acid into 1 l of HPLC-grade water [14]. The use of 0.1% PA is underutilized in HPLC. It is transparent down to 200 nm and is useful for purity methods of raw materials and reagents requiring detection at low UV. However, 0.1% PA, even though it is easily prepared, is not MS-compatible just like phosphate buffers. For critical analysis such as the purity assays of complex pharmaceuticals, a buffer such as 20 mM ammonium formate at pH 3.7 appears to work well for many RPC columns.

### 2.3.5 Ion-Pairing Reagents and Chaotropic Agents

Ion-pairing reagents are detergent-like molecules added to the mobile phase to provide retention of acidic or basic analytes [1, 2, 7]. Long-chain alkyl sulfonates (C5 to C12) combine with basic solutes under acidic pH conditions to form neutral "ion pairs" that are retained in RPC. Retention is proportional to the length of the hydrophobic chain of the ion-pairing agent and its concentration. Note that TFA has some ion-pairing capability and is commonly used in RPC of proteins and peptides. Heptafluorobutyric acid (HFBA) is a MS-compatible ion-pairing reagent. The use of ion-pairing reagents has decreased in recent years due to lower efficiencies and slow column equilibration. Also, the availability of high-pH-compatible silica-based columns offers an effective alternative approach to retain basic analytes [15]. For acidic analytes, ion-pairing reagents such as tetraalkylammonium salts are used.

Another class of additives used to increase retention and selectivity of basic analytes is the inorganic chaotropic agents (such as $PF_6^-$, $BF_4^-$, $ClO_4^-$ ions), which form neutral ion pairs under acidic conditions in RPC [1]. Chaotropes are better suited for gradient analysis and produce lower baseline shifts; however, they are not MS-compatible.

### 2.3.6 High-pH Mobile Phases

Prior to the 1990s, the use of a high-pH mobile phase was not feasible with silica-based columns due to the dissolution of the silica support at pH > 8. The development of improved bonding chemistries and hybrid particles now extends the pH range to 2–10 or even wider (see Chapter 3). These newer high-pH-compatible columns offer an attractive approach to the separation of water-soluble basic drugs (e.g. Opioids) [3, 15]. Figure 2.13 illustrates the basis of this approach in the separation of two closely related basic antidepressant drugs, amitriptyline, and nortriptyline. At low pH, both analytes are ionized and coelute with the solvent front. At pH close to the $pK_a$ of the analytes, the analytes become partially ionized and separate with a large selectivity ($\alpha$) value. At high pH, the nonionized solutes are well retained and resolved. The advantages of high-pH separation versus ion pairing are better MS compatibility and sensitivity (symmetrical peak shapes) (see Section 10.3.3) [14].

Figure 2.14 illustrates the effects of mobile phase pH on solute retention and selectivity in the analysis of a mixture of acidic and basic drugs. Component 1 (acetaminophen) is neutral, components 2–5 are weak bases, and component 6 (ibuprofen) is a carboxylic acid. At low mobile phase pH of 2.5, bases are not well-retained while ibuprofen elutes very late. In contrast, at high pH, the reverse is true. The retention of acetaminophen is not affected by mobile phase pH.

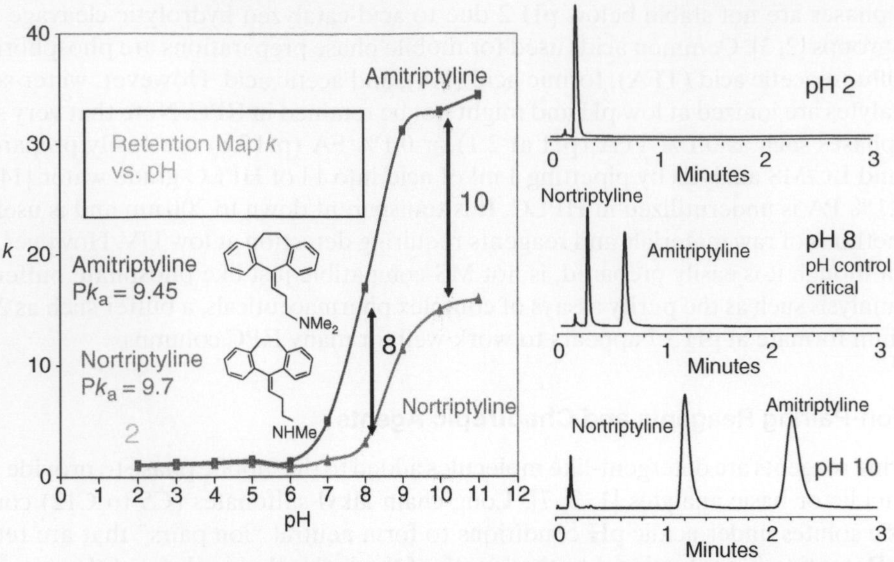


**Figure 2.13.** Retention map and chromatograms of two basic antidepressants using mobile phases at various pH values with the percentage of organic modifier being kept constant. The diagram illustrates the importance of pH in the separation of basic analytes. Source: Reproduced with permission of Waters Corporation.

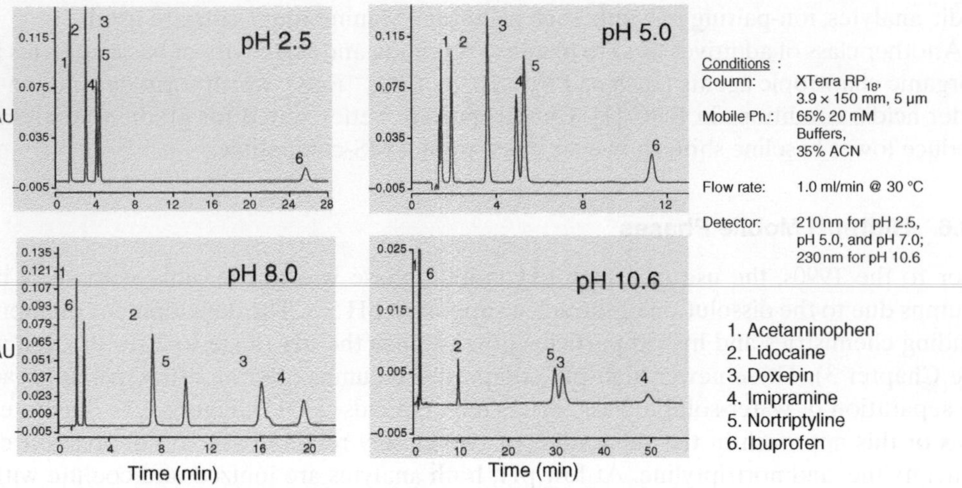


**Figure 2.14.** Selectivity effect of mobile phase pH on six drugs. Source: Courtesy of Waters Corporation.

### 2.3.7 Other Operating Parameters: Flow Rate ($F$) and Column Temperature ($T$)

Typical flow rates ($F$) for analytical columns (4.6 mm i.d.) are 0.5–2 ml/min. Operating at higher flow rates ($F$) increases column back pressure ($\Delta P$) [1, 6] but reduces analysis time:

$$\Delta P = 1000 \frac{F \eta L}{\pi r^2 d_p^2} \tag{2.19}$$

where $L$ = column length, $\eta$ = mobile phase viscosity, $r$ = column radius, and $d_p$ = packing particle diameter.

For isocratic analysis, the flow rate has no impact on $k$ or $\alpha$ since it has the same effect on $t_R$ of each solute. However, the flow rate has a significant effect on $N$, as shown in Section 2.5. Operating flow rates are proportional to the square of the column inner diameter to achieve the same linear velocity through the column. For instance, reducing the column diameter from 4.6 to 2.1 mm, the operating flow rate should be reduced from 1 ml/min to $\{1\ \text{ml/min} \times (2.1^2/4.6^2)\} = 0.21$ ml/min, resulting in significant reduction in solvent consumption. This aspect is important in method conversion from HPLC to UHPLC conditions using columns of 2.1–3.0 mm i.d.

Higher column temperatures ($T$) lower the viscosity of the mobile phase (thus, column back pressure, see Eq. (2.19)) and have significant effects on retention ($k$). Efficiency ($N$) and selectivity ($\alpha$) are also affected to some degree. Some of these effects are discussed further in Chapter 8.

## 2.4 THE RESOLUTION EQUATION

The degree of separation or resolution ($R_s$) between two solutes is dependent on both thermodynamic factors ($k$ and $\alpha$) and kinetic factors ($w_b$ and $N$) [1–3, 6, 7, 12]. Under isocratic conditions, resolution is controlled by three factors (retention, selectivity, and efficiency) as expressed in a quasi-quantitative relationship in the "Resolution Equation": [1]

$$R_s = \left(\frac{k}{k+1}\right) \left(\frac{\alpha - 1}{\alpha}\right) \left(\frac{\sqrt{N}}{4}\right) \tag{2.20}$$

Retention Selectivity Efficiency

To maximize $R_s$, $k$ should be relatively large, though any $k$ values >10 would drive the retention term of $k/(k+1)$ to approach unity. Separation is not achieved if $k = 0$, as $R_s$ becomes zero if k is zero. Selectivity ($\alpha$) must be greater than 1.0 for separation of any closely eluting solutes.

When $\alpha = 1$, $R_s$ would equal to zero. The selectivity is maximized by optimizing column and mobile phase conditions during method development. Figure 2.12 illustrates how resolution can be enhanced by exploiting the selectivity effect of the mobile phase (i.e. by switching from 60% ACN to 60% MeOH). Note that a small change of selectivity can have a significant effect on the resolution as $R_s$ is proportional to $(\alpha - 1)$. Columns of different bonded phases (i.e. C18, phenyl, CN, polar-embedded, see Chapter 3) [2, 6] can also provide different selectivity.

Finally, $N$ should be maximized by using a longer column or a more efficient column packed with smaller particles. Increasing $N$ is a less effective way to achieve resolution since $R_s$ is proportion to $\sqrt{N}$. Doubling $N$ by doubling the column length increases analysis time by a factor of 2 but increases $R_s$ only by $\sqrt{2}$ or by 41%. In contrast, increasing $\alpha$ from 1.05 to 1.10 would almost double the resolution since $R_s$ is proportional to $\alpha - 1$ and $\alpha$ is usually close to one. Nevertheless, for complex samples with many peaks, increasing $N$ can be the most direct approach to increase overall $R_s$. As discussed in Chapter 5, UHPLC is useful for the analysis of complex samples since it allows the use of columns containing small particles and generating higher $N$ per unit column length.

Figures 2.15 and 2.16 show examples of the effect of $k$ and $\alpha$ on peak resolution. In Figure 2.15, seven substituted phenols in a test mixture can be separated by lowering the solvent strength from 45% to 30% ACN. At mobile phase B% (MPB%) of 45% ACN, five

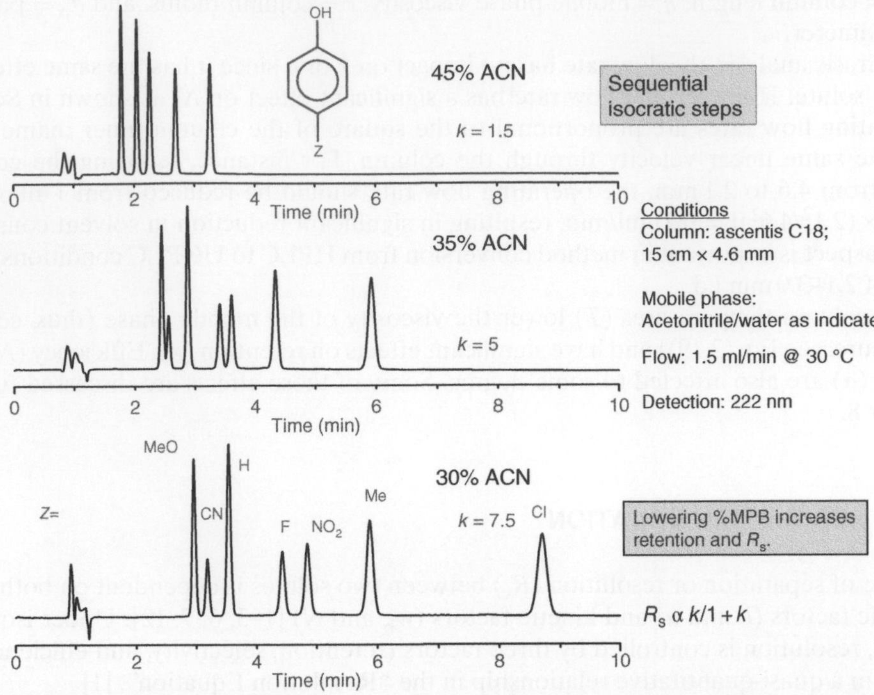


**Figure 2.15.** Examples illustrating the effect of $k$ on the resolution. Both retention ($k$) and resolution typically are increased by using lower solvent strength mobile phase as shown in the separation of seven substituted phenols. This approach is often used to development isocratic methods by lowering the solvent strength known as "Sequential Isocratic Steps." Source: Courtesy of Supelco. (Millipore Sigma).

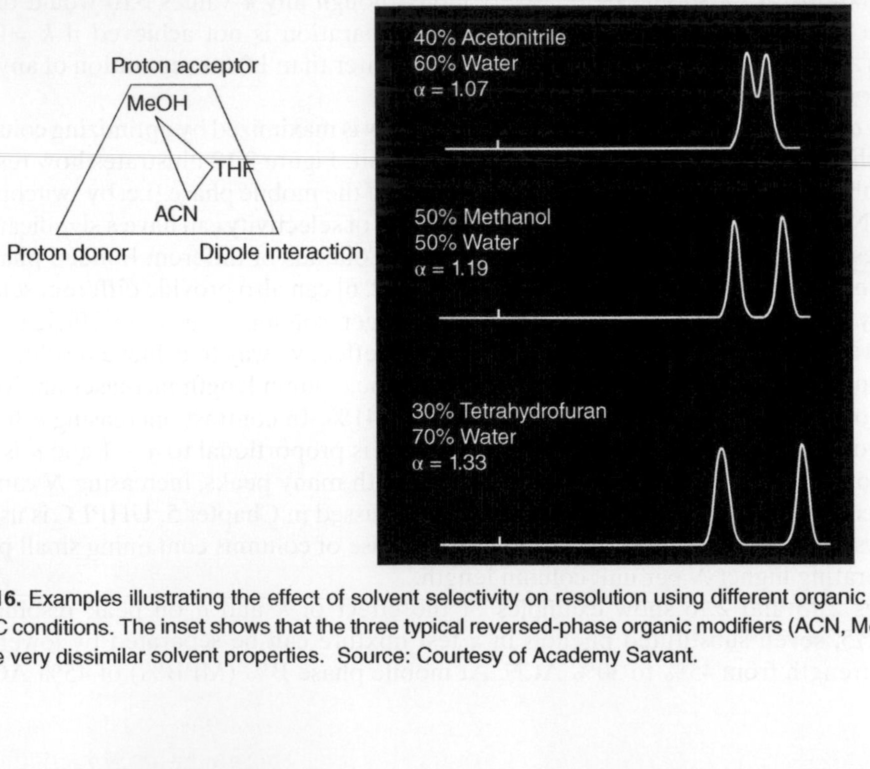


**Figure 2.16.** Examples illustrating the effect of solvent selectivity on resolution using different organic modifiers under RPC conditions. The inset shows that the three typical reversed-phase organic modifiers (ACN, MeOH, and THF) have very dissimilar solvent properties. Source: Courtesy of Academy Savant.

peaks are separated, and the last peak (chlorophenol) elutes at a $k$ of 1.5. At 35% ACN, all seven substituted phenols are separated although the first two peaks are partially resolved. The $k$ of the last peak is 5. At 30% ACN, all seven phenols are baseline-resolved, and the last peak elutes at a $k$ of 7.5. This method development approach by lowering the solvent strength is called "Sequential Isocratic Steps" – a common practice for developing isocratic methods. Typically, $R_s$ and $\alpha$ typically would improve by lowering the solvent strength of the mobile phase and peaks would elute at higher $k$ values.

Figure 2.16 illustrates the effect of changing organic solvent in the mobile phase from acetonitrile to methanol to tetrahydrofuran (THF) through increasing $\alpha$. The elution strength under RPC is MeOH < ACN < THF, so a higher percentage of MeOH is needed to maintain the same $k$ shown in the middle chromatogram. MeOH yields higher $R_s$ since $\alpha$ is 1.19. Since $R_s$ is proportional to $\alpha - 1$ (see the resolution equation), the $R_s$ achieved with MeOH is almost three times higher. The highest $R_s$ is obtained with THF at 30% and a $\alpha$ value of 1.33. Unfortunately, THF is rarely used in RPC due to its reactivity with oxygen and toxicity.

## 2.5 THE VAN DEEMTER EQUATION

The Van Deemter equation was derived in the 1950s to explain band broadening in chromatography. It correlates the HETP or plate height ($H$) with linear flow velocity ($V$) [16]. Figure 2.17 shows a typical Van Deemter curve (HETP vs. $V$), which is a composite curve that is governed by three terms that are largely controlled by stationary phase particle size ($d_p$), and diffusion coefficients ($D_m$) of the solutes [1, 2, 5, 6]. The dip, or minimum point on the Van Deemter curve, marks the minimum plate height ($H_{min}$) found at the optimum flow velocity ($V_{opt}$).

$$\text{The Van Deemter Equation HETP} = A + B/V + CV \qquad (2.21)$$

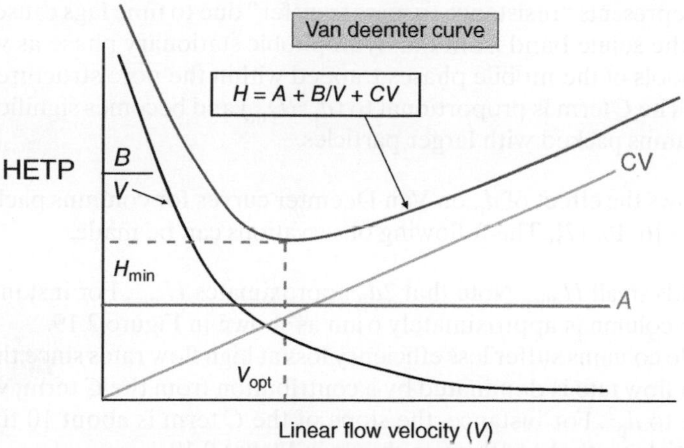


**Figure 2.17.** Van Deemter curves showing the relationship of HETP (H) vs. average linear velocity. The Van Deemter curve has a classic shape and is a composite plot of $A$, $B/V$, and $CV$ terms (plotted below to show their contributions). $H_{min}$ = minimum plate height, $V_{opt}$ optimum velocity.

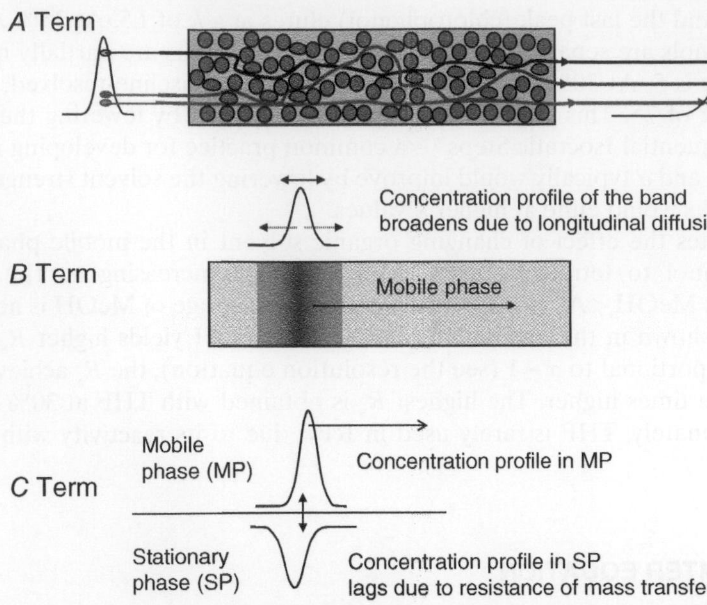


**Figure 2.18.** Diagrams illustrating the Van Deemter terms and the mechanism contributing to chromatographic band broadening.

Figure 2.18 diagrammatically illustrates the three terms in the Van Deemter equation.

- The $A$ term represents eddy diffusion or "multi-path effect" experienced by the solutes as they transverse the packed bed. The $A$ term is proportional to $d_p$ and is approximately equal to $2d_p$ in well-packed columns.
- The $B$ term represents "longitudinal or passive diffusion" of the solute band in the mobile phase and is proportional to $D_m$ of the solute. Note that contribution from the $B$ term is only important at a very low flow rate.
- The $C$ term represents "resistance to mass transfer" due to time lags caused by the slower diffusion of the solute band from the hydrophobic stationary phase as well as from the "stagnant" pools of the mobile phases trapped within the pore structure of the packing material [1]. The $C$ term is proportional to $(d_p^2/D_m)$ and becomes significant at high flow rates for columns packed with larger particles.

Figure 2.19 shows the effect of $d_p$ on Van Deemter curves for columns packed with 10-, 5-, and 3-μm particles [6, 16, 17]. The following observations can be made.

- Small $d_p$ yields small $H_{min}$. Note that $2d_p$ approximates $H_{min}$. For instance, the $H_{min}$ for 3 μm particle column is approximately 6 μm as shown in Figure 2.19.
- Small-particle columns suffer less efficiency loss at high flow rates since the Van Deemter curve at high flow rate is dominated by a contribution from the $C$ term, which is, in turn, proportional to $d_p^2$. For instance, the slope of the $C$ term is about 10 times smaller for the 3 μm particle vs. that of 10 μm as shown in Figure 2.19.

For these reasons, smaller-particle columns (i.e. 3 μm, sub-3 μm, and sub-2 μm) are popular in high-speed applications [17, 18]. For the same reason, columns packed with sub-2 μm particles would require the use of UHPLC equipment as discussed in Chapter 5.

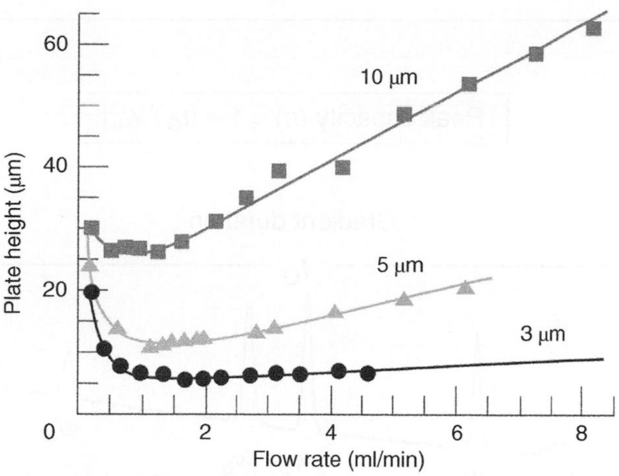


**Figure 2.19.** Van Deemter curves of columns packed with 10-, 5-, and 3-μm particles. Source: Reprinted with permission of Dong and Gant 1984 [17].

## 2.6 ISOCRATIC VS. GRADIENT ANALYSIS

In the early days of HPLC, most separations were performed under isocratic conditions in which the same mobile phase was used throughout the elution of the entire sample. An isocratic analysis is useful for simple mixtures, but a gradient analysis, in which the strength of the mobile phase is increased with time during sample elution, is required for more complex samples containing analytes of diverse polarities [1–5].

Advantages of gradient analysis are as follows:

- Better suited for complex samples that require quantitation of all peaks or multiple analytes of diverse polarities
- Better resolution of early and late eluting peaks
- Peaks have similar peak widths throughout the entire run
- Better sensitivity for late eluting peaks
- Higher peak capacity

Disadvantages are as follows:

- More complex HPLC instrument (i.e. a binary to quaternary pump) is required
- Development gradient methods are more difficult and time-consuming
- Transfer of gradient methods for regulated testing from one lab to another is more difficult
- Longer run times are required due to column equilibration

Several additional parameters are required to be optimized during gradient method development [2, 3]. These are initial and final mobile phase composition, gradient time ($t_G$), and flow ($F$).

### 2.6.1 Peak Capacity (*n*)

Isocratic analysis produces peaks that are broadened with elution time. In contrast, peaks have similar peak widths in gradient elution since they are eluted with an increasingly

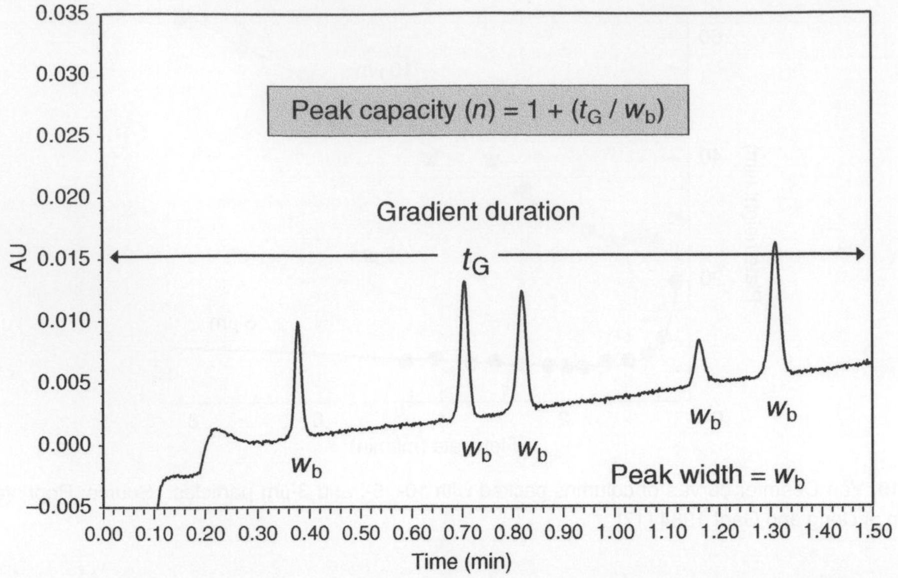


**Figure 2.20.** Chromatogram illustrating peak capacity (n), which is the maximum number of peaks that can be accommodated in a chromatogram with a resolution of 1. Source: Courtesy of Waters Corporation.

stronger mobile phase. Thus, $N$ cannot be measured using Eqs. (2.11) and (2.12). Peak capacity is a useful concept for comparing column performance under different gradient conditions. Peak capacity ($n$) [2, 19] is the maximum number of peaks that can fit in a chromatogram with a resolution value of one. Peak capacity can be calculated using Eq. (2.22).

$$\text{Peak Capacity}, n = 1 + t_G/w_b \tag{2.22}$$

Figure 2.20 illustrates how peak capacities can be approximated by $t_G/w_b$, where $t_G$ is the gradient time (must be linear), and $w_b$ is the average peak width. Higher peak capacities of 100–200 are possible in HPLC gradient analysis versus 50–100 for obtained using isocratic analysis. An important benefit of UHPLC is its higher peak capacities, which are in the range of 400–800 and will be discussed in Chapter 5.

### 2.6.2 Gradient Parameters (Initial and Final Solvent Strength, Gradient Time ($t_G$), and Flow Rate)

Gradient methods are more difficult to develop because the separation is controlled by several additional parameters, such as initial and final solvent strength, flow rate ($F$), and gradient time ($t_G$). The concept of retention factor $k$ is also less intuitive in gradient analysis and is best represented by an average $k$ or $k^*$ [2, 20, 21], which can be calculated using Eq. (2.23).

$$k^* = \frac{t_G F}{1.15 S \Delta\Phi V_M} \tag{2.23}$$

here $k^*$ = average $k$ under gradient conditions, $\Delta\Phi$ = change in volume fraction of strong solvent in RPC, $S$ = a constant that is close to 5 for small molecules, $F$ = flow rate (ml/min), $t_G$ = gradient time (min), and $V_M$ = column void volume (ml).

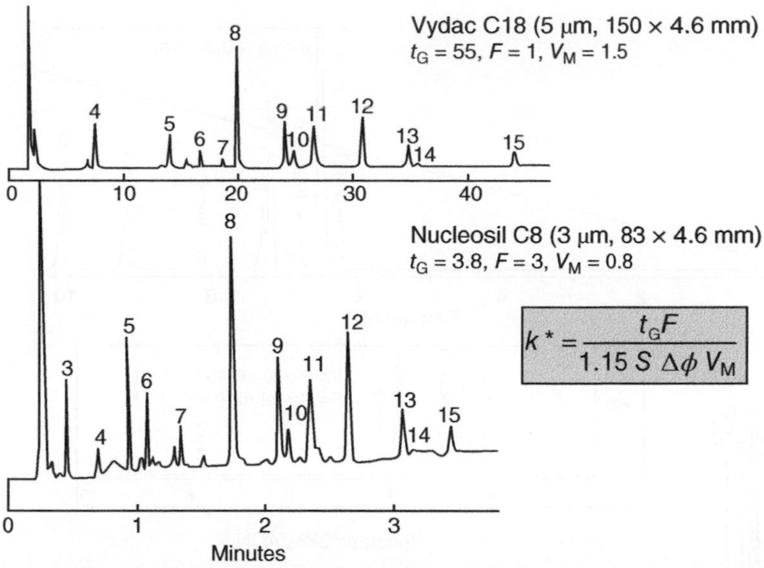


**Figure 2.21.** Two HPLC gradient chromatograms (tryptic maps of lysozyme) illustrating the dramatic effect of flow rate ($F$), gradient time ($t_G$), and holdup volume ($V_M$) on analysis time. Source: Reproduced with permission of Marcel Dekker Inc [21].

Unlike for isocratic analysis, $F$ has a dramatic influence on retention ($k^*$) in gradient analysis. For instance, operating at higher $F$ increases $k^*$, since a higher volume of a lower-strength mobile phase is pumped through the column. This is equivalent to operating at a lower $F$ or a longer $t_G$. Figure 2.21 compares chromatograms for two gradient HPLC peptide maps of a lysozyme protein digest [21] showing how a shorter analysis time can be achieved by reducing $t_G$, increasing $F$, and using a smaller column (lower $V_M$) packed with small $d_p$. Today, high-throughput screening (HTS) assays using ballistic gradients ($t_G$ of 1–2 minutes) are routinely performed using short columns (e.g. 50×2.1 mm) packed with sub-2-μm particles at high $F$ (e.g. 1.5 ml/min) [22].

### 2.6.3 The 0.25 $\Delta t_G$ Rule: When Is Isocratic Analysis More Appropriate?

Isocratic methods are typically used for the quantitation of a single analyte in the sample (or several analytes of similar polarities). Gradient methods are often required for multicomponent assays of complex samples or screening new samples of unknown composition. A rule of thumb called the "0.25 $\Delta t_G$ Rule" is used to determine whether a sample run under gradient conditions can be more effectively handled by isocratic analysis [1]. In this approach, the sample is first analyzed under RPC using a full broad gradient (e.g. 5%–95%B). If all analyte peaks elute within a time span of 0.25 $\Delta t_G$, then an isocratic analysis is preferred. Above 0.40 $\Delta t_G$, a gradient is necessary. Between 0.25 and 0.4 $\Delta t_G$, a gradient is likely preferred. Figure 2.22 illustrates this rule. Here all six impurities from "imp1" to "imp6" plus the main component and a precursor elute within 0.35 $\Delta t_G$. An isocratic analysis is possible at 38%B, but the elution time of the last two impurities 5 and 6 are 20 and 29 minutes, respectively, with broad peaks for these two late elutors. However, if only the first six components in the sample are considered, which elute within 0.2 $\Delta t_G$, an isocratic analysis is preferred as shown in the inset of Figure 2.22 where all six components are baseline-resolved.

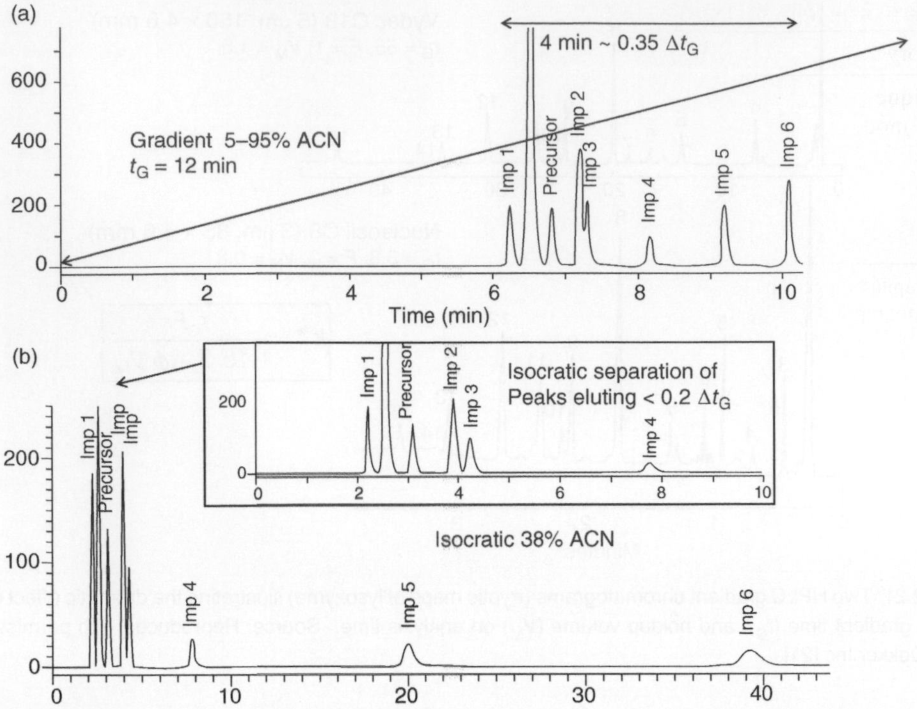


**Figure 2.22.** An example illustrating the 0.25 Δ$t_G$ rule used to indicate when a gradient separation can be more effectively carried out under isocratic conditions. (a) The chromatogram is obtained under gradient conditions where all eight peaks elute in four minutes or 0.35 Δ$t_G$. (b) The chromatogram shows the separation of these eight components under isocratic conditions. If all peaks elute <0.25 Δ$t_G$ in the gradient chromatogram, isocratic separation is preferred (as shown in the inset where the first six peaks are well separated under isocratic conditions of 38% ACN. If all peaks eluting between 0.25 and 0.4 Δ$t_G$, then both isocratic and gradient elution can be used. If all peaks elute in a time span >0.4 Δ$t_G$, gradient separation is preferred.

## 2.7 THE CONCEPT OF ORTHOGONALITY AND SELECTIVITY TUNING

How can one be assured that all the components in the sample are resolved? This issue is particularly important in critical assays such as pharmaceutical impurity testing to ensure that no impurities are coeluting with other components or are hidden under the active pharmaceutical ingredient peak (API or main component). The standard practice is the use of an "orthogonal" separation method [1] to demonstrate that all impurities are accounted for. Ideally, an orthogonal method is one based on a different separation mechanism from the primary method. Some examples of orthogonal methods are shown in Table 2.4. While true orthogonal separation modes based on different retention mechanisms can be used (HPLC vs. CE, GC, or SFC [supercritical fluid chromatography]) or different HPLC modes (RPC vs. NPC, IEC [ion-exchange chromatography], or HILIC [hydrophilic interaction chromatography]), most practitioners prefer to use different variants of RPC such as RPC at different pH's, or using different columns (C18 vs. phenyl, pentafluorophenyl, or polar-embedded). These approaches of changing $\alpha$ (called selectivity tuning) rather than the separation mechanism works well in practice since the overall sample chromatographic profile is preserved but with some changes in the elution pattern [3].

**Table 2.4. Examples of "Orthogonal" Separation Techniques**

| Category | Primary | Orthogonal |
|---|---|---|
| Technique | HPLC | CE, GC, SFC |
| HPLC mode | RPC | IEC, NPC, HILIC |
| Variants of RPLC | RPC | RPC with ion pairing |
| | RPC at low pH | RPC at high pH |
| | RPC with C8 or C18 columns | RPC with phenyl, PFP, or polar-embedded columns |

CE = capillary electrophoresis, SFC = supercritical fluid chromatography, HILIC = hydrophilic interaction liquid chromatography, GC = gas chromatography, PFP = pentafluorophenyl.

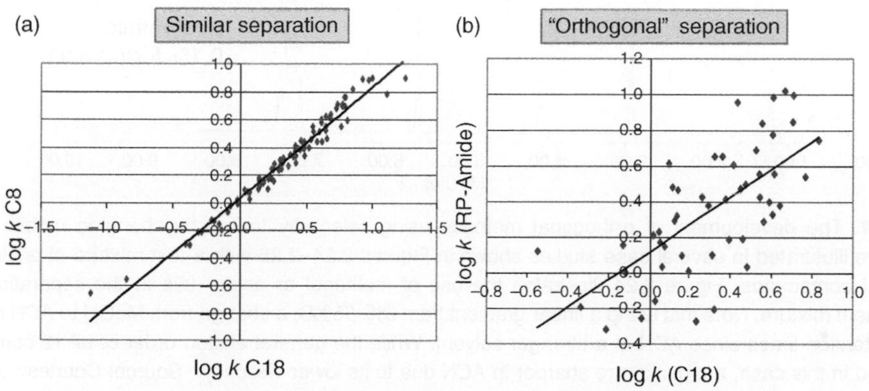


**Figure 2.23.** The concept of orthogonality as shown by retention plots of two sets of columns for a variety of different analytes. (a) Since the log *k* data of the two columns (C8 and C18) are well correlated for most analytes, these two columns are expected to yield similar elution profiles. (b) The selectivity differences of a C18 and a polar-embedded phase (amide) column lead to a very poor correlation with their respective retention data. Methods using a C18 and a polar-embedded column are therefore termed "orthogonal" and hence are expected to yield very different chromatographic profiles. Source: Courtesy of Supelco (Millipore Sigma).

Figure 2.23 shows two retention charts to illustrate the concepts of a pair of similar columns (C18 and C8) and "orthogonal" columns (C18 and amide). A set of 30 pharmaceutical compounds are eluted under identical gradient conditions, and the log *k* data for each component of each column is plotted against each other. Since C8- and C18-bonded phases are very similar, the log *k* data for each component on the two columns are well correlated for all analytes. Thus, methods using C8 and C18 columns are expected to yield similar elution profiles, and coeluting peaks on one column are not likely to be resolved using the other similar column. In contrast, the selectivity differences of a C18 and a polar-embedded phase (amide) column lead to a very poor correlation. A pair of coeluting peaks on the C18 column is likely to be resolved when analyzed on the amide column. RPC methods using a C18 and a polar-embedded column are termed "orthogonal" and expected to yield very different profiles. Another example of "orthogonal" separation can be found in Figure 10.6.

The development orthogonal methods using selectivity tuning by changing mobile phases are illustrated in several case studies shown in Figure 2.24 (MeOH and ACN), Figure 2.25 (C18 and pentafluorophenyl column), and Figure 2.26 (low and high pH mobile phase) using a test mixture consisting of basic, acidic, and neutral compounds. Selectivity tuning is the

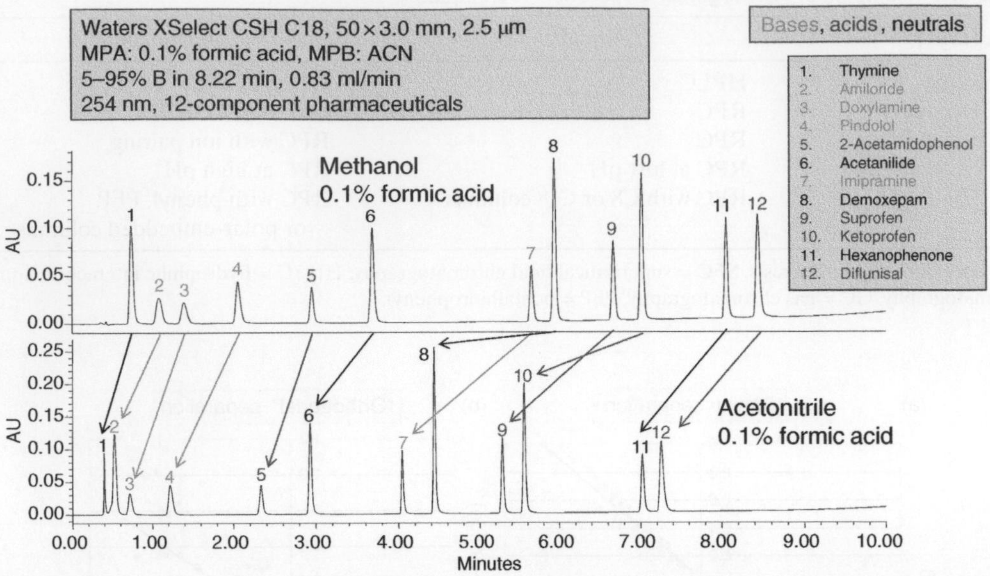


**Figure 2.24.** The development of orthogonal methods using selectivity tuning by changing mobile phases, columns are illustrated in several case studies shown in Figures 2.24–2.26 with a test mixture of acidic, basic, and neutral compounds. Figure 2.24 illustrates the use of methanol or acetonitrile in the separation of the 12-component mixture. Note that using a linear gradient from 5%–95%B, a change from MeOH to ACN generally reduces retention times since ACN is a stronger solvent. While the general elution order of all 12 components is preserved in this case, the peaks are sharper in ACN due to its lower viscosity. Source: Courtesy of Waters Corporation.

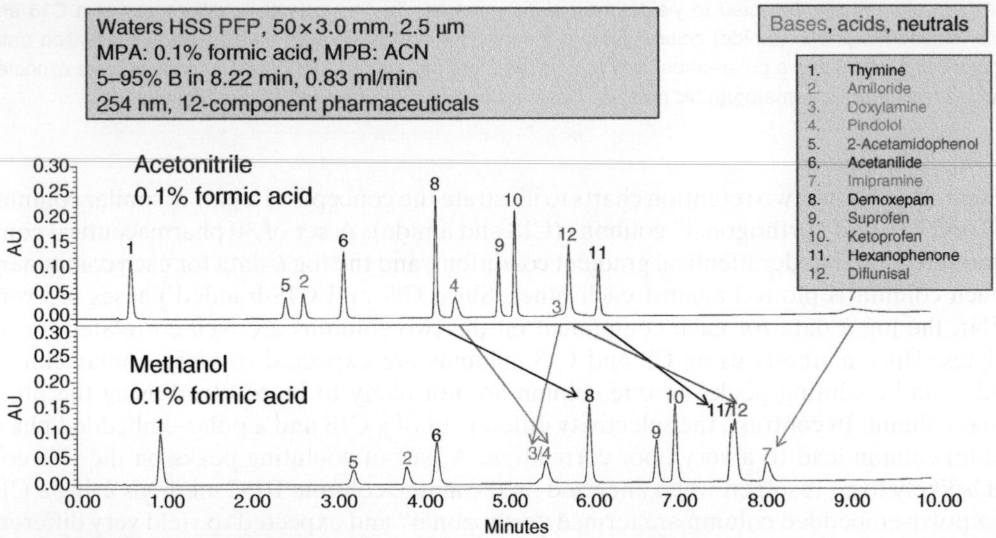


**Figure 2.25.** The elution profile of the 12-component test mix of a pentafluorophenyl column using 0.1% formic acid and ACN or MeOH. Note the changes of elution order in the PFP column that of the C18 column shown in Figure 2.24 and the different selectivity when the two different organic solvents are used as mobile phase B. Source: Courtesy of Waters Corporation.

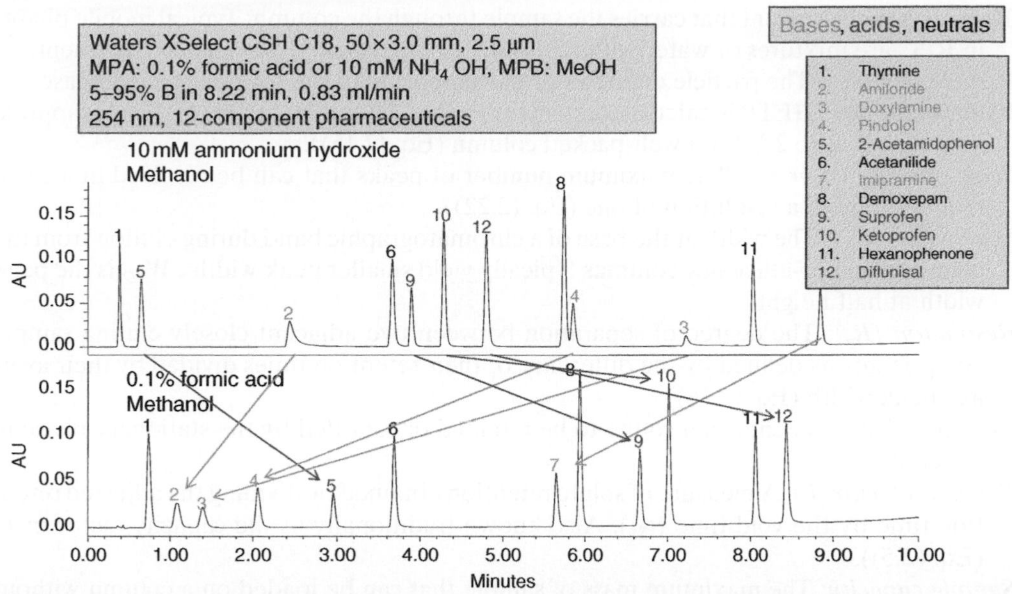


**Figure 2.26.** The use of high- and low-pH mobile phase A in the gradient elution profiles of the 12-component test mix. Note that under acidic pH (0.1% formic acid), all the basic solutes (light gray) have lower retention times since they are ionized while all the acidic solutes (dark gray) have higher retention time since they are now nonionized. The neutral solutes (black) are not affected by the pH of the mobile phase as expected. Source: Courtesy of Waters Corporation.

foundation of HPLC method development of stability-indicating analyses as discussed further in Chapter 10.

## 2.8 SAMPLE CAPACITY

Sample capacity or loading capacity of the column is the maximum amount of solute in milligrams per gram of packing material that can be injected without a significant reduction in column efficiency [1, 2] (e.g. by no more than 10%). Since the amount of packing material is proportional to the column volume, sample capacity is proportional to the square of column diameter and the column length. It is therefore essential in preparative work to maximize the yield of purified material in each sample run by increasing the column size (i.e. column diameter) [1].

## 2.9 GLOSSARY OF HPLC TERMS

*Asymmetry ($A_s$)*: Describes the shape of a chromatographic peak (Eq. (2.17)).

*Band broadening*: The process of increasing peak width via dilution of the solute as it travels through the column.

*Column*: A tube or cylinder that contains the stationary phase. Typical HPLC columns are stainless steel tubes packed with porous silica-based bonded phases.

*Efficiency or plate number (N)*: A measure of column performance. *N* is calculated from retention times and peak widths under isocratic conditions (Eq. (2.10) and (2.11)).

*Mobile phase*: A solvent that carries the sample through the column. Typical mobile phases in RPC are mixtures of water with acetonitrile or methanol. Also called the eluent.

*Particle size ($d_p$)*: The particle diameter of the support material for the bonded phase.

*Plate height ($H$)*: HETP is calculated by dividing the column length by $N$. $H_{min}$ is approximately equal to $2d_p$ for a well-packed column (Eq. (2.15)).

*Peak capacity ($n$ or $P_c$)*: The maximum number of peaks that can be resolved in a chromatogram with a resolution of one (Eq. (2.22)).

*Peak width ($w_b$)*: The width at the base of a chromatographic band during elution from the column. Higher-efficiency columns typically yield smaller peak widths. $W_{1/2}$ is the peak width at half height.

*Resolution ($R_s$)*: The degree of separation between two adjacent closely eluting sample components as defined by the difference of their retention times divided by their average peak width (Eq. (2.16)).

*Retention*: The tendency of a solute to be retained or retarded by the stationary phase in the column.

*Retention factor ($k$)*: A measure of solute retention obtained by dividing the adjusted retention time by the void time ($t_M$). Also known traditionally as the capacity factor or $k'$ (Eq. (2.5)).

*Sample capacity*: The maximum mass of sample that can be loaded on a column without decreasing column efficiency.

*Separation factor or selectivity ($\alpha$)*: The ratio of retention factors ($k$) of two adjacent peaks.

*Stationary phase*: The immobile phase responsible for retaining the sample component in the column. In RPC, this is typically the layer of hydrophobic groups bonded to solid silica support.

*Tailing factor ($T_f$)*: It is a similar term for peak symmetry as defined by the USP [12]. $T_f$ is calculated using the peak width at 5% peak height ($W_{0.05}$) (Eq. (2.18))

*Void volume ($V_M$)*: The total volume of liquid held up in the column. It can be approximated as 65% of the volume of the empty column (Eq. (2.3)).

A complete glossary of liquid-phase separation terms is found in Ref. 10.

## 2.10 SUMMARY AND CONCLUSION

This chapter provides an overview of basic terminologies and essential concepts in HPLC including retention, selectivity, efficiency, resolution, and peak symmetry as well as their relationships with column and mobile phase parameters. Resolution and the Van Deemter equations are discussed along with the concepts of peak capacity, method orthogonality, and gradient parameters. An abbreviated glossary of common HPLC terms is listed. Most concepts and terms in this chapter can be applied to UHPLC, which is predicated on the same principles. Additional discussion on theories of UHPLC can be found in Chapter 5.

## 2.11 QUIZZES

**1.** The correct symbol for retention factor is
(a) $k'$
(b) $k$
(c) $K$
(d) $\alpha$

**2.** Selectivity or $\alpha$ between two adjacent peaks is the ratio of
(a) retention times
(b) $k$
(c) $N$
(d) peak widths

**3.** The minimum resolution for the baseline resolution of two adjacent peaks is equal to
(a) 1.5
(b) 2.0
(c) 1.0
(d) 0.8

**4.** A column void volume ($V_m$) can be estimated by this:
(a) $k$
(b) flow rate
(c) $0.65\,\pi r^2 L$
(d) sample size

**5.** Which relationship is true for the Linear Solvent Strength Theory?
(a) $t_R$ is proportional to %MPB
(b) Log $k$ is proportional to solvent strength
(c) Band broadening is proportional to $k$
(d) Log $k$ is inversely proportional to %MPB

**6.** Which mobile phase additive is NOT MS-compatible?
(a) Ammonia
(b) Phosphate
(c) TFA
(d) Acetate

**7.** Most pharmaceuticals are
(a) acidic
(b) neutral
(c) don't know
(d) basic

**8.** $pK_a$ of acetic acid is
(a) 3.8
(b) 4.8
(c) 0.3
(d) 12

**9.** Peak capacity ($n$) of conventional HPLC under optimum gradient conditions with a high-efficiency column is about
(a) 40 000
(b) 1000
(c) 100
(d) 200

**10.** Peak capacity is dependent on these two factors:
(a) $t_G$ and $W_b$
(b) $\alpha$ and $N$

(c) $k$ and $N$
(d) $t_G$ and $T$

### 2.11.1 Bonus Quiz

From Figure 2.6 and using previous data of $N(1981)$, $\alpha(1.3)$, and $k(3.9)$, calculate the $R_s$ using the resolution equation and compare the calculated value to the actual measured value. How well do the values compare?

$$R_s = \left(\frac{k}{k+1}\right) \left(\frac{\alpha-1}{\alpha}\right) \left(\frac{\sqrt{N}}{4}\right)$$

*Calculated $R_s$ is = 3.9/(3.9 + 1) × (1.3—1)/1.3 × square root (1981)/4 = 2.0, which compares well to measured $R_s$ of 1.8.*

## 2.12 REFERENCES


1. Snyder, L.R. and Kirkland, J.J. (2010). *Introduction to Modern Liquid Chromatography*, 3e. Hoboken, NJ: Wiley. Chapters 2 and 7.
2. Snyder, L.R., Kirkland, J.J., and Glajch, J.L. (1997). *Practical HPLC Method Development*, 2e. New York: Wiley-Interscience.
3. Ahuja, S. and Dong, M.W. (eds.) (2005). *Handbook of Pharmaceutical Analysis by HPLC*. Amsterdam, the Netherlands: Elsevier/Academic Press.
4. Meyer, V.R. (2004). *Practical HPLC*, 5e. New York: Wiley.
5. Dong, M.W. (2006). *Modern HPLC for Practicing Scientists*. Hoboken, NJ: Wiley.
6. Neue, U.D. (1997). *HPLC Columns: Theory, Technology, and Practice*. New York: Wiley-VCH.
7. Stout, T.H. and Dorsey, J.G. (2001). *Handbook of Pharmaceutical Analysis* (ed. L. Ohannesian and A.I. Streeter), 87. New York: Marcel Dekker.
8. Separation Science. Provides Online Training, Webinars, and Conferences in Separation Science, United Kingdom. http://www.sepscience.com/Techniques/LC (accessed 12 January 2019).
9. LC Resources, Training Courses, Lafayette California. http://www.lcresources.com/training.html.
10. Majors, R.E. and Carr, P.W. (2001). *LCGC* 19 (2): 124.
11. Tijssen, R. (1998). *Handbook of HPLC* (ed. E. Katz, R. Eksteen, P. Schoenmakers and N. Miller). New York: Marcel Dekker.
12. (2018). *United States Pharmacopeia 41 – National Formulatory 36*. Rockville, MD: The United States Pharmacopeial Convention.
13. Snyder, L.R. and Dolan, J.W. (2006). *High-Performance Gradient Elution: The Practical Application of the Linear-Solvent-Strength Model*. Hoboken, NJ: Wiley.
14. Dong, M.W. and Boyes, B.E. (2018). *LCGC North Am.* 36 (10): 752.
15. Dong, M.W., Miller, G., and Paul, R. (2003). *J. Chromatogr.* 987: 283.
16. Van Deemter, J.J., Zuidenweg, F.J., and Klinkenberg, A. (1956). *Chem. Eng. Sci.* 5: 271.
17. Dong, M.W. and Gant, J.R. (1984). *LCGC* 2 (4): 294.
18. Guillarme, D. and Dong, M.W. (2013). *Amer. Pharm. Rev.* 16 (4): 36.
19. Neue, U.D. and Mazzeo, J.R. (2001). *J. Sep. Sci.* 24: 921.
20. Snyder, L.R. and Stadalius, M.A. (1986). *High-Performance Liquid Chromatography: Advances and Perspectives*, vol. 4 (ed. C. Horvath), 195. New York: Academic Press.
21. Dong, M.W. (1992). *Advances in Chromatography*, vol. 32 (ed. P. Brown), 21. New York: Marcel Dekker.
22. Wong, M., Murphy, B., Pease, J.H., and Dong, M.W. (2015). *LCGC North Am.* 33 (6): 402.

# 3

# HPLC COLUMNS AND TRENDS

## 3.1 SCOPE

The high-performance liquid chromatography (HPLC) column is the heart of an HPLC system. It holds the support medium for the stationary phase that can provide differential retention of sample components. This chapter offers a concise overview of HPLC columns and their modern trends. Basic HPLC column parameters such as column types (silica, polymer), modes (NPC, RPC, IEC, SEC, SFC, HILIC, HIC, MMC, chiral), dimensions (preparative, analytical, narrow-bore, micro-LC), and packing characteristics (particle and pore size, bonding chemistry) are presented with recent column trends (high-purity silica, hybrid particles, superficially porous particles [SPPs], novel bonding chemistries, and ultra-high-pressure liquid chromatography [UHPLC] columns packed with sub-3- and sub-2-μm particles).

The benefits of shorter and narrower columns packed with smaller particles (Fast LC, UHPLC, micro-LC) are described. Guard columns and specialty columns are discussed together with overall column selection recommendations. For more detailed discussions on column technologies and fundamentals, the reader is referred to Uwe Neue's comprehensive book on HPLC columns [1], other textbooks [2–7], and journal articles [8, 9]. Other useful resources include manufacturers' websites that describe their current column products. As in other chapters, the focus is on reversed-phase HPLC columns for small-molecule drug analysis, which represents a major fraction of all HPLC applications. A glossary of terminologies regarding column and related instrumental aspects is included here. Information on column operation is found in Chapter 7. Additional HPLC and UHPLC applications are found in Chapters 5, 9, 12, and 13 including those for large biomolecules.

### 3.1.1 Glossary and Abbreviations

A list of common terms and abbreviations associated with HPLC columns and related instrumental concepts is shown as follows:

---

*HPLC and UHPLC for Practicing Scientists*, Second Edition. Michael W. Dong.

| | |
|---|---|
| Column support | Spherical/irregular, porous particles used as column packing material |
| Bonded phases | Stationary phase; support material modified with functional groups |
| Packing material | A term that can refer to both the support or the bonded phase |
| $L$, $d_c$, $d_p$, $d_{pore}$, $r$ | Column length; inner diameter; particle size; pore size; radius (inner) |
| HETP, H | Height equivalent to a theoretical plate; plate height |
| $H_{min}$, $v_{opt}$ | Minimum plate height; optimum velocity |
| $N$, $\alpha$, $k$, $R_s$ | Plate count; selectivity = $k_2/k_1$; retention factor; resolution |
| $V_m$, $t_M$ | Column void volume; retention time of void marker |
| $F$, $t_G$, $P_c$, $\Delta P$ | Flow rate; gradient time; peak capacity; backpressure |
| $W_b$, $W_{1/2}$ | Peak width at the base; peak width at half height |
| HILIC, HIC, SFC | Hydrophilic interaction, hydrophobic interaction, supercritical fluid chromatography |
| IBW | Instrumental bandwidth or system dispersion |
| Hybrid phases | Support synthesized from both inorganic and organic materials |
| SPP | Superficially porous particle; also called solid core, core–shell, or fused-core |
| Dwell volume | Volume measured from the point of the solvents mixing in the HPLC system to the head of the column |
| $\lambda_{max}$, $\varepsilon$ | Wavelength of maximum absorbance; molar absorptivity coefficient |

## 3.2 GENERAL COLUMN DESCRIPTION AND CHARACTERISTICS

A typical HPLC column is a stainless steel tube filled with a packing material of porous particles used for sample separations. HPLC columns can be categorized in many ways:

- *By column hardware*: Standard or cartridge, with column hardware made from stainless steel, polyether ether ketone (PEEK), titanium, PEEK-lined stainless steel, or fused silica
- *By chromatographic modes*: Normal-phase chromatography (NPC), reversed-phase chromatography (RPC), ion-exchange chromatography (IEC), size-exclusion chromatography (SEC), hydrophilic interaction chromatography (HILIC), supercritical fluid chromatography (SFC), mixed-mode chromatography (MMC)
- *By dimensions*: Preparative, semipreparative, analytical, Fast LC, mini-bore, narrow-bore, microbore, micro-LC, and nano-LC (capillary)
- *By support types*: Silica, polymer, hybrid, zirconia, porous graphitic carbon; totally porous, nonporous, superficially porous
- *By pressure rating*: UHPLC, HPLC, medium or low-pressure
- *By applications*: Small molecules, bioseparations, synthetic polymers, chiral, SFC, LC/MS, application-specific

The general characteristics of a typical analytical RPC column in use today are included as follows with the most common values shown in parenthesis:

- *Column length* ($L$): 20–300 mm long (150 mm)

- *Column inner diameters and particle sizes* ($d_c$ *and* $d_p$): 2.1–4.6 mm i.d. packed with sub 2–5-μm particles, silica-based bonded phases (4.6 mm, 3 μm C18), or monoliths
- *Flow rate* ($F$): 0.5–2 ml/min using a mobile phase of methanol or acetonitrile in water or aqueous buffer (1 ml/min)
- *Sample loading*: 1 ng to 1 mg (1 μg)
- *Operating pressure* ($\Delta P$): 500–20 000 psi (3000 psi or 200 bar)
- *Plate count* ($N$): 3000–35 000 plates (20 000 plates)
- *Cost*: $150–$2000 ($800)
- *Column lifetime*: 500–2000 injections (2000 injections). Highly dependent on sample type and level of clean-up as well as mobile phase pH and column temperature.

### 3.2.1 Column Hardware – Standard vs. Cartridge Format

Figure 3.1 shows a picture of HPLC columns of various sizes – from large preparative to smaller and shorter analytical columns. Figure 3.2a shows a schematic of the column inlet and the end-fitting of a "standard" column. A typical analytical column is made from a stainless steel tube of 1/4″ outer diameter (o.d.), with end-fittings to allow connections to 1/16″ o.d. tubing (see Chapter 7 for details on fittings and procedures on column connections).

The packing material is held in place inside the cylinder as a packed bed with a set of 0.5- or 2.0-μm porous end frits. For bioseparations or ion-chromatography, the stainless steel column hardware can be replaced by titanium or PEEK for higher corrosion resistance to mobile phases with high salt contents. Micro-LC columns (<0.5 mm i.d.) are packed in fused silica capillaries coated outside with polyimide for flexibility and durability. Figure 3.2b shows a cartridge column design in which a column with two pressed-in frits is placed inside a cartridge holder. This design has a cost advantage since the actual column has no end-fittings, though a cartridge holder must be purchased for a first-time user. Cartridge columns can be changed by unscrewing the knurled nuts at both ends without disconnecting the attached tubing to the end-fittings of the holder. This convenience feature of wrench-free connections of cartridge columns is no longer an advantage as most HPLC column connections are made with reusable finger-tight PEEK fittings.

## 3.3 COLUMN TYPE

An HPLC column may be categorized in many different ways based on the chromatographic mode, dimension, or application.

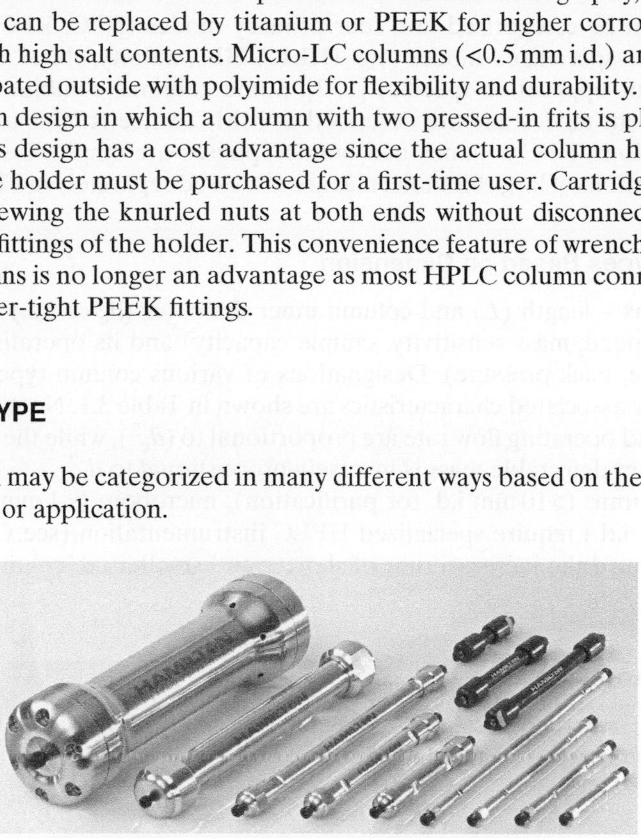

**Figure 3.1.** Picture of HPLC columns of various sizes (preparative, semipreparative, and analytical columns). Source: Courtesy of Hamilton Company.

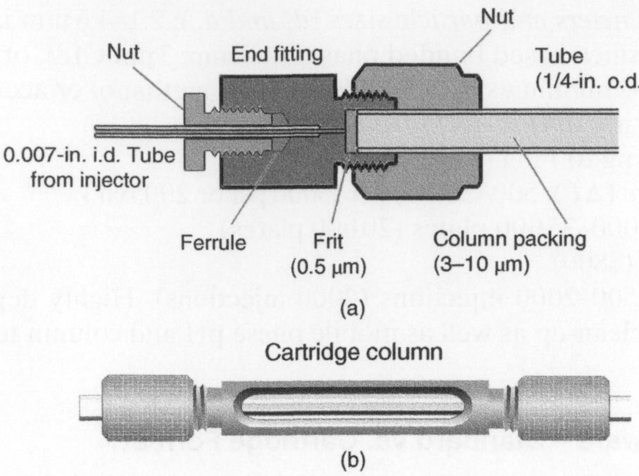


**Figure 3.2.** (a) A cross-sectional diagram of the inlet of a standard conventional HPLC column showing details on the inlet 1/16″ o.d. tube with compression screw and ferrule, the end-fitting, and the inlet 0.5-μm frit. (b) Diagram of a Brownlee cartridge column placed inside an aluminum cartridge holder with finger-tight end-fittings. This was one of the first cartridge column designs. Source: Courtesy of PerkinElmer, Inc.

### 3.3.1 Types Based on Chromatographic Mode

HPLC columns can be categorized into four primary modes (NPC, RPC, IEC, and SEC) in addition to other specialized modes (e.g. HILIC, HIC, affinity, chiral, supercritical fluid extraction [SFE], or application-specific) [1, 3]. Some columns combine the attributes of two or more different modes and are described as MMC. Since RPC is used in 60–80% of all HPLC applications, RPC columns and bonded phases are emphasized in this chapter. Columns for separations of large biomolecules are discussed separately in Chapters 12 and 13.

### 3.3.2 Column Types Based on Dimension

Column dimensions – length ($L$) and column inner diameter ($d_c$ or i.d.) – control column performance ($N$, speed, mass sensitivity, sample capacity) and its operating characteristics (optimum flow rate, back pressure). Designations of various column types based on inner diameters and their associated characteristics are shown in Table 3.1. Note that void volume, sample capacity, and operating flow rate are proportional to ($d_c^2$), while the limit of detection (LOD) for minimum detectable mass is inversely proportional to $d_c^2$.

Preparative columns (>10 mm i.d. for purification), microbore (<1 mm i.d.), and micro-columns (<0.5 mm i.d.) require specialized HPLC instrumentation (see Chapter 4). There is a clear trend toward the increased use of shorter and smaller i.d. columns, which will be explored later.

### 3.3.3 Column Length (*L*)

Column length ($L$) determines column efficiency ($N$), analysis speed ($t_R$ of the last peak), and pressure drop ($\Delta P$) (see Chapter 2):

$$\text{Plate number}, N = L/H$$

Figure 3.3 shows five chromatograms of a test mixture using columns of various lengths ($L = 15, 30, 50, 75$, and 150 mm) packed with 3.5 μm C18 bonded phase, illustrating the effect

**Table 3.1. HPLC Column Types Based on Inner Diameters (i.d.)**

| Column[a] type | Typical[a] i.d. (mm) | Typical flow rate (ml/min) | Void volume (ml) | Typical sample loading |
|---|---|---|---|---|
| Preparative | >25 | >30 | >50 | >30 mg[b] |
| Semi-prep | 10 | 5 | 7.5 | 5 mg[b] |
| Conventional | 4.6 | 1–2 | 1.6 | 100 μg |
| Solvent saver | 3.0 | 0.4–1.0 | 0.7 | 40 μg |
| Mini-bore or narrow-bore | 2.1 | 0.2–1.0 | 0.3 | 20 μg |
| Micro-bore | 1.0 | 0.05–0.2 | 0.075 | 5 μg |
| Micro-LC – capillary | <0.5 | ~10 μl/min | 20 μl | 1 μg |
| Nano-LC – capillary | <0.1 | ~0.5μl/min | 1 μl | 0.05 μg |

[a]Designations of these column types are not universally accepted and may vary with manufacturers. Column i.d. is typical. Void volumes are based on lengths of 150 mm.
[b]Sample loading for preparative applications under overload conditions.

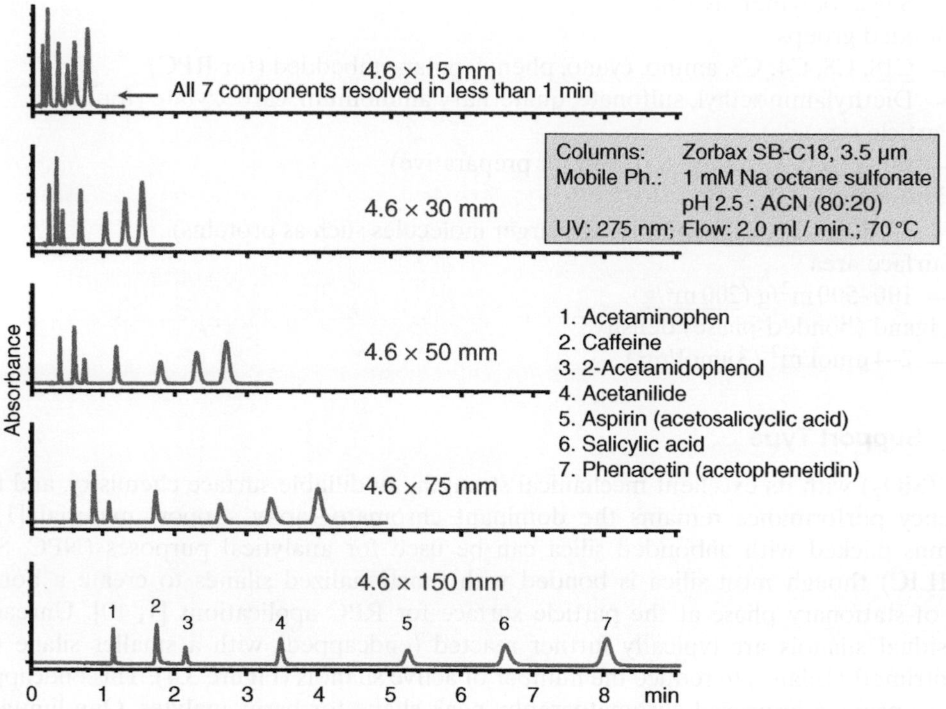


**Figure 3.3.** Chromatograms illustrating the effects of column length on efficiency, analysis time, and resolution. Source: Courtesy of Agilent Technologies.

of $L$ on $N$, analysis time, and $R_s$. Note that longer columns have higher plate counts ($N$ is proportional to $L$) and yield better resolution ($R_s$ is proportional to $N^{1/2}$ or with $L^{1/2}$). Analysis time and $\Delta P$ are also proportional to $L$. For simple sample mixtures or potency assays, short columns are used to yield sufficient resolution in minimum analysis times.

Since plate numbers are additive, columns can be connected in series to produce higher efficiency and resolution at the expense of longer analysis time and lower mass sensitivity due to more sample dilution. Typical column length ranges from 50 to 250 mm, though the use of shorter columns (30–50 mm) packed with small particles is growing rapidly to increase throughput.

## 3.4 COLUMN PACKING CHARACTERISTICS

The nature and characteristics of a column packing material and its associated stationary phase are essential to the column performance and success of the intended applications [1, 2, 6]. Common types and characteristics of column packings in use today for small-molecule separations are summarized as follows. The most common parameters are underlined.

- Support type (totally porous, superficially porous, nonporous particles, TPP, SPP, or NPP)
  — Silica, polymer, hybrid
- Bonded groups
  — C18, C8, C4, C3, amino, cyano, phenyl, polar-embedded (for RPC)
  — Diethylaminoethyl, sulfonate, quaternary ammonium, carboxylate (for IEC)
- Particle size ($d_p$)
  — sub-2, sub-3, 3, 5 μm (≥10 μm for preparative)
- Pore size ($d_{pore}$)
  — 60–300 Å (100 Å) (≥300 Å for larger molecules such as proteins)
- Surface area
  — 100–500 $m^2/g$ (200 $m^2/g$)
- Ligand (bonded-phase) density
  — 2–4 $\mu mol/m^2$ (3 $\mu mol/m^2$)

### 3.4.1 Support Type

Silica ($SiO_2$) with its excellent mechanical strength, modifiable surface chemistry, and high efficiency performance remains the dominant chromatography support material [1, 5]. Columns packed with unbonded silica can be used for analytical purposes (NPC, SFC, or HILIC) though most silica is bonded with functionalized silanes to create a bonded layer of stationary phase at the particle surface for RPC applications [1, 10]. Unreacted or residual silanols are typically further reacted (endcapped) with a smaller silane (e.g. chlorotrimethylsilane) to reduce the number of active silanols (Figure 3.4). This endcapping tends to provide improved chromatography peak shape for basic analytes. One limitation of silica-based bonded phases is the operating pH range of 2–8. At pH values <2, the bonded phase can be hydrolyzed from the silica support leading to retention changes and poor peak shape. At pH values >8, the silica support begins to dissolve leading to column voiding and a total loss of efficiency. Efforts to extend this silica operating range limitation include the development of silica hybrids, novel bonding chemistries, and advanced coating processes.

Polymer support materials such as cross-linked polystyrene-divinylbenzene, polyethers, and polymethacrylates have been used successfully, mostly in IEC, SEC, and bioseparations [1, 3, 6]. Their strength and performance have improved in recent years, though they still lag behind silica in chromatographic efficiency. Major advantages are a wider operating pH range

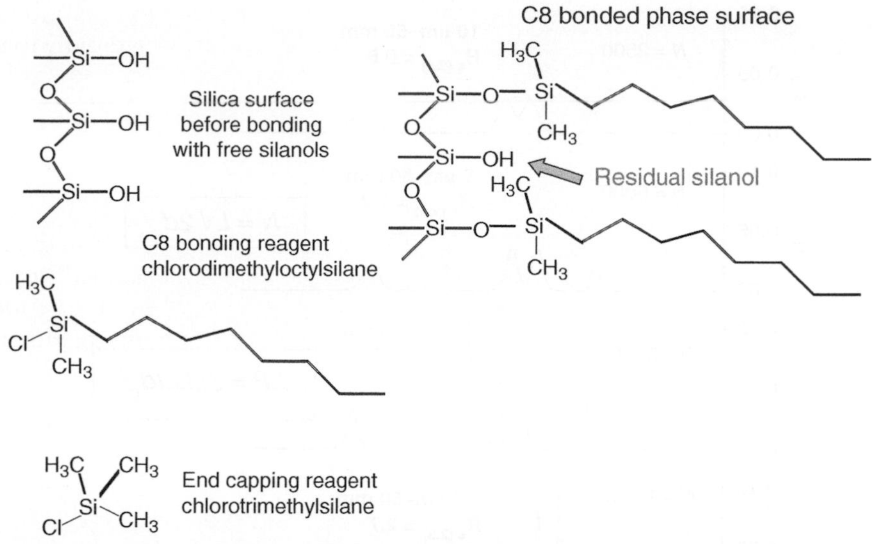


**Figure 3.4.** Structures of the traditional monofunctional silane bonding and endcapping reagents and diagrams of the surfaces of bare silica before bonding and bonded silica with residual silanols. Source: Courtesy of Waters Corporation.

(1–14 for polystyrene-divinylbenzene) and an absence of silanol groups. Other materials used include alumina (occasionally used for NPC or preparative use), titania, and zirconia ($ZrO_2$) [1, 11], which can be coated or bonded to yield stable packing with unique selectivity and high-temperature stability.

### 3.4.2 Particle Size ($d_p$)

Particle size, size distribution, and its mechanical strength are the primary determinants of efficiency and back-pressure of the column [1–3]. The effect of $d_p$ on $H$ is discussed in Chapter 2. For a well packed column, $H_{min}$ is approximately equal to $2d_p$. Also, since the Van Deemter C term is proportional to $d_p^2$, columns packed with small particles have much less efficiency loss at higher flow rates and a wider optimum operating flow range [12]. Despite this advantage, column back-pressure is inversely proportional to $d_p^2$, columns packed with sub-2-μm particles usually require UHPLC equipment.

Figure 3.5 shows the comparative chromatograms of a four-component sample run using four 50-mm-long columns packed with 10, 5, 3.5, and 1.7 μm packings. Note that reducing $d_p$ increases $N$, $R_s$, and peak height or sensitivity. As the $N$ increases sixfold from 2500 (10 μm) to 14 700 plates (1.7 μm), $R_s$ between components 2 and 3 also improves from 0.8 to 2.7. Pressure drop, $\Delta P$, however, increases about 36-fold ($6^2$) as it is inversely proportional to $d_p^2$.

### 3.4.3 Surface Area and Pore Size ($d_{pore}$)

Most chromatographic supports are porous to provide a high surface area to maximize the interaction of the solutes with the bonded stationary phase. Packings for small-molecule separations have pore sizes ranging from 60 to 300 Å and surface areas of 100–500 $m^2/g$. In

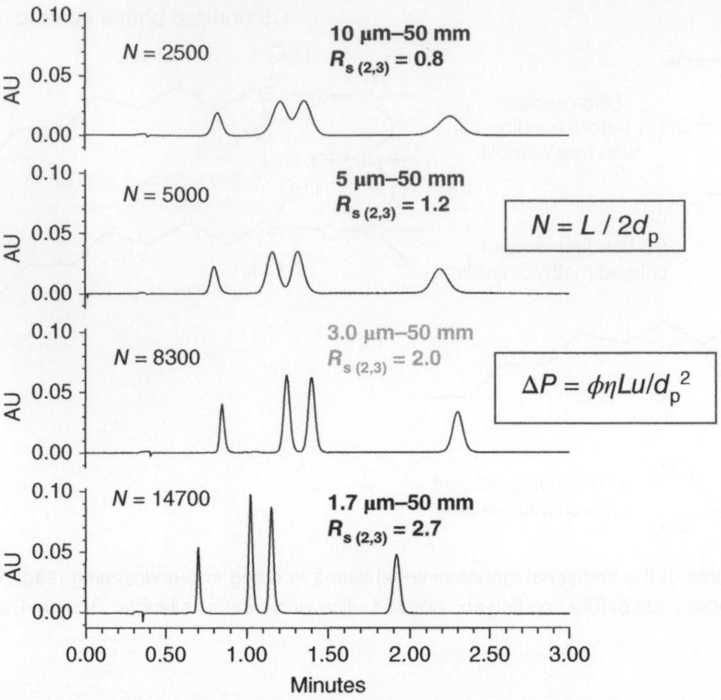


**Figure 3.5.** Comparative chromatograms of four 50-mm-long columns packed with 10, 5, 3.5, and 1.7 μm particles illustrating the enhanced efficiency and resolution performance of columns packed with smaller particles. Source: Courtesy of Waters Corporation.

general, high surface area supports can yield bonded phases with higher bonding density and thus higher retention. Note that small-pore packings are problematic for large biomolecules, which can become entangled in the pores leading to excessive band broadening (see example in Figure 13.27). Thus, wide-pore materials (>300 Å) are used for the separation of large biomolecules [13].

### 3.4.4 Bonding Chemistries

Figure 3.4 shows the traditional bonding chemistry by reacting the surface silanols with a monofunctional C8 reagent such as dimethyloctylchlorosilane to form a layer of hydrophobic C8 stationary phase [1–3, 10]. Since steric hindrance only allows ~50% bonding efficiency [1], the remaining residual silanols are further reacted or "endcapped" with a smaller silane such as trimethylchlorosilane. Some vendors might perform double or triple endcapping to reduce the residual silanol activity. Similarly, other bonding reagents can be used to yield a large variety of bonded phases for RPC, NPC, and IEC. Figure 3.6 shows the most common bonded phases in use today. For RPC, C18 remains the most popular phase supplemented by others such as C8, phenyl, cyano, and polar-embedded phases [8]. For NPC, unbonded silica, amino, and cyano phase are common. For IEC, there are strong and weak IEC phases for both cationic and anionic exchangers.

Figure 3.7 shows eight chromatograms from the analysis of a seven-component sample test mix using columns packed with eight different bonded phases from the same manufacturer.

| | | |
|---|---|---|
| **Reversed-phase (RPC)** | | |
| C18 | Octadecyl | $Si(CH_3)_2$-$(CH_2)_{17}$-$CH_3$ |
| C8 | Octyl | $Si(CH_3)_2$-$(CH_2)_7$-$CH_3$ |
| CN | Cyano | $Si(CH_3)_2$-$CH_2$-$CH_2$-$CH_2$-CN (with propyl linker) |
| ϕ | Phenyl | $Si(CH_3)_2$-$(CH_2)_6$-$C_6H_5$(with hexyl linker) |
| Polar-embedded | | $Si(CH_3)_2$-$(CH_2)_3NHCO(CH_2)_{14}CH_3$ (amide group) |
| **Normal-phase (NPC)** | | |
| Si | Silica | Si-OH |
| $NH_2$ | Amino | $Si(CH_3)_2$-$CH_2$-$CH_2$-$CH_2$-$NH_2$ (with propyl linker) |
| **Ion-exchange (IEC)** | | |
| SP | Sulfopropyl | $Si(CH_3)_2$-$CH_2$-$CH_2$-$CH_2$-$SO_3^-$ |
| CM | Carboxymethyl | $Si(CH_3)_2$-$CO_2^-$ |
| DEAE | Diethylaminoethyl | $Si(CH_3)_2$-$CH_2$-$CH_2$-$CH_2$-$NH^+(C_2H_5)_2$ |
| SAX | Triethylaminopropyl | $Si(CH_3)_2$-$CH_2$-$CH_2$-$CH_2$-$N^+(C_2H_5)_3$ |

**Figure 3.6.** Some common bonded phases for RPC, NPC, and IEC.

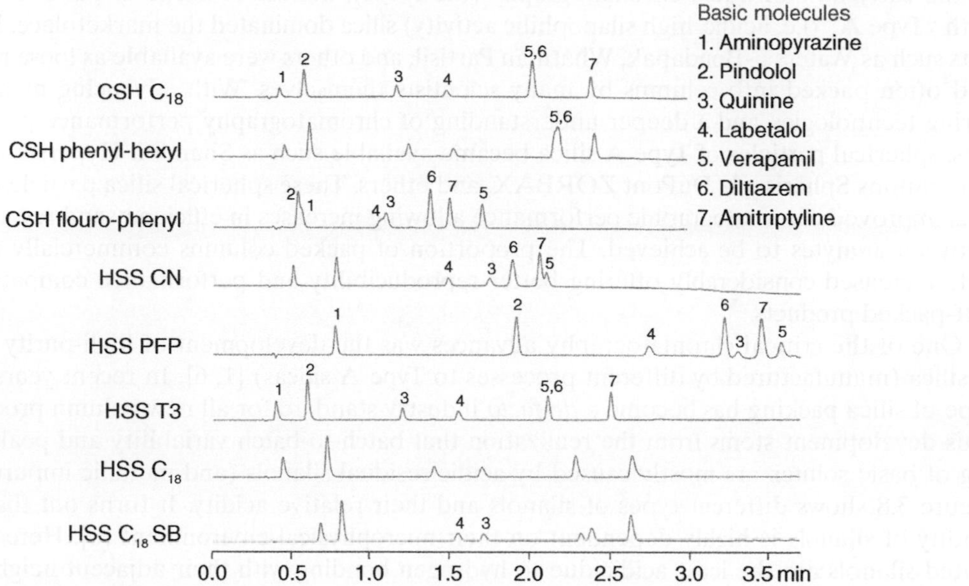


**Figure 3.7.** Comparative chromatograms of the gradient separations of a test mix containing seven pharmaceuticals, illustrating the effect of selectivity of C18 and other bonded phases. All columns are 50 × 2.1 × mm, 1.7 μm, with 10 mM ammonium formate pH 3 and MeOH gradient. Source: Courtesy of Waters Corporation.

Note the differences in elution profiles and selectivity variations particularly for the non-C18 bonded phases with mixed-mode interactions such as the PFP phases. These newer bonding chemistries for improving column performance (i.e. selectivity and peak shape) are discussed in a later section.

## 3.5 MODERN HPLC COLUMN TRENDS

Understanding modern column trends is essential for selecting newer products with higher performance and consistency [8, 9]. Past innovations have targeted the traditional problem areas of silica-based RPC columns. They are as follows:

- Column batch-to-batch inconsistency
- Peak tailing of basic analytes
- Short column lifetime
- pH limitations (2–8)

Many of these problems have been resolved or minimized through developments in the last three decades. This section discusses modern trends of high-purity silica, hybrid and SPPs, and novel bonding chemistries. The benefits of using shorter and narrower columns packed with small particles for fast LC, UHPLC, and micro-LC are discussed with their salient characteristics and applications.

### 3.5.1 Silica Support Material

In the early days of liquid chromatography (the 1970s), the use of irregular particle shape with "Type A" (i.e. acidic, high silanophilic activity) silica dominated the marketplace. Products such as Waters μ-Bondapak, Whatman Partisil, and others were available as loose media and often packed into columns by many scientists themselves. With advancing manufacturing technologies and a deeper understanding of chromatography performance parameters, spherical particles of Type A silica became available such as Shandon Hypersil, Phase Separations Spherisorb, DuPont ZORBAX, and others. These spherical silica particle products improved chromatographic performance allowing increases in efficiency and resolution between analytes to be achieved. The proportion of packed columns commercially available increased considerably offering better reproducibility and performance compared to self-packed products.

One of the critical chromatography advances was the development of high-purity Type B silica (manufactured by different processes to Type A silicas) [1, 6]. In recent years, this type of silica packing has become a *de facto* industry standard for all new column products. This development stems from the realization that batch-to-batch variability and peak tailing of basic solutes are mostly caused by acidic residual silanols (and metallic impurities). Figure 3.8 shows different types of silanols and their relative acidity. It turns out that the acidity of silanols is highly dependent on their microphysical environment [6]. Here associated silanols are the least acidic due to hydrogen bonding with their adjacent neighbors, followed by geminal and free silanols. The worst culprits are the acidic silanols (so-called metal activated or chelated silanols) adjacent to metallic oxide impurities present in the silica particle. Many older silica materials have high metallic contents such as >1000 ppm total metals (e.g. Spherisorb, Hypersil) and therefore a higher proportion of very acidic metal activated silanols. Separations with these stationary phases often require the use of basic additives (triethyl, trimethylamines) in the mobile phase to improve the peak shape of basic analytes.

The proportion and inherent variability of these active (acidic) silanols are also partially responsible for the poor batch-to-batch reproducibility. This inconsistency is a major complaint in many industries (e.g. pharmaceutical analysis) where reproducible separations must

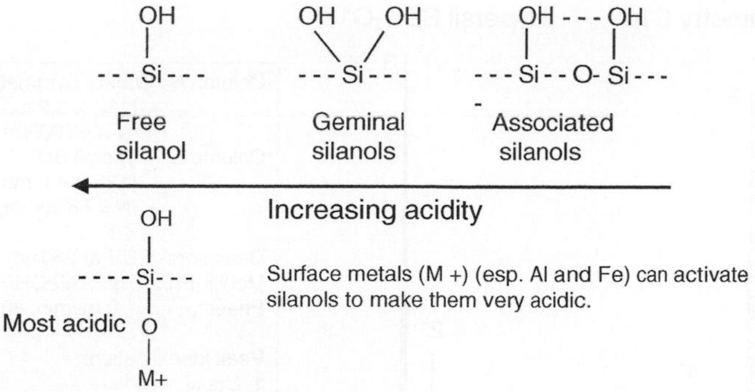


**Figure 3.8.** Schematic diagram of various types of silanol and their relative acidities. Note that a silanol can be "activated" (made more acidic) by an adjacent metal ion.

be consistently achieved in quality control applications. The development of high-purity silica (99.995%) with <50 ppm of total metals in the 1990s resulted in dramatic improvements of batch reproducibility and peak shapes of basic analytes [6]. Other factors are improved and more controlled bonding chemistries with tighter quality control procedures.

Figure 3.9 shows chromatograms that compare the column performance of two C18 columns and their dramatic differences in peak shape performance of basic analytes. While the two columns show similar efficiency performance of ~66 000 plates/m for neutral solutes, the tailing factors of basic analytes (i.e. pyridine and dimethylaniline) are much worse for the Hypersil BDS-C18 column due to the high metallic contents of its base silica. Note that peak tailing of basic analytes is not caused by the number of residual silanols but rather the activity of the silanols [1, 6]. Vydac, Waters Symmetry, ZORBAX (Rockland Technologies/HP/Agilent), and GL Sciences Inertsil were some of the first-generation bonded phases made from high-purity silica.

Figure 3.10 lists some popular RPC columns in use today, with most products based on high-purity silica or hybrid particles. Several older brands packed with non-high-purity silica are included here as a reference. The use of these older columns in new method development activities is not recommended due to their high silanophilic activity and lower batch-to-batch reproducibility. Hybrids are in bold letters while SPP products are highlighted in red and italicized. These packings are discussed in the next sections.

### 3.5.2 Hybrid Particles

An innovative approach to improve silica support materials was the development of hybrid particles [3]. In the synthesis of the first commercial hybrid particles (Waters XTerra introduced in 1999), tetraethoxysilane (the traditional monomer) is mixed with methyltriethoxysilane in a 2 : 1 ratio resulting in hybrid silica support with a claimed 33% fewer surface silanols (Figure 3.11). Consequently, hybrid particles tend to have less residual silanol activity and improved peak tailing for basic analytes. Another advantage is the wider usable pH range of 2–12 for mobile phases. An improved second-generation hybrid particle displaying higher stability and efficiency based on bridged ethylene hybrid (BEH) chemistry was introduced by Waters Corporation in 2004 (Figure 3.12) [14]. These BEH particles are used in Waters ACQUITY, XBridge, XSelect, and charged surface hybrid (CSH) lines of column products. Other manufacturers have also developed and introduced columns based on hybrid particles (e.g. Phenomenex Gemini and YMC-Triart).

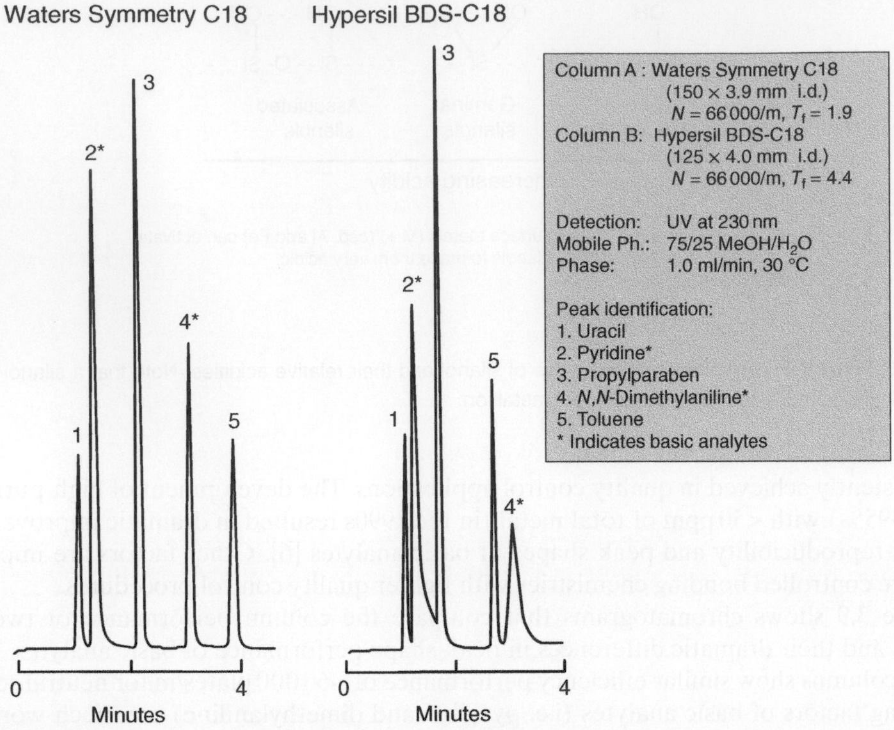


**Figure 3.9.** Comparative chromatograms showing the effect of silica purity on the peak shapes of basic analytes. Note that the column on the right shows significant silanol activity causing considerable peak tailing of basic analytes due to the higher metallic contents of its silica support.

- **Waters:** Symmetry, SunFire, ***XTerra*** *, ***ACQUITY* (BEH, CSH, HSS)***, ***XBridge*****, ***XSelect**** Atlantis, NovaPak, μ-Bondapak, Spherisorb, *CORTECS*
- **Agilent:** Zorbax StableBond, Eclipse XDB, *Extend* C18*, Bonus, *Poroshell*
- **Phenomenex:** *Luna**, Prodigy, *Synergi**, Onyx, ***Gemini*****, *Kinetex*
- **Millipore sigma (supelco):** Discovery, Ascentis, *Ascentis Express*, Supelcosil
- **Thermo/dionex:** HyPURITY, Hypersil, Prism, *Hypersil Gold* *, *Accucore*, Acclaim, Acclaim PA, *Acclaim PA2**
- **Advance chromatography technology (ACT)** ACE, *UltraCore*
- **Advanced materials technologies (AMT)** *Halo*
- **YMC:** YMCbasic, Pack Pro, ***Triart hybrid***, *Meteoric core*
- **Eka chemicals:** ***Kromasil Eternity****
- **GL sciences:** Inertsil
- **Macherey nagel:** Nucleosil, *Nucleoshell*
- **Merck KGaA:** Chromolith (Monolith)
- **Bischoff:** *ProntoSIL**
- **Grace/hichrom:** Vydac, Platinum (Alltech)
- **Restek:** Pinnacle, *Raptor*
- **Nacalai:** Cosmosil, *SunShell*

**Figure 3.10.** A list of popular silica-based columns including some newer product offerings. Columns based on non-high-purity silica are underlined. Columns packed with hybrid particles are in bold and italicized. Phases stable at high pH are italicized and marked with asterisks; SPP are italicized in red.

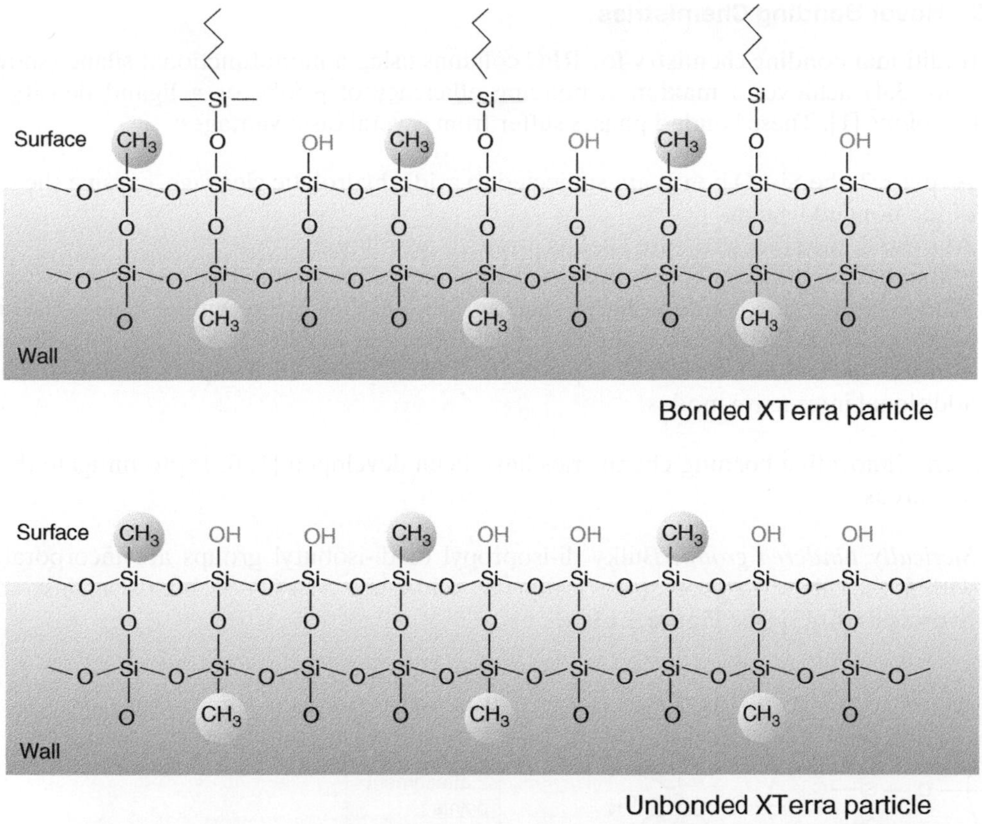


**Figure 3.11.** Diagram illustrating the surface of (b) unbonded and (a) bonded Waters XTerra hybrid particles. Note that XTerra bonded phases contain 33% fewer residual silanols than conventionally bonded phases by the nature of the hybrid chemistry. Source: Courtesy of Waters Corporation.

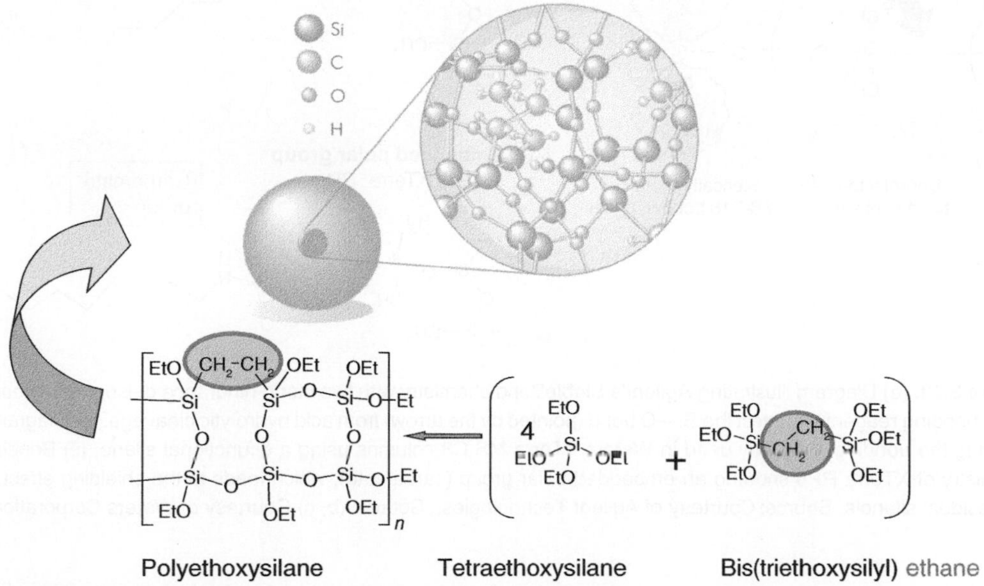


**Figure 3.12.** Diagram illustrating the synthetic route of a second-generation hybrid particle (ethylene bridged hybrid or BEH). Source: Courtesy of Waters Corporation.

### 3.5.3 Novel Bonding Chemistries

The traditional bonding chemistry for RPC columns using a monofunctional silane (shown in Figure 3.4) achieves a maximum bonding efficiency of ~50% or a ligand density of ~3–4 μmol/m$^2$ [1]. These bonded phases suffer from several disadvantages:

- At pH < 2, the Si—O bonds are subjected to acidic hydrolytic cleavage, causing the loss of the bonded ligands.
- At pH > 8, the silica structure itself is prone to dissolution.
- Unbonded acidic silanols may lead to peak tailing of basic analytes.
- While simple alkyl bonded phases such as C18 and C18 are retentive, consistent and highly reliable, the primary retention mechanism for analytes is hydrophobic interaction via dispersion forces. The separation of closely related compounds may require additional interactive modes.

Several innovative bonding chemistries have been developed [1, 6, 15] to mitigate these problem areas:

- *Sterically hindered group*: Bulky di-isopropyl or di-isobutyl groups are incorporated within the silane to protect sterically the acid-labile Si—O—Si bonds against acid hydrolysis [6, 15] (see Figure 3.13a).

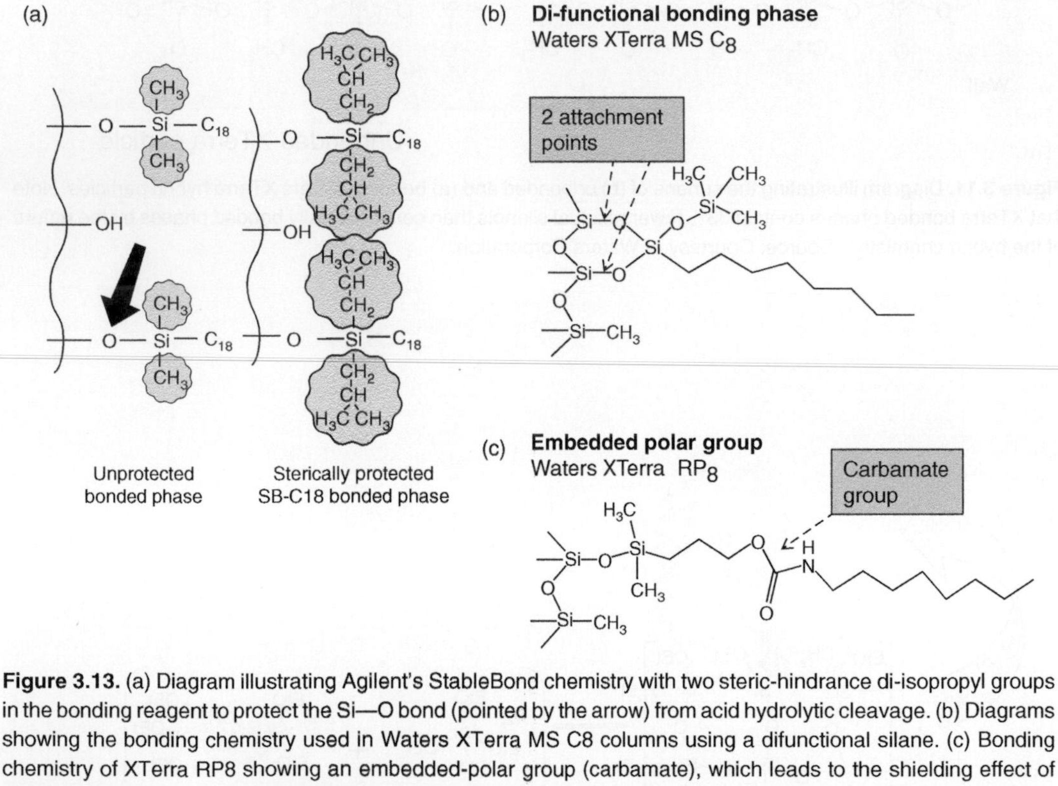


**Figure 3.13.** (a) Diagram illustrating Agilent's StableBond chemistry with two steric-hindrance di-isopropyl groups in the bonding reagent to protect the Si—O bond (pointed by the arrow) from acid hydrolytic cleavage. (b) Diagrams showing the bonding chemistry used in Waters XTerra MS C8 columns using a difunctional silane. (c) Bonding chemistry of XTerra RP8 showing an embedded-polar group (carbamate), which leads to the shielding effect of the residual silanols. Source: Courtesy of Agilent Technologies. Source: (b, c) Courtesy of Waters Corporation.

- *Polyfunctional silane chemistry*: Difunctional or trifunctional silanes are used to create bonded groups with two or three attachment points leading to phases with greater stability at low pH and lower bleed for LC/MS (see Figure 3.13b). Note that polyfunctional silane bonding chemistry is more difficult to control due to a tendency for crosslinking, which may reduce batch-to-batch consistency.
- *Other bonded groups*: Other bonded phases utilize additional interactions or retention modes and yield different selectivity: phenyl ($\pi$–$\pi$ interaction), pentafluorophenyl (PFP) ($\pi$–$\pi$ interaction, dipole–dipole, and H-bonding interactions), and cyano (dipole–dipole, H-bonding). PFP appears to work well for the separation of isomeric compounds.
- *Polar-embedded groups*: The incorporation of a polar group (e.g. carbamate, amide, urea, ether, sulfonamide) in the hydrophobic chain of the bonding reagent has led to a new class of popular phases with different selectivity (polar interaction, H-bonding) and better peak shape for basic analytes due to "shielding" of any residual silanols by the polar-embedded group (see Figure 3.13c). These phases have found use in the food and beverage industry in particular and are useful for the separation of polar analytes where standard C18 phases may not provide adequate retention. These phases can be used with 100% aqueous mobile phase without the problem of "phase collapse" as will be described later.

These polar-embedded columns are less retentive than their C18 or C8 counterparts for neutral and basic analytes. They are used for method development for difficult separations as shown in some case studies in Chapter 9. Examples are Waters ACQUITY BEH Shield RP18 (carbamate), Agilent ZORBAX Bonus-RP (amide), Supelco Ascentis RP-Amide, Thermo Fisher Scientific Acclaim PA (sulfonamide) and PA2, Phenomenex Synergy Fusion-RP, ES Industries Chromegabond ODS-PI (urea), Agilent Polaris C18-Ether, and ACE C18-Amide. Note that these complex bonding chemistries are more difficult to control in the manufacturing process resulting in lesser batch-to-batch consistency in comparison with that of traditional monofunctional C18 chemistry.

***3.5.3.1 Bonded Phases for Retention of Polar Analytes*** Separations of very polar analytes are traditionally difficult in RPC due to lack of retention of these compounds. Lowering the organic strength of the mobile phase to increase retention is effective only for a few percentages of the organic modifiers (e.g. 2–10% methanol or acetonitrile). When the mobile phase is highly aqueous, and the flow is stopped and restarted, the column may show reduced retention due to extrusion of the mobile phase from the hydrophobic pores of the stationary phase (known as pore dewetting or phase collapse; Figure 3.14a,b) [1, 6].

Polar-embedded phases are not prone to phase collapse due to the hydrophilicity of the polar-embedded group (Figure 3.13b). An alternate approach is to use a lower ligand coverage to create a more hydrophilic bonded phase (Figure 3.15). This type of bonded phase is useful for separations of water-soluble compounds and can be used with a 100% aqueous mobile phase. Examples are Grace Alltech Platinum EPS C8 and Waters Atlantis T3. Other approaches employed are the use of polar endcapping or no endcapping. An "AQ" designation of the bonded phases means that the stationary phase is compatible with 100% aqueous mobile phases. Examples are Agilent ZORBAX SB-Aq, YMC Pack ODS-AQ, and Thermo Scientific Hypersil Gold AQ, and Accucore AQ.

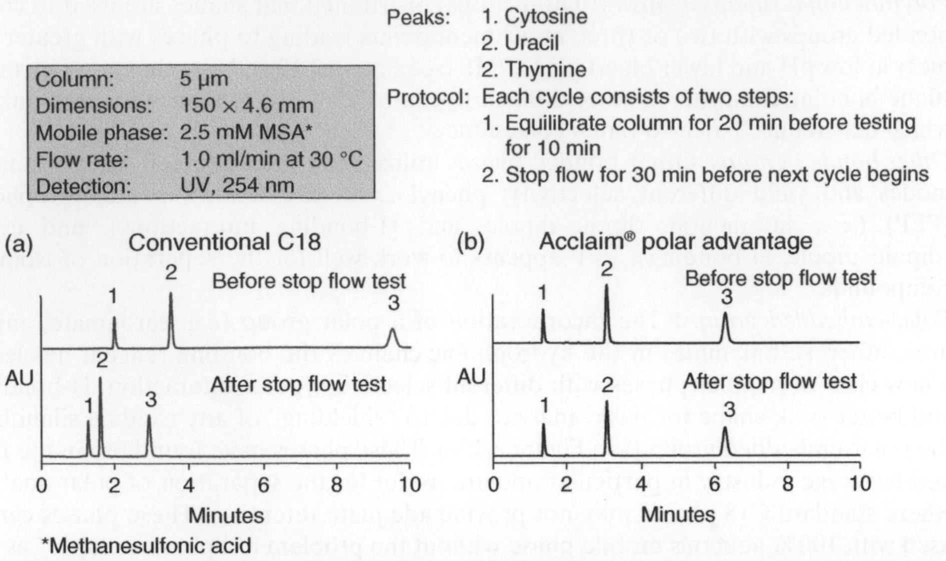


**Figure 3.14.** (a) Chromatograms illustrate the phenomenon of loss of retention after the flow of a highly aqueous mobile phase is stopped and then restarted due to a "phase collapse" phenomenon for a conventional C18 phase. (b) Chromatograms illustrate that polar-embedded phases are less prone to phase collapse (due to the hydrophilicity of the polar groups). Source: Courtesy of Thermo Scientific.

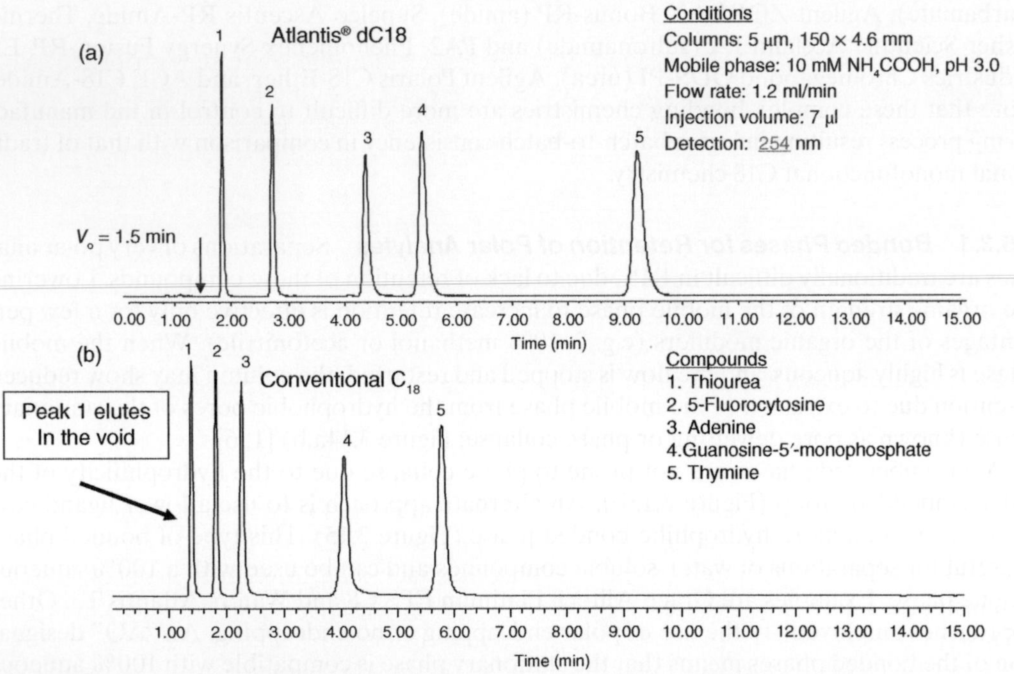


**Figure 3.15.** Comparative chromatograms for two columns under identical conditions in the separation of polar analytes. (a) A column packed with a low-coverage C18 bonded phase designed for the analysis of polar compounds. (b) A column packed with a conventional high-coverage C18 bonded phase showing less retention of polar analytes. Source: Courtesy of Waters Corporation.

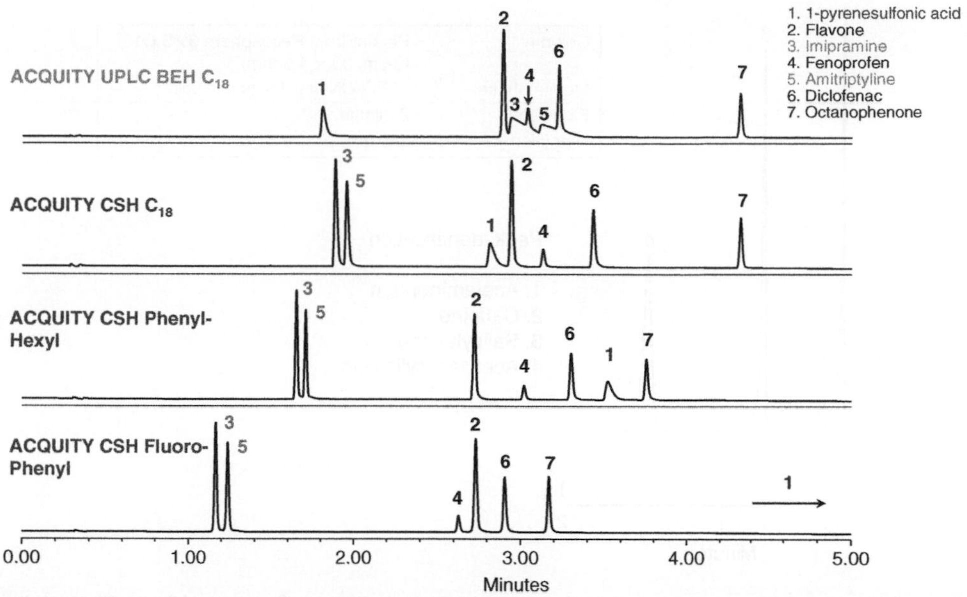


**Figure 3.16.** Comparative gradient chromatograms illustrating the advantage of better peak shapes of basic analytes (peaks 3 and 5) for Charged Surface Hybrid (CSH) columns when used with low ionic strength mobile phases. All columns are 50 × 2.1 mm, 1.7 μm used with 0.1% formic acid and an ACN gradient. Source: Courtesy of Waters Corporation.

***3.5.3.2 Charged Surface Hybrid (CSH)*** Peak tailing of basic analytes is problematic in the pharmaceutical analysis since most drugs are nitrogenous bases, which can interact with residual silanols. The use of high-purity silica has mitigated this issue in well-buffered mobile phases, though less so in low ionic strength mobile phases such as 0.1% formic acid. One innovative approach used in Waters CSHs is the bonding of a slight positive charge (likely an amine) to the packing leading to improved peak shapes for highly basic analytes (e.g. imipramine and amitriptyline) as shown in Figure 3.16. Note that CSH columns may cause less desirable peak shapes for acidic analytes as can be seen for 1 pyrenesulfonic acid (peak 1) in Figure 3.16. A universal generic gradient method based on the CSH column is discussed in Section 10.5.

### 3.5.4 Shorter and Narrower Columns Packed with Small Particles

One prominent trend in modern HPLC is the increased use of shorter and narrower columns packed with smaller particles [6, 8, 9]. The "standard" column for purity assays in the 1980s was a 250 × 4.6 mm column packed with 5 or 10 μm particles. There was a continued trend to a 150 × 4.6 mm column with 3 μm particles and further toward 50–150 mm × 3.0 or 2.1 mm columns packed with sub-3 or sub-2 μm particles for significant increases of analysis speed and resolution [8].

***3.5.4.1 Fast LC*** Fast LC, which was developed in the 1980s, uses short columns (30–50 mm × 4.6 mm i.d.) that are packed with 3 μm stationary phases used with HPLC systems equipped with small UV flow cells [16, 17]. The benefits are fast analysis (isocratic

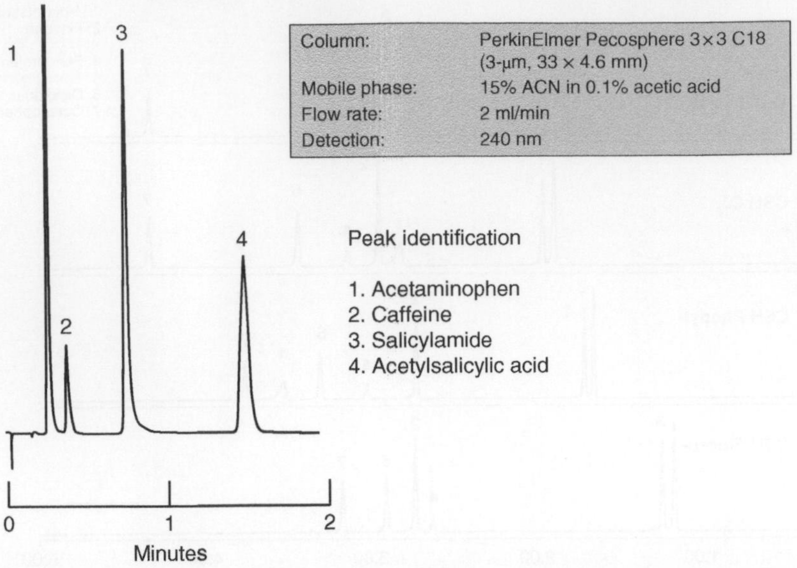


**Figure 3.17.** Fast LC separation of an analgesic tablet extract. Source: Courtesy of PerkinElmer, Inc.

1–3 minutes and gradient 2–10 minutes), rapid method development, and lower solvent consumption. Figure 3.17 shows an example of Fast LC in the separation of an over-the-counter drug product [17]. Figure 3.18 illustrates the flexibility of Fast LC, which can operate in a high-speed or a high-resolution mode for the separations of four paraben preservatives [17].

**3.5.4.2 *UHPLC*** Fast LC was supplanted in 2004 by the commercial availability of UHPLC instruments capable of separations at pressures up to 1000 bar, allowing even faster analysis or very high-resolution of complex samples with minimized system dispersion effects [7, 18, 19]. Figure 3.19 shows the Van Deemter curves from columns packed with 5.0, 3.5, and 1.8 μm particles, which illustrates that sub-2-μm particles can further improve efficiencies by giving a lower $H_{\min}$ and a smaller C term. However, sub-2 μm particles generate much higher back-pressure and necessitate the use of instruments with a higher pressure rating [7, 19]. Issues with viscous heating also mandate the use of smaller i.d. columns (3.0 or 2.1 mm) to lessen the effect of radial thermal gradients [19]. Chapter 5 provides a further overview of UHPLC including descriptions of its benefits, best practices, and potential issues. UHPLC columns are packed at very high pressures with high-strength particles to allow reliable operation at pressures up to 20 000 psi.

### 3.5.5 Micro-LC and Nano-LC

Micro-LC and nano-LC have these benefits: [20]

- Increased mass sensitivity due to lower sample dilution (Figure 3.20).
  - — Useful for samples of limited availability (proteins/peptides in proteomics studies, biomarker studies, pharmacokinetic samples)

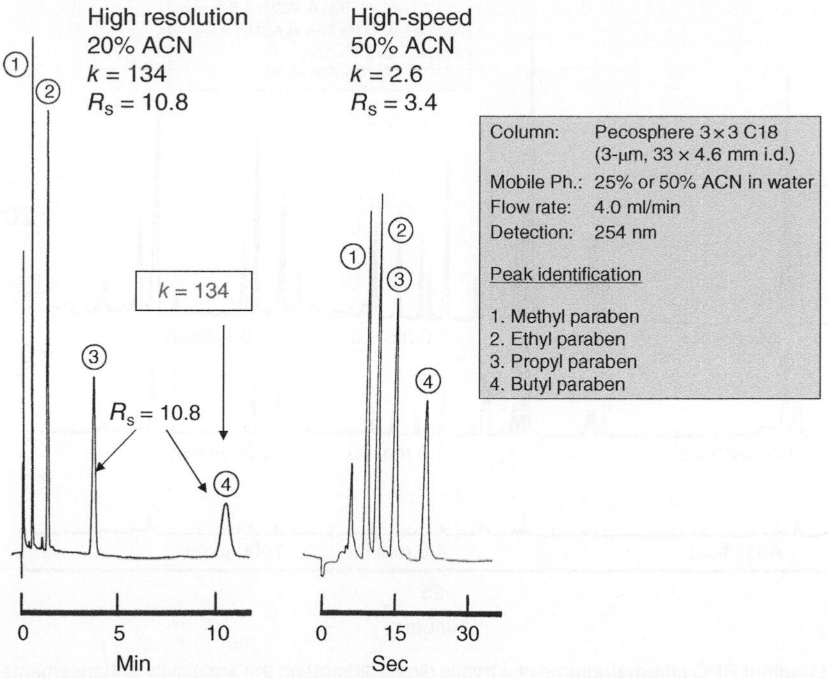


**Figure 3.18.** Diagram illustrating the versatility of fast LC in the high-resolution separation and high-speed of four antimicrobial parabens. Source: Reprinted with permission of Dong and Gant 1984 [17].

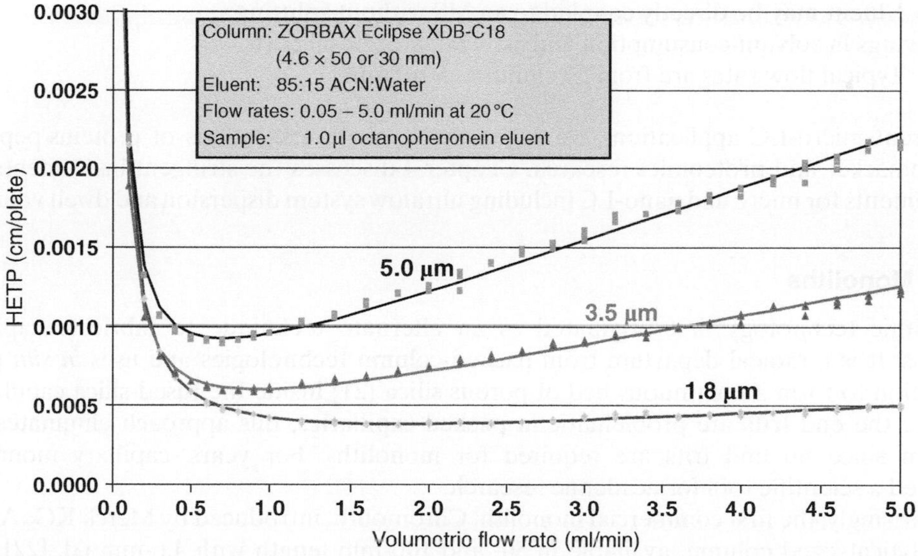


**Figure 3.19.** Van Deemter curves produced using columns packed with 5-, 3.5-, and 1.8-μm particles showing the overall lower $H$ and lesser efficiency loss at higher flows for smaller particles. Source: Courtesy of Agilent Technologies.

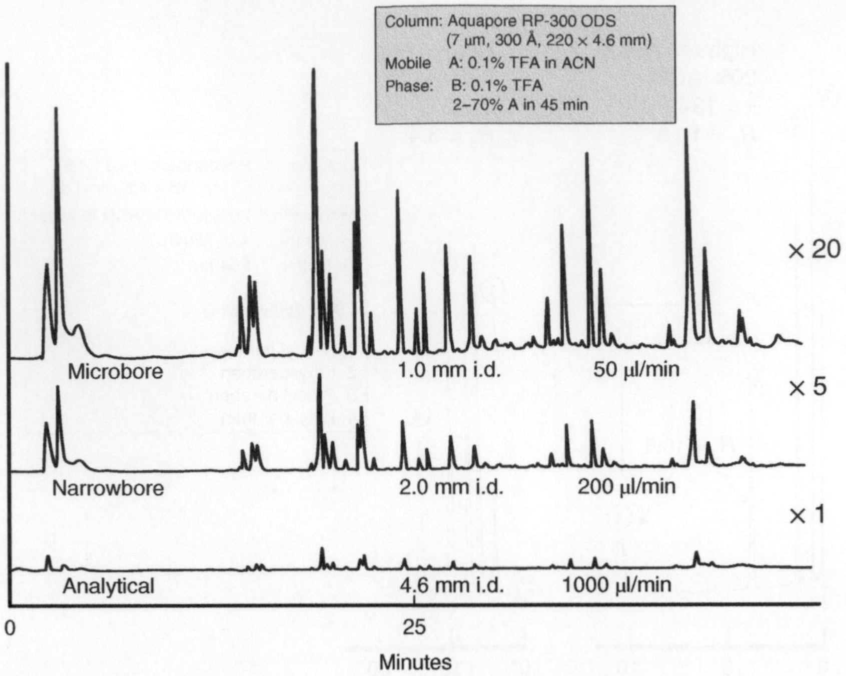


**Figure 3.20.** Gradient RPC chromatograms of a tryptic digest illustrating the sensitivity enhancements of columns with smaller inner diameters by injecting the same small sample amount on each column. Sensitivity enhancements are shown on the right-hand side. Source: Courtesy of PerkinElmer, Inc.

- Better compatibility with LC/MS (e.g. electrospray interface).
  — Eluent may be directly coupled to an MS without splitting
- Savings in solvent consumption and associated disposal costs.
  — Typical flow rates are from 50 nl/min to 50 μl/min

Current micro-LC applications are primarily in micropurifications of proteins/peptides and biomarker and proteomics research. Chapter 4 discusses the stringent instrumentation requirements for micro and nano-LC including ultralow system dispersion and dwell volumes.

### 3.5.6 Monoliths

Monolithic technology first originated as an alternate technique to fabricate capillary columns. It is a radical departure from packed-column technologies and uses *in situ* polymerization to form a continuous bed of porous silica [21] inside the fused-silica capillaries. Because the end frits are problematic in packed capillaries, this approach eliminates this problem since no end frits are required for monoliths. For years, capillary monoliths remained a scientific tool for academic research.

Surprisingly, the first commercial monolith, Chromolith, introduced by Merck KGaA, was an analytical-sized column, available in 50- and 100-mm length with 4.6-mm i.d. [22]. The advantage of the Chromolith is the lower pressure drop, which was claimed by the manufacturer to have the performance of 3-μm particles and pressure drop of 5-μm particles. One application is for in-process control (IPC) testing in process chemistry laboratories because of monolith columns' resistance to sample plugging. Attempts to manufacture smaller-diameter

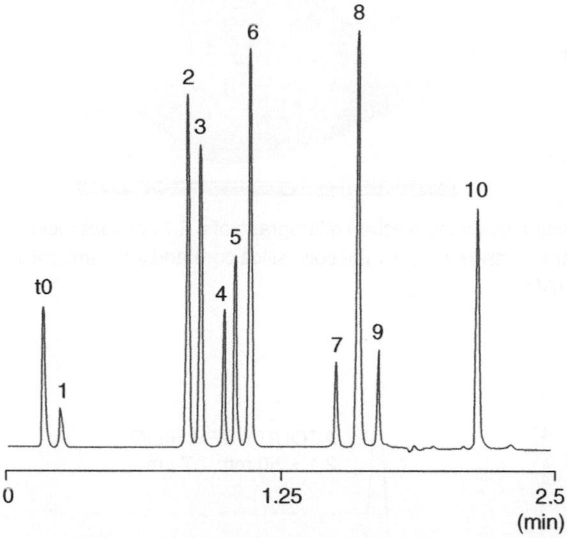

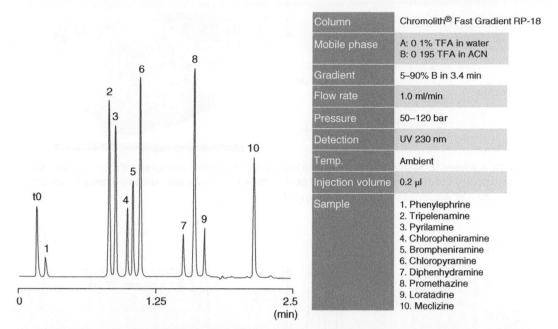

| | |
|---|---|
| Column | Chromolith® Fast Gradient RP-18 |
| Mobile phase | A: 0 1% TFA in water<br>B: 0 195 TFA in ACN |
| Gradient | 5–90% B in 3.4 min |
| Flow rate | 1.0 ml/min |
| Pressure | 50–120 bar |
| Detection | UV 230 nm |
| Temp. | Ambient |
| Injection volume | 0.2 μl |
| Sample | 1. Phenylephrine<br>2. Tripelenamine<br>3. Pyrilamine<br>4. Chloropheniramine<br>5. Brompheniramine<br>6. Chloropyramine<br>7. Diphenhydramine<br>8. Promethazine<br>9. Loratadine<br>10. Meclizine |

**Figure 3.21.** Chromatogram illustrating the gradient performance of a second-generation silica monolith. C18 phase, 50 × 2.0 mm. Source: Courtesy of Millipore Sigma.

columns have been difficulties with the "cladding" or the external bonding of these preformed silica rods with PEEK. However, both 3-mm and 2-mm i.d. Chromoliths are currently available, with a performance that the manufacturer claimed to rival those of sub-3 μm SPP. Figure 3.21 shows the performance of a second-generation Chromolith (50 × 2.0 mm) operating at 1 ml/min at 50–120 bar in the separation of a 10-component sample in less than three minutes. It remains to be seen whether monolith technology can be competitive with the best of packed-column technologies in terms of efficiencies. Polymer-based monoliths are used routinely in the purification of proteins, although their performance appears to be lower than those from silica monoliths.

### 3.5.7 Superficially Porous Particles (SPP)

Superficially porous particles, a.k.a. solid core, core–shell, or fused-core are rapidly becoming the preferred HPLC packing materials due to their higher chromatographic performance (efficiency) vs. those of totally porous particles (TPPs). SPP are made of a solid, nonporous core surrounded by a porous shell layer. The "fused-core" term was introduced by Advanced Materials Technology (AMT), which described the manufacturing procedure that "fuses" a porous silica layer onto a solid silica particle as shown in Figure 3.22 [23–26].

Figures 3.23 and 3.24 show comparisons of the performance of SPP vs. TPP of comparable $d_p$ under isocratic (HILIC) and RPC gradient conditions, respectively. Increases of $N$ in the range of 20–40% are typical, leading to enhanced $R_s$ or peak capacity in the range of 10–20%. It is believed that this advantage of SPP is due to the reduction of Van Deemter A and B terms, rather than the C term [25, 26]. Examples of current column products packed with SPP include AMT HALO, Supelco Ascentis Express, Phenomenex Kinetex, Agilent Poroshell, Thermo Fisher Accucore, Waters CORTECS, Nacalai Sunshell, and ACE UltraCore.

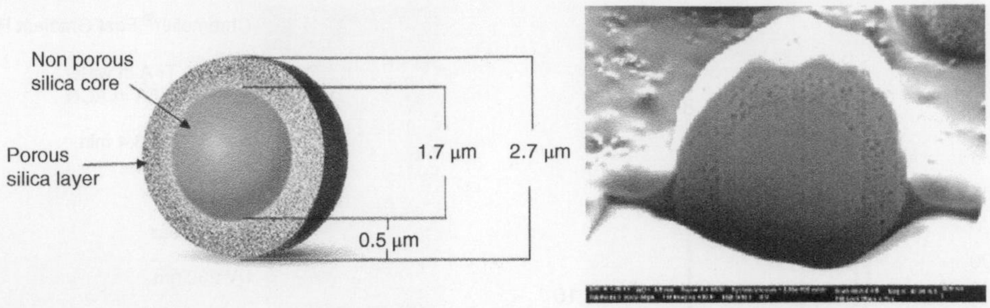


**Figure 3.22.** A schematic diagram and a focused-ion beam scanning electron micrograph of a 2.7 μm superficially porous particle (SPP) of the first commercial HALO particle, showing a 1.7 μm solid silica core and a 0.5 μm shell. Source: Courtesy of Advanced Materials Technology (AMT).

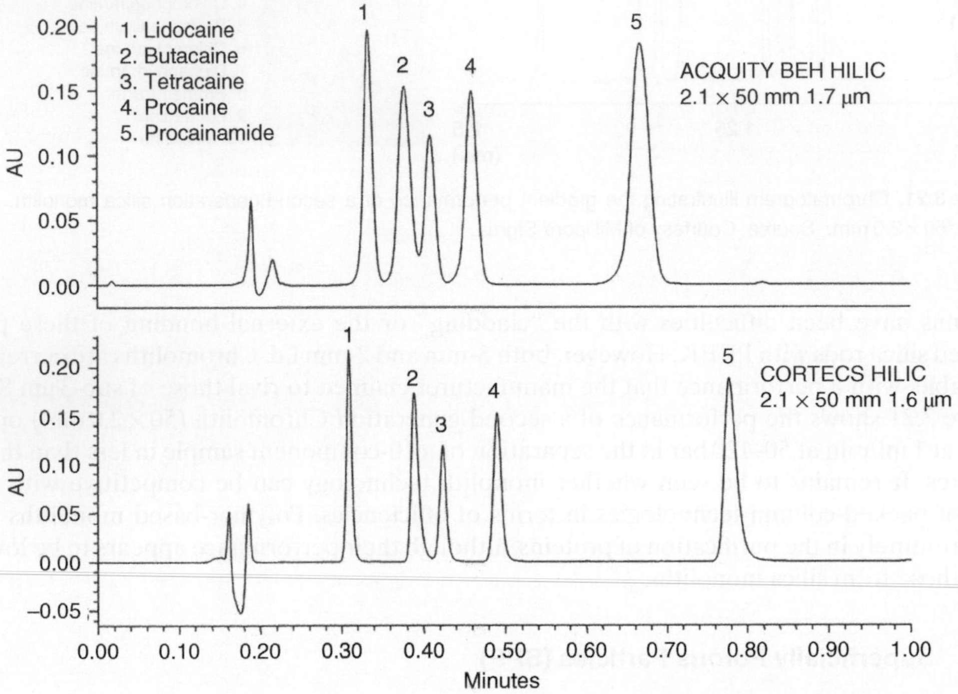


**Figure 3.23.** Comparative separations of five local anesthetics by HILIC using columns packed with TPP vs. SSP. Columns: (TPP) ACQUITY BEH HILIC 50 × 2.1 mm, 1.7 μm, and (SPP) CORTECS UPLC HILIC 50 × 2.1 mm, 1.6 μm. MPA – 50 : 50 ACN/10 mM ammonium formate with 0.125% formic acid, MPB – 90 : 10 acetonitrile ACN/10 mM ammonium formate with 0.125% formic acid, isocratic at 99.9% MPB for 1 minute, then gradient from 99.9% to 0.1% in 1.6 minutes. Flow rate: 0.8 ml/min at 30 °C, detection: 245 nm, injected amount: 5 μl. A Waters ACQUITY I-Class UPLC system was used. Source: Courtesy of Waters Corporation.

The use of SPP is expected to increase as more new phases and applications become available.

***3.5.7.1 Kinetic Plots Demonstrating the Superiority of SPP*** While the Van Deemter plot describes the efficiency performance with respect to flow rate, no consideration is given

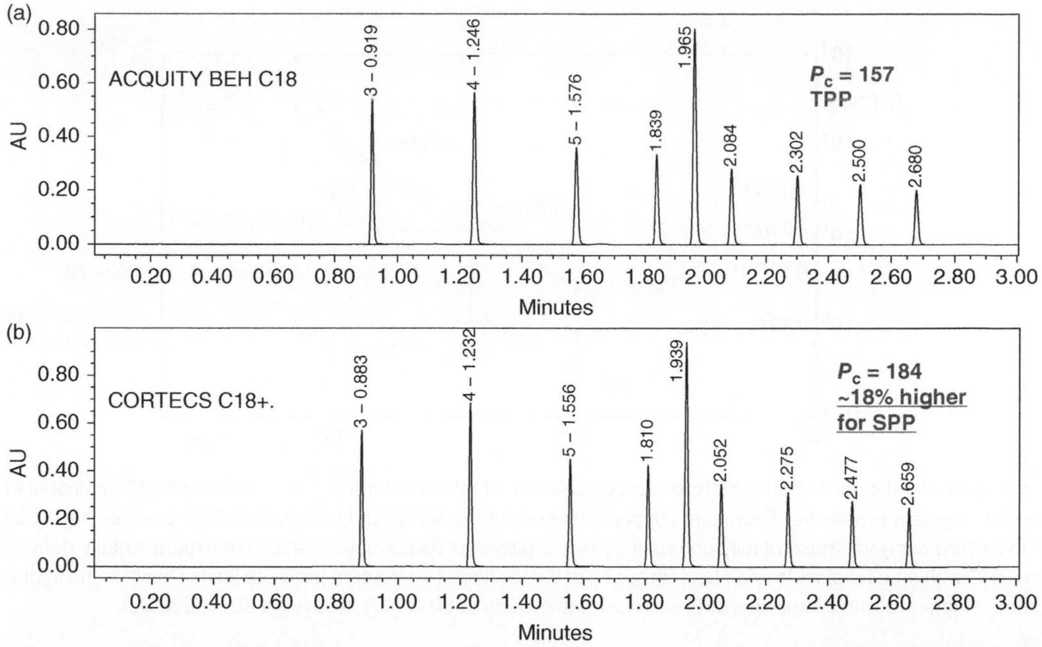


**Figure 3.24.** Comparative performance of columns packed with TPP vs. SPP showing 18% higher peak capacity using SPP. Column: (TPP) ACQUITY BEH C18, (1.7 µm, 50 × 2.1 mm) (a) and (SPP) CORTECS C18+, (1.6 µm, 50 × 2.1 mm) (b), water and ACN, 30–100%B in 3 minutes, 1.0 ml/min at 45 °C, 254 nm with data rate at 40 pt/s Sample mixture containing nine alkylphenones.

to analysis time and system pressure limit. A better comparative model that encompasses these factors is the kinetic plots described by Gert Desmet and coworker [27]. Figure 3.25 shows comparative kinetic plots of fully porous HPLC packings (black dots) and SPP or core–shell particles (red dots). This figure demonstrated the superior kinetic performance of small-particle SPP. Best performance in ultrafast separation is achievable with sub-2 µm SPP (1.3 or 1.6 µm), and high-efficiency analysis in 5–60 minutes range is obtained with sub-3 µm SPP (2.7 µm). The demands on LC hardware, however, are increasing. To use these small SPP (<2 µm) and obtain the efficiency values promised by theory, LC systems and engineering must further evolve to minimize performance losses from system dispersion.

### 3.5.8 Micropillar Array Chromatography (µPAC)

One truly innovative column design that has recently been commercialized by PharmaFluidics is termed micropillar array chromatography (µPAC). These perfectly ordered silica pillar structures (7.5 µm wide and 20 µm deep) shown in the insets of Figure 3.26 are made by microlithography technology and can be packaged in microfluidics devices or LC chips. They are compatible with any capillary or micro-LC equipment at flow rates of 50–1500 nl/min at pressures <100 bar. Two comparative peptide maps of an originator monoclonal antibody biologics drug and that of its biosimilar counterpart are shown in Figure 3.26. Peak capacities in excess of 600–800 and reduced plate heights of 0.75 were reported [28].

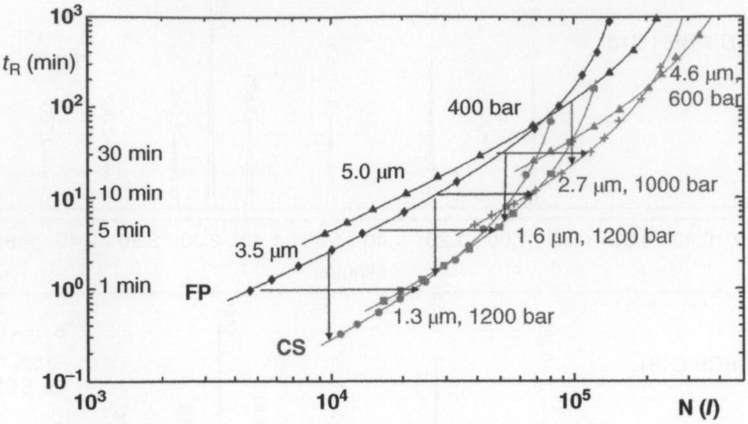


**Figure 3.25.** Kinetic performance plots for various columns of retention time ($t_R$) vs. plate counts ($N$) attainable at maximum system pressures. This figure supports the current superiority of small-particle SPP or core–shell (CS) columns (red dots) vs. those of fully porous (FP) HPLC packings (black dots). Data were experimentally derived from RPC with small molecule analytes with $t_R$ at $k$ of 9 collected at instrument pressure limits shown in the figure. Source: Reprinted with permission of Broeckhoven and Desmet 2014 [27]. Copyright 2014, Elsevier.

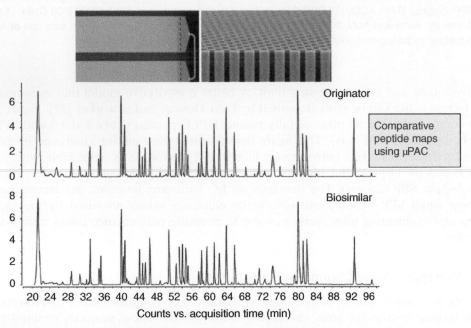


**Figure 3.26.** Comparative peptide maps of an originator monoclonal antibody drug and that of a biosimilar counterpart using a micropillar array chromatography in a microfluidic device on micro-LC with MS detection. The inset shows pictures of the microfluidic device and the structures of the micropillar array. Source: Reprinted with permission of Dr. Koen Sandra of RIC from Ref. [28].

## 3.6 GUARD COLUMNS

A guard column is a small packed cartridge/column placed before an analytical column to protect it from particles or contaminants from the samples [29]. Ideally, guard columns are packed with the same materials as analytical columns and should not cause a significant increase in pressure or performance degradation. Guard columns are commonly used in applications involving "dirty samples" – environmental or bioanalytical samples where more extensive sample clean-up may not be feasible. Some users prefer the use of an in-line filter, although it does not have the sample capacity of a well-designed guard column. Note that the use of guard columns is not prevalent in pharmaceutical analysis laboratories dealing with cleaner drug substance and drug product samples.

## 3.7 SPECIALTY COLUMNS

This section provides a brief overview of HPLC columns involving specialized chromatographic modes (chiral, SFC, HILIC, mixed-mode), bioseparations, and other application-specific columns (environmental and food analysis). Case studies for bioseparations and QC of recombinant biologics are described in Chapters 12 and 13. Note that most modern HPLC columns have little bleeding issues and should be entirely compatible with LC/MS analysis. Nevertheless, it is always astute to assess this characteristic of a column before finalizing your LC/MS method.

### 3.7.1 Bioseparations Columns

The HPLC of large biomolecules such as proteins, monoclonal antibodies, and DNAs often requires specialized columns packed with wide-pore particles [10, 13]. The use of wide-pore SPP is rapidly increasing for these applications [26]. Some of these columns are available in PEEK or titanium hardware to allow the use of high-salt mobile phases and to prevent possible protein denaturing by metallic leachates. Others analytes such as peptides, oligonucleotides, amino acids, and glycans may require columns for specific retention or selectivity characteristics. Further examples can be found in Chapters 12 and 13.

### 3.7.2 Chiral Columns

As many new drugs are chiral compounds, accurate assessments of chiral purity are expected regulatory requirements. The standard methodology for the determination of enantiomers is HPLC using chiral stationary phases (CSPs). Common CSPs include low-molecular-weight selectors (Pirkle type), macrocyclic selectors (cyclodextrins, crown-ethers, macrocyclic antibiotics), and macromolecular chiral selectors (proteins, molecular imprinted polymers, polysaccharides) [30, 31]. Figure 3.27 shows chromatograms on the separation of four chiral drugs using immobilized CSPs columns Daicel Chiralpak columns IA-3 and IC-3. These derivatized polysaccharide phases are very popular CSPs due to their ability to separate many chiral drugs under NPC or RPC conditions. The development of immobilized phases further increases the versatility of these columns (vs. coated phases) for use with more aggressive solvents [31]. Chiral columns are also trending toward the use of smaller particles for fast analysis. Also, SFC is often used for chiral method development and purification applications, although HPLC is still dominant for chiral analysis in quality control (QC) applications.

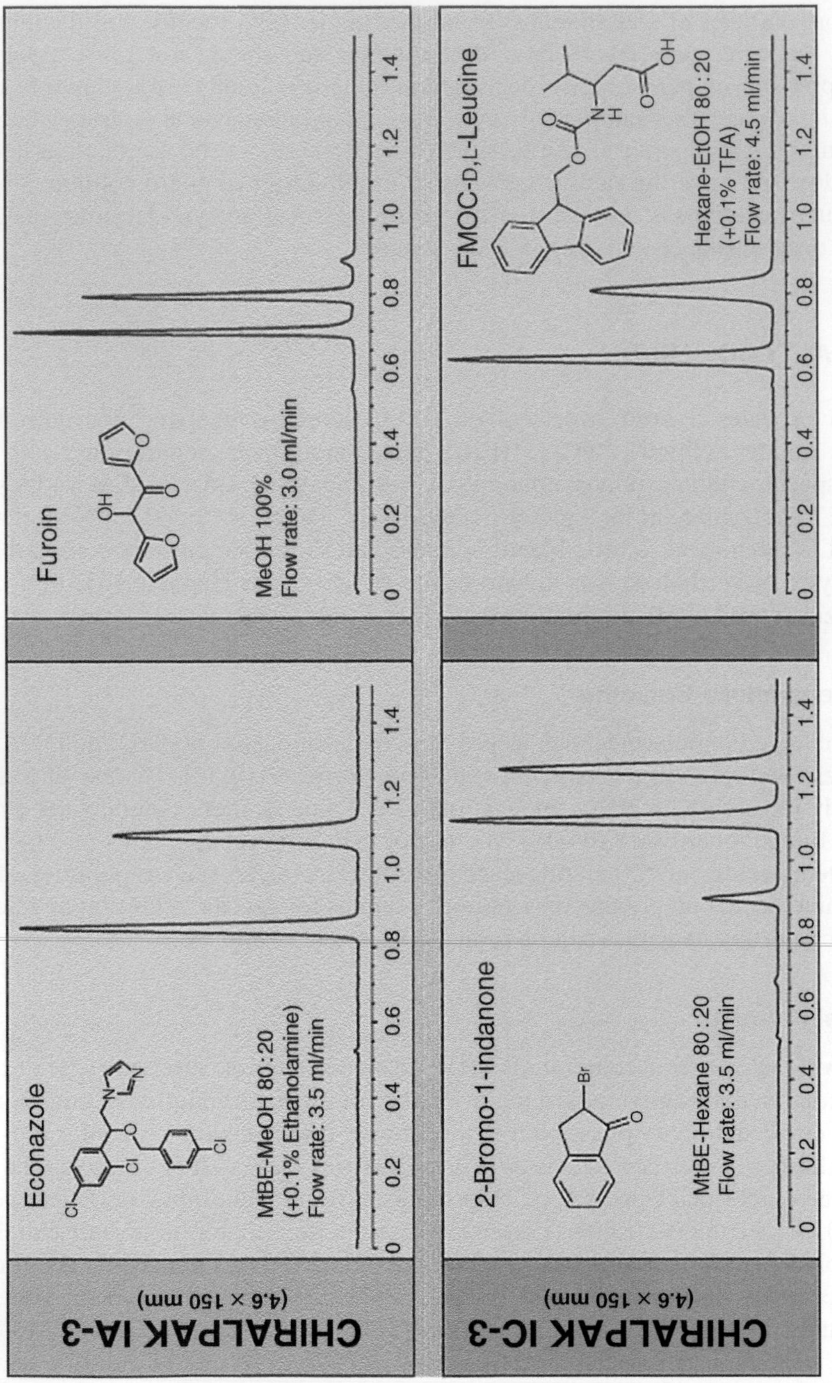


**Figure 3.27.** Chiral HPLC using immobilized polysaccharide phases Daicel Chiralpak 1A and 1C. Source: Courtesy of Chiral Technologies.

### 3.7.3 Supercritical Fluid Chromatography (SFC) Columns

In the last 10 years, SFC is emerging as a viable analytical technique made possible by improved instrumentation from manufacturers such as Waters, Agilent, and JASCO. Improved instrumentation has remedied past limitations of SFC, such as low UV sensitivity, poor reliability, and low precision performance. SFC is now the preferred technique for rapid column/mobile phase screening in chiral method development and high-throughput purification in the micro to preparative scales due to the ease of sample recovery. SFC offers superior kinetic performance (because of the low viscosity of supercritical $CO_2$). It is particularly suited for ultrafast and ultrahigh-resolution for both chiral and achiral compounds [32].

SFC operates under an NPC-like condition with supercritical $CO_2$ substituting for non-polar solvents such as hexanes as the primary mobile phase component. Figure 3.28 shows chromatograms of four SFC separations under gradient conditions on four different SFC columns. Common SFC columns for achiral applications are those packed with unbonded silica or bonded phases such as amino, diol, amide, cyano, or PFP. Several SFC-only bonded phases gaining popularity in pharmaceutical applications are 2-ethylpyridine (EP), pyridylamine, and diethylamino propyl (DEAP) phases [33]. C18 AQ type phases have also been used successfully.

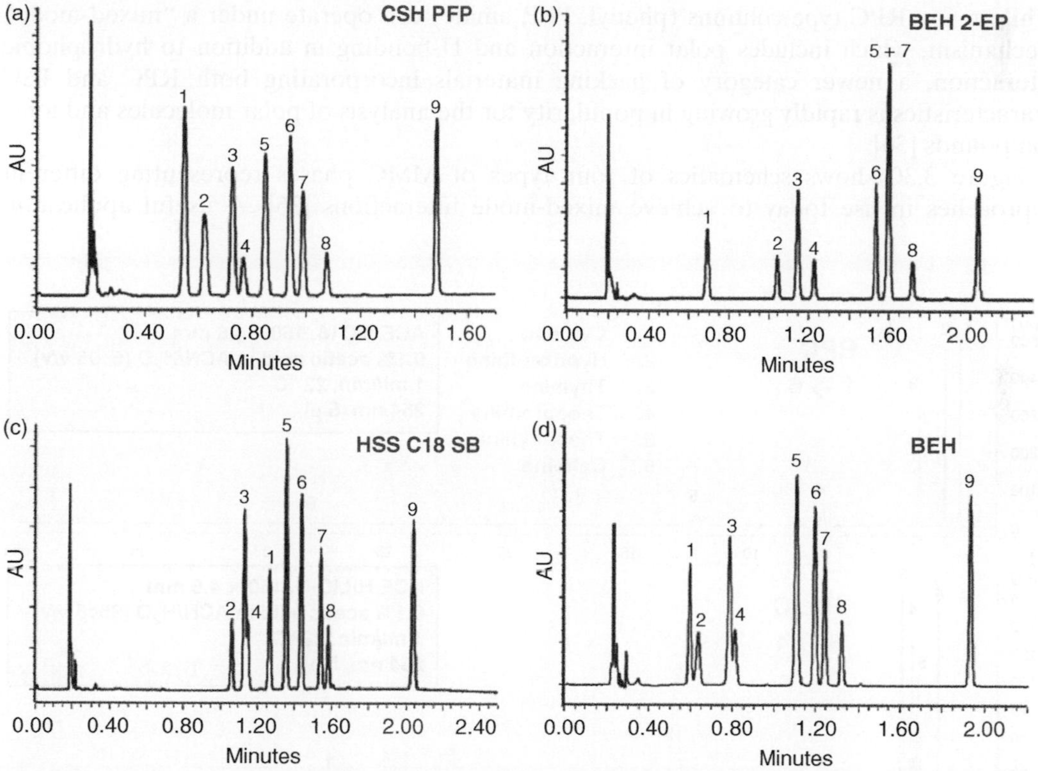


**Figure 3.28.** SFC chromatograms obtained using several different HPLC columns: The separation of steroidal estrogens on four stationary phases: (a) Acquity UPC$^2$ CSH PFP, (b) ACQUITY UPC$^2$ BEH 2-EP, (c) ACQUITY UPC$^2$ HSS C18 SB, (d) ACQUITY UPC$^2$ BEH. Compounds tested: (1) methoxyestradiol, (2) estradiol-17-acetate, (3) estrone, (4) Δ-estrone, (5) $\alpha$-estradiol, (6) ethinylestradiol, (7) $\beta$-estradiol, (8) Δ-estradiol, and (9) estriol. Source: Desfontaine et al. 2015 [32]. Reprinted with permission of Elsevier.

### 3.7.4 Hydrophilic Interaction Liquid Chromatography (HILIC) Columns

The HILIC mode, first described by Andy J. Alpert in the early 1990s [34], uses a hydrophilic stationary phase (silica, diol, cyano, amide, zwitterionic) with an RPC-like aqueous buffer and acetonitrile mobile phases. HILIC is now commonly used for the analysis of polar drugs, secondary drug metabolites, amino acids, peptides, neurotransmitters, oligosaccharides, carbohydrates, nucleotides, and nucleosides. The retention mechanism of HILIC is mainly "partitioning" of analyte molecules to the water layer adhering to the hydrophilic bonded groups or silica (along with hydrogen bonding, dipole–dipole, and ion-exchange interactions). Prominent benefits of HILIC include "orthogonal" selectivity to RPC using similar sample preparation and diluents, higher electrospray ionization sensitivity for MS (5- to 15-fold), and lower operating pressures vs. RPC. Figure 3.29 shows the chromatograms of a pharmaceutical sample obtained under RPC and HILIC conditions demonstrating orthogonal selectivity [35]. Recent work has shown that HILIC columns require longer times to equilibrate (formation of the adsorbed water layer) than RPC. This water layer seems to be an important and necessary feature of the technique for reproducible HILIC analyses.

### 3.7.5 Mixed-Mode Chromatography (MMC) Columns

While many RPC type columns (phenyl, PFP, amide) can operate under a "mixed-mode" mechanism, which includes polar interaction and H-bonding in addition to hydrophobic interaction, a newer category of packing materials incorporating both RPC and IEC characteristics is rapidly growing in popularity for the analysis of polar molecules and ionic compounds [36].

Figure 3.30 shows schematics of four types of MMC phases representing different approaches in use today to achieve mixed-mode interactions [36]. A useful application

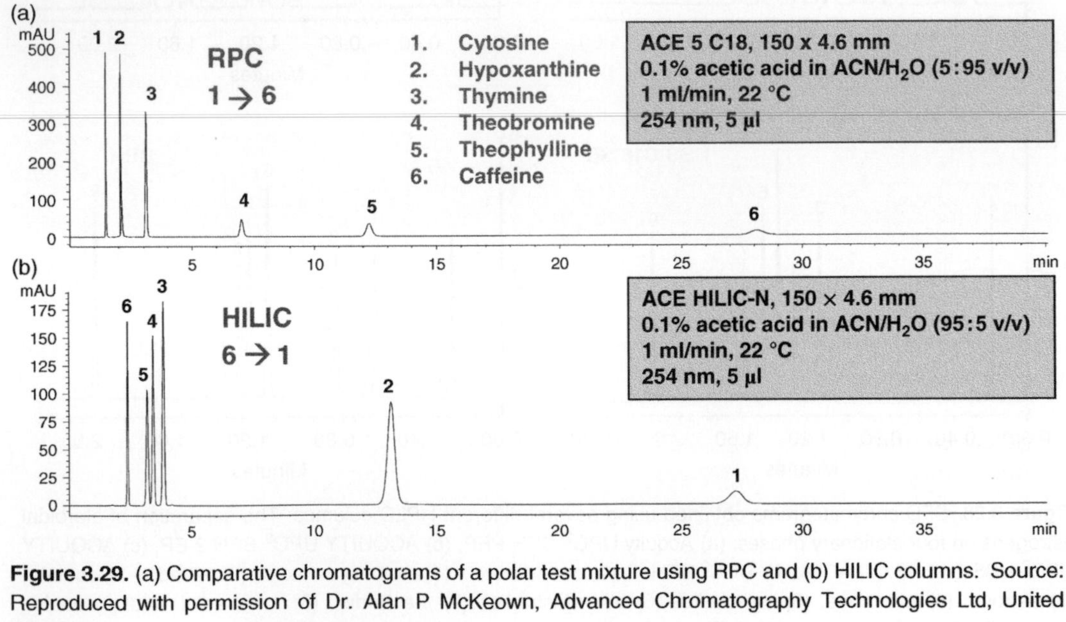


**Figure 3.29.** (a) Comparative chromatograms of a polar test mixture using RPC and (b) HILIC columns. Source: Reproduced with permission of Dr. Alan P McKeown, Advanced Chromatography Technologies Ltd, United Kingdom.

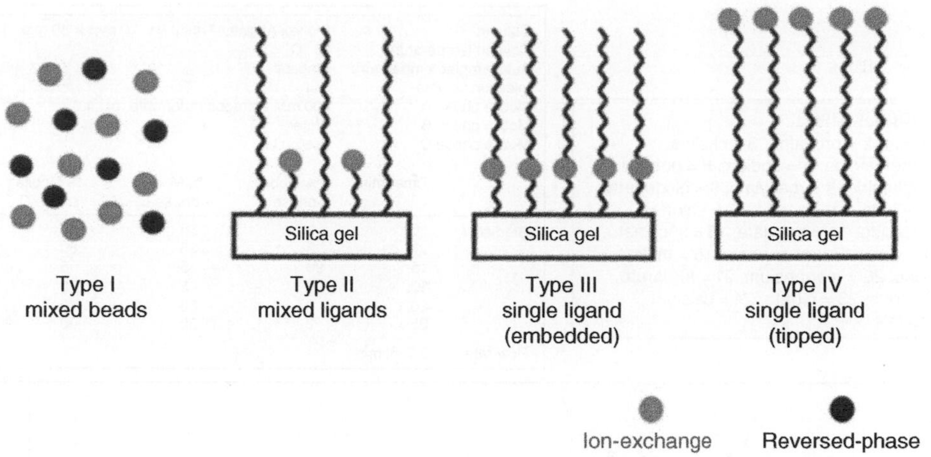


**Figure 3.30.** Schematic diagram of four types of mixed-mode chromatography (MMC) packings using different approaches in combining RPC and IEC. Source: Zhang et al [36]. Reprinted with permission of Elsevier.

demonstrating the unique characteristics of MMC is shown in Figure 3.31 in the separation of common cations and anions of pharmaceutical interest in less than 25 minutes [37]. This methodology has been adopted by the European Pharmacopoeia in the QC of pharmaceutical counterions in drug products [37].

### 3.7.6 Application-Specific Columns

While most columns are used in general-purpose analysis, some columns are marketed for specific applications. Examples include columns for environmental analysis (carbamates, polynuclear aromatic hydrocarbons) or food testing (amino acids, organic acids, sugars). These columns are often shipped with chromatograms demonstrating the performance of the specific application. Others may include a kit containing reagents and procedures for derivatization before analysis. Examples of other specific applications and gel permeation chromatography (GPC) columns for polymer characterization are described in Chapter 13.

## 3.8 RPC COLUMN SELECTION GUIDES

Column selection during HPLC method development often reflects the personal preferences or prior experience of the individual scientist [1, 4, 6]. Nevertheless, some general guidelines based on the best available technologies, literature references, and consensus of experienced chromatographers can be found here.

- Select columns packed with 3- or sub-3 μm high-purity silica-bonded phases from reputable manufacturers. Hybrids and SPP are highly recommended.
  - — 50 × 3.0 mm packed with sub-3 μm particles for rough potency assays, IPC, or cleaning verification testing
  - — 100–150 × 3.0 mm packed with sub-3 μm particles for purity testing or more complex samples
  - — 30–50 × 2.1 mm columns packed with sub-2 μm for high-throughput screening (HTS)

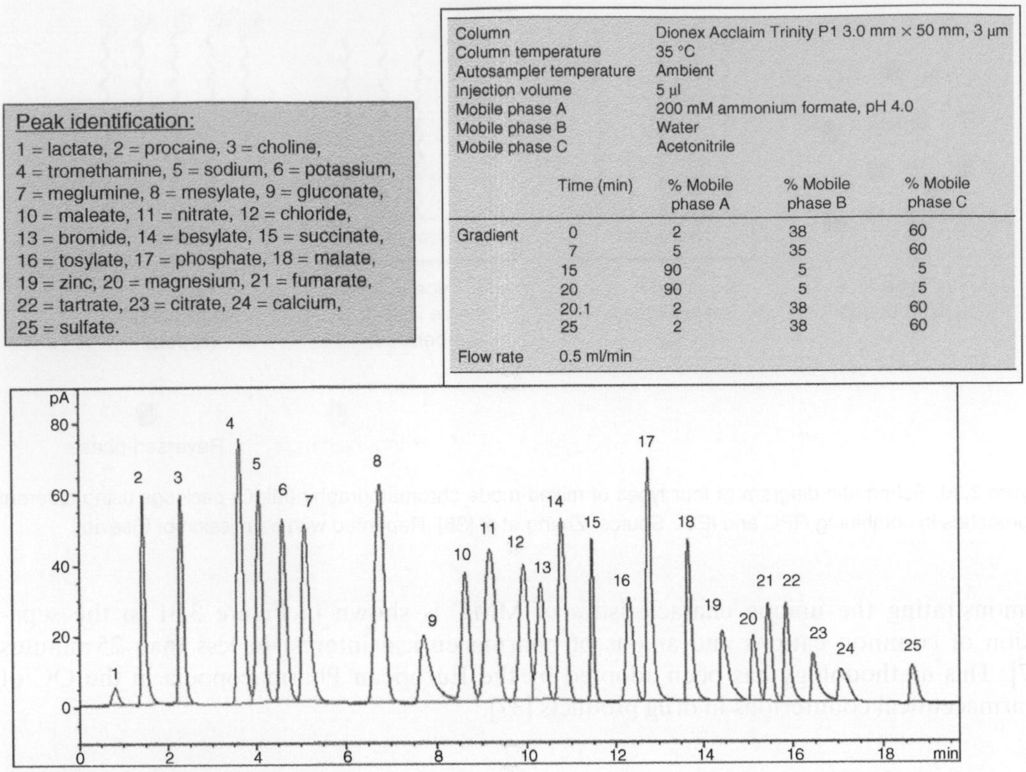
Peak identification:
1 = lactate, 2 = procaine, 3 = choline, 4 = tromethamine, 5 = sodium, 6 = potassium, 7 = meglumine, 8 = mesylate, 9 = gluconate, 10 = maleate, 11 = nitrate, 12 = chloride, 13 = bromide, 14 = besylate, 15 = succinate, 16 = tosylate, 17 = phosphate, 18 = malate, 19 = zinc, 20 = magnesium, 21 = fumarate, 22 = tartrate, 23 = citrate, 24 = calcium, 25 = sulfate.
Column Dionex Acclaim Trinity P1 3.0 mm × 50 mm, 3 μm
Column temperature 35 °C
Autosampler temperature Ambient
Injection volume 5 μl
Mobile phase A 200 mM ammonium formate, pH 4.0
Mobile phase B Water
Mobile phase C Acetonitrile
Time (min) % Mobile phase A % Mobile phase B % Mobile phase C
Gradient 0 2 38 60
7 5 35 60
15 90 5 5
20 90 5 5
20.1 2 38 60
25 2 38 60
Flow rate 0.5 ml/min
pA
min

**Figure 3.31.** A chromatogram shows the concurrent determination of positive and negative pharmaceutical counterions using mixed-mode chromatography coupled with charged aerosol detection. Source: Zhang et al 2010 [37]. Reprinted with permission of Elsevier.

- For resolution of coeluting analytes on C18 columns, explore selectivity differences of polar-embedded, phenyl, PFP, or cyano phases.
  - For very low-pH applications (pH < 2.0), select columns resistant to acid hydrolysis of the bonded groups.
  - For high-pH applications (pH > 8), select columns stable at high pH (hybrids, C18 with trifunctional silanes, or organic polymer particles)
  - For retention of polar compounds or metabolites, choose C18 with hydrophilic characters, such as those with low ligand density and C18 AQ or use HILIC columns.

For new method development projects, the selection of columns packed with bonded phases on high-purity silica is recommended. Columns packed with SPP are highly recommended (see Figure 3.10). Modern columns have improved batch-to-batch reproducibility, stability, and versatility to ensure better success in method transfer.

Figure 3.32 shows a column selectivity chart of many popular reversed-phase bonded phases based on studies by Neue et al. A current version of this chart is available at www.waters.com. The hydrophobicity of the sorbent is measured by the log $k$ of a nonpolar solute (acenaphthene) and plotted on the $x$-axis. The silanophilic activity as indicated by the ratio of log $k$ of amitriptyline vs. acenaphthene is plotted on the $y$-axis. This comparative chart is useful for selecting equivalent or dissimilar bonded phases or columns from various vendors.

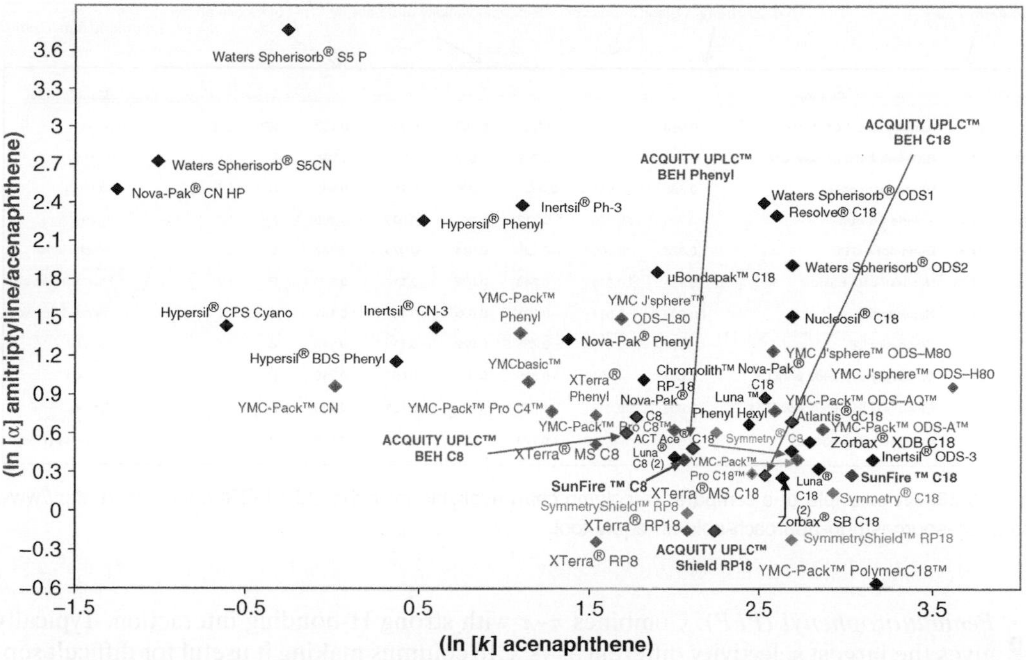


**Figure 3.32.** RPC column selectivity chart. The *x*-axis is an indication of the hydrophobicity and *y*-axis represents the silanophilic activity of the bonded phase. This chart can be useful for selecting equivalent or dissimilar columns from various vendors. Source: Courtesy of Waters Corporation.

Another more detailed classification of HPLC column is based on some extensive experimental studies of hundreds of commercially available reversed-phase columns by Snyder et al. [38] These data are available as a free United States Pharmacopeial (USP) termed PQRI database (http://www.usp.org/resources/pqri-approach-column-equiv-tool). An example of a comparative chart from the website is shown in Figure 3.33. The user can type in any available column in search of equivalent columns based on measured indices of hydrophobicity, steric interaction, hydrogen bonding acidity, and basicity.

### 3.8.1 Some General Guidelines for Bonded Phase Selection

Some general guidelines for selecting RPC-bonded phases are summarized as follows [4, 6].

- *C18*: Very hydrophobic, retentive, and stable phase. It is the first choice for most separations.
- *C8*: C8 has similar selectivity as C18 but is less retentive.
- *C3 or C4*: Less retentive than C8 or C18. Mostly used for protein separations on wide-pore supports.
- *Cyano (CN)*: Less retentive and different selectivity than C8. Note that many CN phases are less stable than C18. This phase is typically a propylcyano group with a C3 linker.
- *Phenyl*: A less retentive phase with unique selectivity for aromatics with $\pi$–$\pi$ in addition to hydrophobic interaction. The phenyl group may be directly attached to the silicon using a C3 to C18 alkyl spacer.

Similarity index — Hydrophobicity — Steric interaction — H-bonding acidity and basicity — pH of mobile phase

| Rank | F | Column | H | S | A | B | C(2.8) | C(7.0) | Type | USP designation | Manufacturer |
|---|---|---|---|---|---|---|---|---|---|---|---|
| 0 | 0 | Acquity UPLC CSH C18 | 0.968 | 0 | −0.355 | 0.199 | 0.065 | 0.157 | EP | L60 | Waters |
| 21 | 22.59 | BIOshell A400 protein C18 | 0.94 | −0.052 | −0.202 | 0.063 | 0.177 | 0.261 | B | L1 | Supelco |
| 22 | 23.39 | Ultrasphere octyl | 0.896 | 0.016 | 0.003 | 0.086 | 0.157 | 0.547 | B | L7 | Hichrom |
| 23 | 23.46 | Polaris C8-Ether | 0.705 | 0.023 | −0.312 | 0.04 | 0.095 | 0.269 | EP | L7 | Agilent/Varian |
| 24 | 24.32 | Bondclone C18 | 0.824 | 0.056 | −0.125 | 0.044 | 0.078 | 0.347 | B | L1 | Phenomenex |
| 25 | 24.73 | Hypersil beta basic-8 | 0.834 | 0.016 | −0.248 | 0.029 | 0.11 | 0.115 | B | L7 | Thermo/Hypersil |
| 26 | 24.74 | Hyputity C8 | 0.833 | 0.011 | −0.201 | 0.035 | 0.157 | 0.161 | B | L7 | Thermo/Hypersil |
| 27 | 24.86 | Discovery C8 | 0.832 | 0.011 | −0.238 | 0.029 | 0.119 | 0.143 | B | L7 | Supelco |
| 28 | 24.9 | Discovery BIO wide pore C5 | 0.654 | −0.019 | −0.305 | 0.029 | 0.091 | 0.219 | B | | Supelco |
| 29 | 24.91 | Discovery BIO wide pore C18 | 0.836 | 0.014 | −0.254 | 0.028 | 0.121 | 0.119 | B | L1 | Supelco |
| 30 | 24.97 | Nucelosil 100-5-C8 HD | 0.865 | −0.008 | −0.174 | 0.029 | 0.045 | 0.188 | A | L7 | Macherey Nagel |

**Figure 3.33.** An example of a comparative column chart available from the USP PQRI database in http://www.usp.org/resources/pqri-approach-column-equiv-tool.

- *Pentafluorophenyl (PFP)*: Combines $\pi$–$\pi$ with strong H-bonding interaction. Typically gives the largest selectivity differences vs. C18 columns making it useful for difficult separations of isomeric compounds.
- *Polar-embedded phases*: A polar group such as amide, carbamate, or ether can be embedded in the hydrophobic chain for mixed-mode separation of coeluting pairs.
- *Others*: Other useful phases for difficult separations include biphenyl, pentabromo, naphthyl, and cholesterol phases.

## 3.9 SUMMARY

This chapter provides an overview of essential column concepts, including a general description, types, packing materials, modern trends, and a selection guide. While the most popular column in use today is a 150 × 4.6-mm i.d. column packed with 3- or 5-μm C18 bonded phases, there is a definitive trend with excellent rationales to migrate toward the use of shorter and smaller inner diameter columns packed with sub 3-μm or sub 2-μm particles. The dominance of silica as the primary support material appears unchallenged and significant performance enhancements have been made possible with high-purity silica, hybrid particles, and SPP. The reader is urged to take advantage of these advances by developing more consistent methods using these improved columns.

## 3.10 QUIZZES

**1.** Most modern HPLC packings Do NOT have this feature.
(a) Spherical particles
(b) Irregular particles
(c) Porous particles
(d) High-purity silica

2. The minimum plate height of a column packed with 2.0 μm particles is about
   (a) 6 μm
   (b) 40 μm
   (c) 4 μm
   (d) 0.4 μm

3. The plate count of a 150-mm-long column packed with 3 μm particles is about
   (a) 25 000
   (b) 45 000
   (c) 2500
   (d) ~5000

4. Which column is NOT with bonded phases from high-purity silica?
   (a) Phenomenex Gemini
   (b) Waters Novapak
   (c) Waters X-Bridge
   (d) ZORBAX StableBond

5. The factor most responsible for higher batch-to-batch reproducibility of modern HPLC columns is the use of
   (a) high-purity silica
   (b) novel bonding chemistry
   (c) smaller particles
   (d) endcapping

6. Which silanol is the most acidic?
   (a) Gemini
   (b) Associated
   (c) Silanols activated by metallic oxides
   (d) Free silanols

7. What is the typical usable pH range for mobile phases for conventional silica-based C18 bonded phases?
   (a) 1–12
   (b) 3–6
   (c) 0–14
   (d) 2–8

8. Hybrids are an excellent choice vs. conventional silica support because of their
   (a) higher efficiency
   (b) smaller $d_p$
   (c) wider pH range
   (d) spherical particles

9. SPP is preferred for this performance advantage vs. totally porous particles.
   (a) Efficiency
   (b) $\Delta P$
   (c) Peak shape
   (d) Retention

10. What is the preferred particle size if one desires to have the best performance on both HPLC and UHPLC equipment in QC applications?
   (a) sub 3-μm

(b) sub 2-μm
(c) 1.7 μm
(d) 5 μm

### 3.10.1 Bonus Quiz

1. In your own words, describe which bonded phase you would select to retain some polar analytes using RPC and explain why.
2. Which RPC bonded phases would you select if your pharmaceutical sample contains some difficult analytes such as isomers and diastereomers, which are not resolved with the C18 column? Explain why.

## 3.11 REFERENCES


1. Neue, U.D. (1997). *HPLC Columns: Theory, Technology, and Practice*. New York: Wiley-VCH.
2. Poole, C. (2003). *The Essence of Chromatography*. Amsterdam, the Netherlands: Elsevier.
3. Unger, K. (1990). *Packings and Stationary Phases in Chromatographic Techniques*. New York: Marcel Dekker.
4. Ahuja, S. and Dong, M.W. (eds.) (2005). *Handbook of Pharmaceutical Analysis by HPLC*. Amsterdam, the Netherlands: Elsevier. Chapter 4.
5. Unger, K. (1979). Porous silica. *J. Chromatogr. Libr.* 16. Elsevier Scientific Publishing Co., New York.
6. Snyder, L.R. and Kirkland, J.J. (2010). *Introduction to Modern Liquid Chromatography*, 3e. Hoboken, NJ: Wiley. Chapter 5.
7. Guillarme, D. and Veuthey, J.-L. (eds.) (2012). *UHPLC in Life Sciences*. Cambridge, United Kingdom: Royal Society of Chemistry. Chapter 2.
8. Majors, R. (2015). *LCGC North Am.* 33 (11): 818.
9. Majors, R. (2015). *LCGC North Am.* 33 (12): 886.
10. Neue, U. (2000). *Encyclopedia of Analytical Chemistry* (ed. R.A. Meyer), 11450–11472. Chichester, United Kingdom: Wiley.
11. Dunlap, C., McNeff, C.V., Stoll, D., and Carr, P.W. (2001). *Anal. Chem.* 73 (21): 591A.
12. Van Deemter, J.J., Zuiderweg, F.J., and Klinkenberg, A. (1956). *Chem. Eng. Sci.* 5: 271.
13. Cunico, R.L., Gooding, K.M., and Wehr, T. (1998). *Basic HPLC and CE of Biomolecules*. Richmond, CA: Bay Bioanalytical Laboratory.
14. Wyndham, K.D., O'Gara, J.E., Walter, T.H. et al. (2003). *Anal. Chem.* 75 (24): 6781.
15. Kirkland, J.J., Adams, J.B. Jr., van Straten, M.A., and Claessens, H.A. (1998). *Anal. Chem.* 70: 4344.
16. DiCesare, J.L., Dong, M.W., and Ettre, L.S. (1981). *Introduction to High-Speed Liquid Chromatography*. Norwalk, CT: Perkin-Elmer.
17. Dong, M.W. and Gant, J.R. (1984). *LCGC* 2 (4): 294.
18. Tolley, L., Jorgenson, J.W., and Moseley, M.A. (2001). *Anal. Chem.* 73 (13): 2985.
19. Guillarme, D. and Dong, M.W. (eds.) (2014). UHPLC: where we are ten years after its commercial introduction. *Trends Anal. Chem.* 63: 1–188. (Special issue).
20. Ishii, D. (1988). *Introduction to Microscale High-Performance Liquid Chromatography*. New York: VCH Publishers Inc.
21. Tanaka, N., Kobayashi, H., Nakanishi, K. et al. (2001). *Anal. Chem.* 73: 420A.
22. Cabrera, K., Lubda, D., Eggenweiler, H.M. et al. (2000). *High Resol. Chromatogr.* 23: 93.
23. Kirkland, J.J., Langlois, T.J., and DeStefano, J.J. (2007). *Am. Lab.* 39: 18.
24. Kirkland, J.J., Schuster, S.A., Johnson, W.L., and Boyes, B.E. (2013). *J. Pharm. Anal.* 3: 303.
25. Fekete, S., Oláh, E., and Fekete, J. (2012). *J. Chromatogr. A* 1228: 57.
26. Fekete, S., Guillarme, D., and Dong, M.W. (2014). *LCGC North Am.* 32 (6): 420.

27. Broeckhoven, K. and Desmet, G. (2014). *Trends Anal. Chem.* 63: 65.
28. Sandra, K., Vandenbussche, J., Vandenheede, I. et al. (2018). *LCGC Eur.* 31 (3): 155–166.
29. Dong, M.W., Gant, J.R., and Perrone, P.A. (1985). *LCGC.* 3: 786.
30. Heyden, Y.V., Mangelings, D., Matthijs, N., and Perrin, C. (2005). *Handbook of Pharmaceutical Analysis by HPLC* (ed. S. Ahuja and M.W. Dong). Amsterdam, the Netherlands: Elsevier. Chapter 18.
31. Zhang, T., Franco, P., Nguyen, D. et al. (2012). *J. Chromatogr. A* 1269: 178.
32. Desfontaine, V., Guillarme, D., Francotte, E., and Nováková, L. (2015). *J. Pharm. Biomed. Anal.* 113: 56.
33. Nováková, L., Chocholouš, P., and Solich, P. (2014). *Talanta* 121: 178.
34. Alpert, A.J. (1990). *J. Chromatogr. A* 499: 177.
35. Wang, X., Li, W., and Rasmussen, H.T. (2005). *J. Chromatogr.* 1083: 58.
36. Zhang, K. and Liu, X.D. (2016). *J. Pharm. Biomed. Anal.* 128: 73.
37. Zhang, K., Dai, L., and Chetwyn, N. (2010). *J. Chromatogr. A* 1217: 5776.
38. Snyder, L.R., Dolan, J.W., and Carr, P.W. (2007). *Anal. Chem.* 79: 3255–3261.

# 4

# HPLC/UHPLC INSTRUMENTATION AND TRENDS

## 4.1 INTRODUCTION

### 4.1.1 Scope

High-performance liquid chromatography (HPLC) is a versatile analytical technique using sophisticated equipment refined over the last five decades. An understanding of the working principles and trends is useful for effective applications of the technique. This chapter provides the reader with a concise overview of HPLC and ultra-high-pressure liquid chromatography (UHPLC) instrumentation, operating principles, recent advances, and trends. The focus is on analytical-scale HPLC systems and modules (pumps, autosamplers, detectors, and data systems). Concepts of system dwell volume and instrumental bandwidth (IBW) are discussed with their impacts on the use of small-diameter columns. Specialized HPLC systems and their attributes are described. Guidelines for selecting HPLC equipment are reviewed. Readers are referred to books [1–5], review articles [6, 7], and manufacturers' literature/websites for more details. UHPLC instruments are described briefly here and are further discussed in Chapter 5.

### 4.1.2 HPLC Systems and Modules

Today's liquid chromatographs are well-engineered instruments with excellent performance and reliability. A typical HPLC system consists of a number of modules – a pump, injector or autosampler, column oven, detector(s), and a data system (see schematic diagram in Figure 4.1). Modern HPLC systems are likely to resemble those in Figure 4.2, consisting of a multisolvent pump, an autosampler, a column oven, a UV/visible absorbance or a photodiode array (PDA) detector, and a chromatography data system (CDS), which also controls the entire HPLC system besides managing its data.

HPLC systems can be modular, integrated, or a mix of both instrument types. Figure 4.2a shows a modular system consisting of separate but stackable modules. A modular system is more serviceable than a fully integrated system because each module can be maintained,

*HPLC and UHPLC for Practicing Scientists*, Second Edition. Michael W. Dong.

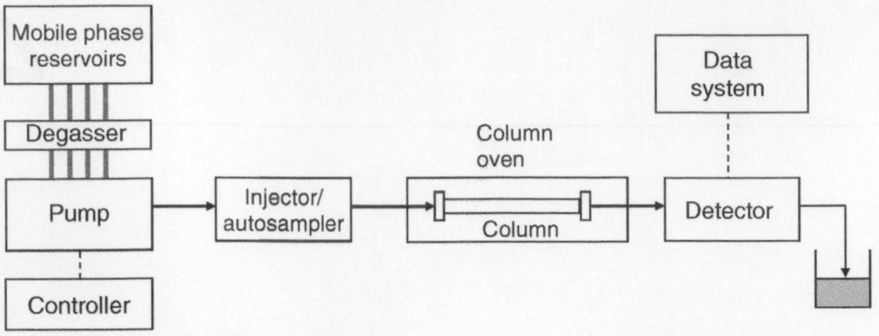


**Figure 4.1.** A schematic of an HPLC system showing all the major modules or components.

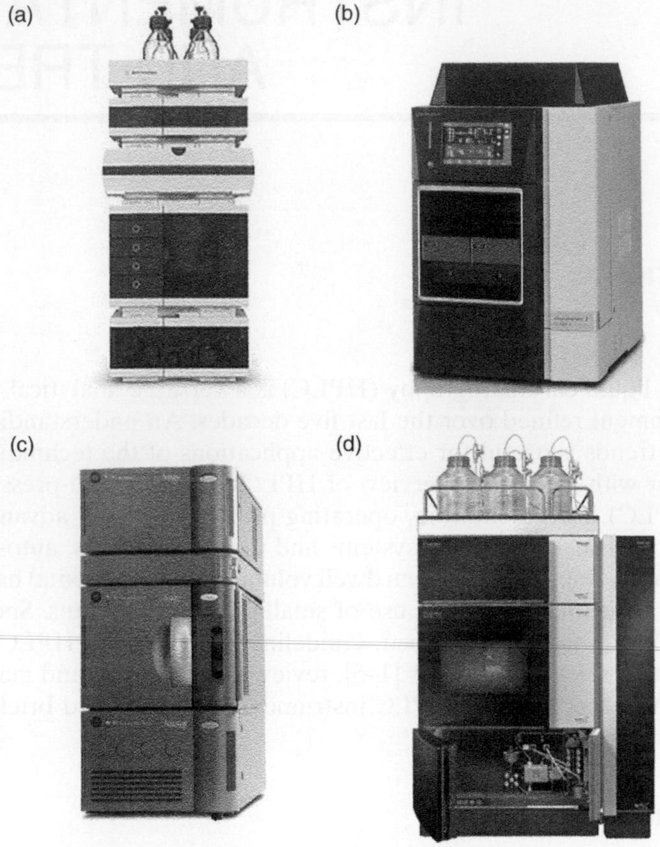


**Figure 4.2.** Examples of modular, integrated, and mixed HPLC systems. Modular: (a) Agilent 1290 Infinity II (binary). Integrated: (b). Shimadzu Prominence-I (quat., HPLC). Mixed systems: (c) Waters Acquity UPLC H-Class (quat.). (d) Thermo Vanquish Flex (quat., biocompatible). Source: Courtesy of Agilent, Shimadzu, Waters, and Thermo Fisher Scientific.

repaired, or exchanged individually if it malfunctions. Figure 4.2b shows an integrated system whose modules are built inside a single housing, typically with a front control panel (Figure 4.2b). There are also mixed systems (Figure 4.2c,d), in which a pump and autosampler are within a main integrated unit, and a column oven and detector(s) can be stacked or placed alongside the main unit. In modern HPLC systems, modules are connected (internally or externally using networking cables), so they can be controlled by a control panel, a workstation, or client-server network. Single-point control by the CDS allows a single method to execute instrument control of all the modules.

Another designation for HPLC systems used for the analysis of biomolecules, which often require the use of high-salt mobile phases that are corrosive to stainless steel is called "biocompatibility." The term biocompatible HPLC systems typically means systems with a "bioinert" fluidic flow paths constructed from titanium, MP35N alloy, or polyetheretherketone (PEEK)-coated stainless steel. Other components such as the pump seal, injector rotor, and sampling needle may also be made from a more corrosion-resistant material.

### 4.1.3 Ultra-High-Pressure Liquid Chromatography (UHPLC)

UHPLC is the most important breakthrough in modern HPLC [6, 7]. Research pioneered by James Jorgenson and coworkers [8] in 1997, and later on by Milton Lee, led to the introduction of the first commercial UHPLC system (Waters Acquity UPLC) in 2004 [9, 10]. Today, all major manufacturers have product offerings in UHPLC, allowing more efficient and faster separations [10–12]. HPLC and UHPLC models from four major manufacturers are listed in Table 4.1 with their pressure rating and associated mass spectrometers (MS) and CDSs.

## 4.2 HPLC AND UHPLC SOLVENT DELIVERY SYSTEMS

The solvent delivery system consists of one or more pumps, solvent reservoirs, and a degasser. Typical specifications of an analytical HPLC pump are as follows:

- Provides precise and pulse-free delivery of solvents
  - *Typical flow rates range*: 0.01–10 ml/min (HPLC), 0.001–2–5 ml/min (UHPLC)
  - *Pressure limits*: 6000 psi (HPLC), 15 000–22 000 psi (UHPLC), 9 000–12 000 psi (intermediate UHPLC or dual-path systems)
- Compatible with common organic solvents, buffers, and salts
- Reliable operation with long pump seal life
- Accurately blends solvents for isocratic or gradient operation
- Easy to maintain and service

Figure 4.3a shows the schematic of a reciprocating pump mechanism. Here, a motorized cam drives a piston to deliver a solvent through a set of inlet and outlet check valves. Since only the inward piston stroke delivers the liquid, a pulse dampener is used to reduce flow fluctuations. All components in the fluidic path are made from inert materials (e.g. stainless steel pump heads, ruby balls/sapphire seats in check valves, sapphire pistons, and fluorocarbon pump seals). This simple and low-cost reciprocating pump design has undergone numerous innovations, as summarized in Table 4.2. Note that gradient pump performance at low flow rates (<200 μl/min) is historically difficult to achieve for pumps with a "standard" piston size of 100 μl. Innovations such as variable pump stroke mechanisms and micro pistons have extended acceptable low-flow performance down to 10–100 μl/min.

Figure 4.3b shows a very common design for enhancing the flow precision and compositional accuracy called a dual-piston in-series pump. The two pistons are connected in series and typically driven by direct screw drives from separate digital motors. Many modern high-end pumps use this design including the Waters Alliance and Acquity, Agilent 1100/1200/1260/1290, and most Shimadzu pumps.

For micro- or nano-LC, either flow-splitting or a syringe pump is required. [13]

**Table 4.1. HPLC and UHPLC Systems from Four Major Manufacturers**

| Manufacturer | Series/model | Type | Max pressure (psi) | CDS | MS |
|---|---|---|---|---|---|
| Agilent | 1290 Infinity II | Binary, quat | 19 000 | OpenLAB ChemStation | SQ, TOF, TQ, QTOF, IMS |
| | 1260 Infinity II, Prime | Quat, binary, bio | 9000, 11 500 | | |
| | 1220 Infinity II | Integrated, L-P mixing Binary | 9000 | | |
| Shimadzu | Nexera X2 | Binary | 19 000 | LabSolutions | SQ, TQ, IT-TOF |
| | Nexera Quat | Quat, bio | 19 000 | | |
| | Prominence | Modular HPLC | 6000 or 9000 | | |
| | i-Series | Integrated HPLC | 6000 | | |
| Thermo Scientific | Vanquish Horizon | Binary bio | 22 000 | Chromeleon | SQ, TQ, IT, Orbitrap |
| | Vanquish Flex, Duo | Quat bio, Dual ternary | 15 000 | | |
| | UltiMate 3000 | Binary, quat, dual, bio | 9000 or 15 000 | | |
| Waters | Acquity H-Class | Quat, bio | 15 000 | Empower | QDa, SQ, TQ, TOF, QTOF, IMS |
| | Acquity I-Class | Binary | 18 000 | | |
| | Acquity UPLC Classics | Binary | 15 000 | | |
| | Acquity Arc | Quat, dual-path | 9500 | | |
| | Alliance | Quat, integrated | 5000 | | |

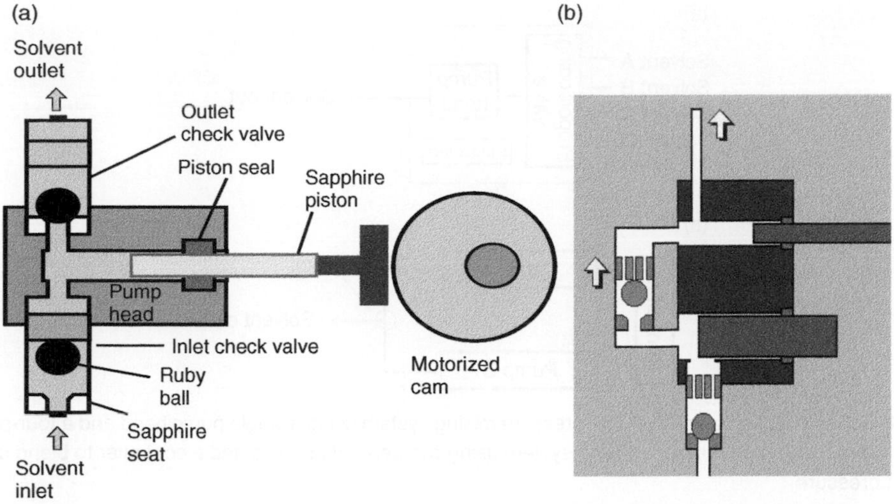


**Figure 4.3.** (a) A schematic of a single-piston reciprocating pump. (b) A schematic of an HPLC pump based on a dual-piston in-series design. Source: Courtesy of Savant Academy.

**Table 4.2. Innovations for Enhancing Pump Performance**

| Pump performance | Innovations |
|---|---|
| Ultra-high-pressure | Pump designed to operating reliability up to 15 000–22 000 psi |
| Piston seal life | Spring-loaded seal, self-aligning piston, piston seal wash |
| Blending accuracy | Dual-piston in-series or in parallel design, direct screw-drive mechanism for each pump, high-speed proportioning valve, auto compensation for compressibility and $\Delta V_{mixing}$ |
| Reduced pulsation | Rapid-refilling pump mechanism, pulse dampener, direct screw-drive, dual-piston in-parallel |
| Better low-flow performance | Micro-pistons, variable stroke length, active check valve, syringe pump |
| Biocompatibility | Titanium or PEEK based system, high-pH compatible seals |
| Multi-solvent pump | Low-pressure mixing system, high-speed proportionating valve |
| Reduced dwell volume | High-pressure mixing system, low-volume mixers |

### 4.2.1 High-Pressure and Low-Pressure Mixing Designs in Multisolvent Pumps

Multisolvent pumping systems are classified by how solvent blending is achieved. All quaternary (quat) pumps use a low-pressure mixing design where a single pump head draws mobile phases from a four-port proportioning valve (a white Teflon block with four solenoid valves, each connecting to a solvent line, as shown in Figure 4.4a). The pump's microprocessor controls the solvent composition of each intake piston stroke by the timed opening of each solvent port. Note that solvent blending occurs inside the pump at low pressures. Solvent degassing is mandatory to prevent outgassing of dissolved air during blending. In a low-pressure mixing design, the pump head, the pulse dampener, the drain valve, the in-line filter, and the pressure transducer all contribute to the internal liquid hold-up (delay) volumes or dwell volumes. However, the major advantages of low-pressure mixing pumps are the simplicity and lower cost.

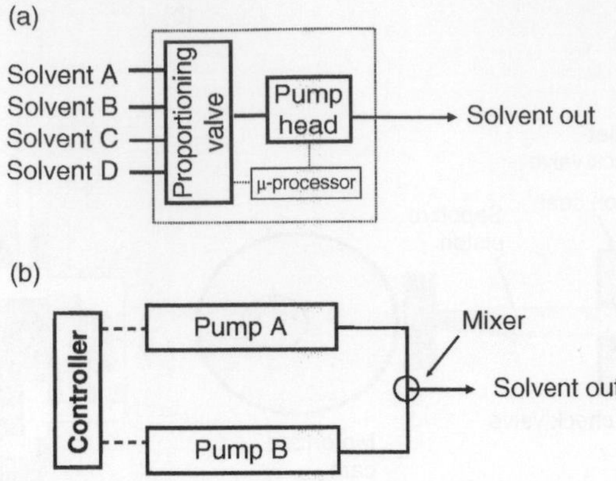


**Figure 4.4.** Schematic diagrams: (a) A low-pressure mixing system using a single pump head and a four-port proportioning valve. (b) A high-pressure mixing system using two separate pumps and a controller to blend solvents under high pressure.

In high-pressure mixing systems, two (and occasionally three) separate pumps are used to mix solvents at high pressures (Figure 4.4b). A flow rate change of each pump is used to generate different isocratic blends or gradient profiles. An external mixer is required to ensure adequate mixing of the two solvents. Binary pumps cost more but have the advantage of lower dwell volumes for applications using small-diameter columns. Added solvent selection valves are used to increase the number of solvent pairs (e.g. A1 or A2 and B1 or B2) though only binary solvents can be delivered at one time.

### 4.2.2 System Dwell Volume

System dwell volume is the liquid hold-up volume of the HPLC system from the point of solvent mixing to the inlet of the column. For low-pressure mixing systems, the dwell volume includes the additive volumes from the proportioning valve, pump head, mixer and any other internal pump volume (e.g. pulse-dampener, pressure transducer, back-pressure device, and filter), injector, sample loop, and all fluidic connection tubing to the column inlet. The typical dwell volume of a low-pressure mixing HPLC system is ~1 ml (e.g. Waters Alliance and Agilent 1100/1200 systems) [2] and can range from 0.5 to 2 ml. High-pressure mixing systems have inherently lower dwell volumes because the point of mixing is external to the pumps. Note that dwell volume is inconsequential in isocratic analysis but becomes important in gradient analyses, since it adds gradient delay time to the analysis. It becomes critical for gradient applications at low flow rates. For instance, a dwell volume of 1 ml represents a 5-minute gradient delay at 200 μl/min or a 20-minutes delay at 50 μl/min for microbore (1-mm i.d.) columns. For this reason, high-pressure mixing systems are used for high-throughput screening (HTS) and micro-LC systems.

The system dwell volume can be measured using a UV detector as follows:

- Disconnect the column and replace it with a zero-dead-volume union.
- Place 0.5% acetone in water in reservoir B and water in reservoir A.
- Set a linear gradient program from 0% to 100% B in 10 minutes at a flow rate of 1.0 ml/min and a detection wavelength of 254 nm.
- Start the solvent gradient and the data system to record the detector signal.
- Measure the intersection point of the extrapolated absorbance trace with the baseline to record the dwell time, as shown in Figure 4.5. Multiply dwell time by flow rate to obtain the dwell volume.

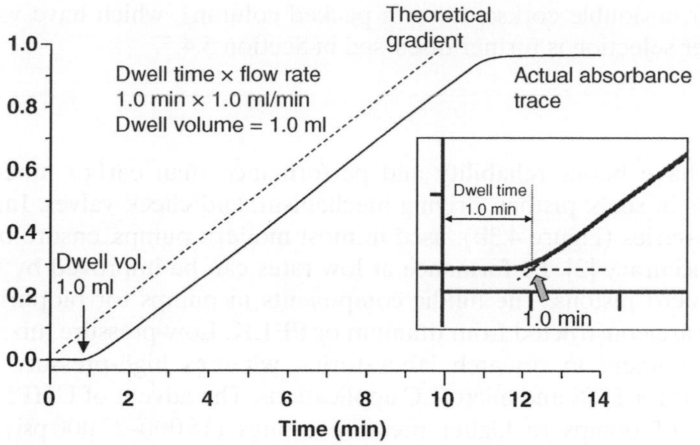


**Figure 4.5.** The absorbance gradient trace used to measure the system dwell volume of a Waters Alliance system. The inset shows the intersection point marking the gradient onset. The system dwell volume is equal to the dwell time times the flow rate.

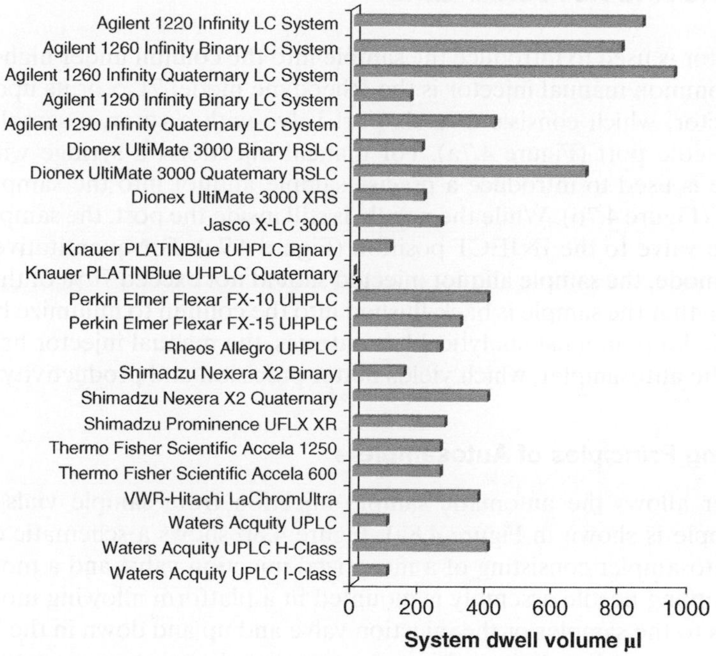


**Figure 4.6.** Chart of system dwell volumes of HPLC and UHPLC system. Source: Fekete et al. 2014 [14]. Reprinted with permission of Elsevier.

***4.2.2.1 Dwell Volumes of UHPLC Systems*** UHPLC systems typically have much lower dwell volumes than those of conventional HPLC. Dwell volumes of binary systems range from 0.1 to 0.4 ml, depending on the choice of mixers, while those of quaternary systems range from 0.4 to 0.8 ml (see Figure 4.6) [14]. External online mixers are required to ensure adequate mixing efficiency for binary pumps for high-sensitivity UV applications. Each manufacturer may offer different optional mixers catering to specific applications (e.g. a jet-weaver, a double corkscrew, or a packed column), which have various volumes (30–450 μl). Mixer selection is further discussed in Section 5.4.5.

### 4.2.3 Trends

Modern pumps have better reliability and performance than earlier models because of improved designs in seals, pistons, driving mechanism, and check valves. Innovations such as dual-piston in-series (Figure 4.3b), used in most modern pumps, ensure better compositional and flow accuracy [2]. Performance at low rates can be improved by variable stroke mechanism or micro pistons. The fluidic components in pumps for biopurification or ion chromatography are constructed from titanium or PEEK. Low-pressure mixing quat pumps are standard equipment in research laboratories, whereas high-pressure mixing binary pumps are popular for HTS and micro-LC applications. The advent of UHPLC has elevated the performance of pumps to higher pressure ratings (15 000–22 000 psi) and precision performance. Dual-piston in-parallel designs are returning for some UHPLC pumps (e.g. Thermo Vanquish) to yield flows with lower pulsations.

## 4.3 INJECTORS AND AUTOSAMPLERS

An HPLC injector is used to introduce the sample into the column under high-pressure flow conditions. A common manual injector is the Rheodyne model 7125 or its updated replacement 7725 injector, which consists of a six-port valve with a rotor, a sample loop, and a front-loading needle port (Figure 4.7a). For manual injections, a syringe with a 22-gauge blunt-tip needle is used to introduce a precise sample aliquot into the sample loop at the LOAD position (Figure 4.7b). While the needle is still inside the port, the sample is delivered by switching the valve to the INJECT position (Figure 4.7c). For quantitative analysis in a partial loop fill mode, the sample aliquot injected should not exceed 70% of the sample loop volume [2]. Note that the sample is back-flushed into the column to minimize band broadening in the sample loop. In most analytical laboratories, the manual injector has been mostly supplanted by the autosampler, which yields better precision and productivity.

### 4.3.1 Operating Principles of Autosamplers

An autosampler allows the automatic sample injection from sample vials or microtiter plates (an example is shown in Figure 4.8a). Figure 4.8b shows a schematic diagram of an "*XYZ*"-type autosampler consisting of a motorized injection valve and a moving sampling needle. The sampling needle assembly is mounted in a platform allowing movement in the "*XY*" directions to the samples or the injection valve and up and down in the "Z" direction. It is connected to a sampling syringe driven by a precise digital stepper motor. The syringe can draw samples or flush solvents depending on the positions of a solenoid valve.

The injection sequence mimics the operation of a manual injection. First, the needle cleans itself by drawing the flush solvent and empties it into a drain port. Then, it moves to the sample vial, withdraws a precise sample aliquot, and delivers it to the injector valve. The motor then turns the valve to inject the sample. Small air gaps are used to segment the sample aliquot from the flush solvent inside the sampling needle. This basic design allows rapid sampling from a variety of sample vials or microtiter plates.

Another popular design is the integrated-loop autosampler, where a fixed sampling needle forms part of the sampling loop [4]. Sample vials are brought underneath the sampling

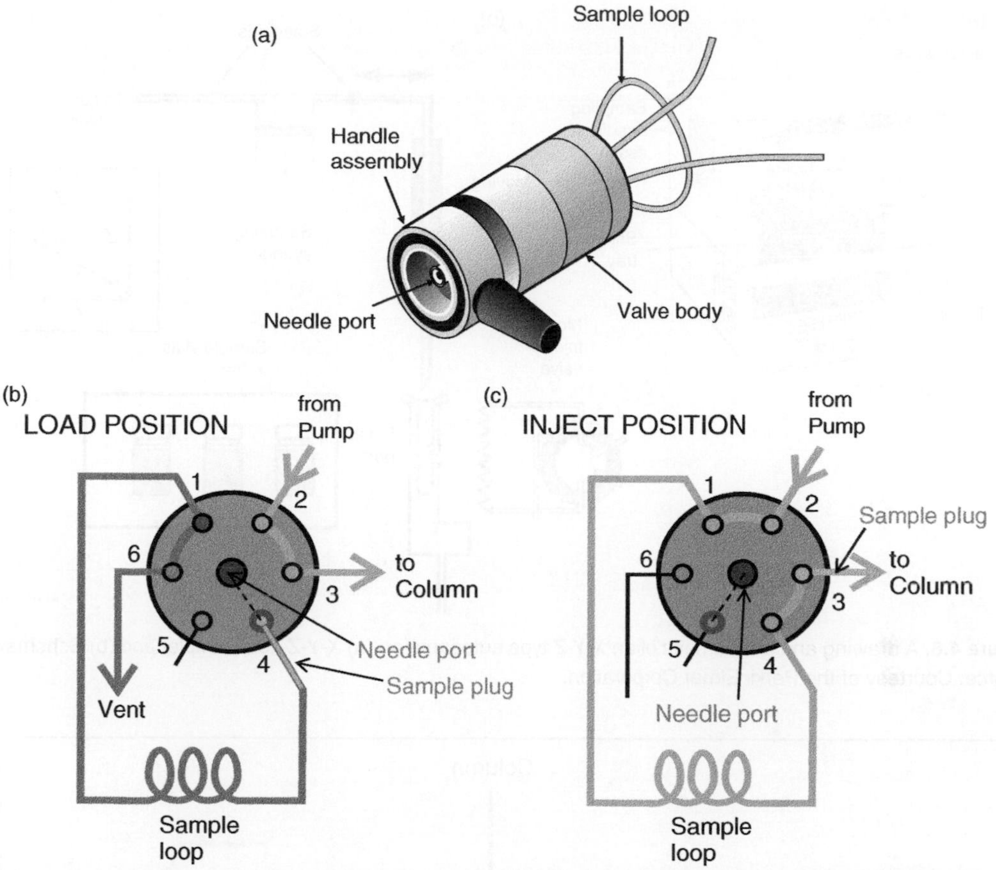


**Figure 4.7.** (a) A diagram of a manual injector valve (Rheodyne 7125). (b, c) A schematic of a Rheodyne 7125 injector valve during the LOAD and INJECT cycle under a partial-loop injection mode.

needle by a set of robotic arms or a carousel turntable system. No flush solvent is needed as the mobile phase acts as the carrier fluid to withdraw the sample through a valving system and a metering device. A high-pressure needle seal is used to seal the sample needle during injection. This design is gaining popularity because of its excellent precision and carryover characteristics. Figure 4.9 shows a schematic of an integrated-loop design used in several popular autosamplers.

The autosamplers in UHPLC must have a high-pressure rating, lower dispersion, and excellent precision at small injection volumes to accommodate the use of small-diameter columns.

### 4.3.2 Performance Characteristics and Trends

Autosamplers have made significant progress in reliability and performance. Primary performance criteria are sampling precision and carryover. Precision levels of <0.2% relative standard deviation (RSD) for peak area and a low carryover of <0.05% are routinely achievable in most HPLC autosamplers. "Carryover" refers to the percentage of the previous sample that is "carried over" to the next sample and must be minimized for most applications

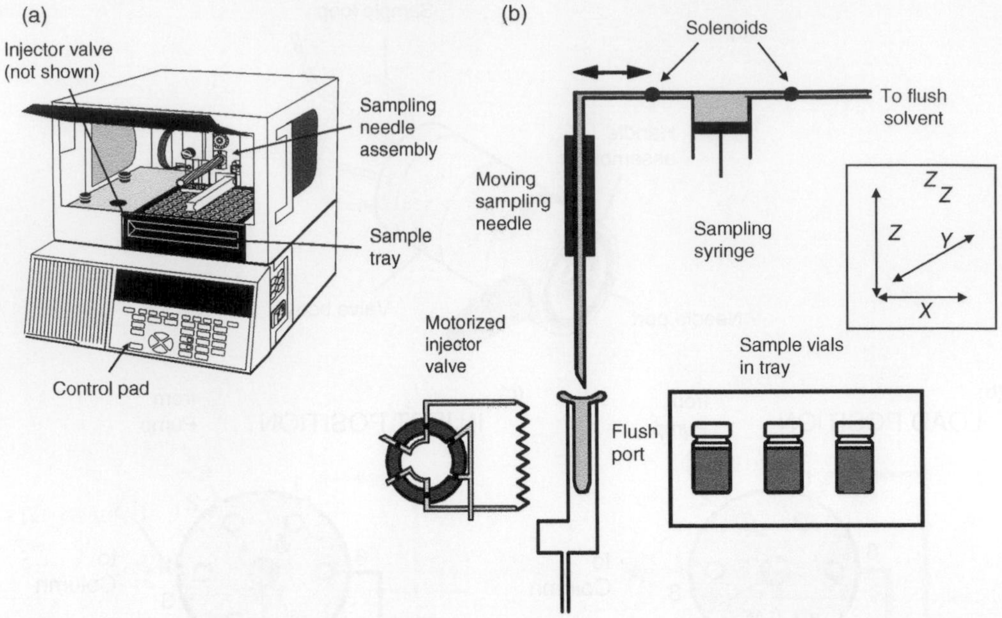


**Figure 4.8.** A drawing and a schematic of an *X-Y-Z* type autosampler. (a) *X-Y-Z* auto sampler and (b) Schematic. Source: Courtesy of the PerkinElmer Corporation.

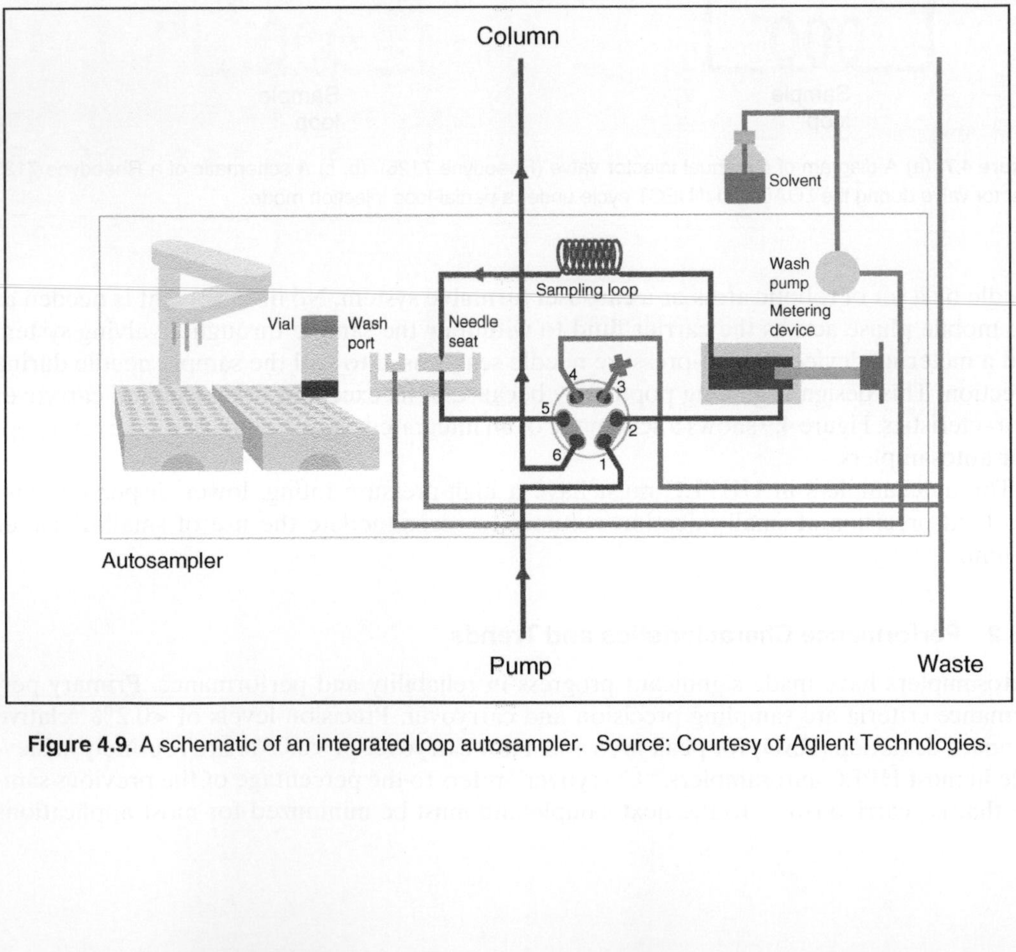


**Figure 4.9.** A schematic of an integrated loop autosampler. Source: Courtesy of Agilent Technologies.

(e.g. bioanalytical analysis). Fast operation (up to 3 injections/min) and sampling from 96-well microplates are desirable for HTS assays. Many have optional Peltier coolers for sample trays and additional liquid handling capabilities for dilution, standard addition, and derivatization [2]. Autosamplers for HTS handle a large number of microplates by using large *X*–*Y* platforms or robotic plates feeders.

UHPLC has pushed modern autosamplers to higher levels of precision (<0.1% RSD) even for very small sampling volumes (<2 μl). Historically, the minimum injection volume for precise quantitation by HPLC is ~ 5 μl. This level is now ~1 μl for UHPLC. Not surprisingly, the equipment cost for UHPLC pumps and autosamplers is higher due to the more complex engineering designs.

## 4.4 DETECTORS

An HPLC detector measures the concentration (or mass) of eluting analytes by monitoring one of their inherent physicochemical properties, such as UV absorbance [2–4]. A detector can be "universal" and respond to all analytes or "specific" to particular classes of analytes. Common detectors and their attributes are listed in Table 4.3. Early HPLC detectors were spectrometers equipped with small flow cells, but all modern HPLC UV detectors are designed solely for this purpose. The ubiquitous UV/visible variable wavelength absorbance and the PDA detectors are covered here together with a universal detector (charged aerosol detector, CAD). Mass spectrometers (MS) are briefly described here with a more in-depth discussion of LC/MS principles, equipment, and applications in Chapter 6.

**Table 4.3. Common HPLC Detectors and Attributes**

| Detector | Analyte/attributes | Sensitivity |
|---|---|---|
| UV/Vis absorbance (UV/Vis) | Specific: Compounds with UV Chromophores | ng–pg |
| Photo diode array (PDA) | Specific: Same as UV/Vis detectors also provides UV spectra | ng–pg |
| Fluorescence (FLD) | Very specific: Compounds with native fluorescence or with fluorescent tag | fg–pg |
| Refractive index (RID) | Universal: polymers, sugars, triglycerides, organic acids, excipients; not compatible with gradient analysis | 0.1–10 μg |
| Evaporative light-scattering (ELSD) | Universal: non-volatile or semi-volatile compounds, compatible with gradient analysis | 10 ng |
| Corona charged aerosol (CAD) | Universal: use nebulizer technology like ELSD and detection of charges induced by a high-voltage corona wire | Low ng |
| Chemiluminescence nitrogen (CLND) | Specific to N-containing compounds based on pyro-chemiluminescence | <0.1 ng of nitrogen |
| Electrochemical (ECD) | Very specific: Electro-active compounds (Redox) | pg |
| Conductivity | Specific to anions and cations, organic acids, surfactants | ng or ppm–ppb |
| Radioactivity | Specific, radioactive-labeled compounds | Low levels |
| Mass spectrometry (MS), MS/MS | Both universal and specific, structural identification<br>Very sensitive and specific | ng–pg<br>pg - fg |
| Nuclear magnetic resonance (NMR) | Universal, for structure, elucidation and conformation | mg–ng |

## 4.5 UV/VIS ABSORBANCE DETECTORS

The single-wavelength UV/Vis absorbance detector used to be the most common HPLC detector, but its prominence has been superseded by the PDA detector.

### 4.5.1 Operating Principles

The principle for UV/Vis absorption is the Beer's Law [1], where

$$\text{Absorbance } (A) = \text{molar absorptivity } (\varepsilon) \times \text{path length } (b) \times \text{concentration } (c)$$

Absorbance is defined as the negative logarithm of transmittance, which is equal to the ratio of transmitted light intensity and the incident light intensity. Note that absorbance is equal to 1.0 if 90% of the light is absorbed and 2.0 if 99% of incident light is absorbed. Most UV absorption bands correspond to transitions of electrons in the analyte molecules from $\pi \rightarrow \pi^*$, $n \rightarrow \pi^*$, or $n \rightarrow \sigma^*$ molecular orbitals [1]. Table 4.4 lists some common organic functional groups with chromophoric (light absorbing) properties.

The UV/Vis absorbance detector monitors the absorption of UV or visible light in the HPLC eluent by measuring the energy ratio of the sample beam against that of a reference beam. It is a very common detector since many analytes of interest (e.g. pharmaceuticals) contain one or more chromophoric groups. A UV/Vis detector consists of a deuterium lamp, a monochromator, and a small flow cell (Figure 4.10a). A monochromator consists of a grating that can be turned to select a specific wavelength through an exit slit. Dual-beam optical design is common. Here, the light source is split into a sample and a reference beam, and the intensity of each beam is monitored by a separate photodiode that transforms light energy into an electronic signal. Only the sample beam passes through the sample flow cell. A flow cell (Figure 4.11a) has typical volumes of 2–10 μl and path lengths of 10 mm with quartz lenses or windows at both ends of the flow cell.

**Table 4.4. Common UV Chromophores**

| Chromophore | $\lambda_{max}$ (nm) | Molar absorptivity ($\varepsilon$) |
|---|---|---|
| Alkyne | 225 | 160 |
| Carbonyl | 280 | 16 |
| Carboxyl | 204 | 41 |
| Amido | 214 | 41 |
| Azo | 339 | 5 |
| Nitro | 280 | 22 |
| Nitroso | 300 | 100 |
| Nitrate | 270 | 12 |
| Olefin conjugated | 217–250 | >20 000 |
| Ketone | 282 | 27 |
| Alkylbenzenes | 250–260 | 200–300 |
| Phenol | 270 | 1450 |
| Aniline | 280 | 1430 |
| Naphthalene | 286 | 9300 |
| Styrene | 244 | 12 000 |

Source: Data extracted from Ref. [1].

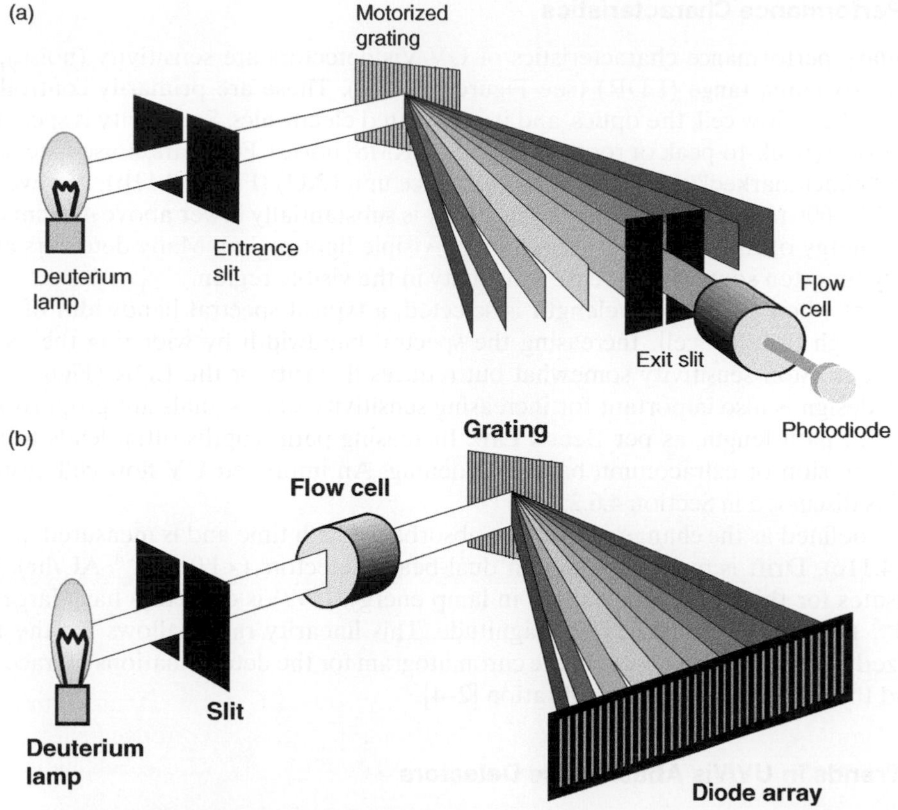


**Figure 4.10.** A schematic of (a) a UV–Vis absorbance detector, (b) a Photodiode array (PDA) detector.

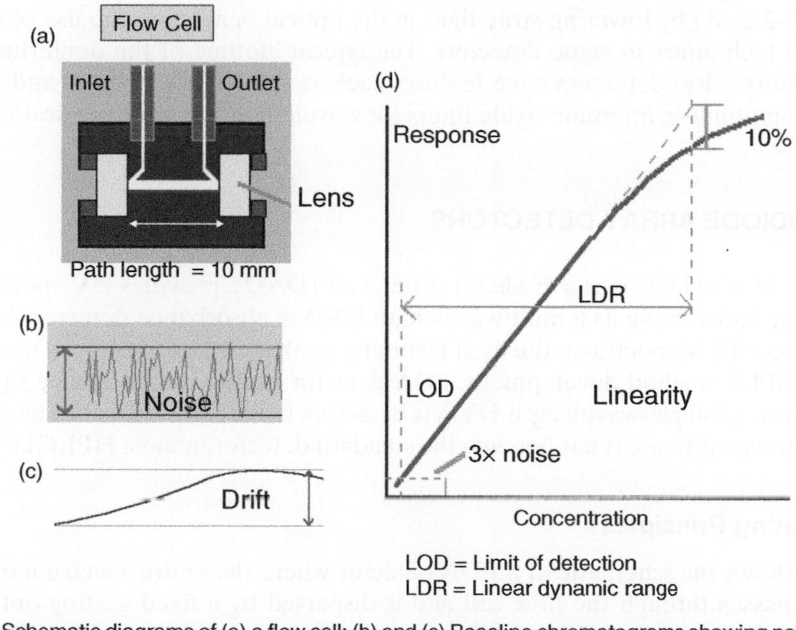


**Figure 4.11.** Schematic diagrams of (a) a flow cell; (b) and (c) Baseline chromatograms showing noise (magnified) and drift, (d) Chart of UV response vs. concentration of the analyte injected. Linearity range is from the limit of detection (LOD) to the point deviating 10% from a linear response.

### 4.5.2 Performance Characteristics

The primary performance characteristics of UV/Vis detectors are sensitivity (noise), drift, and linear dynamic range (LDR) (see Figure 4.11b–d). These are primarily controlled by the design of the flow cell, the optics, and its associated electronics. Sensitivity is specified by baseline noise (peak-to-peak or root mean square [RMS] noise). For years, noise specification has been "benchmarked" at $\pm 1.0 \times 10^{-5}$ absorbance unit (AU) (Figure 4.11b). A wavelength range of 190–600 nm is typical, though sensitivity is substantially lower above 400 nm due to a lack of energy of the deuterium source in the visible light region. Many detectors allow a secondary tungsten source to increase sensitivity in the visible region.

Note that when a single wavelength is selected, a typical spectral bandwidth of 5–8 nm passes through the flow cell. Increasing the spectral bandwidth by widening the exit slits improves detection sensitivity somewhat but reduces linearity or the LDR (Figure 4.11d). Flow cell design is also important for increasing sensitivity since signals are proportional to the flow cell path length, as per Beer's Law. Increasing path lengths often leads to higher system dispersion or extracolumn band broadening. An improved UV flow cell design for UHPLC is discussed in Section 4.6.2.

Drift is defined as the change of baseline absorbance with time and is measured in AU/hr (Figure 4.11c). Drift is typically lower in dual-beam detectors ($<1.0 \times 10^{-4}$ AU/hr), which compensates for the long-term changes in lamp energy. UV/Vis detectors have large LDR from $10^{-5}$ to ~2 AU or 5 orders of magnitude. This linearity range allows for the use of normalized peak area percentages in the chromatogram for the determinations of trace impurities and the use of single-point calibration [2–4].

### 4.5.3 Trends in UV/Vis Absorbance Detectors

Sensitivity and linearity performance has improved steadily in recent decades. The benchmark noise level of $\pm 1 \times 10^{-5}$ AU/cm is surpassed by many modern UV/Vis detectors (e.g. to $\pm 0.3 \times 10^{-5}$ AU). The high end of the LDR has been extended from a typical level of 1–1.5 AU to 2–2.5 AU by lowering stray-light in the optical bench and the use of electronic compensation techniques in some detectors. The typical lifetime of the deuterium lamp is now ~2000 hours. Most detectors have features such as self-aligned sources and flow cells, leak sensors, and built-in holmium oxide filters for wavelength accuracy verification.

## 4.6 PHOTODIODE ARRAY DETECTORS

A PDA detector, also known as a diode array detector (DAD), provides UV-spectra of eluting peaks while functioning as a multiwavelength UV/Vis absorbance detector. Because of its ability to record UV-spectra on the fly, it facilitates peak identification and is the preferred detector for HPLC method development. PDA detector sensitivity used to be significantly lower than those of single-wavelength UV/Vis detectors in earlier models but has improved significantly in recent years. It has become the standard detector in most HPLC laboratories.

### 4.6.1 Operating Principles

Figure 4.10b shows the schematic of a PDA detector where the entire spectrum of the deuterium lamp passes through the flow cell and is dispersed by a fixed grating onto a diode array element that measures the intensity of light at each wavelength. Most PDAs use a

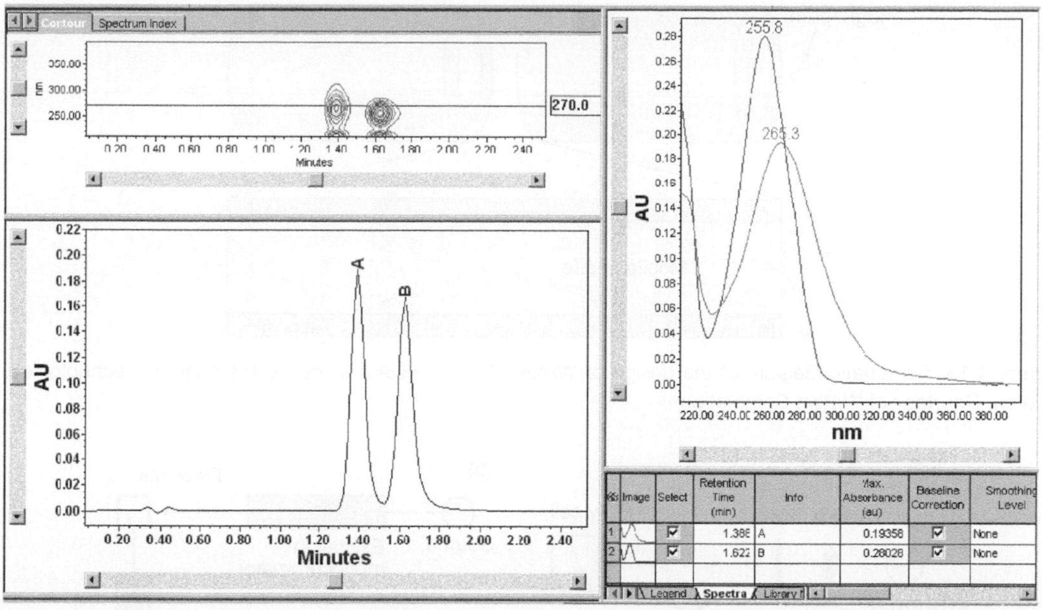


**Figure 4.12.** Waters Empower 3 (CDS) screenshot showing a UV spectral contour map, a chromatogram at 270 nm showing the separation of nitrobenzene and propylparaben. UV spectra of these two components are shown in the right-hand pane annotated with their respective $\lambda_{max}$.

charge-coupled diode array with 512–1024 diodes (or pixels), capable of a spectral resolution of about 1 nm. Spectral evaluation software allows the display of both chromatographic and spectral data of all the peaks in the sample (Figure 4.12). These features are integrated into the CDS and can include automated spectral annotations of $\lambda_{max}$ and display of UV-spectra in reports, contour maps, peak matching, library searches, and peak purity evaluation. Peak purity evaluation works by comparing the up-slope, apex, and down-slope spectra and can detect a coeluting impurity with different UV spectral characteristics.

### 4.6.2 Trends in PDA Detectors

Modern PDAs have sensitivity performance close to the benchmark level of $\pm 1 \times 10^{-5}$ AU for UV/Vis detectors. One innovation for UHPLC is in flow cell design using light pipe (fiber optics) technology to extend the pathlength without increasing noise or chromatographic dispersion. By constructing the light-pipe with a reflective polymer to allow total internal deflection, very small flow cells (e.g. 0.5 µl) can be made with long path lengths (e.g. 10 mm) to minimize peak dispersion (see Figure 4.13). Another development is a programmable slit width design, which allows the user to select via software for either high detector sensitivity (wider slit) or high spectral resolution (narrower slit).

## 4.7 OTHER DETECTORS

This section describes the operating characteristics and salient features of several universal (RID, ELSD, CAD, conductivity) and specific (FLD, CLND, ECD and radiometric) detectors.

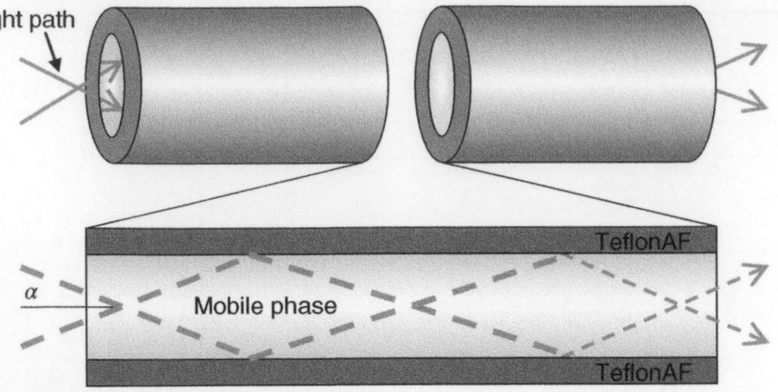


**Figure 4.13.** Schematic diagram of the design of Waters Acquity PDA flow cell with LightPipe Technologies. Source: Courtesy of Waters Corporation

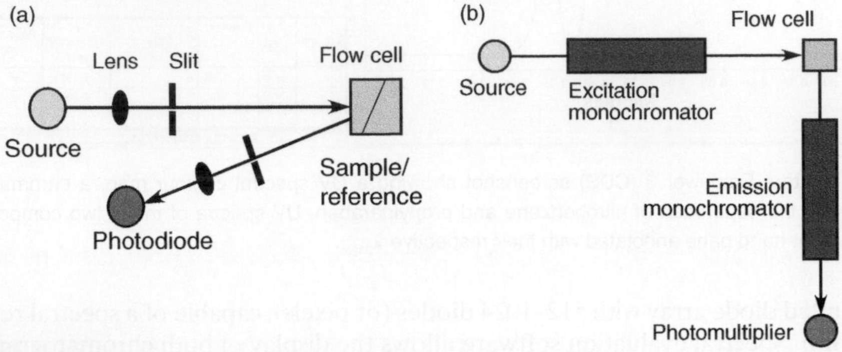


**Figure 4.14.** (a) Schematic diagram of a refractive index detector (deflection type). (b) Schematic diagram of a dual-monochromators fluorescence detector with a square flow cell.

### 4.7.1 Refractive Index Detector (RID)

A refractive index detector (RID) measures the refractive index changes between the sample cell containing the eluting analyte and the reference cell purged with pure eluent (see Figure 4.14a). It has lower sensitivity (0.01–0.1 μg) vs. the UV detector and is prone to temperature and flow fluctuations [1–4]. RID offers universal detection and is commonly used for analytes of low chromophoric activities such as sugars, triglycerides, organic acids, pharmaceutical excipients, and polymers. It is the standard detector in GPC [2, 4]. Modern RID is mostly a differential deflection type (Figure 4.14a). Baseline stability has improved significantly in recent years by better thermostatting the flow cell. However, its lower sensitivity and incompatibility with gradient elution are the biggest disadvantages.

### 4.7.2 Evaporative Light Scattering Detector (ELSD)

An evaporative light scattering detector (ELSD) nebulizes the HPLC eluent to droplets and then eliminates mobile phase solvent by evaporation until a stream of dried particles is formed by semi- and nonvolatile analytes. Finally, scattered light of a laser beam that crosses this particle stream of all nonvolatile analytes in the eluent is measured. [2] Compared with RI detection, ELSD has higher sensitivity (<10 ng) and compatibility with gradient elution. ELSD is used for analytes of low UV absorbance (e.g. sugars, triglycerides, pharmaceutical excipients), polymers, and as a supplemental detector to UV and MS in HTS (Figure 4.15).

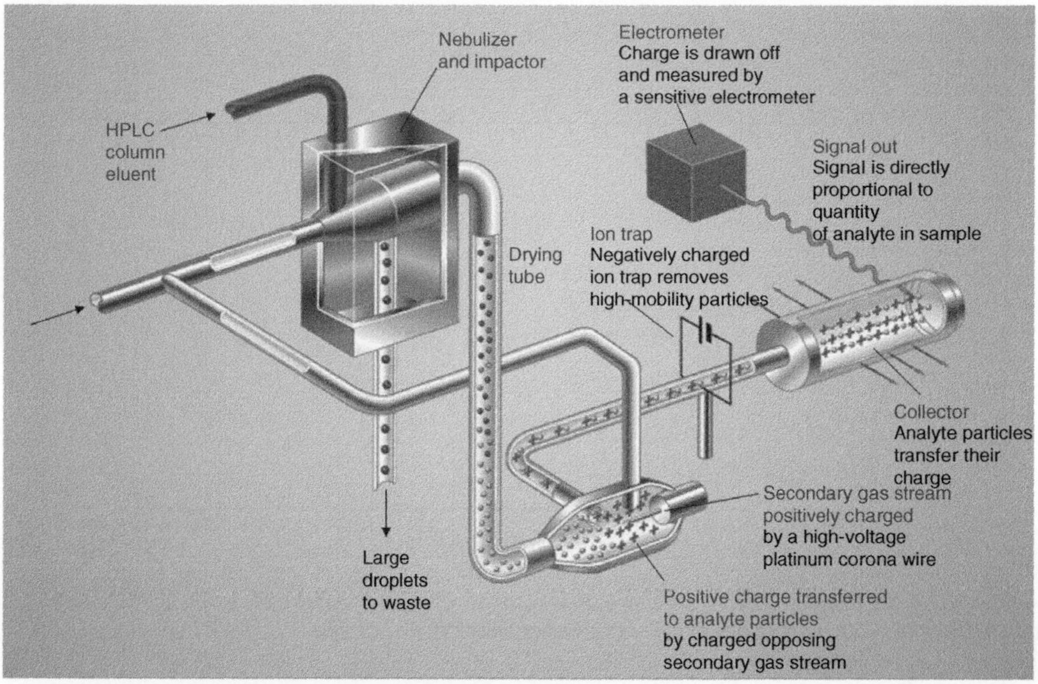


**Figure 4.15.** Schematic diagram of a charged aerosol detector (CAD). Source: Adapted from ESA (Thermo Fisher Scientific).

### 4.7.3 Charged Aerosol Detector (CAD)

A corona CAD is a newer universal detector introduced by ESA Bioscience, Inc. in 2004. It uses a nebulizer technology similar to ELSD, which creates a stream of dried particles that collide with a stream of positively charged gas from the corona charged needle kept at high voltage. The charge is then transferred to the particles, which are quantitatively collected and measured by an electrometer [15]. Sensitivity is in the lowest ng to pg range. Compared with the ELSD, CAD has the advantages of higher sensitivity, easier operation (fewer parameters to optimize), and more consistent mass response factors to diverse analytes ($<\pm10\%$ under isocratic conditions). Figure 4.16 shows an HPLC/CAD chromatogram of a sample of acetaminophen with a 0.005% level of glucose. The detector is used in HTS quantitation studies with an absolute calibration curve in drug discovery research [16, 17].

### 4.7.4 Conductivity Detector (CD)

A conductivity detector is a universal and sensitive detector for ions at ppm levels by measuring changes in the electrical conductivity of the HPLC eluent stream. It is the primary detector for ion chromatography (IC) [18] when used with a suppressor to deliver detection at ppb levels (see example in Figures 13.23 and 13.24).

### 4.7.5 Fluorescence Detector (FLD)

A fluorescence detector (FLD) monitors the emitted fluorescent light of the analytes in the HPLC eluent in the flow cell with irradiation with excitation light at a right angle (see schematic diagram in Figure 4.14b). It is selective and extremely sensitive (pg to fg level) but

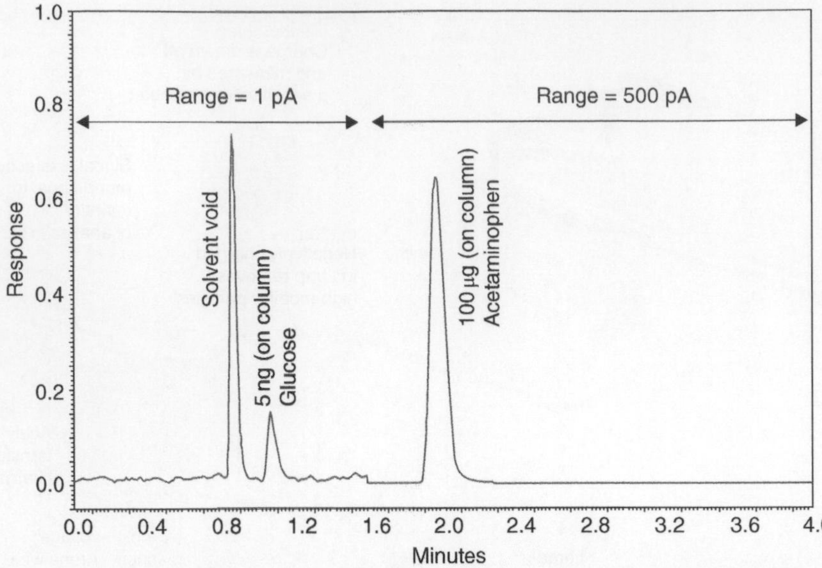


**Figure 4.16.** Chromatogram showing a sample of acetaminophen with a glucose peak at 0.005% using HPLC with CAD detection. Source: Courtesy of Thermo Fisher Scientific.

is limited to compounds with innate fluorescence or those labeled with fluorescent tags [1, 4]. A FLD consists of a xenon source, an excitation monochromator, an emission monochromator, a square flow cell, and a photomultiplier for amplifying the emitted photons. The xenon lamp can be a continuous source or a lower-wattage pulsed source. A continuous source is more sensitive than a pulsed source, though the latter allows the measurement of phosphorescence, chemiluminescence, and bioluminescence. More expensive units have a double monochromator, which can be time-programmed to optimize detection for multiple components (see example in Figure 13.17). Filters are used instead of monochromators in lower-cost units.

***4.7.5.1 Postcolumn Reaction Technique*** A postcolumn reaction unit is an online derivatization system that supplies reagents to the column eluent into a heated coil to convert the analytes into more chromophoric forms for sensitive detection. Some common applications of postcolumn reaction systems are an amino acid analysis using ninhydrin (with visible detection of the reaction products at 550 nm, see Figure 1.7) and a carbamate pesticide analysis using *o-phthalaldehyde* (with fluorescence detection according to Environmental Protection Agency [EPA] method 531.2). Knitted coiled tubing is often used to reduce band broadening for larger reaction coils.

### 4.7.6 Chemiluminescence Nitrogen Detector (CLND)

Chemiluminescence nitrogen detector (CLND) is a nitrogen-specific detector based on pyro-chemiluminescence technology where nitrogen-containing compounds are oxidized to nitric oxide. The nitric oxide generated then reacts with ozone and emits light of a specific wavelength. It is highly specific and sensitive (<0.1 ng of nitrogen) and yields equimolar response to the nitrogen atoms in the compound.

It is useful for the determination of relative response factors of unknown impurities and degradants and the quantitation of new chemical entities with an absolute calibration curve with a single reference material such as caffeine [17]. Note that CLND cannot be used with any nitrogen-containing mobile phase components, such as acetonitrile or ammonia.

### 4.7.7 Electrochemical Detector (ECD)

An electrochemical detector (ECD) measures the electrical current generated by electroactive analytes in the HPLC eluent between electrodes in the flow cell. [2, 4] It offers sensitive detection (pg levels) of catecholamines, neurotransmitters, reducing sugars, glycoproteins, and compounds with phenolic, hydroxyl, amino, diazo, or nitro functional groups. ECD can be the amperometric, pulsed-amperometric, or coulometric type. Common electrode materials are carbon, silver, gold, or platinum, operated in the oxidative or reductive mode. It is capable of high sensitivity and selectivity, though the fouling of electrodes can be a common problem. Many of these traditional analyses using HPLC/ECD have been superseded by LC/MS analysis.

### 4.7.8 Radiometric Detector

A radiometric detector is used to measure the radioactivity of radioactive analytes in the HPLC eluent passing through a flow cell. Most are based on liquid scintillation technology to detect phosphors caused by the radioactive nuclides. A scintillating liquid can be added post-column with a pump, or a permanent solid-state scintillator can be used around the flow cell. This detector is specific only to radioactive compounds and can be extremely sensitive and selective. This detector is commonly used for experiments using tritium or C-14 radiolabeled compounds in toxicological, metabolism, distribution, or degradation studies.

## 4.8 HYPHENATED AND SPECIALIZED SYSTEMS

Hyphenated systems such as LC/MS and 2D-LC are briefly described here although the importance of LC/MS deserves a more systematic treatment in Chapter 6. This section also describes briefly preparative HPLC, micro- and nano-LC, and other application-specific systems.

### 4.8.1 LC/MS and LC/MS/MS

LC/MS, combining the separation power of an HPLC with the sensitivity and specificity of an MS, is emerging as the dominant analytical platform technology for complex samples [19, 20]. The importance of MS was recognized by the awarding of the 2002 Nobel Prize in Chemistry to John B. Fenn of Virginia Commonwealth University and Koichi Tanaka of Shimadzu for their research in soft-ionization MS methods for biological macromolecules. The difficulties of combining LC and MS have been the interface (the ionization of analytes in a stream of condensed liquids and the transfer of ions into the high vacuum inside the MS). Two common LC/MS ionization sources are electrospray ionization (ESI) and atmospheric pressure chemical ionization (APCI).

Figure 4.17 shows a schematic diagram of an ESI interface where HPLC eluent is sprayed across a capillary inlet of the MS. During the nebulizing process, solvent molecules are

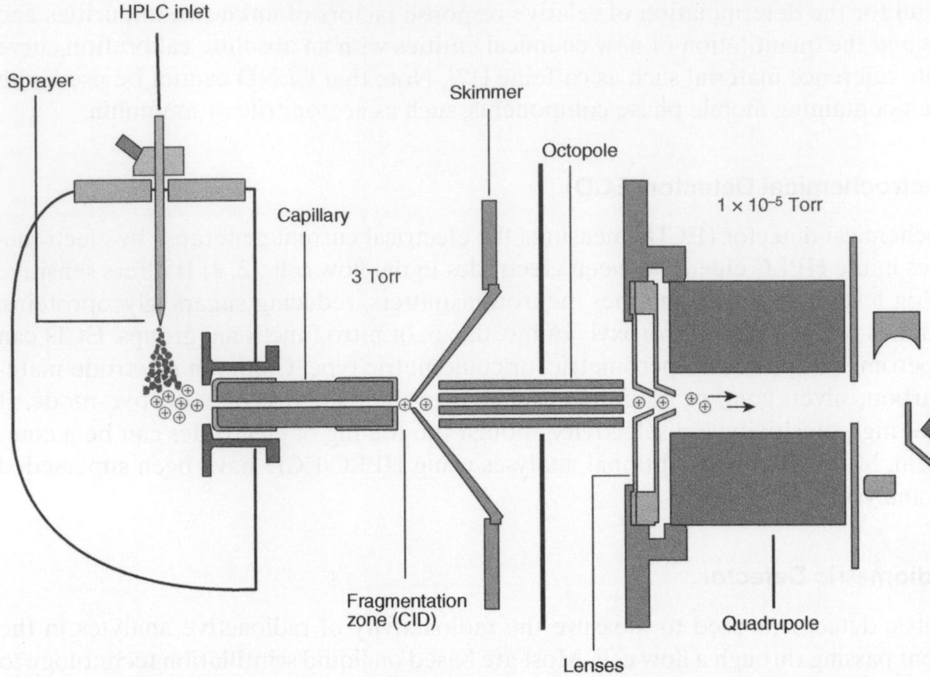


**Figure 4.17.** Schematic diagram of an LC/MS electrospray ionization (ESI) interface showing the nebulization of the LC eluent into droplets, evaporation of the solvent, and the ionization of the analytes, which are pulled inside the MS through a capillary inlet. Source: Courtesy of Agilent Technologies.

removed with the aid of counter-flow of heated gases while the charged analytes are guided into the MS. Common MS types are the single quadrupole (SQ), ion trap (IT), triple quadrupole (TQ), time of flight (TOF), Orbitrap, and Fourier transform mass spectrometry (FTMS). Figure 4.18 shows several models of MS. While most SQ or IT systems have unit mass resolution and a mass range up to 3000, TOF, Orbitraps, FTMS, and TOF systems are capable of a higher mass range and a resolution of >10 000. These high-resolution mass spectrometry (HRMS) systems can often provide enough information to determine the exact chemical formula of an unknown compound. Hybrid MS such as quadrupole-time-of-flight (Q-TOF) or tribrids such as Q-Orbitrap-IT are becoming an effective tool for structure elucidation. Details of MS types are discussed elsewhere [19, 20]. and in Chapter 6. One exciting advance has been the development of low-cost, compact MS detector, which can be used by most chromatographers without expert knowledge.

LC/MS/MS can further increase sensitivity and specificity for trace analysis of complex matrices. Here the precursor ions of the analyte are fragmented, and the product/fragment ions are monitored for quantitative analysis. LC/MS/MS is now the dominant platform methodology for bioanalytical analysis (drugs and metabolites in physiological fluids and tissues) and has become a gold standard in trace analysis (e.g. multiresidue testing of environmental and food samples).

### 4.8.2 LC/NMR

The hyphenated technique of LC/NMR is made possible by instrument enhancements such as probe miniaturization, noise reduction, and innovative interface technologies (see Figure 4.19a) [21]. Nevertheless, many NMR experts still prefer to conduct identifications in the offline mode by isolating the impurities first for more flexibility and better sensitivity.

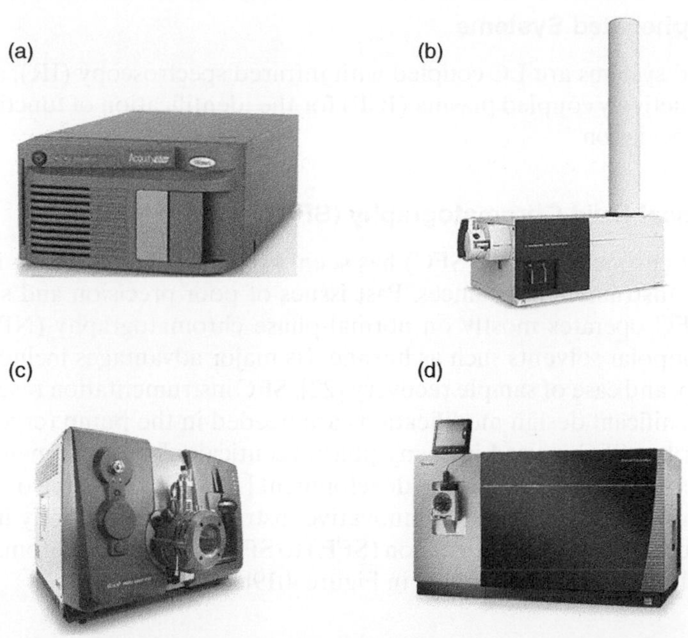


**Figure 4.18.** Images of various MS systems. (a) Waters QDa (compact SQ-MS). (b) Agilent 6500 Q-TOF. (c) Sciex 6500+ TQ, (d) Thermo Fusion Lumos Tribrid (Q-Orbitrap-IT). Source: Courtesy of Waters, Agilent, Sciex and Thermo Fisher Scientific.

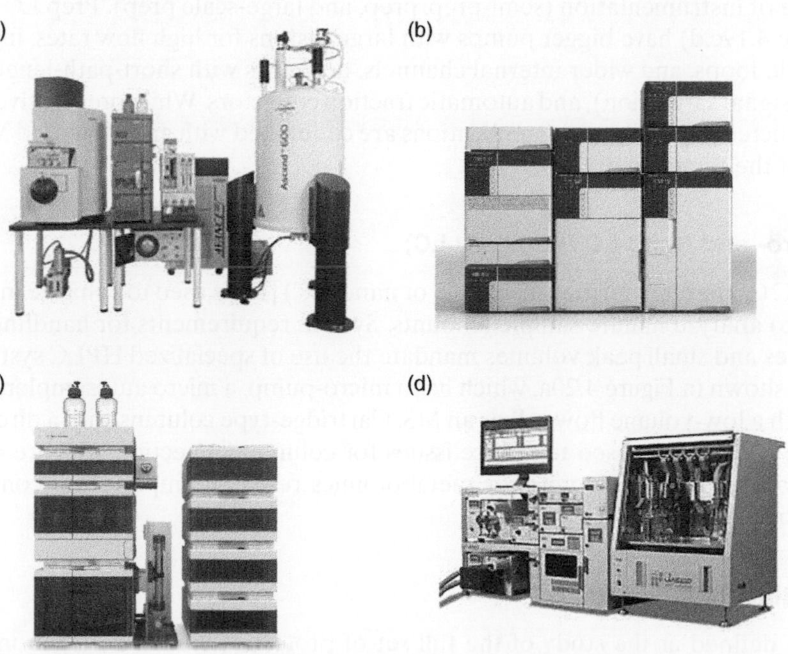


**Figure 4.19.** Images of (a) Bruker LC-NMR/MS, (b) Shimadzu Nexera UC (SFE/SFC), (c) Agilent 1260 Prep LC, and (d) Jasco Prep SFC. Source: Courtesy of Bruker, Shimadzu, Agilent, and Jasco.

### 4.8.3 Other Hyphenated Systems

Other hyphenated systems are LC coupled with infrared spectroscopy (IR), atomic absorption (AA), or inductively coupled plasma (ICP) for the identification of functional groups or studies of metal speciation.

### 4.8.4 Supercritical Fluid Chromatography (SFC)

Supercritical fluid chromatography (SFC) has seen a remarkable comeback in recent years due to significant instrumental advances. Past issues of poor precision and sensitivity have been resolved. SFC operates mostly on normal-phase chromatography (NPC) mode with $CO_2$ replacing nonpolar solvents such as hexane. Its major advantages include fast analysis, low-pressure drop, and ease of sample recovery [22]. SFC instrumentation resembles those of HPLC, though significant design modifications are needed in the pump (cryogenic cooling) and the autosampler. SFC is used in many pharmaceutical labs in automated column and mobile phase screening for chiral method development [16]. SFC is an area of active research for both chiral and achiral analysis. An innovative instrument was recently introduced that directly couples supercritical fluid extraction (SFE) to SFC/MS to allow automated extraction and analysis of labile compounds (shown in Figure 4.19b) [23].

### 4.8.5 Preparative LC and SFC

The goal of preparative LC or SFC is the efficient purification and recovery of purified materials [24]. The quantity of materials for purification (mg, g, kg) dictates the size of the column and the type of instrumentation (semi-prep, prep, and large-scale prep). Prep LC or SFC systems (Figure 4.19c,d) have bigger pumps with large pistons for high flow rates, injectors with larger sample loops, and wider internal channels, detectors with short-path-length flow cells (to prevent signal saturation), and automatic fraction collectors. While preparative LC or SFC can be conducted manually, most applications are conducted with an automated MS-directed collection of the target components [16].

### 4.8.6 Micro- and Nano-LC (Capillary LC)

Capillary LC (sometimes termed micro-LC or nano-LC) [13] is used to enhance mass sensitivity needed to analyze minute sample amounts. System requirements for handling extremely low flow rates and small peak volumes mandate the use of specialized HPLC systems such as an example shown in Figure 4.20a, which has a micro-pump, a micro autosampler, and a PDA detector with a low-volume flow cell or an MS. Cartridge-type columns with a direct interface into MS are increasingly used to reduce issues for column connections (Figure 4.21). Many micro-LCs are used in proteomics or metabolomics research employing a comprehensive 2D-LC-MS mode [25].

### 4.8.7 Multidimensional LC

Proteomics, defined as the study of the full set of proteins encoded by a genome, involves the study of samples too complex to be sufficiently resolved by a single HPLC column. The traditional approach is two-dimensional polyacrylamide gel electrophoresis (2-D-PAGE). A better approach is 2D-LC where the sample fractionated by the first column is subsequently

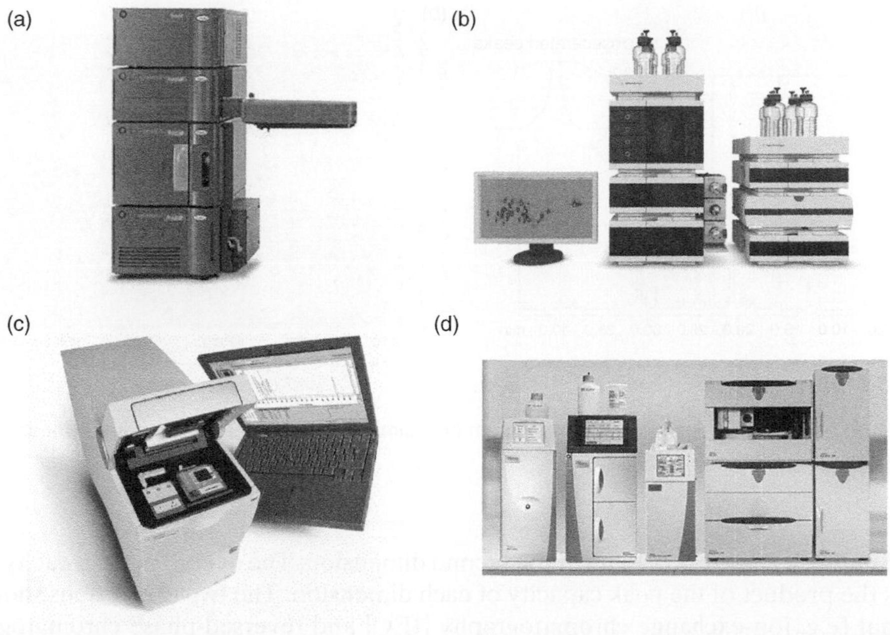


**Figure 4.20.** Images of (a) Waters Acuity UPLC M-Class Micro-LC, (b) Agilent Infinity II-2D-LC solution, (c) Agilent Bioanalyzer 2100, and (d) Thermo scientific ion chromatographs. Source: Courtesy of Waters, Agilent, and Thermo Fisher Scientific.

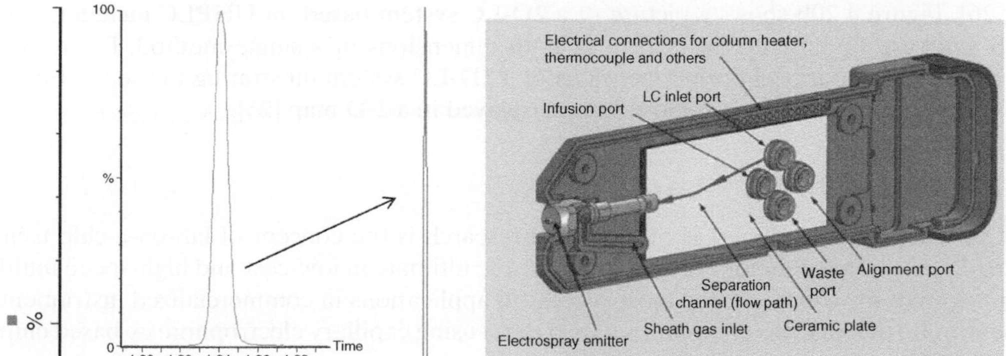

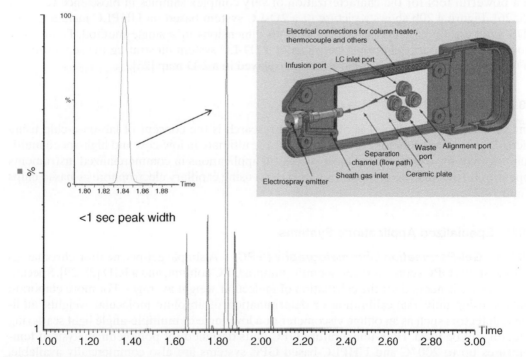


**Figure 4.21.** Picture of Waters iKey Cartridge micro-LC column and a chromatogram demonstrating its efficiency performance. HPLC Conditions: 300 mm × 50 μm iKey, gradient – 5 to 95% B in 1 minute at 22 μl/min; sample – a mixture of small molecules: propranolol, verapamil, and alprazolam, tolbutamide, and fluticasone propionate. Source: Courtesy of Waters Corporation.

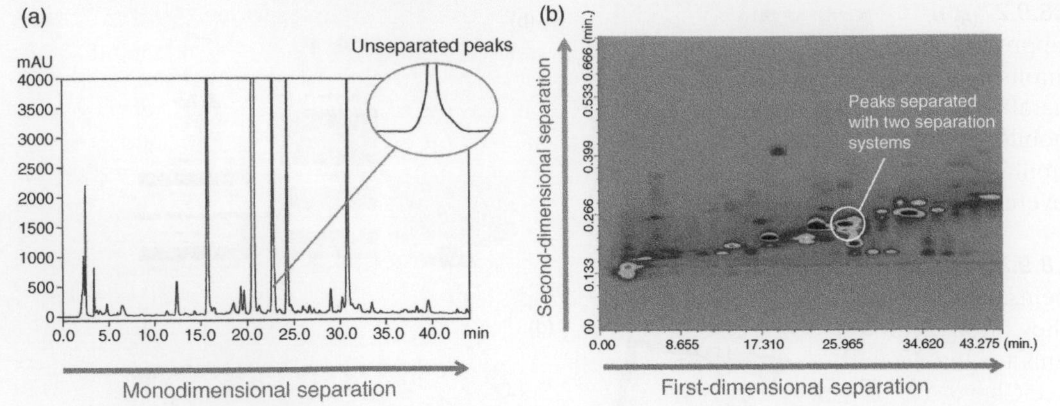


**Figure 4.22.** 2D-LC Chromatograms: a.Chromatogram (first dimension) and 2-D display from Shimadzu 2-D LC system. Source: Courtesy of Shimadzu.

analyzed online by another column in the second dimension. The overall peak capacity of the system is the product of the peak capacity of each dimension. The two dimensions should be orthogonal (e.g. ion-exchange chromatography [IEC] and reversed-phase chromatography [RPC]) to maximize resolving power. For instance, if each dimension has a peak capacity of 400 using UHPLC, a total peak capacity of ~16 000 or 400 × 400 can be predicted for a 2D-LC system. When coupled with HR-MS and bioinformatics software, these systems can be a powerful tool for the characterization of very complex samples in bioscience research [25, 26]. Figure 4.20b shows a picture of a 2D-LC system based on UHPLC including new CDS software for instrument control of both dimensions in a single method. Figure 4.22 shows chromatograms of both dimensions of a 2D-LC system illustrating the separation of an unresolved peak by the second column displayed in a 2-D map [23].

### 4.8.8 Lab-on-a-Chip

One of the most exciting areas of analytical research is the concept of lab-on-a-chip using micro-fabrication technology [27]. It promises the ultimate in low-cost and high-speed multi-channel analysis. Currently, the most successful applications in commercialized instruments appear to be the analysis of biomolecules (DNA) using capillary electrophoresis-based chips (Figure 4.20c).

### 4.8.9 Specialized Applications Systems

***4.8.9.1 Gel-Permeation Chromatography (GPC)*** A simple gel-permeation chromatography system (GPC) consists of an isocratic pump, a GPC column, and a RID [28, 29]. Specialized software is needed for the calculation of molecular weight averages. For more elaborate analysis using universal calibration or determination of absolute molecular weights, additional detectors such as an online viscometer or a low angle or multiple-angle light scattering detector can be used. High-temperature GPC units designed for polyolefin analysis at temperatures up to 200 °C and UHPLC-based GPC systems are also commercially available. Additional examples are shown in Section 13.4.1.

***4.8.9.2 Ion Chromatography (IC)*** In essence, ion chromatography is IEC with suppressed conductivity detection. It is particularly useful in the analysis of low levels of anions and metals in environmental and industrial samples [18]. IC instruments often have metal-free fluidics to enhance sensitivity. Recent advances include the use of reagent-free mobile phase generation, charged detector, and system for higher-pressure microbore or capillary operation to allow for continuous operation. Figure 4.20d shows examples of several modern IC systems.

***4.8.9.3 Application-Specific Systems*** While most HPLCs are general-purpose instruments, a few are dedicated analysis systems packaged with specific columns and reagents. They often come with a guaranteed performance from the manufacturer. Examples are systems for the analysis of amino acids (Figure 1.7), carbamate pesticides, cannabis, or sugars (see Chapter 13 for examples).

## 4.9 HPLC ACCESSORIES

This section describes several common accessories for HPLC systems. Solvent degassers are often built-in with integrated systems, and column ovens can be purchased with optional column selector valves to increase the flexibility of the system.

### 4.9.1 Solvent Degasser

Solvent degassers are necessary accessories for HPLC pumps as dissolved gas in the mobile phase can cause inaccuracy in pump blending or air lock issues in check valves. Past practices of degassing by stirring or ultrasonication under vacuum are inadequate while helium sparging is inconvenient and expensive [2]. Online vacuum degassers are available as separate modules or as built-in units in integrated HPLC systems. Vacuum degassers work by passing solvents through tubes of semiporous polymer membranes inside an evacuated chamber to eliminate dissolved gaseous molecules (Figure 4.23). The reliability and efficiency of modern degassers have improved, and degasser tube volumes have been reduced to <1 ml. Note that helium pressurization systems are still required for pumping volatile solvents (e.g. pentane, methylene chloride) or labile reagents (e.g. ninhydrin).

### 4.9.2 Column Oven

A column oven is required for automated assays to improve retention time precision. Column temperatures of 30–50 °C are most typical in HPLC operation. Temperatures higher than 60 °C are often used to allow higher flow rates without exceeding system pressure limits by reducing solvent viscosity. Column temperatures up to 80 °C are frequently used to reduce peak tailing and enhance recovery in the RPC of proteins and peptides. The subambient operation is used to enhance selectivity or to reduce on-column degradation of labile analytes [4]. Column ovens operate either by circulating heated air or direct contact (clam-shell type). Solvent preheating is achieved by passing the mobile phase through a coiled tube embedded onto the heating element before the column. In UHPLC column ovens, active solvent preheater devices are common. New trends are toward wider temperature ranges (e.g. 4–100 °C) using Peltier devices.

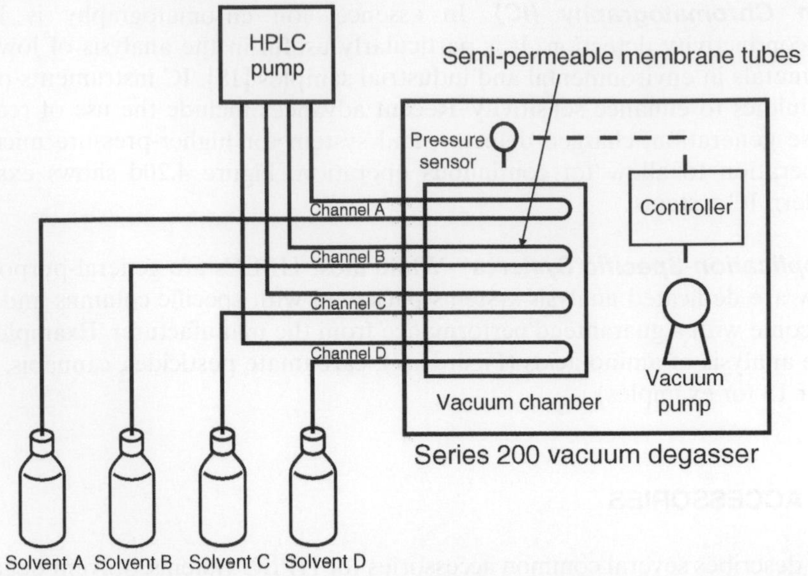


**Figure 4.23.** Schematic diagram of an online vacuum degasser. Older systems have relatively large semipermeable vacuum tubes of ~10 ml. New systems have smaller tubes <1 ml. Note that for effective degassing, all four lines should be filled with solvents. Source: Courtesy of the PerkinElmer.

### 4.9.3 Valves for Column and Mobile Phase Selection

Column selector valves can be added as accessories to allow column switching for multi-dimensional chromatography or automatic column selection to facilitate methods development. Column switching valves are often located inside the column oven.

Low-pressure selection valves increase the number of mobile phases that can be selected for method development. For instance, most binary UHPLC pumps have built-in two or three-way selection valves to increase the solvent selection from A1 or A2 and B1 or B2. An inexpensive six-port selection valve can be added to most HPLC systems and turn it into an automated mobile phase screening system.

## 4.10 CHROMATOGRAPHY DATA SYSTEMS (CDS)

Figure 4.24 shows the historical development of chromatography data handling devices or CDS ranging from a strip chart recorder, an electronic integrator, a PC-based workstation, and a client–server network system [30]. In the last few decades, CDS has seen a progression of increasing sophistication and automation. While the chart recorder requires manual measurement of peak heights, the electronic integrator debuted in the 1980s has built-in algorithms for peak integration, calculation, and report generation. Windows-based PC-based workstations popularized in the 1980s incorporate additional functions of data archiving and single-point system control. For large laboratories with many chromatography systems, a centralized client–server network system is the mandated solution to ensure data security and regulatory compliance. Many CDS data network also offers remote access through the Internet and mobile devices [23]. For many years, Waters Empower CDS has been dominant in laboratories performing regulatory testing. More recently, other CDS such as Thermo Chromeleon 7.2 and Agilent OpenLAB 2.3 have improved their performance in user interfaces and regulatory compliance to be contenders in this area.

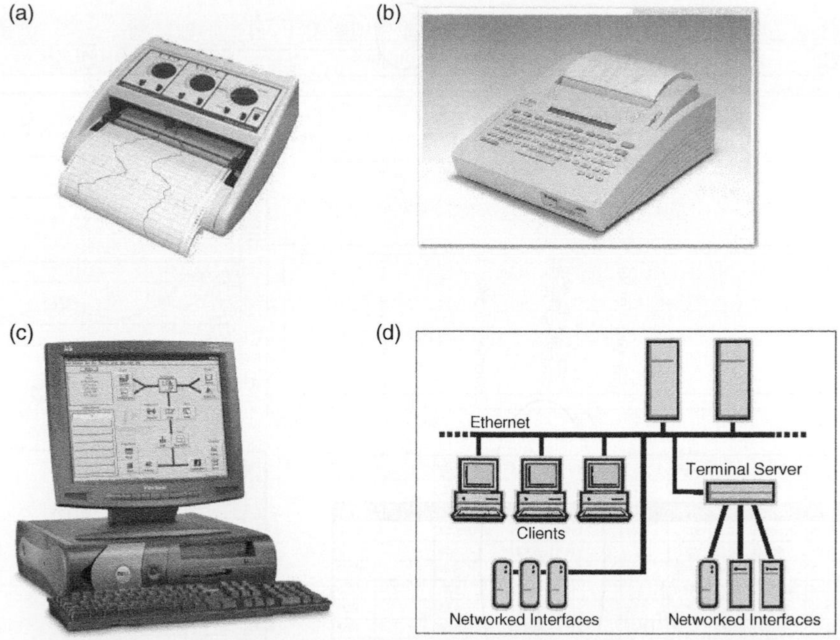


**Figure 4.24.** Diagrams depicting the historical development of chromatography data systems (CDS): (a) Chart recorder, (b) electronic integrator, (c) PC workstation, and (d) client-server CDS network.

### 4.10.1 User Interface and CDS Workflow

Figure 4.25 shows a schematic diagram of the various steps involved in a chromatographic sample analysis using a CDS system [30]. While the terminologies and details may vary with different CDS, the essence of a CDS workflow tends to be similar. Most starts with the setting up an instrumental method for system control and raw data collection, a processing method

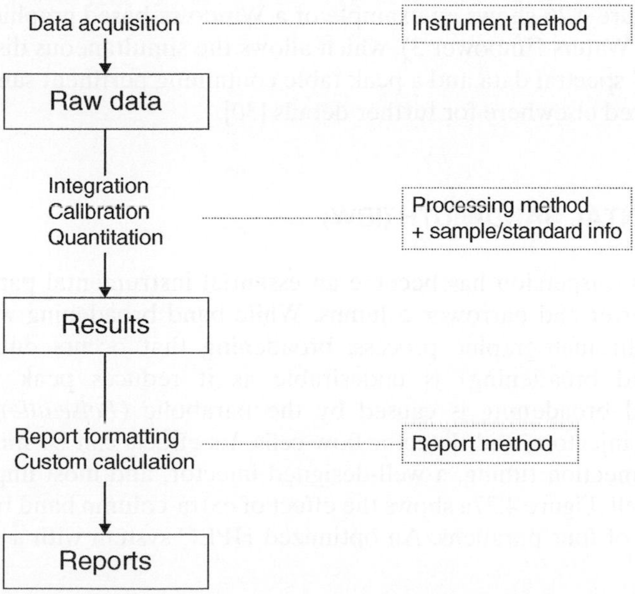


**Figure 4.25.** Schematic diagram showing the workflow in a CDS from data acquisition to the generation of raw data, result files, and formatted reports.

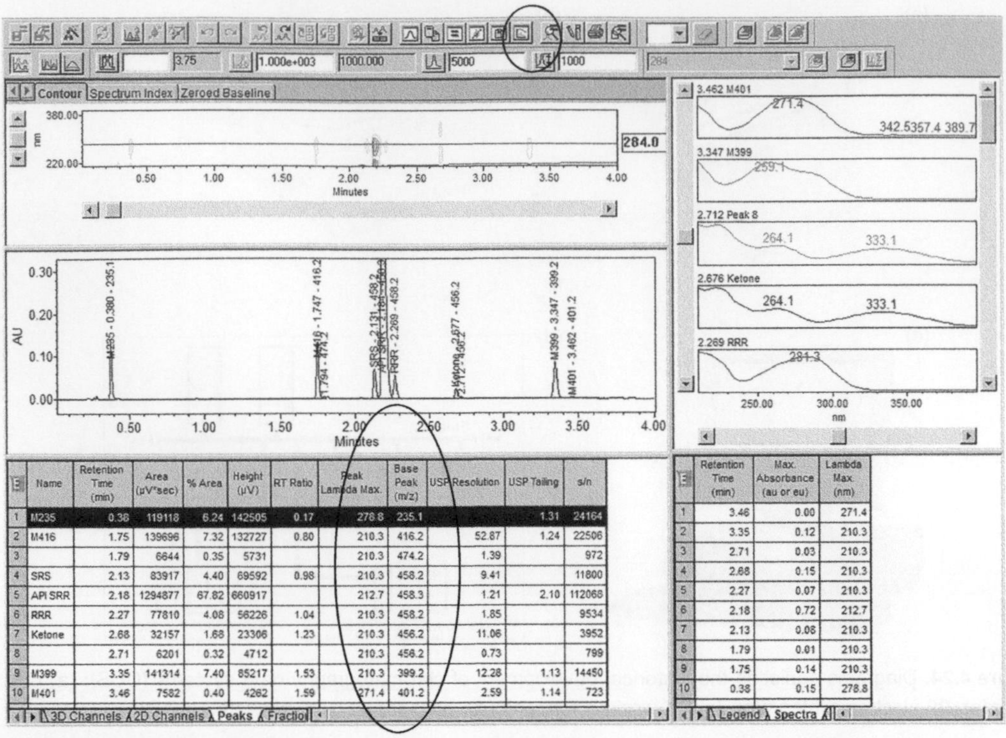


**Figure 4.26.** The graphical user interface of Waters Empower 3 CDS showing a chromatogram with a UV contour map, UV spectra, and a peak table.

for peak integration, calibration, and identification to generate a result file, and a reporting method to format the final report (Figure 4.25). The CDS have evolved into complex software and networking systems requiring major investments in user training, validation, and support efforts. Figure 4.26 shows an example of a Windows-based graphical user interface of a popular CDS (Waters Empower 3), which allows the simultaneous display of the chromatogram with UV spectral data and a peak table containing pertinent sample information. The reader is referred elsewhere for further details [30].

## 4.11 INSTRUMENTAL BANDWIDTH (IBW)

The IBW or system dispersion has become an essential instrumental parameter in recent years for using shorter and narrower columns. While band broadening within the column is innate to the chromatographic process, broadening that occurs outside the column (extra-column band broadening) is undesirable as it reduces peak resolution [2–4]. Extra-column band broadening is caused by the parabolic (*Poiseuille*) flow profiles in connection tubing, injectors, and detector flow cells. Its effects can be minimized by using small-diameter connection tubing, a well-designed injector, and most importantly, a small UV detector flow cell. Figure 4.27a shows the effect of extra-column band broadening on the fast LC separation of four parabens. An optimized HPLC system with a semi-micro 2.4-µl

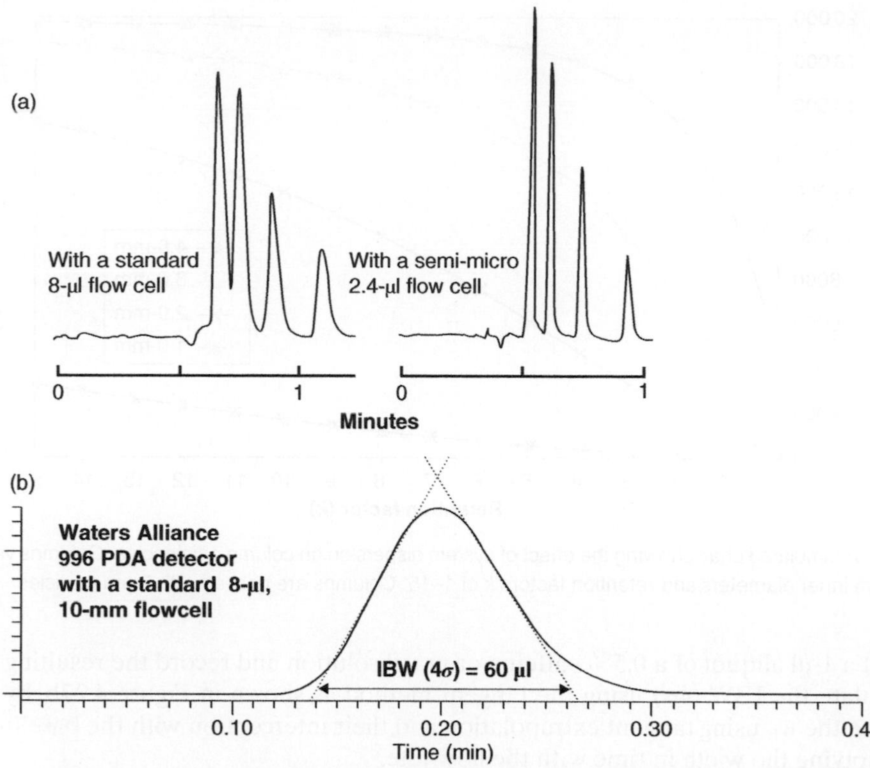


**Figure 4.27.** (a) Comparative chromatograms showing the deleterious effect of instrumental bandwidth on the performance of a fast LC separation of four paraben antimicrobials. (b) The instrumental bandwidth (IBW) of a Waters Alliance HPLC system with a 966 PDA detector with a standard 8-μl flow cell.

flow cell yields sharp peaks and good resolution while the same instrument equipped with an 8-μl standard flow cell shows considerably broader peaks [31].

This effect of system bandwidth can be calculated using the additive relationship of variances where the total variance of the observed peak is equal to the sum of the true peak variances plus the variance of the IBW:

$$\sigma^2_{\text{total}} = \sigma^2_{\text{column}} + \sigma^2_{\text{IBW}}$$

Here, the variance of the IBW is the summation of variances from the injector, the detector, and the connection tubing:

$$\sigma^2_{\text{IBW}} = \sigma^2_{\text{injector}} + \sigma^2_{\text{detector}} + \sigma^2_{\text{connection tubing}}$$

### 4.11.1 How to Measure IBW

This instrumental bandwidth can be measured as follows:

- Replace the column with a zero-dead-volume union
- Set the HPLC system to 0.5 ml/min, UV detection at 254 nm, and a data rate of 10 points per second

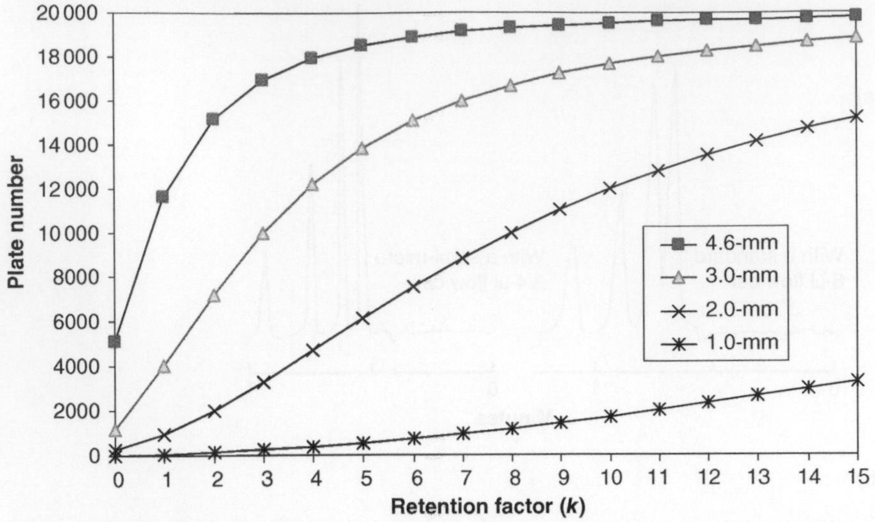


**Figure 4.28.** A simulated chart showing the effect of system dispersion on column efficiencies. Columns vary from 1.0 to 4.6 mm inner diameters and retention factors $k$ of 1–15. Columns are packed with 3-μm particles.

- Inject a 1-μl aliquot of a 0.5% caffeine or uracil solution and record the resulting trace
- Calculate the IBW ($4\sigma$) using the tangent method as shown in Figure 4.27b, by determining the $w_b$ using tangent extrapolation and their interception with the baseline then multiplying the width in time with the flow rate.

An IBW of ~60 μl was found for this HPLC. Figure 4.28 shows a simulation chart plotting the projected column efficiencies of a number of 150-mm long columns with various inner diameters packed with 3-μm particles using an HPLC system with a 60-μl IBW. Note that the efficiency loss from instrumental dispersion can be severe for small inner diameter columns (see Section 2.2.6). Note that peaks of low $k$ values under isocratic conditions are affected more by system dispersion since peak volumes are proportional to $k$.

The IBW of most conventional HPLC systems can be reduced by these modifications:

- Replace connection tubing with shorter lengths of 0.005–0.007″ i.d. tubing
- Replace the UV detector flow cell with a smaller flow cell (1–5 μl)
- Reduce the sample loop volumes <20 μl for isocratic analysis

In gradient RPC analysis, the sample band is reconcentrated at the head of the column, so the dispersive contributions of the injector and tubing before the column is irrelevant. Only the postcolumn dispersion of the detector flow cell and its connection tube to the column outlet are significant.

### 4.11.2 IBW of UHPLC Systems

Typical UHPLC instruments are low-dispersion systems and have substantially smaller IBW than conventional HPLC. Figure 4.29 shows the IBW trace of a UHPLC system with a table of measured values of various systems shown in the inset. Typical IBW values for UHPLC ranged from 10 to 20 μl, depending on the selection of switching valve, sample loop, and inner diameters of the connection tubing.

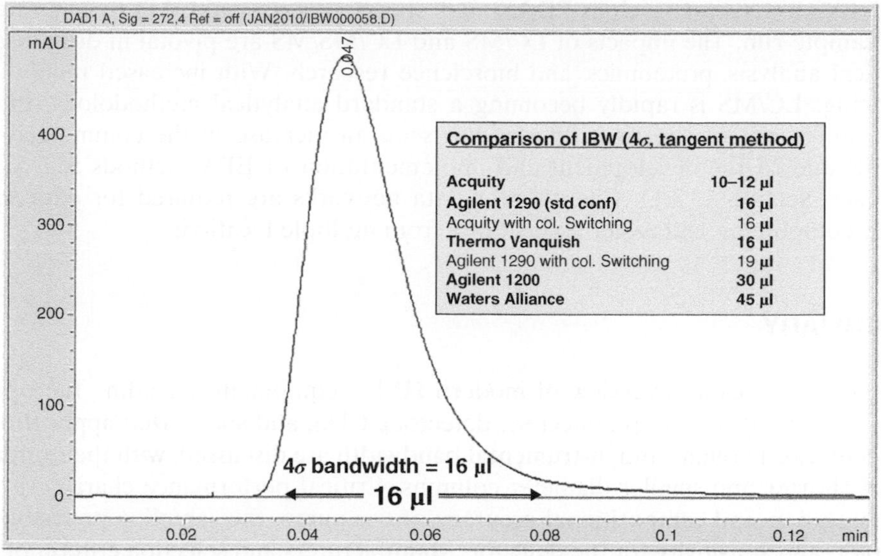


**Figure 4.29.** A UV detector signal trace showing the instrumental bandwidth of a UHPLC system (Agilent 1290 binary). Procedure according to Section 4.11.1 was used except at a flow rate set at 0.4 ml/min. The inset shows data from other UHPLC systems.

## 4.12 MANUFACTURERS AND EQUIPMENT SELECTION

Major manufacturers of HPLC and UHPLC instruments include Waters, Agilent, Thermo Scientific, and Shimadzu. Some instrumental models and pertinent details are listed in Table 4.1. There have been mergers and acquisitions in the last decade resulting in the reduction of the number of instrument brands (e.g. Dionex, Varian, and Eksigent). HPLC is a mature technology, and most manufacturers have reliable products with competitive performance in the marketplace. However, there can be significant differences between the different instrument model on systems (dwell volume, dispersion, initial and maintenance costs), pumps (performance at low flow rates, mixer options), autosamplers (precision, carryover, speed, sample capacity, minimum sample volume), and UV detectors (sensitivity and linearity range). Equipment selection can be based on performance specifications, pricing, features, and the vendor's technical or service support. Major companies tend to purchase HPLC systems from one or two vendors to reduce cost in CDS software, system qualification, equipment service, and operating training. The compatibility to an existing chromatographic data-handling network is likely the dominant factor in the purchase decision for HPLC equipment.

## 4.13 TRENDS IN HPLC AND UHPLC EQUIPMENT

HPLC is a mature analytical technique based on well-engineered instrumentation with a definitive trend toward higher performance, better reliability, and easier maintenance. Demands in life sciences and drug discovery are driving HPLC technologies toward higher resolution, sensitivity, and throughput. Recent innovations in micro-LC, UHPLC, and 2D-LC are direct responses to these demands. Research laboratories are using more information-rich detectors such as PDAs, mass spectrometers, and CAD to generate more data per sample run. The impacts of LC/MS and LC/MS/MS are pivotal in drug discovery, bioanalytical analysis, proteomics, and bioscience research. With increased reliability and lower pricing, LC/MS is rapidly becoming a standard analytical methodology. In recent years, the environmental testing industry has seen an increase in the commercial use of LC/MS/MS due to the development and implementation of EPA methods 537, 539, 544, and 545 (see Section 13.3.1). Client/server data networks are required for data security, regulatory compliance, and easier accessibility from multiple locations.

## 4.14 SUMMARY

This chapter provides an overview of modern HPLC equipment, including the operating principles and trends of pumps, injectors, detectors, CDS, and specialized applications systems. System dwell volume and instrumental bandwidth are discussed, with their impacts on the use of shorter and smaller-diameter columns. Critical performance characteristics are the flow precision and compositional accuracy of the pump, the sampling precision of the autosampler, and sensitivity for the detector. Manufacturers and selection criteria of HPLC equipment from different vendors are reviewed.

## 4.15 QUIZZES

**1.** Most high-end pumps have this design EXCEPT
   (a) dual-pistons in-series
   (b) dual-pistons in-parallel
   (c) low dispersion
   (d) micro piston

**2.** Most pumps used in HPLC method development are
   (a) micro-piston pumps
   (b) low-pressure mixing pumps
   (c) high-pressure mixing pumps
   (d) syringe pumps

**3.** Conventional HPLC systems have dwell volumes around
   (a) 0.2 ml
   (b) 0.5 ml
   (c) 1 ml
   (d) 5 ml

**4.** This type of pumps is preferred in high-throughput screening or LC/MS.
   (a) High-pressure mixing
   (b) low-pressure mixing
   (c) micro-piston pumps
   (d) quaternary pumps

5. The smallest volume in microliters that a conventional HPLC autosampler can inject precisely is about
   (a) 5
   (b) 100
   (c) 20
   (d) 1

6. Which factor is the LEAST important for UV/Vis detector sensitivity?
   (a) The path length of flow cell
   (b) Noise specification
   (c) Spectral bandwidth
   (d) Stray light

7. The typical benchmark noise of a UV detector is
   (a) $\pm 1 \times 10^{-6}$ AU
   (b) $\pm 1 \times 10^{-4}$ AU
   (c) $\pm 1 \times 10^{-3}$ AU
   (d) $\pm 1 \times 10^{-5}$ AU

8. The most common detection wavelength for potency assay of an analyte is
   (a) $\lambda_{max}$
   (b) 254 nm
   (c) 220 nm
   (d) 280 nm

9. A common replacement for the refractive index detector for universal detection is
   (a) MS
   (b) Fluorescence
   (c) CLND
   (d) CAD

10. Chloride ion can be detected by
   (a) UV/Vis
   (b) PDA
   (c) CAD
   (d) CLND

11. The instrumental bandwidth (IBW at $4\sigma$) of a typical HPLC is (e.g. Waters Alliance)
   (a) 6 µl
   (b) 60 µl
   (c) 600 µl
   (d) 2 ml

12. Which MS type is considered to have high-resolution accurate mass?
   (a) Orbitrap
   (b) Ion trap
   (c) Triple-quadrupole
   (d) Single-quadrupole

**13.** The most common ionization technique in LC/MS is
(a) APCI
(b) electron impact
(c) ESI
(d) nebulization

**14.** Most major pharmaceutical companies use this type of CDS.
(a) Client-server network
(b) Electronic integrator
(c) Cloud storage
(d) PC workstations

**15.** The LEAST significant factor affecting extra-column band broadening is
(a) column temperature
(b) system dispersion
(c) *k* of the analyte
(d) the inner diameter of the column

### 4.15.1 Bonus Quiz

1. In your own words, describe the reasons why low-pressure mixing pumps are used in HPLC method development.
2. Why is a UV detector the primary HPLC detector for pharmaceutical analysis?
3. Describe the reasons why LC/MS is emerging as the standard analytical technique in scientific discovery.

## 4.16 REFERENCES

1. Skoog, D.A., Holler, F.J., and Crouch, S.R. (2006). *Principles of Instrumental Analysis*, 6e. Pacific Groves, CA: Brook Cole.
2. Dong, M.W. (2005). *Handbook of Pharmaceutical Analysis by HPLC* (ed. S. Ahuja and M.W. Dong). Amsterdam, the Netherlands: Elsevier. Chapter 4.
3. Fanali, S., Haddad, P.R., Poole, C. et al. (eds.) (2013). *Liquid Chromatography: Fundamentals and Instrumentation*. Amsterdam, the Netherlands: Elsevier.
4. Snyder, L.R., Kirkland, J.J., and Dolan, J.W. (2010). *Introduction to Modern Liquid Chromatography*, 3e. Hoboken, NJ: Wiley.
5. Dong, M.W. (2006). *Modern HPLC for Practicing Scientists*. Hoboken, NJ: Wiley. Chapter 4.
6. Arnaud, C.H. (2016). *Chem.Eng. News* 94 (24): 29–35.
7. Guillarme, D. and Dong, M.W. (2013). *Amer. Pharm. Rev.* 16 (4): 36–43.
8. MacNair, J.E., Lewis, K.C., and Jorgenson, J.W. (1997). *Anal. Chem.* 69: 983.
9. (2004). *Waters Ultra Performance Liquid Chromatography Acquity System*. Milford, MA: Waters Corp.
10. Guillarme, D. and Dong, M.W. (eds.) (2014). UHPLC: where we are ten years after its commercial introduction. *Trends Anal. Chem.* 63: 1–188. (Special issue).
11. Guillarme, D. and Veuthey, J.-L. (eds.) (2012). *UHPLC in Life Sciences*. Cambridge, United Kingdom: Royal Society of Chemistry.
12. Wu, N. and Clausen, A.M. (2007). *J. Sep. Sci.* 30: 1167.
13. Ishii, D. (1988). *Introduction to Micro and Semimicro HPLC*. Weinheim, Germany: Wiley-VCH Verlag.
14. Fekete, S., Kohler, I., Rudaz, S., and Guillarme, D. (2014). *J. Pharm. Biomed. Anal.* 87: 105.

15. Gamache, P.H. (ed.) (2016). *Charged Aerosol Detection and Related Separation Techniques*. Hoboken, NJ: Wiley.
16. Wong, M., Murphy, B., Pease, J.H., and Dong, M.W. (2015). *LCGC North Am.* 33 (6): 402.
17. Lin, B., Pease, J.H., and Dong, M.W. (2015). *LCGC North Am.* 23 (8): 534.
18. Fritz, J.S. and Gjorde, D.T. (2009). *Ion Chromatography*, 4e. Weinheim, Germany: Wiley-VCH.
19. Gross, J.H. (2011). *Mass Spectrometry*, 2e. Heidelberg, Germany: Springer.
20. Zhou, L. (2005). *Handbook of Pharmaceutical Analysis by HPLC* (ed. S. Ahuja and M.W. Dong). Amsterdam, the Netherlands: Elsevier. Chapter 19.
21. Gonnella, N.C. (2013). *LC-NMR: Expanding Limits of Structure Elucidation*. Boca Raton, FL: CRC Press.
22. Webster, G.K. (ed.) (2014). *Supercritical Fluid Chromatography: Advances and Applications in Pharmaceutical Analysis*. Singapore: Pan Stanford.
23. Dong, M.W. (2013). *LCGC North Amer.* 31 (4): 313; 32(4), 270 (2014); 33(4), 254 (2015); 34(4), 262 (2016); 35(4), 246 (2017); 36(4), 256 (2018).
24. Schmidt-Traub, H., Schulte, M., and Seidel-Morgenstern, A. (eds.) (2013). *Preparative Chromatography*, 2e. Weinheim, Germany: Wiley-VCH.
25. Gritsenko, M.A., Xu, Z., Liu, T., and Smith, R.D. (2016). *Proteomics by Mass Spectrometry* (ed. S. Sechi), 237–247. New York: Humana Press.
26. Zhang, K., Wang, J., Tsang, M. et al. (2013). *Amer. Pharma. Rev.* (December issue).
27. Castillo-Leon, J. and Svendsen, W.E. (eds.) (2014). *Lab-on-a-Chip Devices and Micro Total Analysis Systems*. Heidelberg, Germany: Springer.
28. Striegel, A., Yau, W.W., Kirkland, J.J., and Bly, D.D. (2009). *Modern Size-Exclusion Chromatography: Practice of Gel Permeation and Gel Filtration Chromatography*, 2e. Hoboken, NJ: Wiley.
29. Reuter, W.M., Dong, M.W., and McConville, J. (1991). *Am. Lab.* 23 (5): 45.
30. Mazzarese, R. (2005). *Handbook of Pharmaceutical Analysis by HPLC* (ed. S. Ahuja and M.W. Dong). Amsterdam: Elsevier. Chapter 21.
31. Dong, M.W. (2000). *Today's Chemist at Work* 9 (2): 46.

# 5

# UHPLC: PERSPECTIVES, PERFORMANCE, PRACTICES, AND POTENTIAL ISSUES

## 5.1 INTRODUCTION

### 5.1.1 Scope

Ultra-high-pressure liquid chromatography (UHPLC) is the most important innovation in modern HPLC, setting a new performance benchmark for faster and higher-resolution separations with improved precision. This chapter provides an overview of UHPLC systems and applications including historical perspectives, fundamental concepts, benefits, potential issues, and best practices.

### 5.1.2 Glossary and Abbreviations

A list of standard UHPLC terms and abbreviations (acronyms) is shown as follows.

- *UHPLC*: Ultra-high-pressure LC systems with pressure limits >15 000 psi (1000 bar) with lower system dispersion and dwell volumes.
- *UPLC*: Ultra performance liquid chromatography, a trademark by Waters Corporation. It stands for the Waters Acquity UPLC, the first commercial UHPLC system introduced in 2004.
- $V_M$, *column void volume*: The liquid holdup volume of an HPLC column, which roughly equals to 65% of the empty column volume.
- *Peak volume* = $W_bF$ or the volume of eluent containing the peak; $W_b$ is proportional to $V_M$ and $k$.
- $d_p$, *Particle size*: Determines plate height and $\Delta P$ of the column.
- *System dispersion* or instrument bandwidth: Relates to the fluidic volume of the system flow path as seen by the sample including the detector flow cell.

*HPLC and UHPLC for Practicing Scientists*, Second Edition. Michael W. Dong.

- *UV detector flow cell* volume and path length: Primary contributors to system dispersion; pathlength must be maintained at 10 mm for acceptable sensitivity.
- *Extra-column band broadening*: Caused by system dispersion and is particularly deleterious to column efficiencies of smaller i.d. and $d_p$ columns.
- *Dwell volume*: System volume from the point of solvent mixing in the pump to the head of the column; it is responsible for gradient delay time.
- $P_c$ *or n, Peak capacity*: # of peaks in the chromatogram that can be accommodated with $R_s$ of 1. $P_c$ is roughly equal to $t_G/w_b$.
- *Viscous heating* and thermal gradients: Caused by frictional heat generated when mobile phase is pumped through small-particle columns at high flow; radial thermal gradients are detrimental to *N* while longitudinal thermal gradient increases average temperatures of a column by 10–20 °C.
- *Method translation (conversion)*: Conversion of HPLC method to UHPLC conditions or vice versa using geometric scaling.

### 5.1.3 Historical Perspectives: What is UHPLC?

For four decades after the debut of HPLC in the late 1960s, its resolution has been bounded by a system pressure of 6000 psi, limiting practical performance to column efficiency (*N*) of ~20 000 plates or $P_c$ of ~200 [1]. The "revolution" in UHPLC began in 1997 with the proof-of-concept studies by Professor James Jorgenson, who demonstrated the achievement

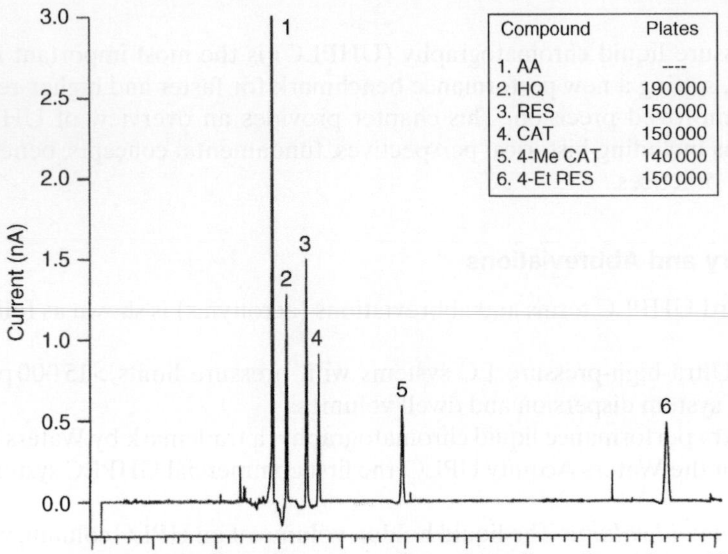


**Figure 5.1.** Chromatogram from a 52-cm-long column with an inlet pressure of 4100 bar (59 000 psi) obtained from a proof-of-concept study on UHPLC published in 1997. Peak identities and plate counts are included in the inset. Source: MacNair et al. 1997 [2]. Copyright 1997. Reprinted with permission of American Chemical Society.

of 190 000 plates in 10 minutes using a capillary column packed with 1.5 μm nonporous particles at the pressure of 59 000 psi (Figure 5.1) [2]. In a follow-on study, research groups from Jorgenson and Professor Milton Lee published extensively on UHPLC including a gradient peptide map with a $P_c$ of >300 in 30 minutes and high-speed separations with improved efficiency [3].

The first commercial UHPLC system, the Acquity UPLC, was introduced by the Waters Corporation in 2004 [4–6]. It was equipped with a binary pump with a pressure limit of 15 000 psi, an autosampler, and a photodiode array (PDA) detector with a 0.5 μl PDA flow cell. The system was introduced with 2.1 and 1.0 mm i.d. C18 and C8 columns packed with 1.7 μm hybrid particles. While the first system was well-received by the research community, universal acceptance of UHPLC for regulatory testing took several more years [1, 5, 6].

Today, the transformation from HPLC to UHPLC is complete with all major manufacturers having UHPLC product offerings capable of system pressures from 15 000 to 22 000 psi [7] (see Figure 4.2; Tables 4.1 and 5.1). There are books [9, 10] and review and research articles [1, 8, 11, 12] on UHPLC. A reader can refer to them for more detailed coverage.

Table 5.1 summarizes the salient features of UHPLC systems. The higher system pressures allow the use of columns packed with small particles (e.g. sub-2 μm) for faster analyses and superior separations of complex samples. All current UHPLC systems have lower system dispersion (5–20 μl at 4$\sigma$ bandwidths) from the use of improved injectors, narrower i.d. connection tubing (<0.005″), and smaller UV detector flow cells (0.5–1 μl) (see Table 5.2). The use of 2.1–3.0 mm i.d. columns packed with sub-2 μm or sub-3 μm particles is typical. Other system characteristics include smaller system dwell volumes (0.1–0.4 ml) and faster detector response/data acquisition (>40 pt/s) for high-throughput screening (HTS) applications [1, 8, 11].

**Table 5.1. Prominent Characteristics of UHPLC**

| System characteristics of UHPLC | Range and comment |
|---|---|
| High pressure limit | 15 000–22 000 psi (1000–1500 bar) with flow rate limits of 2–5 ml/min. Compatible with conventional and sub-2 μm particle columns. |
| Low system dispersion | Instrumental bandwidth of 5–20 μl (4$\sigma$) depending on configuration. System band broadening reduced by using smaller connection tubing <0.005 i.d. and small UV cell (0.5–2 μl). Compatible with columns down to 2.1–3 mm i.d. |
| Low gradient dwell volume | 0.1–0.4 ml for binary pumps (0.4–0.8 ml for quaternary pumps). Compatible to high-throughput screening (HTS). Small dwell (mixing) volumes may negatively impact UV detector noise. |
| Others | Fast injection cycle (~20 s) and detector response, and high acquisition rate (100 pt/s) for HTS. Compatibility to existing HPLC methods desirable (flow range, column oven size, and sample loop are potential issues) |

Source: Adapted from Guillarme and Dong 2013 [8].

**Table 5.2. Pressure Rating and Maximum Flow Rate of Typical UHPLC Systems**

| Company/ *CDS* | Model name | Year of introduction | Pressure (psi) | Flow (ml/min) | Binary or Quat pump | Comment |
|---|---|---|---|---|---|---|
| **Waters** | Acquity UPLC | 2004 | 15 000 | 2.0 | Binary | |
| *Empower* | Acquity H-Class, Plus | 2010<br>2018 | 15 000 | 2.2 | Quat | Bio option |
| | Acquity I-Class, Plus | 2011<br>2018 | 18 000 | 2.0 | Binary | Low dispersion |
| | Acquity M-Class | 2014 | 15 000 | 0.1 | Binary | Micro-LC |
| | Acquity Arc | 2015 | 9 500 | 5.0 | Quat | Dual path |
| **Agilent** | Infinity I/II 1290 | 2010, 2014 | 18 000, 19 000 | 5.0 | Binary, Quat | |
| *OpenLAB* | Infinity I/II 1260<br>Prime | 2010, 2017<br>2018 | 9 000<br>11 500 | 10.0<br>5.0 | Quat, binary | Bio option |
| | Infinity I/II 1220 | 2010, 2017 | 9000 | 10.0 | Isocratic, dual binary gradient | |
| **Thermo** | Ultimate 3000 | 2009 | 9 000<br>15 000 | 10.0 | Quat or Binary | Bio option |
| *Chromeleon* | Vanquish Horizon | 2014 | 22 000 | 5.0 | Binary | |
| | Vanquish Flex | 2015 (2017) | 15 000 | 8.0 | Quat, Binary | Bio |
| | Duo | 2018 | 15 000 | 8.0 | Dual Ternary | Dual path |
| **Shimadzu** | Nexera | 2010 | 19 000 | 5.0 | Binary or Quat | |
| *LabSolutions* | Nexera X2 | 2013 | 19 000 | 5.0 | Binary or Quat | |

Source: Adapted from Dong 2017 [1].

## 5.2 PRACTICAL CONCEPTS IN UHPLC

This section provides a brief review of chromatography concepts that are essential for a better understanding of UHPLC and its applications. Many of these concepts are covered separately in Chapter 2, 3, and 4.

### 5.2.1 Rationale for Higher System Pressure

The upper system pressure of HPLC remained at 6000 psi or 400 bar for about four decades after its debut in the 1960s [13, 14]. This pressure limit was a good match for the standard HPLC columns in use at the period that trended down from 10 to ~3 μm over several decades. While the prediction by A. J. P. Martin ("the highest chromatographic performance is achieved with columns packed with very small particles operating under very high pressure") was published in 1941, its realization took many decades to implement [15].

Since the column back pressure is inversely proportional to $d_p^{\,2}$ according to Eq. (2.19), the need for a higher system pressure was becoming apparent with the advent of sub-2 μm particles as demonstrated by Jorgenson and coworkers in the mid-1990s [2, 3].

$$\Delta P = 1000 \frac{F \eta L}{\pi r^2 {d_p}^2} \tag{2.19}$$

### 5.2.2 Rationale for Low-Dispersion Systems

The 4.6 mm i.d. column had been the "standard" column diameter for analytical HPLC for many decades [14, 16]. The first commercial UHPLC system was introduced with columns packed with sub-2-μm particles in a 2.1 mm i.d. format because of concerns about efficiency loss with viscous heating. However, small i.d. columns generate small peak volumes, which are severely impacted by extra-column band broadening.

The peak volume under isocratic conditions is a function of $V_m$ and $k$ according to Eq. (2.14).

$$\text{Peak Volume} = \frac{4V_R}{\sqrt{N}} = \frac{4V_M(1+k)}{\sqrt{N}} \tag{2.14}$$

For example, the peak volume of an analyte from a "standard HPLC column," 150×4.6 mm packed with 5 μm particles at $k = 3$ is equal to

$$4 \times 1500\,\mu\text{l}\,(1+3)/(15\,000)^{0.5} = 200\,\mu\text{l}$$

In contrast, the peak volume of a "standard" UHPLC column for fast analyses: 50×2.1 mm column packed with 1.7 μm at $k = 3$ is equal to

$$4 \times 100\,\mu\text{l}\,(1+3)/(15\,000)^{0.5} = 13\,\mu\text{l}$$

Therefore, low-dispersion systems equipped with improved injectors, small i.d. connection tubing, and small UV flow cells must be used to prevent efficiency loss. As noted, small flow cells (Figure 4.13) with a 10-mm path length were designed successfully without incurring sensitivity loss. Typical instrumental bandwidth in modern UHPLC systems ranged from 7 to 20 μl (4$\sigma$) as summarized in some measured values of UHPLC systems in Table 5.3. These were substantially reduced from bandwidths of 30–60 μl of conventional HPLC systems [1, 11].

### 5.2.3 Rationale for Low Dwell Volumes

A low system dwell volume is desirable to reduce gradient delay time in HTS and low-flow-rate applications (<0.5 ml/min) [17]. Early UHPLC models were all binary high-pressure mixing systems with small dwell volumes for this reason (e.g. <0.2 ml). Nevertheless, external mixers are needed for efficient solvent blending to reduce baseline perturbation for low UV detection. This issue is discussed further in Section 5.4.5. Many second-generation UHPLC systems are equipped with quaternary low-pressure mixing

**Table 5.3. System Dispersion or IBW (4$\sigma$) of Selected UHPLC**

| Company | Model name | Comments | IBW (4$\sigma$) in μl |
|---|---|---|---|
| Waters | Alliance HPLC | Standard configuration | 45 |
| | Acquity UPLC Classic | Standard configuration | 10–12 |
| | Acquity UPLC classic | With switching valve | 16 |
| | Acquity I-class | Standard configuration | 4–6 |
| Agilent | 1200 HPLC | | 30 |
| | Infinity I 1290 binary | Standard configuration | 14 |
| | Infinity I 1290 binary | With switching valve | 17 |
| Thermo | Vanquish | Standard configuration | 16 |

Source: Adapted from Dong 2017 [1].

pumps with dwell volumes in the range of 0.4–0.8 ml, which may not work well for HTS applications (see Figure 4.6 for a listing of dwell volumes) [12, 17].

### 5.2.4 Other UHPLC Instrumental Characteristics

Requirements for fast separations are fast UV detector responses, higher data sampling rates for both UV and MS, and shorter autosampler cycle time (e.g. three injections per minute). Other important UHPLC system features are the abilities to inject small sample volumes precisely (e.g. 1 μl) and the compatibility to run existing HPLC methods.

## 5.3 BENEFITS OF UHPLC AND CASE STUDIES

Table 5.4 summarizes the benefits of UHPLC including fast separations, high-resolution analysis, rapid method development, flexibility for customizing resolution, and other performance enhancements. Each benefit is described with case studies as follows.

### 5.3.1 Benefit #1: Fast Separations with Good Resolution

*Benefit #1*: "*Fast separation with good resolution*" is perhaps the main reason for most laboratories to consider purchasing UHPLC equipment. UHPLC can increase sample throughput by 3–5-fold versus conventional HPLC while maintaining similar resolution, for example 5 minutes vs. 20 minutes for purity analysis. While this benefit stems from the performance

**Table 5.4. Benefits of UHPLC**

| Benefits of UHPLC | Comment |
|---|---|
| Fast separation with good resolution | Increase throughput by 3–5-fold vs. conventional HPLC while maintaining similar resolution, e.g. 5 minutes vs. 20 minutes purity analysis. |
| High-resolution analysis of complex samples | Increase resolution by up to threefold vs. HPLC, e.g. peak capacities ($P_c$) 400–800 vs. ~200 for HPLC. |
| Rapid method development | Fast analysis with short columns is ideal for rapid column and mobile phase screening and method optimization. |
| Customizable resolution | Flexibility to customize sample resolution by using different column lengths and gradient times. |
| Others | |
| Solvent saving | Typical 5–15-fold reduction vs. HPLC due to shorter analysis time and use of smaller ID columns. |
| Higher mass sensitivity | Three- to tenfold increase of mass sensitivity (reduction of sample amounts injected). Use of long-path-length UV flow cells (25–60 mm) can increase concentration sensitivity up to six times. |
| Higher precision | Significant increase of retention time (two- to threefold) and peak area precision (<0.1% RSD achievable at injection volumes >1 μl) |
| Can be combined with other approaches | UHPLC is compatible with high-temperature LC, 2D-LC, or SPP columns individually or in combination. These are optional rather than alternative approaches. |

Source: Adapted from Guillarme and Dong 2013 [8].

of smaller columns packed with sub-2 or sub-3 μm particles, it is the low-dispersion UHPLC instruments with higher pressure ratings that allow full realization of these performance gains.

This benefit is best illustrated in a method conversion case study using geometric scaling as shown in Figure 5.2 [18]. Chromatograms (A) and (B) show separations of a "standard" HPLC column (150 × 4.6 mm, 5 μm) with a run time of 45 minutes and an UHPLC column

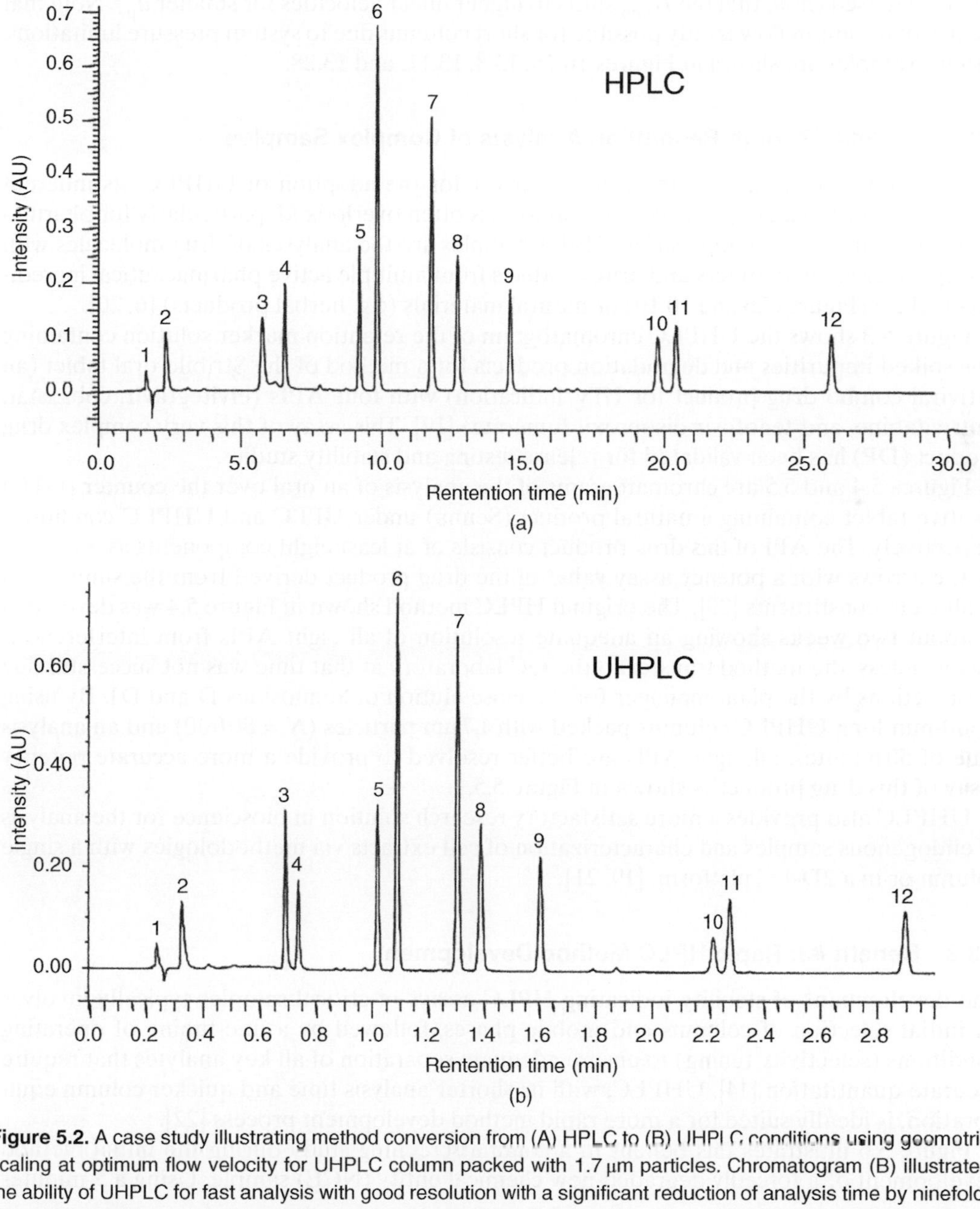


**Figure 5.2.** A case study illustrating method conversion from (A) HPLC to (B) UHPLC conditions using geometric scaling at optimum flow velocity for UHPLC column packed with 1.7 μm particles. Chromatogram (B) illustrates the ability of UHPLC for fast analysis with good resolution with a significant reduction of analysis time by ninefold. Operating conditions: (a) Column C18, 150 × 4.6 mm, 5 μm, $F$ = 1 ml/min, $V_{inj}$ = 20 μl, Runtime, 45 minutes. (b) Column C18, 50 × 2.1 mm, 1.7 μm, $F_{opt}$ = 0.61 ml/min, ($F_{scaled}$ = 0.2 ml/min), $V_{inj}$ = 1.4 μl, Runtime, 5.1 minutes. Source: Guillarme et al. 2008 [18]. Reprinted with permission of Elsevier.

(50×2.1 mm, 1.7 μm) with a run time of five minutes. Note that the separations look remarkably similar even though the UHPLC separation is about nine times faster.

The principle of geometric scaling was used in the method conversion. The ratio of $L/d_p$ was kept the same, so both columns had column efficiency of ~15 000. Gradient time ($t_G$) was scaled to column length while the injection volume was scaled to $V_M$. The flow rate scaled to $r^2$ should be 0.2 ml/min for the 2.1 mm column, though a faster optimum flow of 0.61 ml/min for this $d_p$ was used (note that the $H_{min}$ shifts to higher linear velocities for smaller $d_p$). Note that the use of optimum flow is only possible for short columns due to system pressure limitations. Other examples are shown in Figures 10.26, 13.8, 13.11, and 13.28.

### 5.3.2 Benefit #2: High-Resolution Analysis of Complex Samples

While faster separations are the primary driver for the adoption of UHPLC, its inherent superiority for the analysis of complex samples is often overlooked, particularly for pharmaceutical quality control applications [19]. Examples are the analyses of drug molecules with multiple stereogenic centers and drug products from multiple active pharmaceutical ingredients (APIs) (Figures 7.6 and 10.10) or natural materials (e.g. herbal products) [6, 20].

Figure 5.3 shows the UHPLC chromatogram of the retention marker solution containing 60+ spiked impurities and degradation products for a method of the Stribild oral tablet (an antiviral combo drug product for HIV indication) with four APIs (elvitegravir, cobicistat, emtricitabine, and tenofovir disoproxil fumarate) [19]. This assay of this very complex drug product (DP) has been validated for release testing and stability studies.

Figures 5.4 and 5.5 are chromatograms of the analysis of an oral over the counter (OTC) laxative tablet containing a natural product (Senna) under HPLC and UHPLC conditions respectively. The API of this drug product consists of at least eight components as indicated by the arrows with a potency assay value of the drug product derived from the summation of all eight constituents [20]. The original HPLC method shown in Figure 5.4 was developed in about two weeks showing an adequate resolution of all eight APIs from interferences. Nevertheless, the method transfer to the QC laboratory at that time was not successful due to objections by the plant manager for the close elution of Sennosides D and D1. By using a 300-mm long UHPLC columns packed with 1.7 μm particles ($N = 80\,000$) and an analysis time of 50 minutes, all eight APIs are better resolved to provide a more accurate potency assay of this drug product as shown in Figure 5.5.

UHPLC also provides a more satisfactory research solution in bioscience for the analysis of endogenous samples and characterization of cell extracts via methodologies with a single column or in a 2D-LC platform. [19, 21]

### 5.3.3 Benefit #3: Rapid HPLC Method Development

The development of stability-indicating HPLC assays of critical samples typically involves an initial selection of columns and mobile phases, followed by a fine-tuning of operating conditions (selectivity tuning) to ensure adequate separation of all key analytes that require accurate quantitation [14]. UHPLC, with its shorter analysis time and quicker column equilibration, is ideally suited for a more rapid method development process [22].

Figure 5.6 illustrates this benefit in a column screening study during the initial method development of a forcedly degraded new chemical entity (NCE) sample. Using a 5-minutes

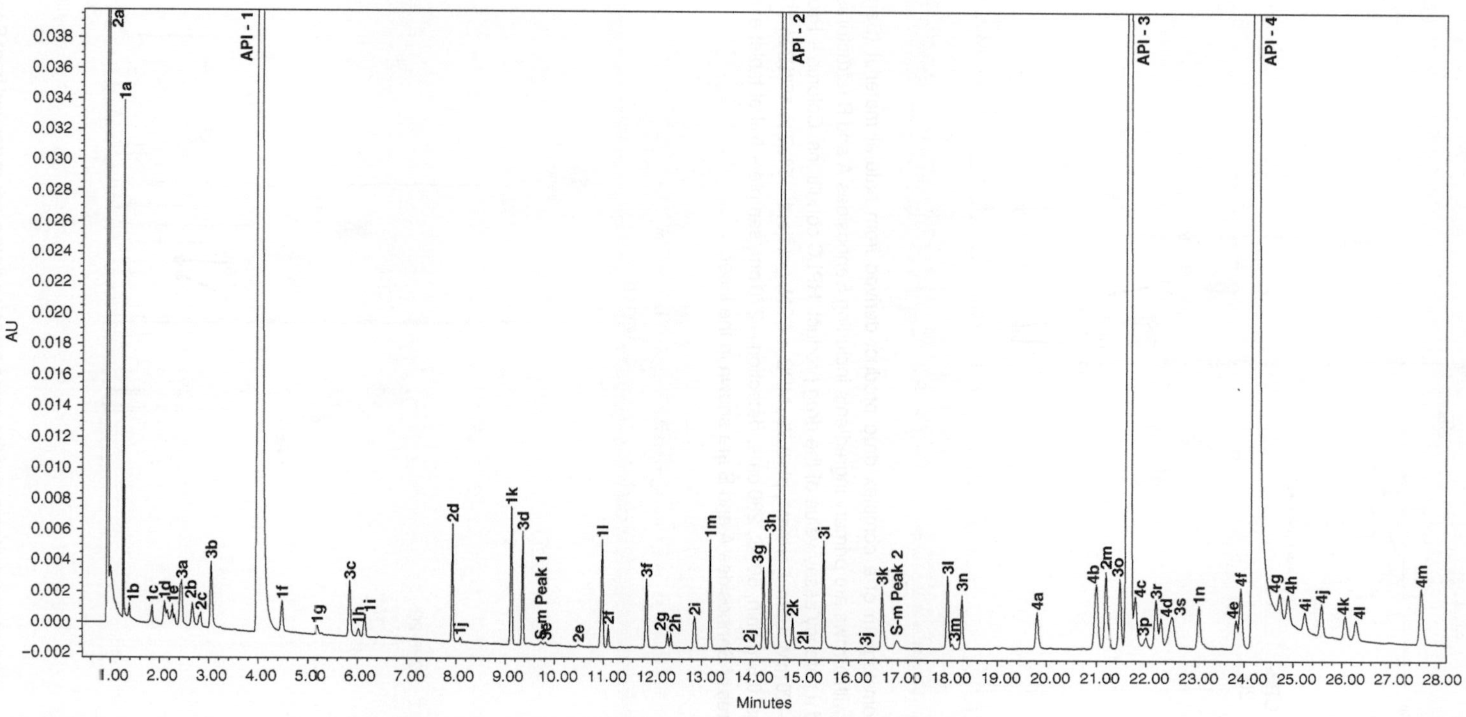


**Figure 5.3.** Chromatogram showing a stability-indicating UHPLC method for a drug product Stribild oral tablet with four APIs (elvitegravir, cobicistat, emtricitabine, and tenofovir disoproxil fumarate). UHPLC method conditions: Column – ACQUITY shield RP18, 1.7 mm, 150 × 2.1 mm, MPA – A: pH 6.2 buffer, MPB – ACN, gradient program – a five-segment gradient program developed using Fusion QbD, flow rate – 0.34 ml/min @ 30 °C, detection – 240 nm, sample – a retention marker solution containing four APIs spiked with reference standards of impurities and degradation products. Source: Reprinted with permission of Dong et al. 2014 [19].

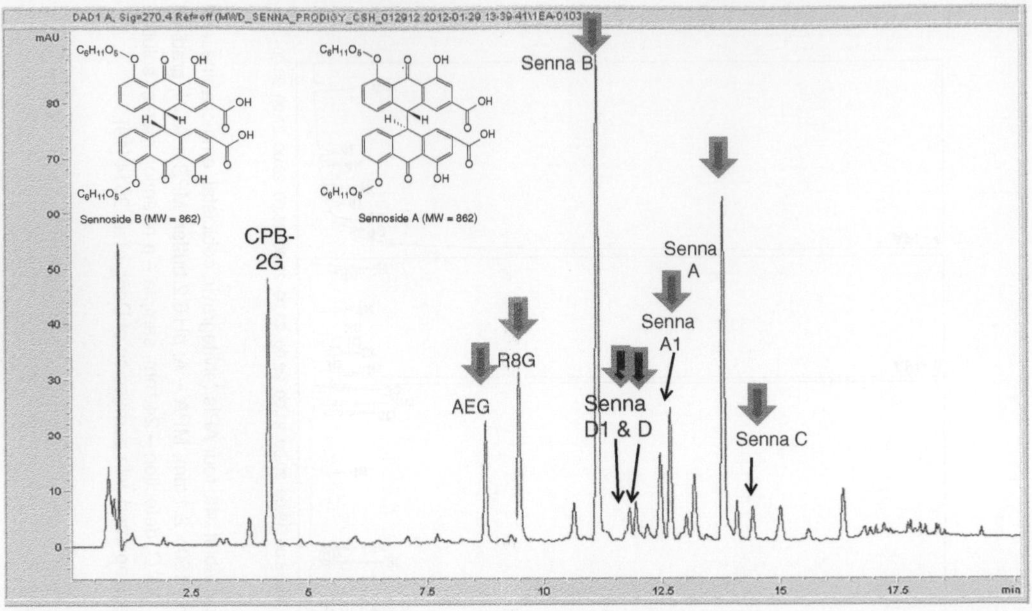


**Figure 5.4.** An HPLC chromatogram of a complex drug product derived from natural material (Senna). The eight constituents marked with arrows are primary ingredients including Sennosides A and B (structures shown), which are summed to yield a potency assay value of the drug product. HPLC conditions: Column – Prodigy C18 150 × 2.1 mm, 3 μm, MPA – 20 mM Amm. Formate, pH 3.75, MPB – ACN, gradient – 5%B to 15%B in 15 minutes, 15% to 85%B in 10 minutes, 0.5 ml/min, 50 °C, 280 bars, detection – 270 nm, sample – 5 μl of tablet extract, runtime – 34 minutes. Structures of Sennosides A and B are shown in the inset.

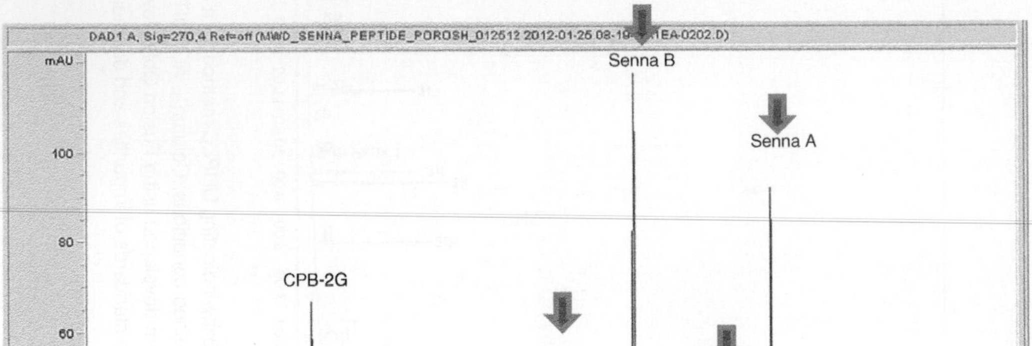

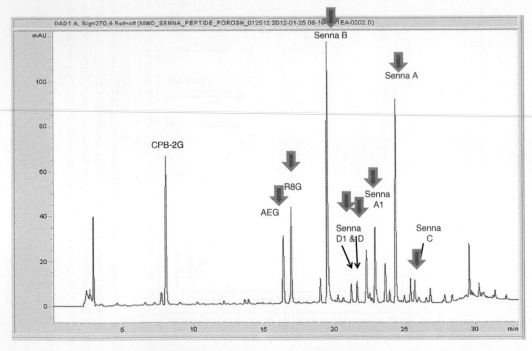


**Figure 5.5.** A UHPLC high-resolution chromatogram of the same complex drug product derived from Senna shown in Figure 5.4 using a 300 mm column packed with 1.7 μm particles with $N = 82000$ plates. UHPLC conditions: Column – ACQUITY BEH130 C18 300 × 2.1 mm, 1.7 μm, MPA: 20 mM Amm. Formate, pH 3.75, MPB – ACN, gradient – 0.3 ml/min, 40 °C, 800 bars, 5%B to 15%B in 25 minutes, 15% to 85%B in 10 minutes, detection – 270 nm, sample – 5 μl of tablet extract, runtime – 50 minutes.

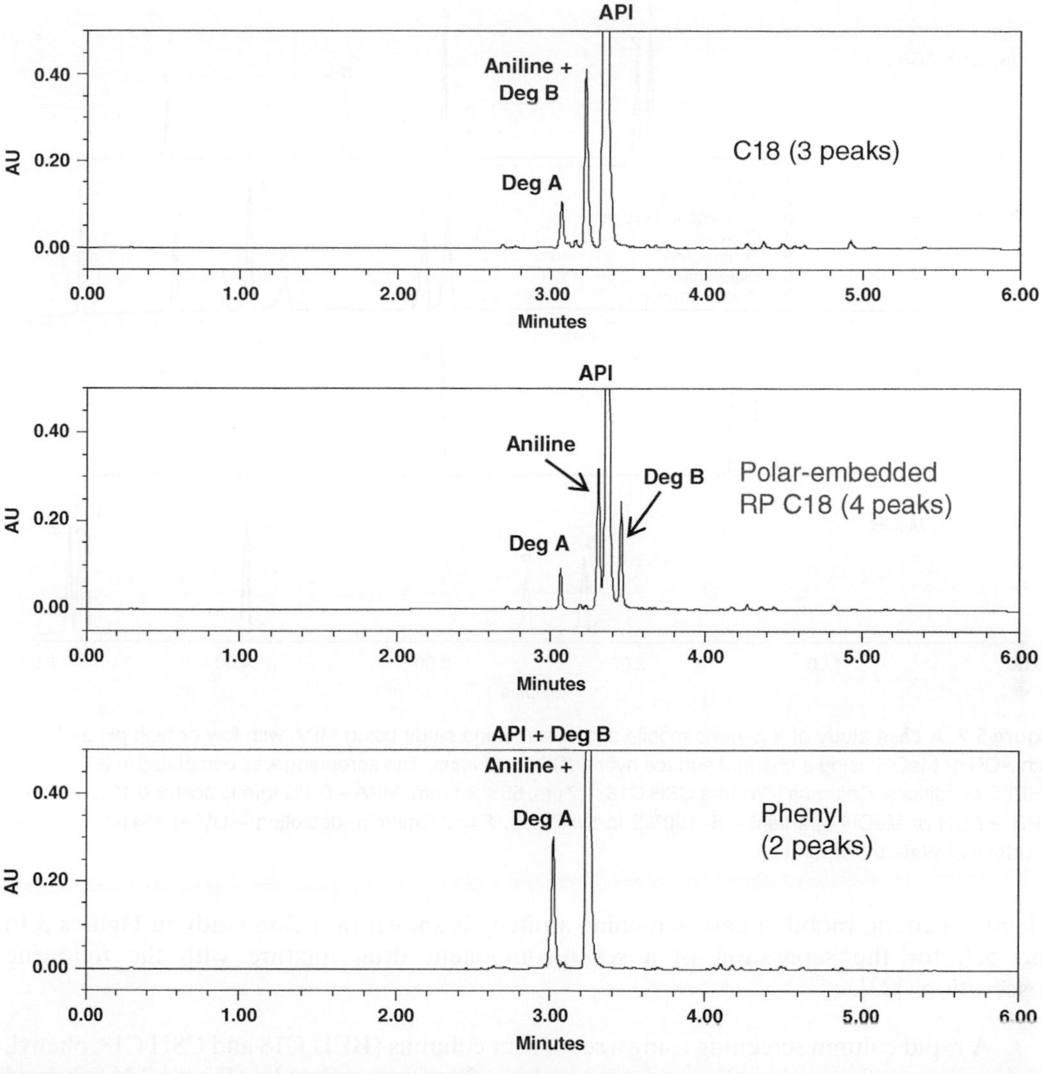


**Figure 5.6.** A case study for rapid HPLC method development by UHPLC demonstrating a column screening study of a forced degradation NCE sample completed in 30 minutes. UHPLC conditions: Column – ACQUITY C18, RP and phenyl, 1.7 μm, 50 × 2.1 mm, MPA – 0.05% formic acid, MPB – ACN, solvent program – 10–90%B in 5 minutes, flow rate – 0.5 ml/min @ 30 °C, 7700 psi, detection – PDA at 254 nm 20 pt/s, instrument – Waters Acquity UPLC, sample – 1 μl of light degraded API at 0.4 mg/ml in ACN spiked with impurities. Source: Dong and Zhang 2014 [22]. Copyright 2014. Reprinted with permission of Elsevier.

broad gradient, column screening using three orthogonal columns was completed in 30 minutes. The best separation of the NCE and its three major impurities was realized by a polar-embedded C18 column (the middle chromatogram). Subsequently, a longer column packed with the same bonded phase was selected for further method development [22].

Column and mobile phase screening (pH, buffer type, and strength, organic solvent) can be a time-consuming task in HPLC method development [14]. An exceptionally rapid but

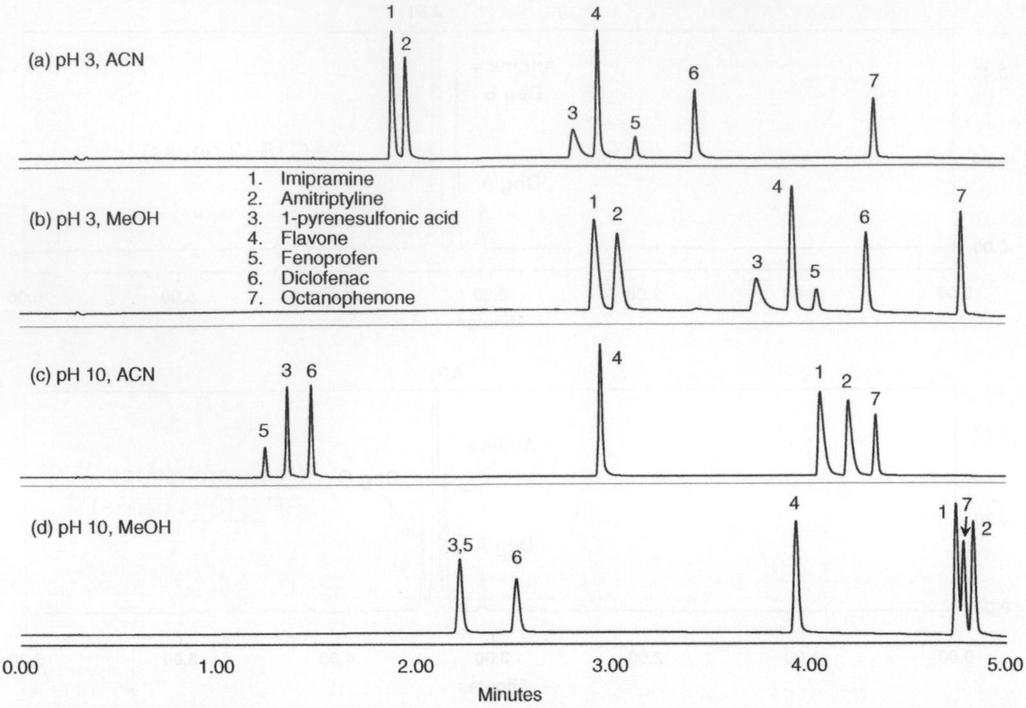


**Figure 5.7.** A case study of a generic mobile phase screening study using MPA with low or high pH and MPB with ACN or MeOH using a charged surface hybrid (CSH) columns. The screening was completed in 30 minutes. UHPLC conditions: Columns – Waters CSH C18, 1.7 μm, 50 × 2.1 mm, MPA – 0.1% formic acid or 0.1% ammonia, MPB – ACN or MeOH, gradient – 5–100%B in 5 minutes, $F = 0.5$ ml/min, detection – UV at 254 nm. Source: Courtesy of Waters Corporation.

effective generic mobile phase screening strategy is shown in a case study in Figures 3.16 and 5.7, for the separation of a seven-component drug mixture with the following observations [23].

1. A rapid column screening using a set of four columns (BEH C18 and CSH C18, phenyl, and pentafluorophenyl) was performed in ~30 minutes. Results (Figure 3.16) showed that the CSH C18 and phenylhexyl columns yielded the best separation. The former was selected for further mobile phase screening.
2. As shown in Figure 5.7, each mobile phase screening run was completed in ~5 minutes using a broad gradient (5–95%B in five minutes) with acidic or basic mobile phase A (pH 3 or 10) coupled with either methanol or acetonitrile as the strong solvent (MPB). The entire mobile phase screening was completed in ~30 minutes.
3. Note that the retention times of the neutral analytes (components 4 and 7) are relatively unaffected by mobile phase pH. In contrast, those of the acidic (components 3, 5, 6) and basic analytes (components 1 and 2) are controlled primarily by the ionization states of the molecules (see Section 2.3.6 for explanation). Retention times using methanol are appreciably longer due to its lower solvent strength compared to acetonitrile.
4. This case study illustrates the use of a generic column and mobile phase protocol (four orthogonal columns for initial screening followed by mobile phase screening with low or high pH and methanol or acetonitrile) that can be completed in one hour on an automated system.

### 5.3.4 Flexibility for Customizing Resolution

A major benefit of UHPLC with increased pressure limits is the versatility for the development of either ultra-fast or high-resolution methods [22]. Figure 5.8a–c shows three chromatograms of the same forcedly degraded sample (shown previously in Figure 5.6) conducted with a narrower gradient range of 20–60%B (with gradient time $t_G$ in 3–20 minutes) to enhance the resolution of the region around the API peak. These examples illustrate the versatility of UHPLC to provide ultra-fast, mid-resolution, and very high-resolution analysis: (i). A fast six-minute analysis with good resolution using a 50-mm column; (ii). A 16-minute analysis with excellent resolution using a 100-mm column; (iii). A 30-minute analysis with

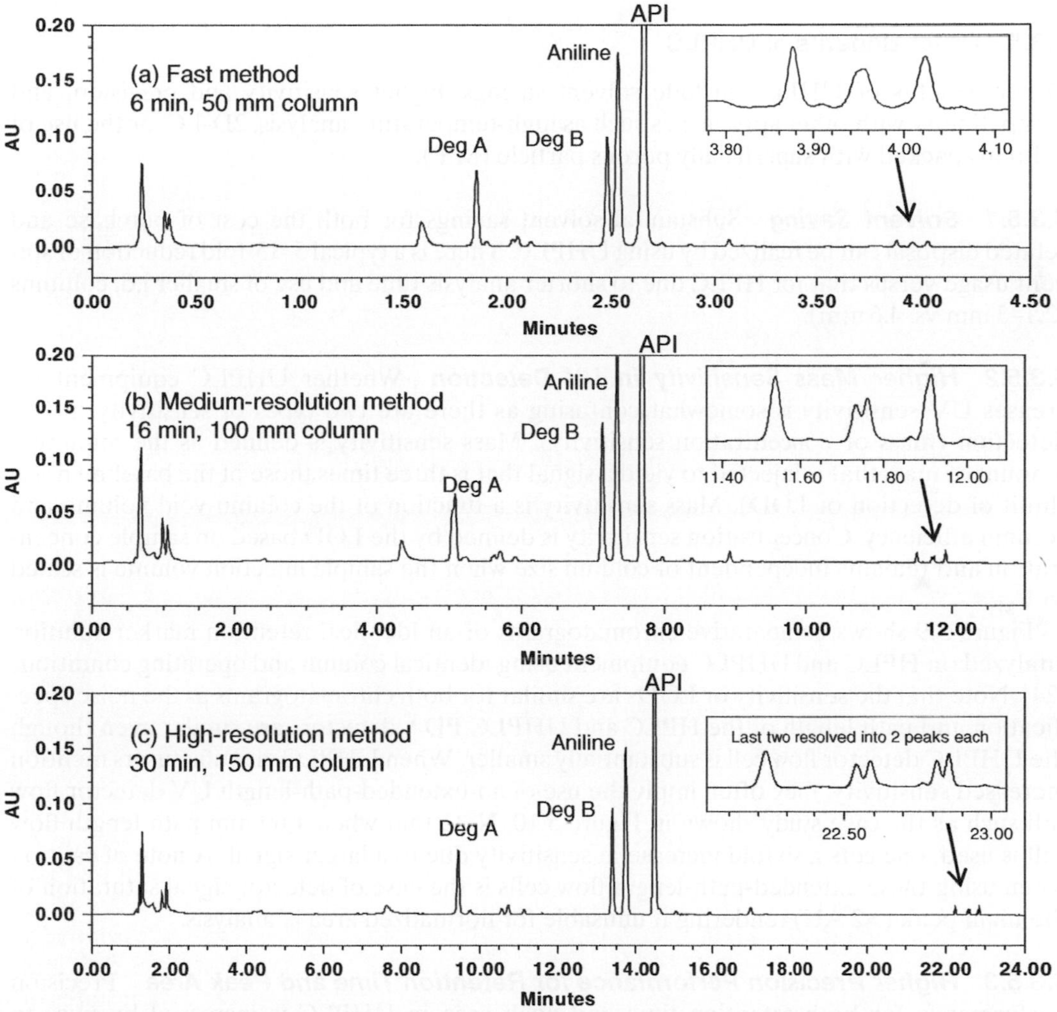


**Figure 5.8.** Three chromatograms demonstrating the ability to customize resolution using UHPC columns of different lengths. UHPLC conditions: MPA = 0.05% formic acid, MPB = ACN, gradient – 20–60%B. (a) Fast method: Acquity RP shield column 1.7 μm, 50 × 2.1 mm, $t_G$ = 6 minutes, $F$ = 0.6 ml/min at 30 °C, 9000 psi. (b) Medium resolution method: Acquity RP shield column 1.7 μm, 100 × 2.1 mm, $t_G$ = 16 minutes, $F$ = 0.4 ml/min at 30 °C, 8000 psi. (c) High-resolution method: Acquity RP shield column 1.7 μm, 150 × 2.1 mm, $t_G$ = 30 minutes, $F$ = 0.3 ml/min at 30 °C, 11 000 psi. Source: Dong and Zhang 2014 [22]. Copyright 2014. Reprinted with permission of Elsevier.

very high resolution using a 150-mm column. The flow rate and gradient times ($t_G$) were adjusted to yield similar gradient volumes for the three chromatograms, normalized to the respective column void volumes. Not surprisingly, the resolutions between the four major peaks increased substantially with the longer column with higher $t_G$. This high-resolution capability is useful for separation of closely eluting peaks (the bottom chromatogram). Note that the last two impurity peaks in the inset of the top chromatogram (fast method) were subsequently resolved into two distinct peaks (possibly isomers) as shown in the inset of the bottom chromatogram (high-resolution method).

UHPLC allows the versatility for the development of a wide variety of methods – from fast analysis to very high-resolution methods for detailed profiling of complex samples, thus providing a higher level of flexibility for different applications.

### 5.3.5 Other Benefits of UHPLC

Other benefits of UHPLC include solvent savings, higher sensitivity and precision, and compatibility with other approaches such as high-temperature analysis, 2D-LC, or the use of columns packed with superficially porous particle (SPP).

***5.3.5.1 Solvent Saving*** Substantial solvent savings for both the cost of purchase and related disposal can be realized by using UHPLC. There is a typical 5–15-fold reduction of solvent usage versus that for HPLC due to shorter analysis time and use of smaller i.d. columns (2.1–3 mm vs. 4.6 mm).

***5.3.5.2 Higher Mass Sensitivity in UV Detection*** Whether UHPLC equipment increases UV sensitivity is somewhat confusing as there are two types of sensitivity in UV detection (mass or concentration sensitivity). Mass sensitivity is defined as the minimum amount of mass that is injected to yield a signal that is three times those of the baseline noise (limit of detection or LOD). Mass sensitivity is a function of the column void volume and column efficiency. Concentration sensitivity is defined by the LOD based on sample concentration and remains independent of column size when the sample injection volume is scaled to $V_M$.

Figure 5.9 shows comparative chromatograms of an identical retention marker solution analyzed on HPLC and UHPLC equipment using identical column and operating conditions [24]. Note that the sensitivity or LODs are similar for both chromatograms as the noise specification and path length of the HPLC and UHPLC PDA detectors are similar even though the UHPLC detector flow cell is substantially smaller. When UHPLC manufacturers mention increased sensitivity, they often imply the use of an-extended-path-length UV detector flow cell such as the case study shown in Figure 5.10. Note that when a 60-mm path length flow cell is used, one gets a sixfold increase in sensitivity due to a larger signal. A note of caution when using these extended-path-length flow cells is the ease of detector signal saturation of the main peak (>2 AU) rendering it unusable for normalized area% analysis.

***5.3.5.3 Higher Precision Performance for Retention Time and Peak Area*** Precision performance for both retention time and peak area in UHPLC is increased by two- to threefold versus those of conventional HPLC. With better-designed pumps, retention time precision is improved for repeated runs on the same day and over an extended period. Many analysts have noticed that calibration tables containing retention time windows for peak identification in UHPLC can often be used without any modifications for many months in extended stability studies. In contrast, similar chromatography data system (CDS)

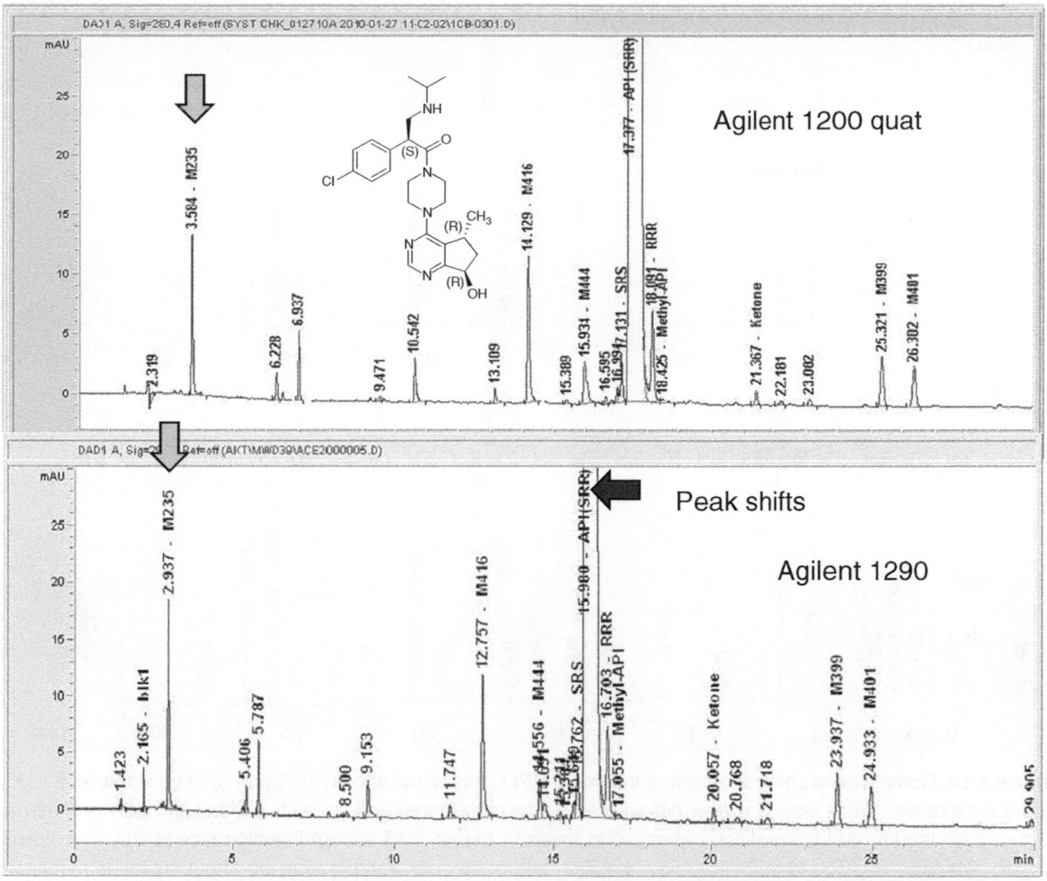


**Figure 5.9.** Comparative chromatogram a retention marker solution of a stability-indicating HPLC assay run on an HPLC (Agilent 1200 quat system) and an UHPLC (Agilent 1290 binary system). HPLC conditions: Column – ACE C18, 3 µm, 150 × 4.6 mm, MPA – 20 mM ammonium formate at pH 3.7, MPB – ACN with 0.05% formic acid, gradient program – 5%B to 15% in 5 minutes, 15%B to 40% in 25 minutes, 40%B to 90% in 3 minutes, run-time = 42 minutes, flow rate – 1.0 ml/min @ 30 °C, detection – PDA at 280 nm 40 pt/s, sample – 10 µl 0.5 mg/ml API, pressure – 160 bar (HPLC), 200 bar (UHPLC). Note that noise = 0.02 mAU, gradient shift of baseline, and *S/N* ratio of most peaks are all comparable except that retention times are shifted by 0.8–1.4 minutes in UHPLC due to the lower dwell volume. Source: Reprinted with permission of Dong 2013 [24].

processing methods on HPLC systems often require multiple adjustments of retention time windows to ensure correct identifications.

Higher peak area precision due to more precise autosamplers is another desirable feature of UHPLC. Most HPLC autosamplers are capable of peak area precision levels of 0.2–0.5% relative standard deviation (RSD) for injection volumes >5 µl. This precision performance is improved to 0.1–0.2% RSD in UHPLC even for injection volumes down to 1 µl. [12] Example data on peak area precision are shown in Section 5.4.4.

***5.3.5.4 UHPLC are Compatible with Other Approaches*** In the early days of UHPLC, one often heard arguments that UHPLC equipment was not needed since higher column temperatures or SPP can be used to reduce pressure without resorting to sub-2 µm particles.

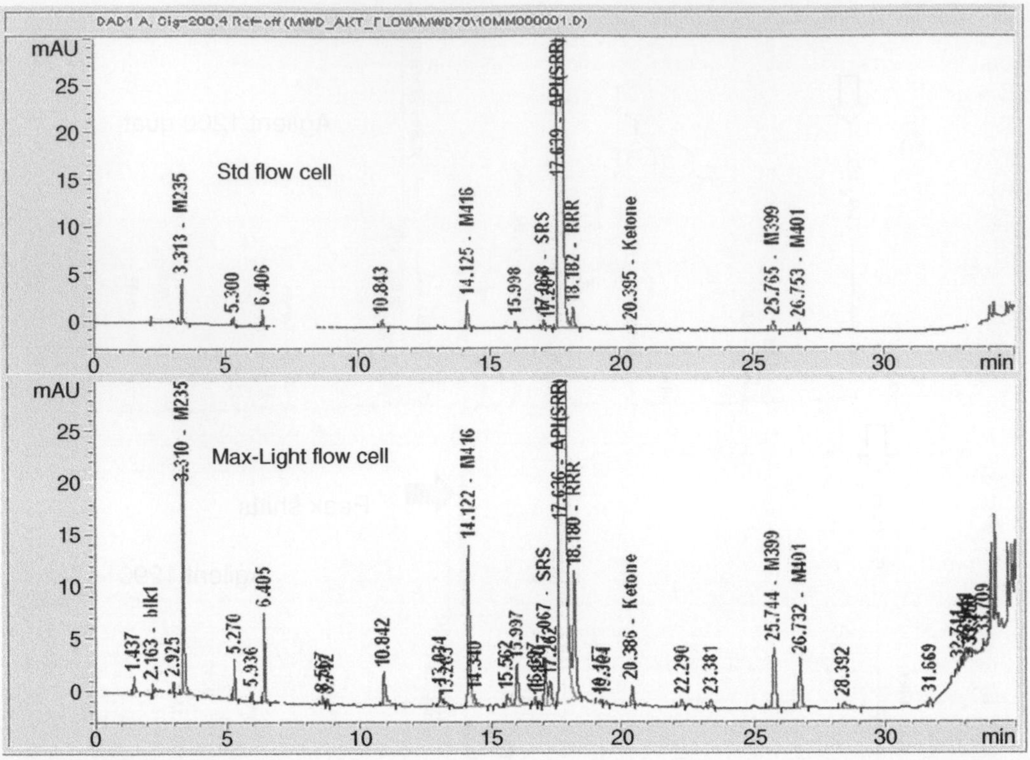


**Figure 5.10.** Comparative chromatogram of the same HPLC assay run shown in Figure 5.13 on an Agilent 1290 binary system with (a) a standard flow cell with a 10-mm path length and (b) with a Max-Light flow cell with a 60-mm path length. HPLC conditions identical to those in Figure 5.13 except injection size is 2 μl. The noise of both detectors was found to 24 μAU with the *S/N* ratio of all peak estimated to be about six times higher in chromatogram. (b) The limits of quantitation or LOQs are calculated to be: (a) 0.05% at 2 μl injection and 0.01% at 10 μl injection; b. 0.01% at 2 μl injection and 0.002% at 10 μl injection. Note that the main API peak saturated the detector at 10 μl injection in chromatogram (b). Source: Reprinted with permission of Dong 2013 [24].

Most realize now that these arguments are not valid since UHPLC is an improved platform that can be used with all these approaches including 2D-LC with even better results [24].

## 5.4 POTENTIAL ISSUES AND HOW TO MITIGATE

The use of UHPLC under high operating pressures with low-dispersion equipment can present potential issues for many new users. Earlier issues have been mitigated with improved instrumentation and the introduction of second-generation UHPLC equipment [5, 25]. Each potential issue is described here along with its mitigation strategies.

### 5.4.1 Safety Issues

Safety concerns of UHPLC were addressed in a study by Professor Milton Lee et al. This study concluded that UHPLC poses little inherent dangers to users in normal operation because of the low flow rates (<1 ml/min) and small compressibility of liquids [26].

### 5.4.2 Viscous Heating

Effects of viscous (frictional) heating caused by pumping liquid under high pressure through small-particle columns have been a favorite research topic in many published papers. [9, 24, 25, 27] The generation of frictional heat and its dissipation are complex phenomena and highly dependent on viscosity, $F$, $d_p$, $L$, column i.d., the thermal conductivity of the mobile phase, and the type of column oven used.

Viscous heating can cause two types of thermal gradients within the column.

1. *Radial thermal gradient,* where the center of the column has higher temperatures since generated heat dissipates mostly from the column wall. Radial gradient causes more band dispersion (a more pronounced parabolic flow profile), which is deleterious to column efficiency. Early UHPLC columns were mostly packed in small i.d. formats (e.g. 2.1 mm) to minimize this effect. However, further studies indicated that the radial thermal effect is not a practical issue in most still air columns due to poor heat dissipation from the stainless steel column wall (vs. the case where the column is kept in a constant temperature water bath or a forced air column oven).
2. *Longitudinal thermal gradients*: The heat generated is cumulative giving rise to *longitudinal thermal gradients* along the length of the UHPLC column where the temperature at column outlet can be 5–20 °C higher than those at the column inlet (Figure 5.11) [27]. So the average column temperature can be substantially higher than that of the set column temperature under high operating pressures. While longitudinal thermal gradient does not affect column efficiency (just the average column temperature), there may be potential method transfer issues for critical pairs in the sample whose selectivity is temperature dependent (case study is shown in Figure 5.12) [28].

For most users, it is essential to acknowledge the existence of viscous heating and its potential ramifications. However, it may not be a practical issue except for columns packed with very small particles (<2 μm) operating at very high pressures (>12 000 psi or 800 bar) or with a forced air oven and if there are critical pairs in the sample whose relative retention times are sensitive to temperature changes.

### 5.4.3 Instrumental and Operating Nuances

While UHPLC has improved performance regarding speed, resolution, and precision, new users may experience some surprises and potential issues. First, there is a cost premium of 10–50% over those of conventional HPLC equipment. Second, the use of smaller i.d. columns packed with small particles used on low-dispersion instruments would require a better understanding of concepts such as $V_M$, peak volumes, dwell volumes, instrumental dispersion, and peak capacity particularly for method development, conversion, or troubleshooting situations. Perhaps additional fundamental and operating training would be helpful for easing the transition to UHPLC.

Third, backward compatibility considerations to existing HPLC methods can be an immediate issue for performing legacy HPLC methods requiring high flow rates, longer columns, or large injection volumes. Many UHPLC systems have an upper flow rate limit of 2 ml/min, a standard injector loop of ≤20 μl, or a column oven size that fits column lengths of less than 150 mm. The implementation of these existing HPLC methods that exceed the instrument limits may require the installation of additional optional equipment (e.g. larger sample loop, larger column oven). Some manufacturers appear to offer UHPLC equipment that has more backward compatibility in this regard.

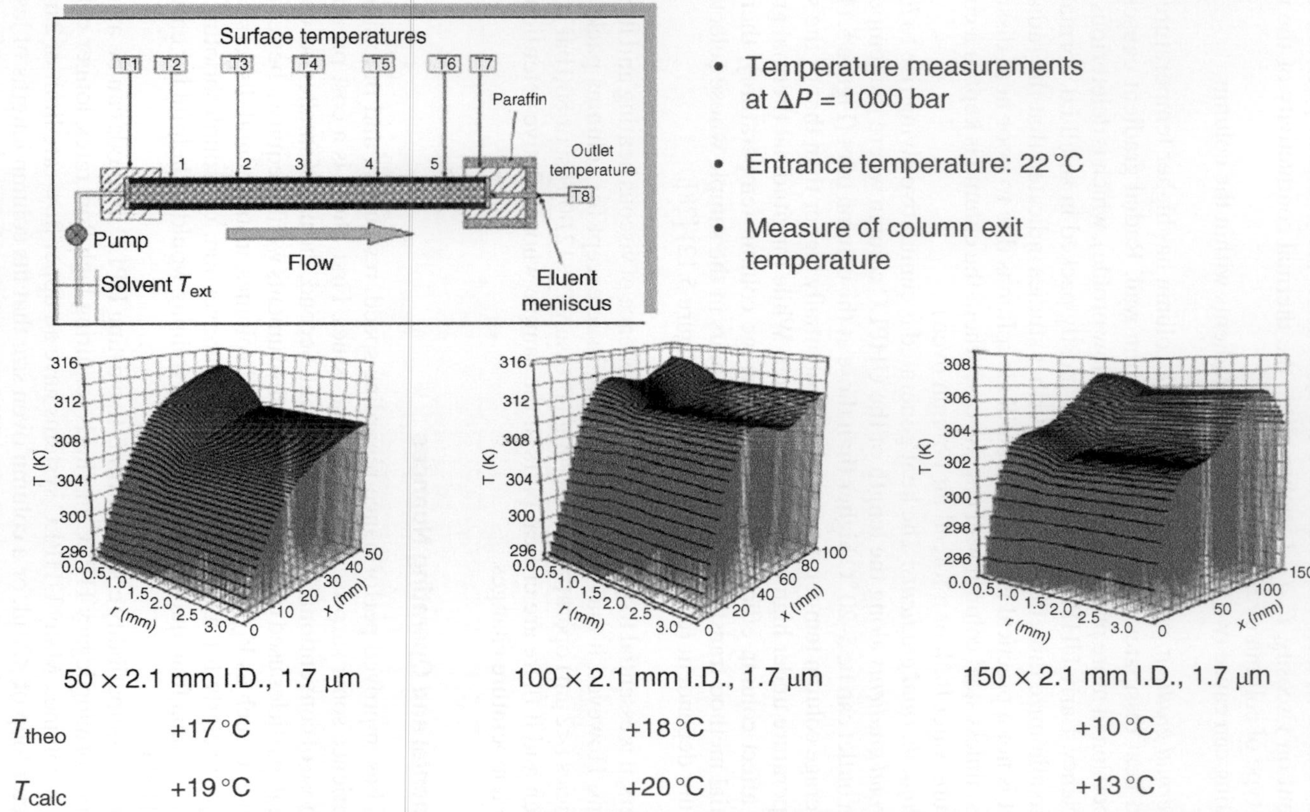


**Figure 5.11.** Diagram showing the experimental setup and thermal gradient profiles of various columns illustrating the effect of longitudinal thermal gradient effects in UHPLC. Source: Gritti and Guichon 2008 [27]. Reprinted with permission of American Chemical Society.

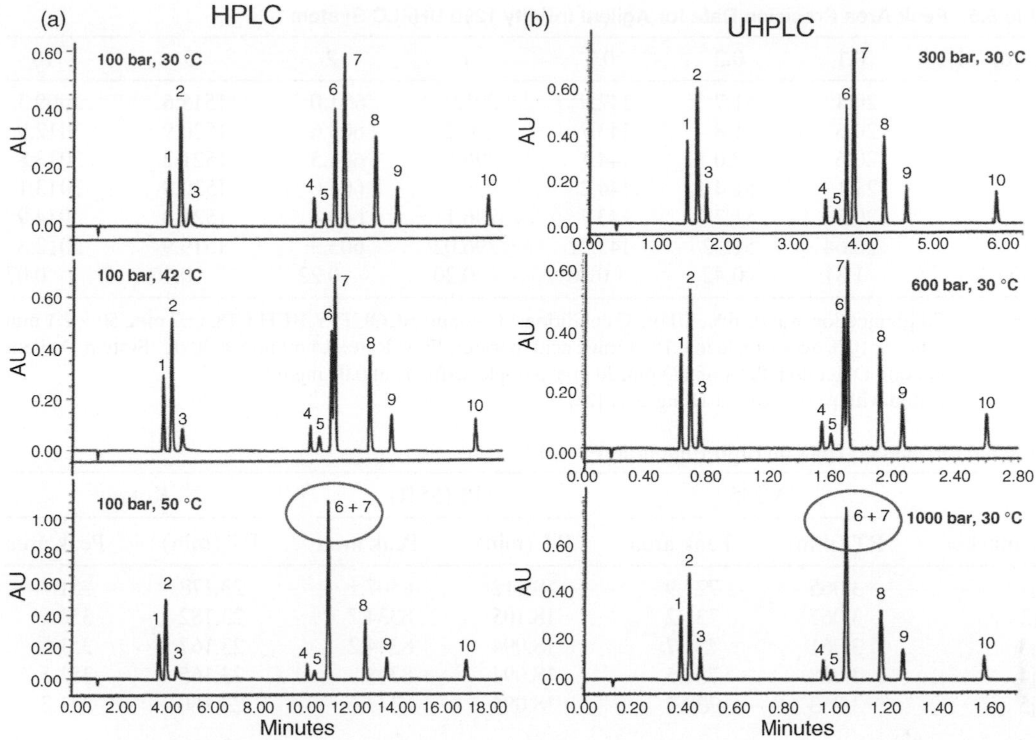


**Figure 5.12.** Comparative chromatogram illustrating the influence of column temperature on peak coelution under HPLC and UHPLC operating conditions. Note that coelution of peaks 6 and 7 occurs under 1000 bar conditions for UHPLC due to the viscous heating effect under higher pressures. Source: Novakova et al. 2011 [28]. Reprinted with permission of Elsevier.

### 5.4.4 Injector Precision

Poor precision performance for autosampler injection was reported in some early UHPLC models under a partial-loop injection mode [5, 29]. With continued improvements of UHPLC autosamplers, the precision performance of peak areas has improved significantly even for small sampling volumes. Recent data indicated superior precision for retention time and peak area even for small injection volumes (i.e. ~0.1–0.2% RSD). Examples of precision performance data are shown in Tables 5.5 and 5.6.

### 5.4.5 UV Detection Noise vs. Mixer Volumes

While the reduction of system dwell volumes is critical for HTS applications, it may be problematic for high-sensitivity UV detection at a low wavelength (<230 nm) if efficient pump blending is not achieved. Figure 5.13 shows a chromatogram of an isocratic analysis of an analgesic drug product extract using a pump-blended mobile phase with UV detection at 227 nm. The assay was performed in 2006 using an early UHPLC system with a binary pump equipped with a "standard" 100-μl external mixer [5]. Note that a periodical baseline perturbation is visible, which was attributable to an inadequate blending of mobile phase A (0.1% acetic acid, which has an absorbance at low UV) and acetonitrile (mobile phase B). Baseline noise perturbation, synchronous with the piston stroke of pump B, can be eliminated by using a premixed mobile phase or by adding mixing volumes to the system as shown in Figure 5.14.

**Table 5.5. Peak Area Precision Data for Agilent Infinity 1290 UHPLC System**

| Inj vol (μl) | 0.1 | 0.2 | 0.5 | 1 | 2 | 5 | 10 |
|---|---|---|---|---|---|---|---|
| 1 | 20.4 | 51.7 | 142.9 | 295.0 | 601.0 | 1515.6 | 3009.3 |
| 2 | 20.6 | 51.8 | 143.7 | 296.2 | 603.6 | 1520.9 | 3012.9 |
| 3 | 20.6 | 52.0 | 144.1 | 296.5 | 604.5 | 1521.2 | 3013.8 |
| 4 | 21.1 | 51.4 | 144.2 | 296.3 | 604.0 | 1520.8 | 3013.1 |
| 5 | 20.5 | 51.7 | 143.7 | 296.1 | 603.4 | 1521.0 | 3014.9 |
| Avg. | 20.64 | 51.72 | 143.72 | 296.02 | 603.3 | 1519.9 | 3012.8 |
| RSD (%) | 1.31 | 0.42 | 0.36 | 0.20 | 0.22 | 0.16 | 0.07 |

Precision of 20 μl injection was 0.06%. UHPLC conditions: Column: ACQUITY BEH C18, (1.7 mm, 50 × 2.1 mm i.d.), Mobile Phase.: 10% acetonitrile in 0.1% formic acid in water, Flow Rate: 0.4 ml/min @ 30 oC, System: Agilent 1290 binary system; Detection: PDA at 273 nm, 20 pt/s; Sample: caffeine at 0.05 mg/ml.
Source: Reprinted with permission of Dong 2017 [25].

**Table 5.6. Precision Data on High-Resolution-UHPLC**

| | M235 | | API (SSR) | | Ketone | |
|---|---|---|---|---|---|---|
| RT marker | RT (min) | Peak area | RT (min) | Peak area | RT (min) | Peak area |
| Inj1 | 3.065 | 727.9 | 18.112 | 8307.1 | 23.178 | 351.7 |
| inj2 | 3.067 | 731.2 | 18.105 | 8334.7 | 23.182 | 351.4 |
| Inj3 | 3.061 | 731.7 | 18.094 | 8344.2 | 23.167 | 351.6 |
| Inj4 | 3.068 | 731.5 | 18.094 | 8348.8 | 23.165 | 351.5 |
| Inj5 | 3.063 | 732.2 | 18.094 | 8347.6 | 23.160 | 351.3 |
| Mean | 3.065 | 730.9 | 18.100 | 8336.5 | 23.170 | 351.5 |
| Stdev. | 0.003 | 0.495 | 0.005 | 6.434 | 0.008 | 0.114 |
| RSD (%) | 0.09 | 0.07 | 0.03 | 0.08 | 0.04 | 0.03 |

Specificity: no interferences from blank, all impurities and deg. separated. Linearity 50–150%" $R = 1.000$, y-intercept 0.4%; Precision of RT and peak area <0.1% RSD; System: Agilent Infinity 1290 binary system.
Source: Reprinted with permission of Dong et al. 2014 [19].

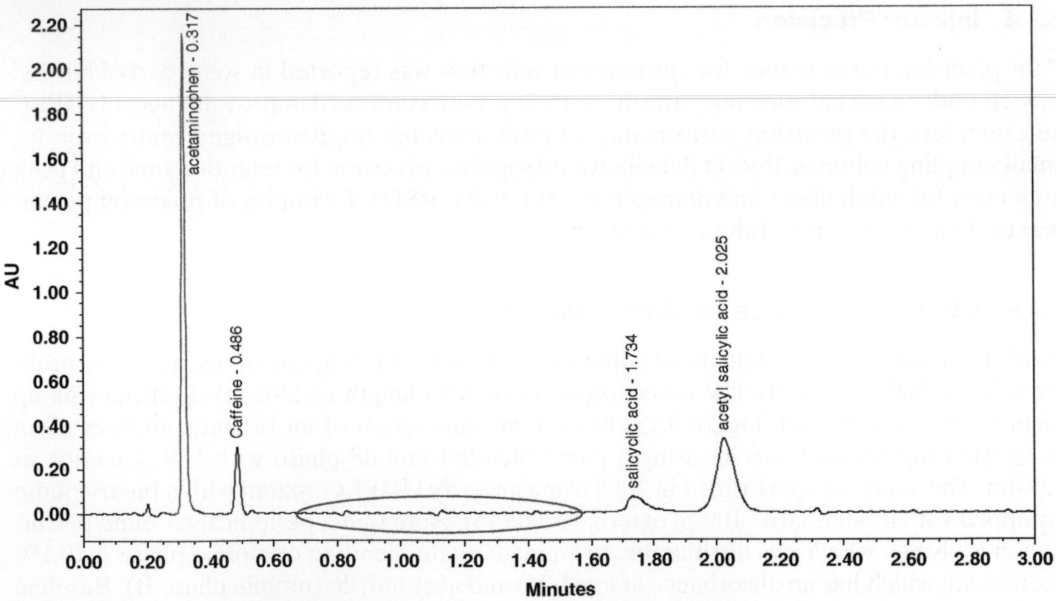


**Figure 5.13.** UHPLC chromatogram of an analgesic tablet extract showing baseline perturbation from an inadequate blending of the two mobile phases using a standard 100-μl mixer. UHPLC condition: Column – XBridge, C18, 50 × 3.0 mm, 3.5 μm, MPA – 1% acetic acid, MPB – ACN, pump blended at 11%B isocratic, $F = 1.5$ ml/min at 30 °C, 2700 psi, detection – UV at 227 nm, 20 pt/s, sample – 1 μl of an Excedrin tablet extract (aged). Source: Reprinted with permission of Dong 2007 [5].

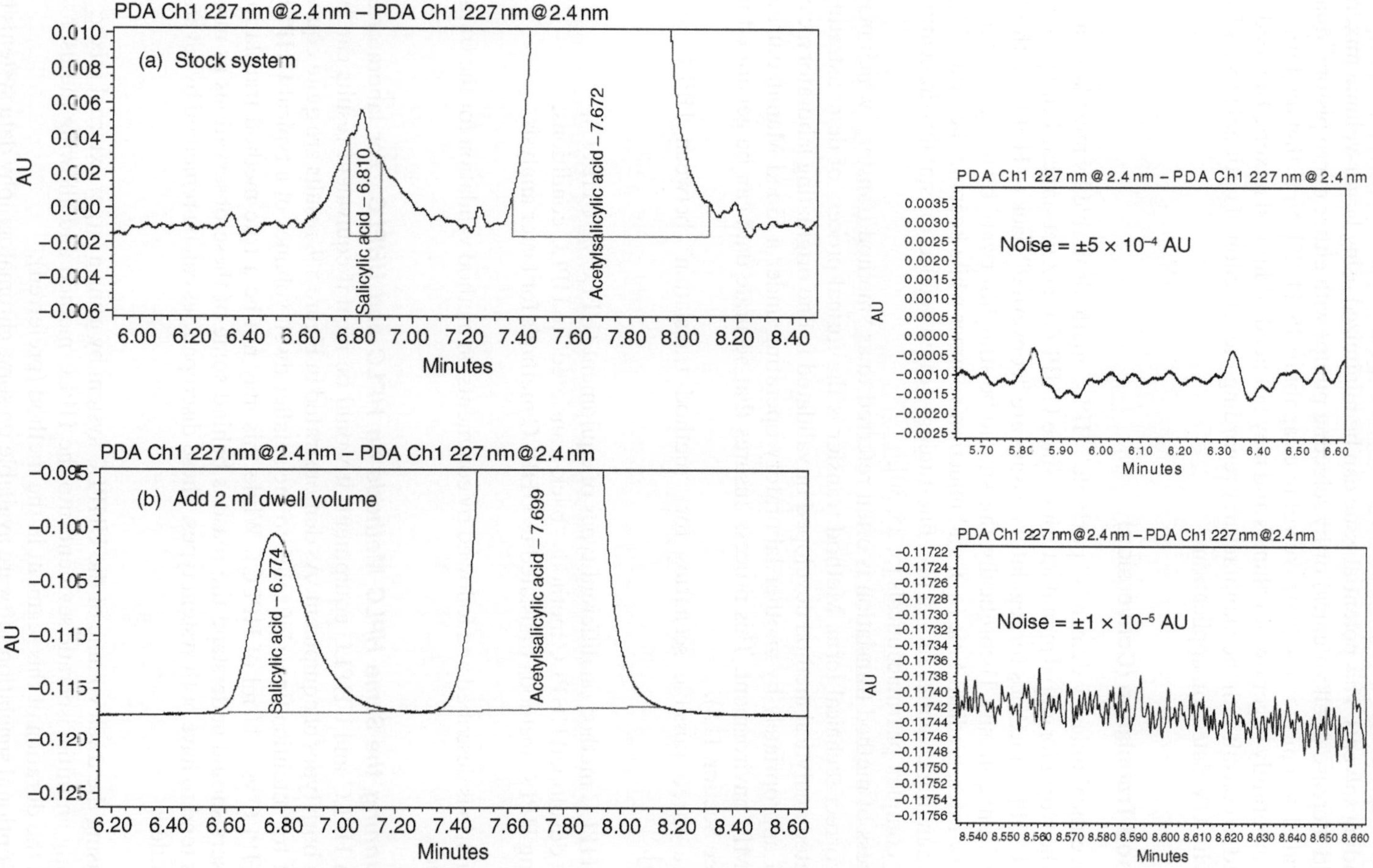


**Figure 5.14.** An expanded view of the UHPLC chromatograms (around the acetylsalicylic acid peak region) of an analgesic tablet extract showing baseline perturbation from an inadequate blending of the two mobile phases using a standard 100 μl mixer shown in Figure 5.14a. Excellent mixing achieved by adding 2 ml of mixing volume before the injector shown in Figure 5.14b. Insets show noise and calculated noise for chromatograms (a) and (b). Source: Reprinted with permission of Dong 2007 [5].

The severity of this UV baseline problem is expected to be a function of the flow rate, the piston stroke volume, the volume and efficiency of the mixer, and the relative absorbance of the two blending solvents. This potential issue can be minimized using larger-volume mixers (which would increase dwell volumes) or by selecting pumps with either micro pistons, dual piston in parallel design, or variable stroke volume capability [5, 11, 25]. Note that quaternary pumps have inherently larger dwell volumes and may not need additional mixers. The reader is encouraged to consult with the manufacturer regarding the selection of optional mixers for high-sensitivity UV detection applications.

### 5.4.6 Method Translation (Conversion)

The pharmaceutical industry demands "portable" HPLC methods usable by most laboratories around the world for global manufacturing. Since UHPLC is not yet standard equipment, many faster UHPLC methods for regulatory assays are "converted" back to HPLC methods using longer 3-μm columns. Theoretically, this should be straightforward by using geometric scaling of flow rates, gradient time ($t_G$), and injection volumes while keeping the same $L/d_p$ ratio when changing columns. In practice, fine-tuning is often needed for stability-indicating methods for complex pharmaceuticals [5, 25, 30].

This process of method translation is often referred to as "method transfer," which may not be the correct technical term. Method transfer is the formal process of demonstrating that a validated analytical method developed or validated in one originating laboratory, can be executed appropriately by another laboratory operating under a Good Manufacturing Practice (GMP) environment. This process ensures that accurate data can be generated in the latter (see Section 11.5).

There are three possible scenarios for "method translation" between HPLC and UHPLC [25].

a. Same HPLC method on different types of equipment (HPLC vs. UHPLC);
b. Newly developed UHPLC methods "back-translated" to HPLC conditions;
c. Existing HPLC methods translated to UHPLC methods for faster analysis.

Each scenario is described and followed by comments on method validation for the translated methods.

**5.4.6.1 *Running the Same HPLC Methods on HPLC and UHPLC*** For laboratories having both HPLC and UHPLC equipment, it would be ideal if equivalent results can be obtained on both types of equipment. As demonstrated in Figure 5.9, results are quite equivalent except for retention time shifts due to the smaller dwell volumes of a typical UHPLC (~0.3 ml UHPLC vs. ~1.0 ml of HPLC). While this may not be a true method translation situation, users should understand the reasons behind some of these observations as most laboratories tend to have both system types. These discrepancies can be remedied by several means if needed:

1. Increasing the dwell volume of the UHPLC system by using a larger external mixer;
2. Building an initial isocratic segment into the HPLC method and allowing the user to adjust the duration of this segment in the method (preferred);
3. Using optional simulation software available on some chromatography data systems to simulate the performance of various equipment by automatic method adjustments or purchasing a dual-path system that converts a UHPLC into an HPLC with a valving switch [7].

***5.4.6.2 Back Conversion of UHPLC Methods to HPLC Method Conditions*** Many laboratories use UHPLC for rapid method development including column/mobile phase screening and method optimization [22] and "back transfer" the optimized UHPLC methods to HPLC conditions using longer columns following the principles of geometric scaling. The converted HPLC method is then validated and serves as the primary regulatory method to support global manufacturing operation. Case studies for these method conversion processes are available elsewhere [5, 22].

***5.4.6.3 Conversion of Existing HPLC Methods to Faster UHPLC Methods*** The primary driver for purchasing UHPLC equipment is the ability to perform faster analysis with "good" resolution. This benefit can be accomplished by geometric scaling with the following ground rules: [24]

1. Column length ($L$) is scaled to particle size keeping $L/d_p$ ratios the same. The bonded phase chemistry must be identical.
   For example, Acquity (1.7 µm, 50 × 2.1 mm) – XBridge (3.5 µm, 100 × 3.0 mm)
2. Flow rate ($F$) is scaled to cross-sectional area of the column.
   For example, 2.1 mm i.d. (0.5 ml/min), 3.0 mm i.d. (1 ml/min), 4.6 mm i.d. (2 ml/min)
3. Gradient time ($t_G$) is scaled to column length.
4. Sample injection volume is scaled to column void volume.
   For example, 2 µl (50 × 2.1 mm) – 10 µl (3.5 µm, 100 × 3.0 mm)

One requirement is that the new UHPLC columns used must contain identical bonded phase materials to eliminate any selectivity differences. Also, the mobile phases used should be identical (the type of buffer, strength, pH, organic modifier). Figure 5.2 provides an example of this method conversion noting that a higher optimum flow rate is also used for the UHPLC column. Programs for geometric scaling are available from various vendors' websites (Waters, Agilent, and Thermo/Dionex) and other sources [24]. This method conversion process may be straightforward for simple assays but can be challenging for complex samples or QC methods for commercial products.

***5.4.6.4 Method Validation Requirements After Method Translation*** For validated HPLC methods, there were numerous discussions on what constitutes a method adjustment versus a method change [31, 32] (see Figure 5.15 for USP guidelines) and to what point a method revalidation is needed. [24] The current consensus including a direct response from an FDA reviewer indicated that a partial method validation (including specificity, intermediate precision, linearity, and robustness) is needed plus data on method equivalency between the two methods.

## 5.5 HOW TO IMPLEMENT UHPLC AND PRACTICAL ASPECTS

### 5.5.1 How to Transition from HPLC to UHPLC

For those considering the purchase of UHPLC equipment, the first consideration is the intended use: "Is the UHPLC system used for method development, routine testing, research, QC, HTS, or LC/MS?" Answers to the application question would likely determine the choice of systems (binary or quat pumps, detectors, and vendors with the best reputation in the intended application).

The next consideration is the existing CDS as a UHPLC system manufactured by the same CDS vendor should provide better interfacing and system control (see Tables 4.1, 5.1, and 5.2). For UHPLC serving as the front-ends of LC/MS or LC/MS/MS, the MS data

| Summary of *USP* adjustment guidelines | | | |
|---|---|---|---|
| Variable | Allowed range | OK for | |
| | | Isocratic | Gradient |
| pH | ±0.2 | Yes | Yes |
| Buffer concentration | ±10% | Yes | Yes |
| Mobile phase | The lesser of ±30% relative or ±10% absolute for minor components | Yes | NR* |
| UV wavelength | 0, but error ±3 nm OK | Yes | Yes |
| Column length and particle size | $L/d_p$ = −25% to +50% or $N$ = −25% to +50% | Yes | No |
| Column diameter | OK if linear velocity constant | Yes | No |
| Flow rate | OK if linear velocity constant, plus an additional ±50%; exceptions | Yes | No |
| Injection volume | OK if performance OK | Yes | Yes |
| Column temperature | ±10 °C | Yes | Yes |

*NR = not recommended, but not explicitly prohibited.

**Figure 5.15.** Summary of USP method adjustment guidelines. Source: Reprinted with permission of Dolan 2017 [32].

system becomes the pivotal consideration. Secondary considerations are the pressure limits, which vary from 15 000 to 22 000 psi for UHPLC systems or 9000–15 000 psi for intermediate or dual-flow-path systems. [7] The upper flow rate limit of the new UHPLC can be essential for performing existing HPLC methods. Other considerations are system performance, cost, reliability, and reputation of the vendor for service and technical support. For systems to be used in ion-exchange chromatography (IEX) or hydrophobic interaction chromatography (HIC) for biomolecule characterization, system biocompatibility (corrosion resistance to salts) can be an overriding requirement. Note that operating training including a review of UHPLC fundamentals would be highly desirable for all operators.

### 5.5.2 End-Fittings

Reliable fittings for UHPLC column connections and leak-free operation were issues during the early days of UHPLC [5, 24]. Gold-plated nuts (to prevent seizing of the threads) and double metallic ferrules were used in the first-generation UHPLC fittings [24]. They have fixed insertion depths and were not universally compatible with columns from different manufacturers. Today, many choices are available including reusable fittings for a finger-tight or wrench-tight operation, which can be resealed many times at 20 000 psi. Examples are found in Figure 5.16. Note that UHPLC fittings remain expensive though the connections from column to detector can use the lower-cost HPLC PEEK fittings [24].

### 5.5.3 A Summary of UHPLC System Performance Tradeoffs

The design of low-dispersion UHPLCs involves many compromises of cost, performance, and reliability tradeoffs [24]. Some of these are summarized here.

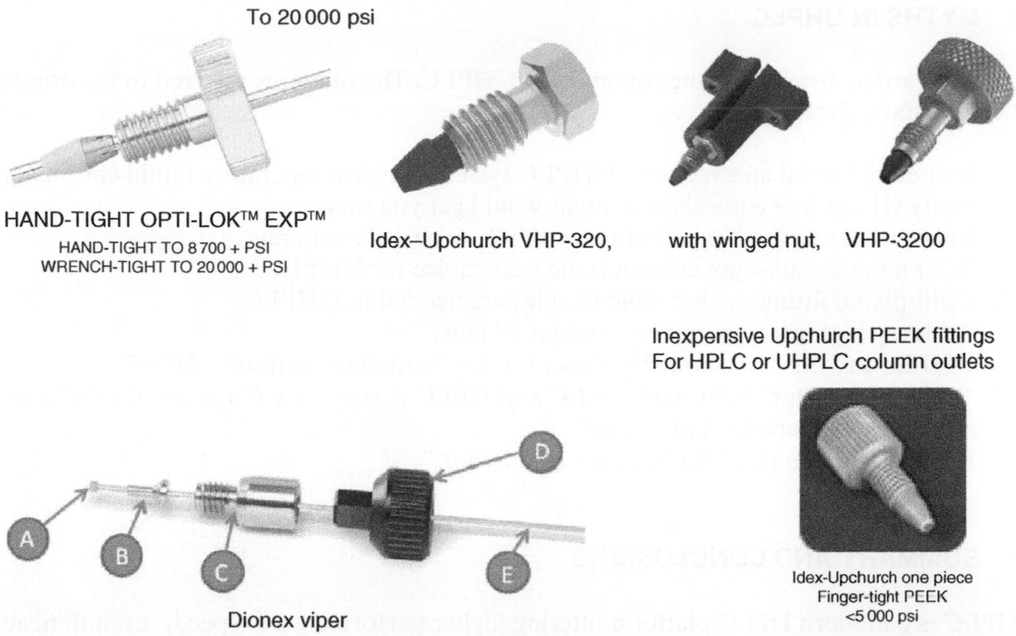


**Figure 5.16.** Examples of UHPLC column end-fittings with upper pressure limits up to 20 000 psi and a low-cost PEEK HPLC end-fitting for connections after the column outlet.

*System dispersion*: Low-dispersion systems are desirable because they allow the use of smaller columns without efficiency loss particularly under isocratic conditions. It is useful to realize that system dispersion before the column (injector, loop, switching valve) is not relevant for gradient analysis since sample bands are refocused at the top of the column. Postcolumn dispersion (tubing from column to detector, detector flow cell or mass spectrometer inlet) is critical since it broadens separated bands.

Low-dispersion systems using very small i.d. connection tubing (<0.003″ i.d.) can generate substantial backpressure at flow rates >1 ml/min without a column and therefore reducing the usable pressure range.

*System precision*: Peak area precision is mostly dependent on autosampler design and the volume of the sampling syringe. In general, an integrated loop design yields better precision and carryover performance. However, an *X-Y-Z* type of autosamplers is preferred for HTS application using microtiter plates. The precision of *X-Y-Z* autosamplers can be improved by using smaller sampling syringes.

*UV Detection Sensitivity*: As mentioned earlier, high-sensitivity UV applications at low UV wavelengths require adequate blending of mobile phases with different absorptivity to eliminate baseline perturbation. A larger mixer is often needed for binary pumps except for those equipped with micro pistons or variable piston stroke capability.

*Backward compatibility with existing HPLC methods*: This backward compatibility is vital for laboratories intending to use the same UHPLC systems to run existing HPLC methods, which may employ longer columns (>150 mm), higher flow rates (>2 ml/min), or higher injection volume (>20 μl). Here, some UHPLC systems may be more suited than others without the need for purchasing additional optional equipment.

## 5.6 MYTHS IN UHPLC

An earlier article listed a number of myths in UHPLC. The reader is referred to the original article for more details [24].

1. You do not need an expensive UHPLC system – high-temperature liquid chromatography (HTLC) or core–shell columns would get you there.
2. Viscous heating is a "huge" issue for sub-2 μm particle columns.
3. A 2.1 mm i.d., sub-2 μm column is the best choice for UHPLC.
4. Gold-plated fittings with double ferrules are needed in UHPLC.
5. A binary high-pressure mixing pump is a "must."
6. UHPLC provides substantially higher sensitivity than conventional HPLC.
7. "Method transfer" between UHPLC and HPLC is very easy ("a piece of cake"), and method revalidation is not needed.
8. Lower-dispersion UHPLC systems are "better."

## 5.7 SUMMARY AND CONCLUSIONS

UHPLC is a modern HPLC platform offering higher performance in speed, resolution, and precision. UHPLC is useful in method development and suitable for regulatory analysis. System compatibility with existing CDS and HPLC methods and the need for "method translation" are essential considerations for first-time buyers. Many design tradeoffs are made by manufacturers including system dispersion/flexibility/backward compatibility, dwell volume/blending efficacies, and higher pressure limit/cost or reliability. Method conversion between UHPLC and HPLC conditions is accomplished following the principles of geometric scaling, and partial method validation is needed after the conversion of validated methods in regulated testing. Purchasing decisions for UHPLCs are based mostly on the intended applications, compatibility with existing CDS or the need to perform legacy HPLC methods, and the reputation of the manufacturer for technical and service support.

## 5.8 QUIZZES

**1.** UHPLC is generally NOT capable of delivering
(a) faster analysis
(b) much higher UV sensitivity
(c) very high peak capacities
(d) faster method development

**2.** The peak volume of a 50 × 2.1 mm column packed with 1.7 μm particles at $k = 2$ is about
(a) 4 μl
(b) 13 μl
(c) 9 μl
(d) 200 μl

**3.** The typical path length of a UHPLC PDA detector flow cell is
(a) 25 mm
(b) 8 μl
(c) 500 nl
(d) 10 mm

4. Geometric scaling used in method conversion includes all these factors EXCEPT
   (a) type of organic solvent in MPB
   (b) injection volume
   (c) gradient time and column length
   (d) column diameter and particle size

5. Very high resolution can be delivered by UHPLC by using
   (a) very small $d_p$ such as 1.5 μm particles
   (b) small $d_p$, short column at optimum flow
   (c) small $d_p$, longer column, and $t_G$
   (d) very small i.d. columns

6. UHPLC is a "greener" HPLC platform because of
   (a) faster analysis time, lower flow rates, and smaller sample size
   (b) greater UV sensitivity using longer-path-length flow cells
   (c) lower system dispersion yielding sharper peaks
   (d) does not use ACN

7. UHPLC facilitates method development on all these processes EXCEPT
   (a) quicker column and mobile phase screening
   (b) faster method optimization
   (c) faster method prequalification
   (d) quicker method transfer of validation methods

8. Which one fact is NOT true for UHPLC vs. conventional HPLC?
   (a) It cost more than conventional HPLC
   (b) Injection volumes are typically lower
   (c) Path lengths of UV flow cell are lower
   (d) UV flow cell volumes are lower

9. To reduced UV baseline noise in UHPLC using binary pumps, one should first try this:
   (a) Change to a larger UV flow cell
   (b) Reduce the pump piston size
   (c) Avoid using low UV detection
   (d) Select a large mixer

10. Some UHPLC systems may have compatibility issues with existing HPLC methods due to this: (select best answer)
    (a) Cannot fit column longer than 150 mm
    (b) All factors mentioned here
    (c) Cannot have flow rates higher than 2 ml/min
    (d) Cannot inject >20 μl sample volumes

11. Method conversions in UHPLC are often performed, but NOT for this reason:
    (a) To support global manufacturing
    (b) To increase lab productivity
    (c) To support faster method development by UHPLC
    (d) As required by the US FDA

12. The top reason for adopting UHPLC in most laboratories is
    (a) to increase resolution
    (b) to buy more expensive equipment
    (c) to reduce analysis times
    (d) to use higher pressures

**13.** Which one is NOT true in UHPLC/HPLC method conversions?
(a) Uses geometric scaling
(b) Keeps $L/d_p$ the same
(c) Uses the same type of bonded phase
(d) Scales injection volumes to particle *size*

**14.** The most important reason for selecting UHPLC equipment from a vendor is
(a) compatibility with current CDS network
(b) pressure ratings and serviceability
(c) equipment specifications
(d) compliance with regulations

**15.** The issue of UV detector noise due to poor pump mixing is NOT related to
(a) piston size
(b) Availability of variable pump strokes
(c) Viscous heating issues
(d) Mixer volume

### 5.8.1 Bonus Quiz

1. In your own words, describe the tradeoff in UHPLC design between system dispersion/flexibility and UV sensitivity/dwell volume.
2. Describe the three important benefits of UHPLC over those of conventional HPLC.

## 5.9 REFERENCES


1. Dong, M.W. (2017). *LCGC North Am.* 35 (6): 374.
2. MacNair, J.E., Lewis, K.C., and Jorgenson, J.W. (1997). *Anal. Chem.* 69: 983.
3. MacNair, J.E., Patel, D., and Jorgenson, J.W. (1999). *Anal. Chem.* 71: 700.
4. Neue, U.D., Kele, M., Bunner, B. et al. (2009). *Advances in Chromatography*, vol. 48, 99–143. Boca Raton, FL: CRC Press.
5. Dong, M.W. (2007). *LCGC North Am.* 25: 89.
6. Guillarme, D. and Dong, M.W. (eds.) (2014). UHPLC: where we are ten years after its commercial introduction. *TrAC, Trends Anal. Chem.* 63: 1–188. (Special issue).
7. Dong, M.W. (2018). *LCGC North Amer.* 36 (4): 252; 35(4), 246 (2017); 34(4), 262 (2016); 33(4), 254 (2015); 32(4), 270 (2014); 31(4), 313 (2013).
8. Guillarme, D. and Dong, M.W. (2013). *Am. Pharm. Rev.* 16 (4): 36.
9. Guillarme, D. and Veuthey, J.-L. (eds.) (2012). *UHPLC in life sciences*. Cambridge, UK: Royal Society of Chemistry.
10. Xu, Q.A. (2013). *UHPLC and Its Applications*. Hoboken, NJ: Wiley.
11. De Vos, J., Broeckhoven, K., and Eeltink, S. (2016). *Anal. Chem.* 88: 262.
12. Dong, M.W. and Guillarme, D. (2017). *LCGC North Am.* 35 (8): 486.
13. Arnaud, C.H. (2016). *Chem. & Eng. News* 94 (24): 29.
14. Dong, M.W. (2006). *Modern HPLC for Practicing Scientists*. Hoboken, NJ: Wiley, Chapters 4 and 8.
15. Martin, A.J.P. and Synge, R.L.M. (1941). *Biochem. J.* 35: 1358.
16. Dong, M.W. (2014). *LCGC North Am.* 32 (8): 552.
17. Fekete, S., Kohler, I., Rudaz, S., and Guillarme, D. (2013). *J. Pharm. Biomed. Anal.* 87: 105.
18. Guillarme, D., Nguyen, D., Rudaz, S., and Veuthey, J.L. (2008). *Eur. J. Pharm. Biopharm.* 68 (2): 430.

19. Dong, M., Guillarme, D., Fekete, S. et al. (2014). *LCGC North Am.* 32 (11): 868–876.
20. Zhou, L. (2005). *Handbook of Pharmaceutical Analysis by HPLC* (ed. S. Ahuja and M.W. Dong). Amsterdam, the Netherlands: Elsevier, Chapter 17.
21. Guillarme, D., Grata, E., Glauser, G. et al. (2009). *J. Chromatogr. A* 1216: 3232.
22. Dong, M.W. and Zhang, K. (2014). *TrAC, Trends Anal. Chem.* 63: 21.
23. Fountain, K.J., Hewitson, H.B., Iraneta, P.C., and Morrison, D. (2010). Waters Corporation, 720003720EN, Milford, Massachusetts.
24. Dong, M.W. (2013). *LCGC North Am.* 31 (10): 868.
25. Dong, M.W. (2017). *LCGC North Am.* 35 (11): 818.
26. Xiang, Y., Maynes, D., and Lee, M.L. (2003). *J. Chromatogr. A* 991: 189.
27. Gritti, F. and Guichon, G. (2008). *Anal. Chem.* 80: 5009.
28. Novakova, L., Veuthey, J.L., and Guillarme, D. (2011). *J. Chromatogr. A* 1218 (44): 7971.
29. de Villiers, A., Lestremau, F., Szucs, R. et al. (2006). *J. Chromatogr. A* 1127 (1–2): 60.
30. Debrus, B., Rozet, E., Hubert, P. et al. (2012). *UHPLC in Life Sciences* (ed. D. Guillarme, J.-L. Veuthey and R.M. Smith), 67–98. Cambridge, UK: Royal Society of Chemistry Publishing.
31. General Chapter <621> "Chromatography" (2017). *United States Pharmacopeia 40 National Formulary 35* (USP 40-NF 35, United States Pharmacopeial Convention), 508–520. Maryland: Rockville.
32. Dolan, J. (2017). *LCGC North. Amer.* 35 (6): 368.

# 6

# LC/MS: FUNDAMENTALS, PERSPECTIVES, AND APPLICATIONS

**CHRISTINE GU**

## 6.1 INTRODUCTION

### 6.1.1 Scope

In recent years, mass spectrometry (MS) has become a prevalent and indispensable analytical methodology used in research and development [1–4]. When coupled with liquid chromatography (LC), it adds an orthogonal detection dimension in sample analysis, providing outstanding performance in sensitivity, selectivity, speed, and definitive mass spectral information.

In this chapter on LC/MS, we will provide an overview of the fundamentals and applications in bioscience research, pharmaceutical development, and multiresidue analysis, supplementing the information described in other chapters (Chapter 4 on instruments, Chapter 7 on pharmaceutical analysis, Chapter 12 on biopharmaceuticals, and Chapter 13 on other HPLC applications).

### 6.1.2 LC/MS Technology and Instrumentation

High-performance liquid chromatography (HPLC) is a physical separation technique that relies on analytes' differences in hydrophobicity or other characteristics between a flowing mobile phase and a stationary phase. MS separates ions by their mass-to-charge ratio (*m/z*) in a high-vacuum [4]. A standard mass spectrometer consists of three major components (Figure 6.1) – an ion source, mass analyzer, and detection system.

Compared to common HPLC detectors, such as ultraviolet/visible (UV/Vis), charged aerosol detector (CAD), or evaporative light scattering detector (ELSD), MS often provides

*HPLC and UHPLC for Practicing Scientists*, Second Edition. Michael W. Dong.

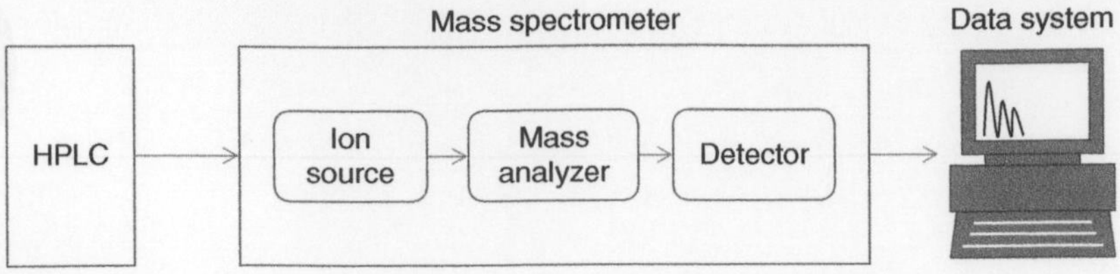


**Figure 6.1.** The basic components of an LC/MS system.

higher sensitivity, specificity, and definitive information on mass and structures of the analytes. In contrast to GC/MS, which is limited to the analysis of volatile and thermally stable analytes, LC/MS is amenable to a wide range of involatile compounds including those with high molecular weights and polarities, ions, proteins, drugs, natural products, and biomolecules.

### 6.1.3 Basic Terminologies and Concepts for MS

A few of the basic concepts of MS are described here briefly in this chapter, and interested readers are referred to textbooks for a thorough discussion of fundamentals and concepts in MS [2–4].

*Resolving power (RP)* is the ability of an MS instrument to perform the separation of adjacent ions. RP is similar to mass resolution, but the term is less widely used. Simpler MS instruments have lower RP than more expensive MS with higher RP with narrower mass peaks.

*Mass spectral resolution (R)* is defined by the mass of a peak divided by its full width at half maximum (FWHM) in Dalton (Da) or atomic mass unit (amu or u) (Eq. (6.1) and Figure 6.2). The concept of "unit resolution" represents the resolving power necessary to separate two adjacent ions one mass unit apart. This is the typical resolution of a lower-cost MS instrument.

$$\text{Mass Resolution} = \frac{\text{Peak mass (Da)}}{\text{Full width at half maximum (Da)}} \tag{6.1}$$

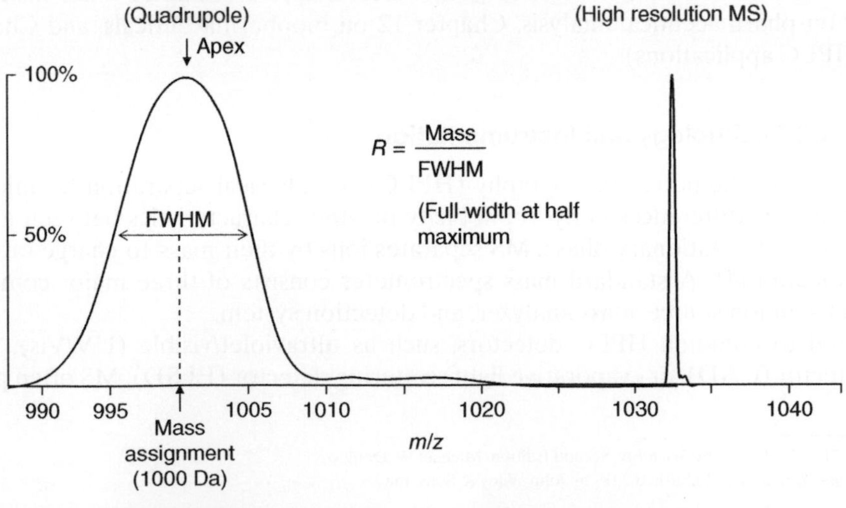


**Figure 6.2.** Comparative resolution between a unit-resolution quadrupole and high-resolution MS.

*Nominal mass* is the sum of nominal (integer) masses of the constituent elements of a compound.

*Average mass*, also known as molecular weight or formula weight, is the weighted average of the molecular masses of the analyte, where atomic masses of each atom are based on the natural abundance of all isotopes of that element. Note that this value is not used for the interpretation of mass spectra. It is more commonly used as a means of calculating the intact mass of larger unisotopically resolved proteins.

*Monoisotopic mass*, also called an exact mass, is an exact molecular mass of the structure, where atomic masses of each atom are based on the most abundant isotope for that element. Note that all structure elucidation and mass spectral interpretation are based on the exact mass of the analyte. It is more commonly used to calculate small molecular weight analytes, <2000 Da, that are isotopically resolved.

*Mass error and mass accuracy* defined in Eqs. (6.2) and (6.3) are used in high-resolution mass spectrometry (HRMS). They indicate the accuracy of the mass information provided by MS in reference to the true exact mass. To calculate the molecular formula effectively, a mass accuracy of less than 5 ppm is typically required of the instrument.

$$\text{Mass error} = \text{theorical mass} - \text{measured mass} \tag{6.2}$$

$$\text{Mass accuracy in parts per million (ppm)} = \frac{\text{mass error}}{\text{theorical mass}} \times 10\text{e}^{6} \tag{6.3}$$

*Mass defect*, defined as the difference between the exact mass and the integer/nominal mass. In the case of $^{16}O$, for example, the isotopic mass is 15.994 915 u, being 5.085 milli-mass units or mmu deficient as compared to the nominal value. With the growth of HRMS, the application of this concept is increasing. Filtering based on known mass defects allows the selective profiling of specific compound classes in complex samples.

*The nitrogen rule* is useful for determining the nitrogen content of an unknown compound in MS. If a compound has an odd number of nitrogen atoms, its molecular weight (MW) will be odd nominally, and its protonated molecule will have an even nominal mass. If a compound has an even number of nitrogen atoms, its MW will be even, and its protonated molecular ion will be odd in nominal mass.

*Isotope abundances of elements* can be classified into three general categories (as shown in Table 6.1), including $A$, $A+1$, and $A+2$ elements. The common elements and their isotopes percentages are summarized in Table 6.2. Multiple $A+2$ elements produce increasingly complex clusters, but with the characteristic spectral fingerprint. If the $(A+2)/(A)$ peak ratio is less than 3%, the analyte peak cannot contain any $A+2$ elements.

*Adduct and cluster ions* are ions formed by the adduction of other ions or molecules to the protonated or deprotonated molecule of a certain analyte in LC/MS. The most commonly observed adduct ions under positive and negative electrospray ionization (ESI) interfaces are listed in Table 6.3.

**Table 6.1. General Categories of Element Types Based on their Isotopic Abundances**

| Category | Description |
|---|---|
| "$A$" elements | Those elements with only one natural isotope |
| "$A+1$" elements | Those elements that have two isotopes, the second of which is one mass unit heavier than the most abundant isotope |
| "$A+2$" elements | Those elements that have an isotope that is two mass units heavier than the most abundant isotope |

**Table 6.2. Common Elements and their Isotopic Abundance Percentages**

| "A" elements | | | "A + 1" elements | | | | | "A + 2" elements | | | | | | |
|---|---|---|---|---|---|---|---|---|---|---|---|---|---|---|
| | *A* | % | | *A* | % | *A* + 1 | % | | *A* | % | *A* + 1 | % | *A* + 2 | % |
| **H** | 1 | 100 | **C** | 12 | 100 | 13 | 1.1 | **O** | 16 | 100 | 17 | 0.04 | 18 | 0.2 |
| **F** | 19 | 100 | **N** | 14 | 100 | 15 | 0.37 | **Si** | 28 | 100 | 29 | 5.1 | 30 | 3.4 |
| **P** | 31 | 100 | | | | | | **S** | 32 | 100 | 33 | 0.79 | 34 | 4.4 |
| **I** | 127 | 100 | | | | | | **Cl** | 35 | 100 | 36 | 0 | 37 | 32 |
| | | | | | | | | **Br** | 79 | 100 | 80 | 0 | 81 | 97.3 |

**Table 6.3. Common Adduct Ions Observed in Electrospray Ionization**

| Positive ion electrospray | | Negative ion electrospray | |
|---|---|---|---|
| $[M+H]^+$ | Protonated molecule | $[M-H]^-$ | Deprotonated molecule |
| $[M+Na]^+/[M+K]^+$ | Alkali metal adduct | $[M-H+S]^-$ | Solvent adduct |
| $[M+NH4]^+$ | Ammonium adduct | $[M+X]^-$ | Anionic adduct from spray solvent |
| $[M+H+S]^+$ | Solvent adduct | | |
| $[2M+H]^+$ | Protonated dimer formed at higher concentrations | | |

### 6.1.4 Interfacing HPLC and MS

The primary difficulty of combining LC and MS has traditionally been the interface. LC uses high pressure to separate analytes, which elute as distinct bands in the eluent under atmospheric pressure. Traditional detectors (UV, refractive Index, and fluorescence) involve flow-through cells, which accommodate a wide range of mobile phase composition and volatility and are virtually insensitive to the buffers and salts content of the eluent. Mass spectrometry detection requires that all components, including the mobile phase and additives therein, be transferred into the gaseous phase. This conversion represents the most significant challenge, as ionic equilibria that exist in solution are not preserved, qualitatively and quantitatively, upon transition of the ions into the gas phase. With the development of atmospheric pressure ionization sources, MS can be directly coupled with an HPLC when using MS-compatible mobile phases with volatile additives, which do not adversely affect the ionization in solution of the analytes of interest (Section 2.3.3).

There has been a major focus on improving the MS scan speed when the liquid chromatography instrument is transitioned from HPLC to ultra-high-pressure liquid chromatography (UHPLC) as these result in faster-eluting peaks. UHPLC/MS often requires high scanning speed to avoid the loss of resolution and chromatographic peak shape. Advances in modern mass analyzers allow a sufficient number of scans (≥10) across the chromatographic peak in full scan modes without compromising resolving power and sensitivity.

## 6.2 LC/MS INSTRUMENTATION

### 6.2.1 Ion Sources

Analytes require ionization in order to be manipulated by the mass spectrometer, and this is achieved through an ion source. Analytes that are not ionized will be expelled through

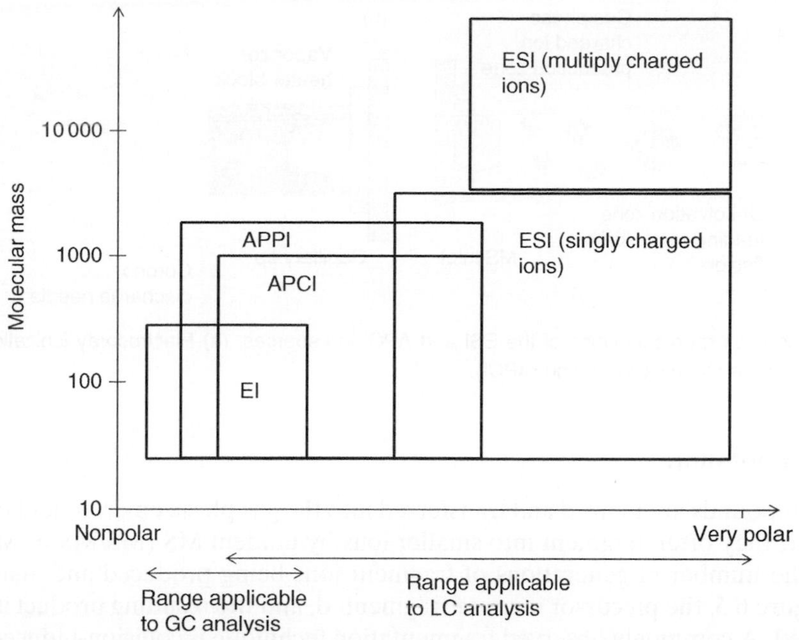


**Figure 6.3.** Common ionization techniques and application areas in MS. EI stands for electron impact and is primarily used in GC/MS.

the exhaust. The most commonly observed forms of ionization modes in LC/MS are ESI, atmospheric pressure chemical ionization (APCI), and atmospheric pressure photoionization (APPI). As illustrated in Figure 6.3, ESI is considered as a "soft" ionization technique and is better suited to higher-molecular-weight and polar compounds (e.g. Biologics), while APCI is more compatible with low- to medium-polarity compounds with a limited upper mass range (<*m*/*z* of 1000). Recent studies showed that APPI expands the range of compounds that can be ionized and has become the third option for nonpolar compounds, such as steroids and polycyclic aromatic hydrocarbons (PAHs) [5]. Additionally, the combination APCI/APPI may be used in a complementary fashion, with the proper source setup. Nanoelectrospray ionization source [6], connected to nano-LC (<500 nl/min), is a minimized-flow ESI source, which may have a different mechanism of ion formation at low flow rates. In contrast to conventional ESI, nanospray technique provides higher ionization efficiency and can greatly increase sensitivity in sample-limited applications such as proteomics and bioscience research.

ESI is the dominant ionization mode used in 80–90% of LC/MS applications. The ion formation in ESI (Figure 6.4a) can be divided into three main steps [7]. (i) The HPLC effluent is pumped through a nebulizing needle and creating an electrically charged spray at the capillary tip under a high electrical potential; (ii) A counter flow of heated nitrogen drying gas shrinks the charged droplets and carries away the uncharged solvent ions to waste; (iii) As the droplets shrink, it leads to coulombic fission and the analyte ions are ultimately desorbed into the gas phase. These gas-phase ions pass through the ion transfer interface into the low-pressure region of MS.

Complementary to ESI, APCI (Figure 6.4b) uses the process of evaporation followed by ionization. The heat rapidly evaporates the spray droplets, and then the gas-phase analyte and solvent molecules are ionized by the discharge from a corona needle kept at a high voltage.

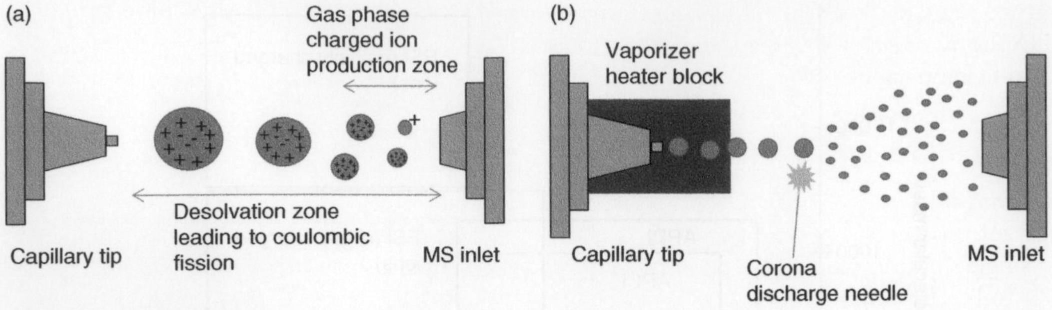


**Figure 6.4.** Schematic representations of the ESI and APCI ion sources. (a) Electrospray ionization (ESI); (b) Atmospheric pressure chemical ionization (APCI).

### 6.2.2 Fragmentation

After the compounds are ionized and transferred into the gas phase environment of the mass spectrometer, they often fragment into smaller ions by tandem MS (MS/MS or MS$^n$, where $n$ refers to the number of generations of fragment ions being produced and analyzed). As shown in Figure 6.5, the precursor ions are fragmented, and the resulting product ions can be mass analyzed. A commonly observed fragmentation technique is collision-induced dissociation (CID), which is sometimes termed a collision-activated dissociation (CAD). The ions are exposed to ion/molecule collisions with neutral atoms, for example, inert gases argon, helium, or nitrogen. One limitation of using CID is the low mass cutoff using an ion trap (also called the "one-third rule"), which doesn't allow for trapping of fragment masses below 28% of the precursor mass. An alternative fragmentation technique is called high-energy collisional dissociation (HCD). Due to higher energy used in dissociation than in CID, HCD enables a wider range of fragmentation pathways and provides more structural information. Electron transfer dissociation (ETD) has emerged as a newer technique to employ on multiply

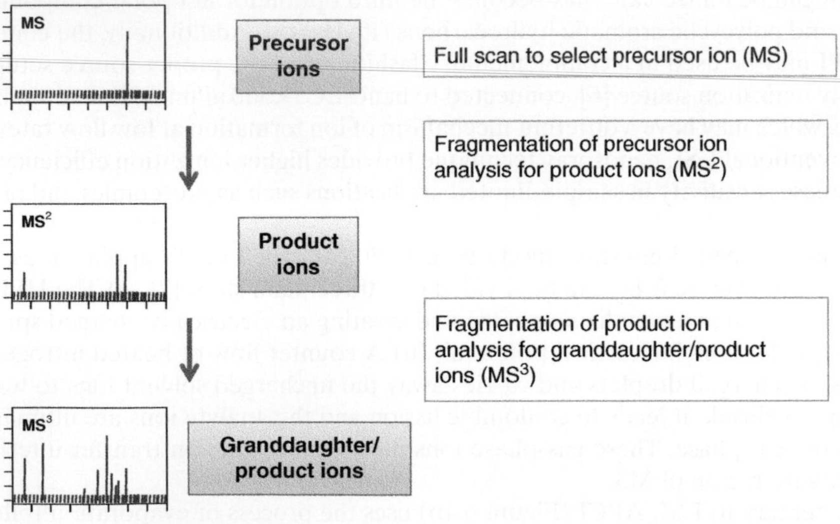


**Figure 6.5.** Schematics showing a precursor ion, product ions in MS$^2$, and granddaughter/product ions on MS$^3$. These ions are commonly found using sequential fragmentation with an ion trap MS.

charged analytes, such as biomolecules including peptides, proteins, and oligonucleotides. ETD is often used within biopharma characterization as it is complementary to CID as it doesn't cleave labile posttranslational modifications (PTMs) and cleaves disulfide bonds.

### 6.2.3 Mass Analyzers

Following their passage through the source interface, once ions are in the gas phase, the mass analyzer separates the ions according to their *m/z* values. The most common mass analyzer is a single quadrupole (SQ) type (Figure 6.6a), which contains two sets of identical stainless steel rods. The combined direct current (DC) and radio frequency (RF) potentials on these rods can be set to pass only a range of *m/z's* ions. Any ion outside of the selected range will have an unstable trajectory and will collide with the rods. The single-quadrupole instrument is widely adopted in different areas because of its compactness, ease of operation/maintenance, high sensitivity, fast scan rate, low cost, and moderate vacuum requirements (~$10^{-5}$ torr).

When three quadrupoles are connected together, it is called a triple-quadrupole (TQ) instrument, also known as a tandem-in-space instrument or QqQ or MS/MS (Figure 6.6b). The first and third quadrupoles act as mass filters and the second set is RF-only quadrupoles in which the fragmentation of ions occurs through interaction with a collision gas (nitrogen or argon). The most common method used in the quantitative analysis is called selected reaction monitoring (SRM) or multiple reactions monitoring (MRM). It is performed by a TQ system, where a precursor ion selected in Q1, is fragmented in Q2, and its unique product ion is monitored in Q3. LC/MS/MS analysis using TQ is currently the standard platform for trace quantitation in bioanalytical or multiresidue applications. Other scan modes used by quadrupole MS and their applications are summarized in Table 6.4.

An ion trap (IT), also known as a tandem-in-time instrument, uses RF and DC voltages as well but traps ions in a linear or three-dimensional quadrupole (Figure 6.6c). The ions can be dynamically stored for isolation or fragmentation and are then driven from the

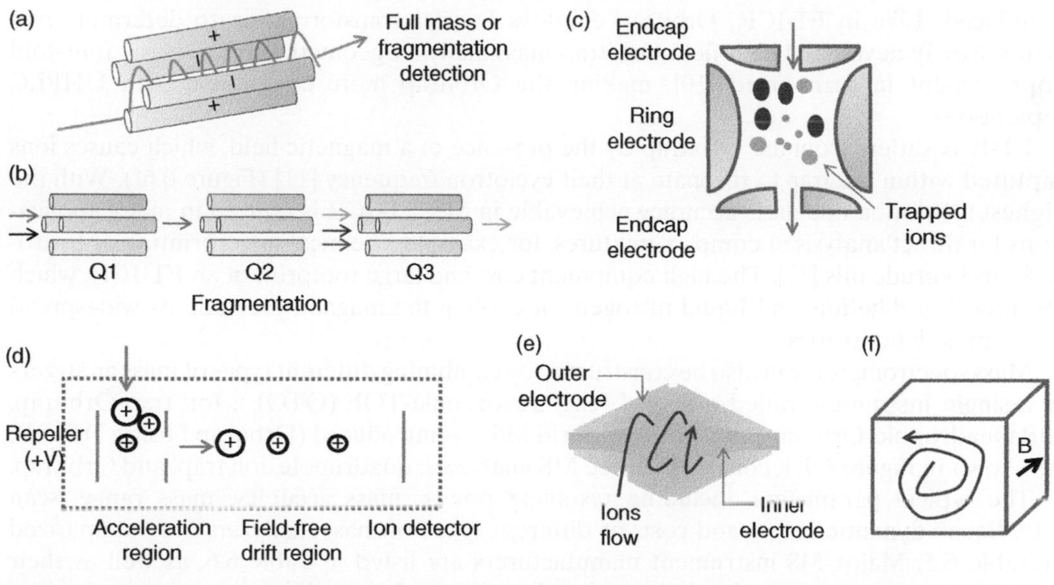


**Figure 6.6.** Schematics of common mass analyzers. (a) Single-quadrupole (SQ); (b) Triple-quadrupole (TQ); (c) Ion trap (IT); (d) Time-of-flight (TOF); (e) Orbitrap; (f) Fourier transform ion cyclotron resonance (FT-ICR).

**Table 6.4. Scan Modes Available in Quadrupole Mass Spectrometer**

| | Scan mode | Q1 | Q2 | Q3 | Purpose |
|---|---|---|---|---|---|
| Single-quadrupole MS (SQMS) | Full-scan | Scanning | Pass All | Pass All | MW Information |
| | SIM | Fixed *m/z* | Pass All | Pass All | Quantitation |
| Triple-quadrupole MS (TQMS) | Product | Fixed *m/z* | Pass All (+CE) | Scanning | Structural Information |
| | SRM | Fixed *m/z* | Pass All (+CE) | Fixed *m/z* | Targeted Quantitation |
| | Neutral loss | Scanning | Pass All (+CE) | Scanning | Analyte screening |
| | Precursor | Scanning | Pass All (+CE) | Fixed *m/z* | Analyte screening |

cell in a low-to-high *m/z* sequence. A key advantage of an ion trap is its ability to perform multistage MS fragmentation experiments such as $MS^n$, in which ion fragmentation stages can be performed sequentially (Figure 6.5). It generates a compound-specific ion spectral tree and provides useful structural information. However, ion traps can have a limited dynamic range, rendering it primarily for qualitative applications and less useful for quantitation.

A time-of-flight (TOF) mass analyzer (Figure 6.6d) determines an ion's *m/z* ratio based on the length of time it takes to travel down the instrument's flight tube and reach the detector. A TOF MS allows for accurate mass measurements, with up to 100 000 RP in modern instruments. It also provides the highest practical mass range and fastest scan speed among all MS analyzers rendering it desirable for UHPLC/MS.

In addition to TOF, Orbitrap and Fourier transform ion cyclotron resonance (FT-ICR) are the other two high-resolution ion trap mass analyzers (Figure 6.6e,f). Both types of MS provide ultrahigh RP, for example, up to 1 000 000 RP for Orbitrap and up to 20 000 000 RP for FT-ICR. Orbitrap is the newest mass analyzer on the market; it was introduced in 2005 [8, 9]. Compared to other ion traps, Orbitrap operates on an electrostatic field only, by trapping ions around an inner spindle-like electrode in an orbital motion. An outer barrel-like electrode is coaxial with the inner electrode where the image current from the trapped ion is induced. Like in FT-ICR, Orbitrap employs Fourier transformation to determine *m/z* ratios. Newly developed high-field Orbitrap mass analyzer geometry also allows a four-fold improvement in scan speed [10], making the Orbitrap more compatible with UHPLC separations.

FT-ICR differs from the Orbitrap by the presence of a magnetic field, which causes ions captured within the trap to resonate at their cyclotron frequency [11] (Figure 6.6f). With the highest resolution and mass accuracy achievable in MS, FT-ICR is utilized in many applications for direct analysis of complex mixtures, for example, chemical fingerprinting of hydrocarbons in crude oils [12]. The high equipment cost and large footprint of an FT-ICR, which requires liquid helium and liquid nitrogen for cooling the magnet, preclude its widespread use in most laboratories.

Mass spectrometers can also be constructed by combining different types of mass analyzers in a single instrument called a hybrid (e.g. Quadrupole-TOF (QTOF), Ion trap-Orbitrap, and Quadrupole-Orbitrap). Recently, a tribrid MS was introduced (Orbitrap Fusion Tribrid). As shown in Figure 6.7, it consists of three MS analyzers: quadrupole, ion trap, and Orbitrap.

The typical parameters, including resolving power, mass accuracy, mass range, scan rate, linear dynamic range, and cost, of different types of mass analyzers are summarized in Table 6.5. Major MS instrument manufacturers are listed in Table 6.6, as well as their available mass analyzers and associated data handling and control software.

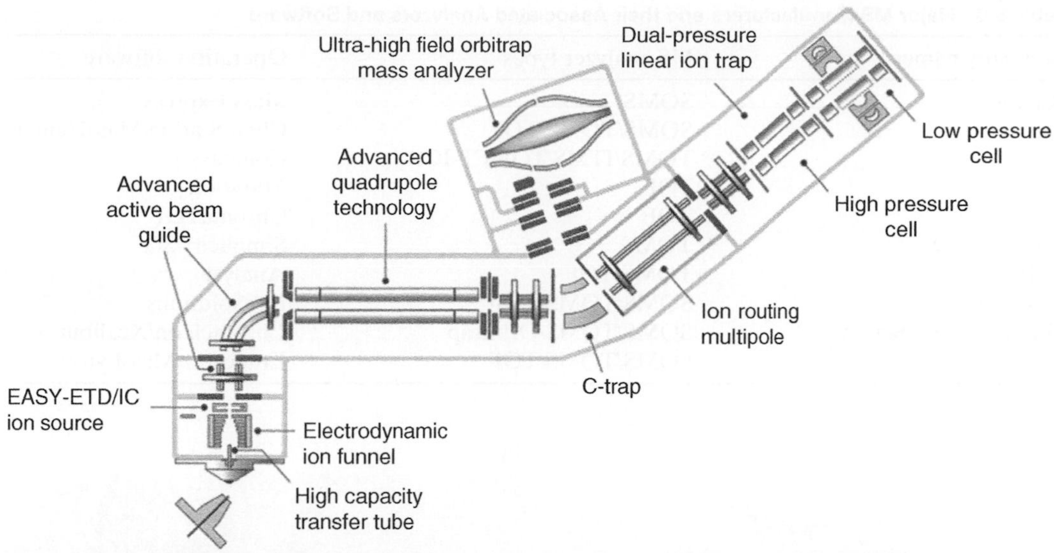


**Figure 6.7.** A scheme of Orbitrap Tribrid mass spectrometer, which consists of quadrupole, Orbitrap, and ion trap. Source: Reproduced with permission of Thermo Fisher Scientific.

**Table 6.5. Performance Parameters of Different Mass Analyzers**

| Parameters | Quadrupole | Ion trap | TOF | Orbitrap | FT-ICR |
|---|---|---|---|---|---|
| Resolving power (FWMH) | Up to 10 000 Unit resolution | Up to 10 000 Unit resolution | 20 000–80 000 | Up to 1 000 000 | Up to 20 000 000 |
| Mass accuracy (ppm) | 50–100 | 50–100 | 1–5 | 1–5 | 0.3–1 |
| Mass range (*m*/*z*) | Up to 4000 but usually below 2000 | Up to 4000 but usually below 2000 | Over 100 000 | Up to 80 000 | Up to 10 000 |
| Scan rate (Hz) | 2–10 | 2–10 | 10–100 | 1–40 | 0.5–2 |
| Linear dynamic Range | $10^5$–$10^6$ | $10^4$–$10^5$ | $10^4$–$10^6$ | $5 \times 10^3$ | $10^4$ |
| Cost | Low | Low | Moderate | Moderate | High |

### 6.2.4 Detectors

The ion detection system in MS is used to detect the presence of ion signals emerging from the mass analyzer of a mass spectrometer [2]. The simplest ion collector and detector is a Faraday cup (Figure 6.8a), which is still in use to measure abundance ratios with the highest accuracy in isotopic ratio mass spectrometry (IR/MS). The most common detectors in quadrupole mass spectrometer are electron multipliers (Figure 6.8b,c), which multiply the numbers of electrons emitted from the surface by impingement of ions to increase the signal intensity. According to its construction, it can be classified into discrete-dynode electron multiplier (DEM) (Figure 6.8b) and the continuous-dynode electron multiplier (also referred

**Table 6.6. Major MS Manufacturers and their Associated Analyzers and Software**

| Company names | MS analyzer types | Operation software |
|---|---|---|
| Advion | SQMS | Mass Express |
| Agilent | SQMS/TQMS/TOF | ChemStation/MassHunter |
| Bruker | TQMS/ITMS/TOF/FT-ICR | Compass |
| JEOL | TOF | MStation |
| LECO | TOF | ChromaTOF |
| PerkinElmer | TQMS | Simplicity 3Q |
| SCIEX | TQMS/TOF | Analyst |
| Shimadzu | SQMS/TQMS/TOF | LabSolutions |
| Thermo Fisher Scientific | SQMS/TQMS/Orbitrap | Chromeleon/Xcalibur |
| Waters | SQMS/TQMS/TOF | Empower/MassLynx |

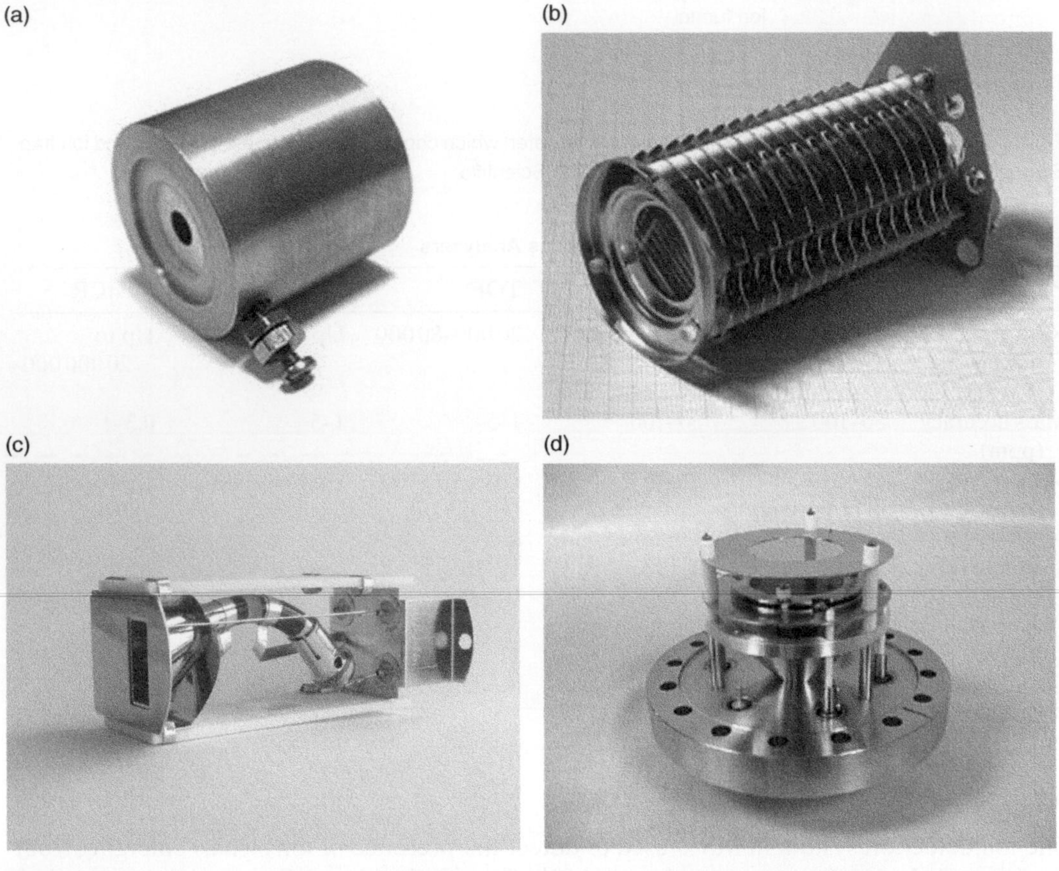

**Figure 6.8.** Common mass detectors. (a) Faraday cup; (b) Discrete dynode electron multipliers; (c) Channeltron multipliers; (d) Microchannel plates.

to as a channel electron multiplier or CEM) (Figure 6.8c). Microchannel plates (MCPs) is another ion counting system, which puts millions of tubes together in a "bundle" and yields a CEM array (Figure 6.8d). MCP is mainly used in TOF-MS to detect the electron current every nanosecond while the ions are arriving at the detector. Ion counting detectors are not used in FT-ICR or Orbitrap instrument where image current detection yields superior results.

## 6.3 SMALL-MOLECULES DRUG RESEARCH AND DEVELOPMENT

The role of separation science (including LC/MS) in small-molecule drug discovery and development has been described in a series of four white papers recently published [13–16]. Several prominent LC/MS applications such as mass measurement, structure elucidation, and trace quantitation are discussed here and in Chapter 7.

### 6.3.1 Mass Measurement and Elemental Composition Determination

During drug discovery, rapid HPLC/UV/MS methods that provide nominal molecular weight data along with isotopic distribution patterns are used in high-throughput purification and characterization [13]. As shown in Figure 6.9, the most abundant ions for small-molecule drugs (which are mostly weak bases) are the singly charged protonated molecular ions $[M+H]^+$ in a positive ionization mode. In addition to the molecular ions, adduct ions are commonly observed (Table 6.3), including those from alkali metal ions (Na, K), ammonium ions $[M+NH_4]^+$, adducted solvent molecules $[M+H+CH_3OH]^+$, in a positive ionization mode. Ions such as $Na^+$ are pervasive in the mobile phases, glassware, and samples and are difficult to eliminate. These adduct ions add complexity to the interpretation of unknowns, though they could be useful to identify the molecular ions when their signals are weak. The dimer ions of the main component, that is $[2M+H]^+$, may be observed at higher concentrations of the analyte due to their combination in the gas phase at the ion source.

During early-phase drug development, accurate mass data are used to determine the elemental compositions of the active pharmaceutical ingredient (API) and its impurities and

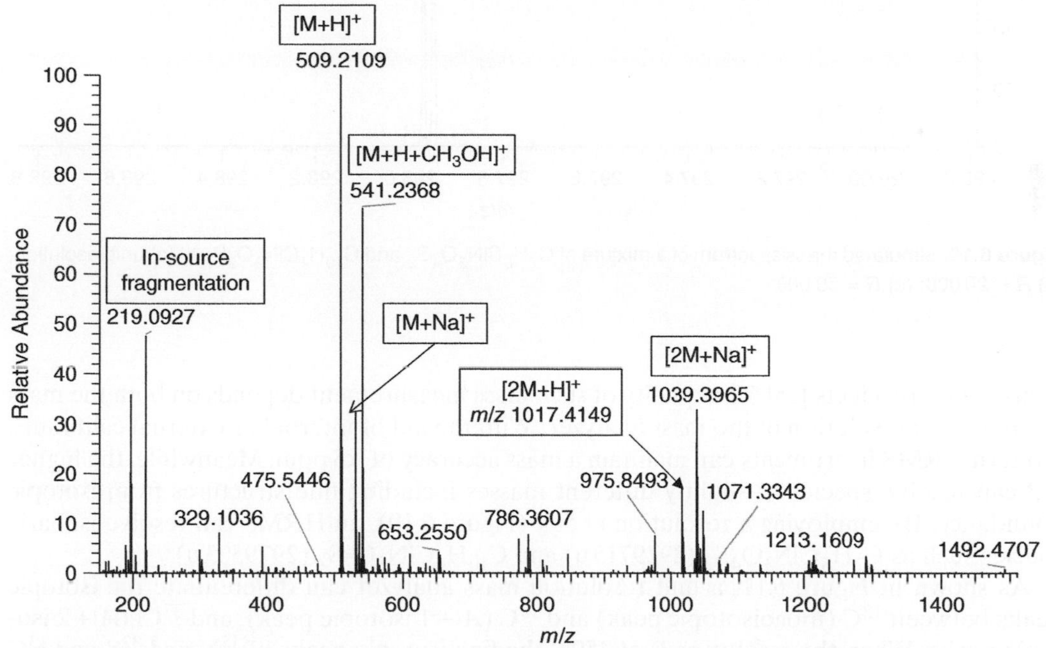


**Figure 6.9.** An example of a full-mass spectrum at positive ESI mode showing precursor ion, product, adduct ions, and doubly charged ion.

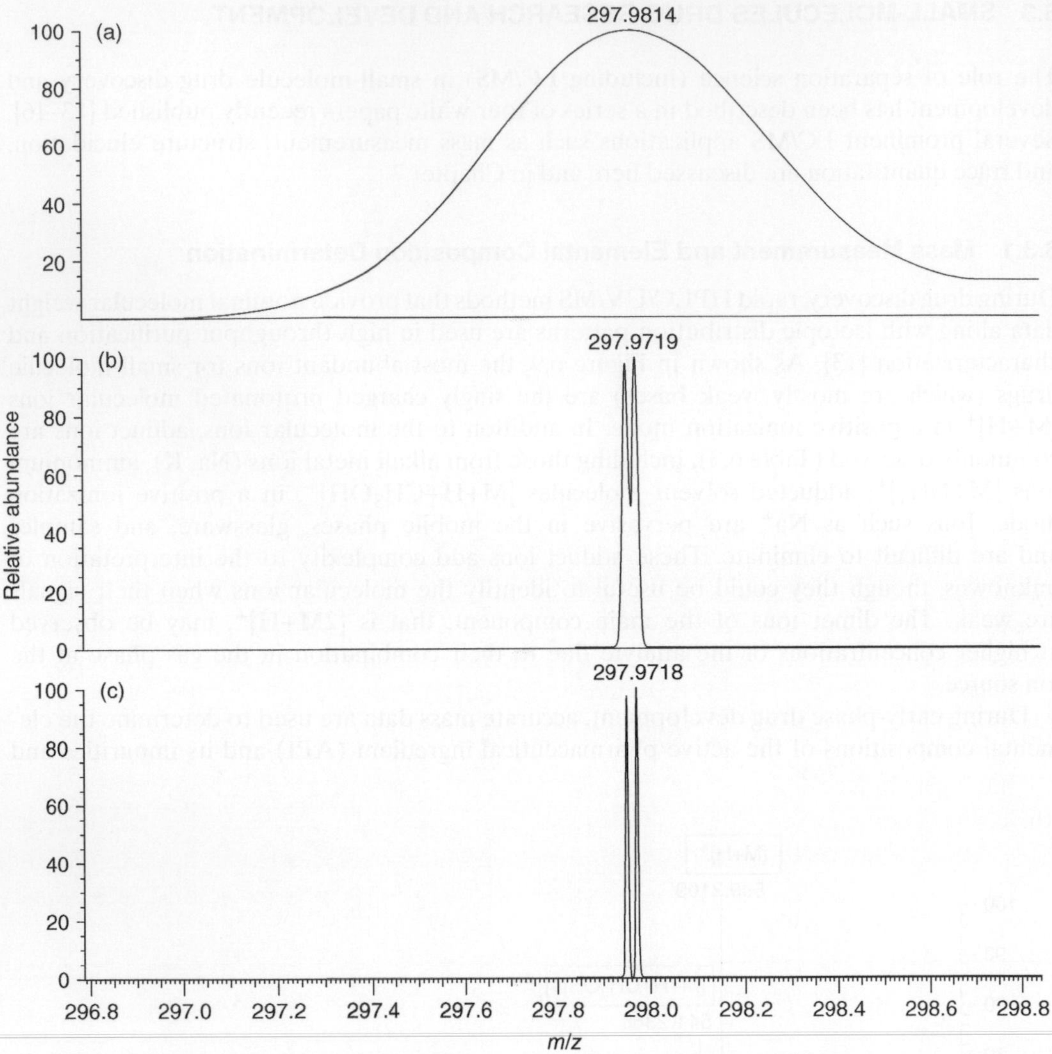


**Figure 6.10.** Simulated mass spectrum of a mixture of $C_7H_8ClN_3O_4S_2$ and $C_{10}H_4ClN_3O_2S_2$ at (a) unit resolution; (b) $R = 20\,000$; (c) $R = 50\,000$.

degradation products [15]. The quality of such mass measurement depends on both the mass accuracy and resolution of the mass analyzer. With the aid of internal or external calibrants, modern HRMS instruments can maintain a mass accuracy of <5 ppm. Meanwhile, the higher RP can resolve species of slightly different masses including fine structures from isotopic abundance. By employing a resolution at 50k (Figure 6.10), an HRMS can resolve isobaric species such as $C_7H_8ClN_3O_4S_2$ (297.9715 u) and $C_{10}H_4ClN_3O_2S_2$ (297.9503 u).

As shown in Figure 6.11, a unit resolution mass analyzer can differentiate the isotopic peaks between $^{12}C$ (monoisotopic peak) and $^{13}C$ ($A + 1$ isotopic peak), and $^{37}Cl$ ($A + 2$ isotopic peak). When the resolution is at 450k, the fine isotopic peaks of $^{15}N$ and $^{32}S$ and $^{13}C$ can be well resolved (Figure 6.12a). For $A + 2$ isotopic peaks (Figure 6.12b), $^{34}S$ and $^{37}Cl$ and $^{15}N^{13}C$ can be separated at an RP of 450k. In principle, the numbers of each element can be

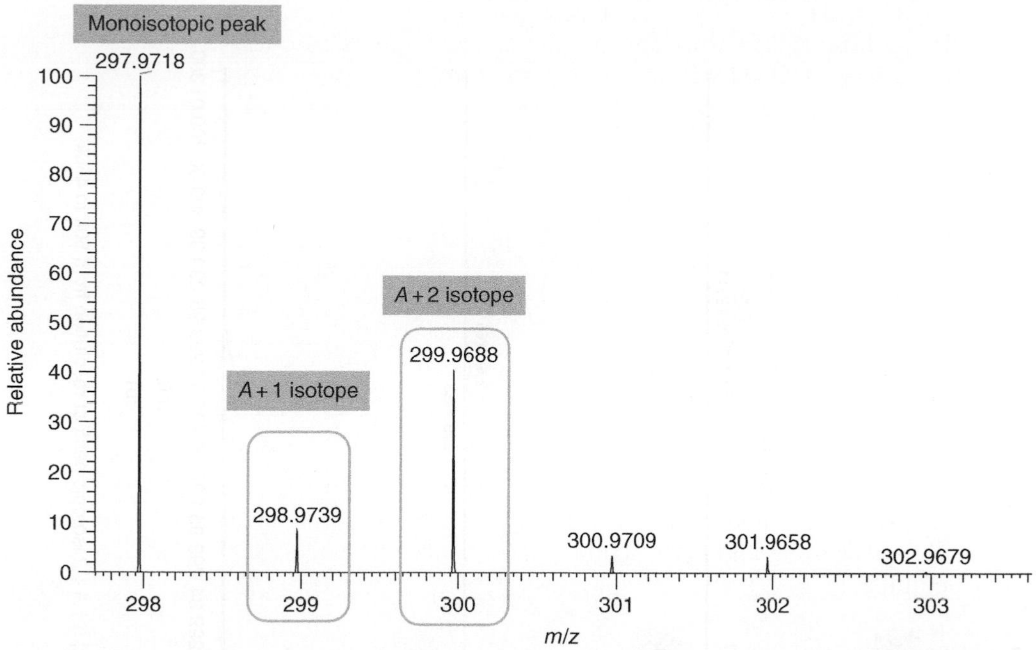


**Figure 6.11.** Simulated mass spectra of $C_7H_8ClN_3O_4S_2$ with the isotopic distribution at unit resolution.

readily calculated using the relative isotopic abundance (RIA) and the natural abundance of the corresponding isotopes, as shown in the following equations [17]:

$$P_m^N : P_m^N \times \binom{N}{1} P_i = A_m : A_i$$

$$N = \frac{P_m}{P_i} \times \frac{A_i}{A_m} \tag{6.4}$$

where $A_m$ and $A_i$ indicate the monoisotopic and isotopic peak area and $P_m$ and $P_i$ indicate the natural abundance percentages of the corresponding isotopes, respectively.

The elemental composition of an unknown compound, obtained from its exact mass measurement combined with the isotopic envelops and nitrogen rules, provides a good starting point for structural elucidation. An elemental calculator function for suggesting the potential molecular formulae of unknowns is available in most data handling software of HRMS listed in Table 6.6.

### 6.3.2 Structural Elucidation

Additional structural information of various chemical/functional groups of target molecules can be obtained from MS fragmentation data, which are acquired by tandem-in-time/space MS instruments, such as ion trap, TQ, and Q-Trap systems. In Figure 6.13, the multiple-stage mass spectra ($MS^n$) is constructed into "spectral trees" that illustrate the compound product ion spectra in its respective sequential mass spectral stage. The ion spectral information defines the structure of the molecule and is considered as footprints of the molecule. The structures of unknown compounds with similar structures as the main component in the

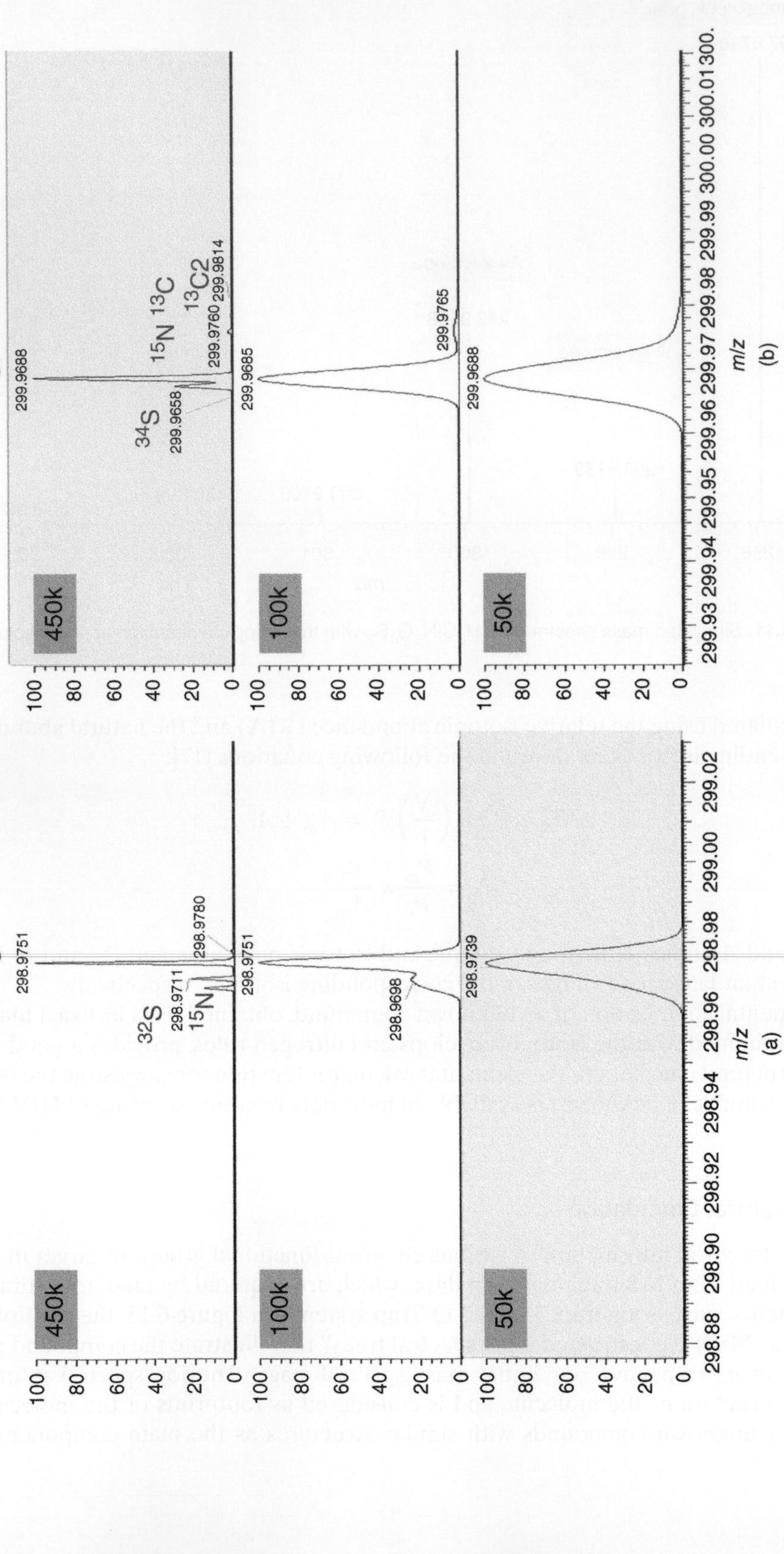


**Figure 6.12.** Simulated mass spectra of $C_7H_8ClN_3O_4S_2$ at (a) A + 1 isotopic peak; (b) A + 2 isotopic peak at resolution from 50k to 450k.

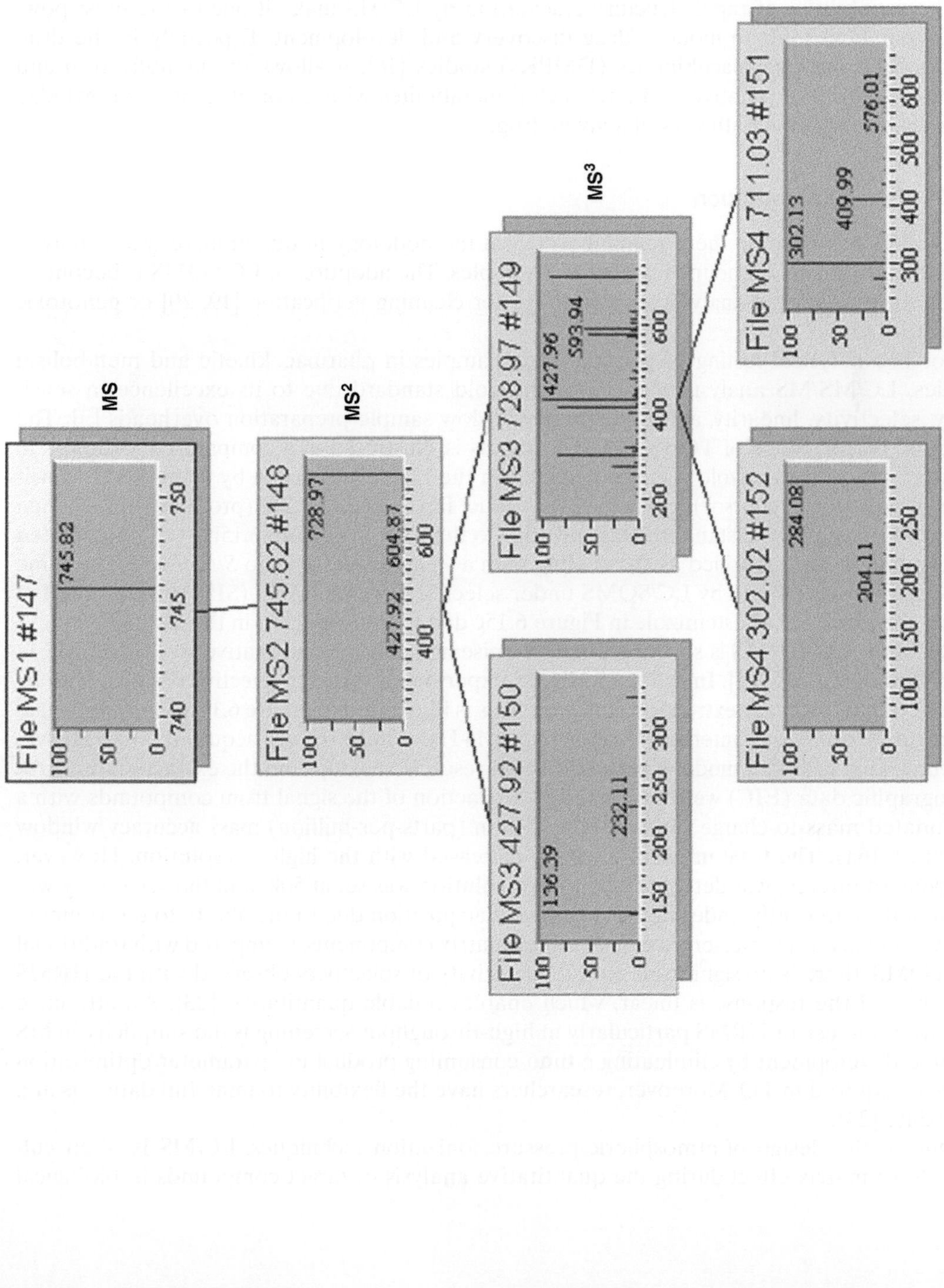


**Figure 6.13.** Spectral trees generated from multiple-stage mass spectra ($MS^n$).

sample can often be deduced from these spectral tree data, like gallocatechol (GC) and an unknown whose mass spectra are shown in Figure 6.14a,b. The unknown has the same ion at $MS^3$ spectra as GC, (139.10 *m/z*), indicating possession of the same core structure, triol-methylbenzene of GC. As a result, a putative structure of epigallocatechin gallate (EGCG) is deduced as a tentative structure that is in the same class of catechin as GC. Nevertheless, a definitive structure elucidation required sign-off of the identity of an API, the gold standard is still nuclear magnetic resonance spectrometry (NMR), supplemented by HRMS data, which can be acquired quickly on a minimum sample amount.

The capabilities of rapid structure elucidation by LC/MS make it one of the most powerful analytical tools in modern drug discovery and development. Especially in the drug metabolism and pharmacokinetics (DMPKs) studies [18], it allows the identification and characterization of putative and unexpected metabolites with no or little prior knowledge of biotransformation pathways of a given drug.

### 6.3.3 Trace Quantitation

HPLC/MS has become the dominant platform methodology in quantitative trace analysis of pharmaceutical compounds in complex samples. The adoption of LC/SQMS is becoming common in regulated analysis such as those for cleaning verification [19, 20] or genotoxic impurities [21].

For bioanalytical testing of plasma/serum samples in pharmacokinetic and metabolism studies, LC/MS/MS analysis with TQ is the gold standard due to its excellence in sensitivity, selectivity, linearity, and robustness with low sample preparation overheads [3]. The superior performance of TQ vs. SQ instruments is illustrated in a comparative example in Figure 6.15a–c. Astemizole was not detected in the blank rat plasma by using SRM transitions at *m/z* 459 (precursor) and *m/z* 135 (product ion 1) and *m/z* 218 (product ion 2). When 1.5 ng/ml of astemizole standard was spiked into the plasma (Figure 6.15b), the peak eluted at 1.3 minutes was identified as the analyte with a signal-to-noise ratio $S/N > 100$. The same spiked sample, analyzed by LC/SQMS under selected ion monitoring (SIM) mode, failed to detect the presence of astemizole in Figure 6.15c due to interferences in the sample.

Recently, LC/HRMS is showing great promise as a feasible alternative to LC/MS/MS in quantitative studies [22]. In full-scan HRMS experiments, greater selectivity is achieved by the use of narrow mass-extraction windows. This is illustrated in Figure 6.16b where the mass spectrum of analyte of interest ($C_{14}H_{20}ON_3$, $[M+H]^+$ 246.1601) was acquired on an HRMS instrument in full-scan mode with the different resolution setup, and the extracted ion chromatographic data (EIC) were processed by extraction of the signal from compounds with a protonated mass-to-charge ratio within a 5 ppm (parts-per-million) mass accuracy window (Figure 6.16a). The total intensity was not increased with the higher resolution. However, the peak of interest was detected when the resolution was set at 50k, and the selectivity was improved significantly under higher resolution acquisition due to the ability to discriminate analyte signals from interferences in complex matrix components. Compared with traditional LC/TQMS, there is no significant drop in sensitivity or selectivity observed with the HRMS system, and the response is linear, which enables reliable quantitation [23]. An attractive benefit of the use of HRMS particularly in high-throughput screening is the simplicity in MS method development by eliminating a time-consuming product ion parameter optimization process required in TQ. Moreover, researchers have the flexibility to mine full data sets at a later date [24].

Due to the design of atmospheric-pressure ionization techniques, LC/MS is often vulnerable to matrix effect during the quantitative analysis of target compounds in biological

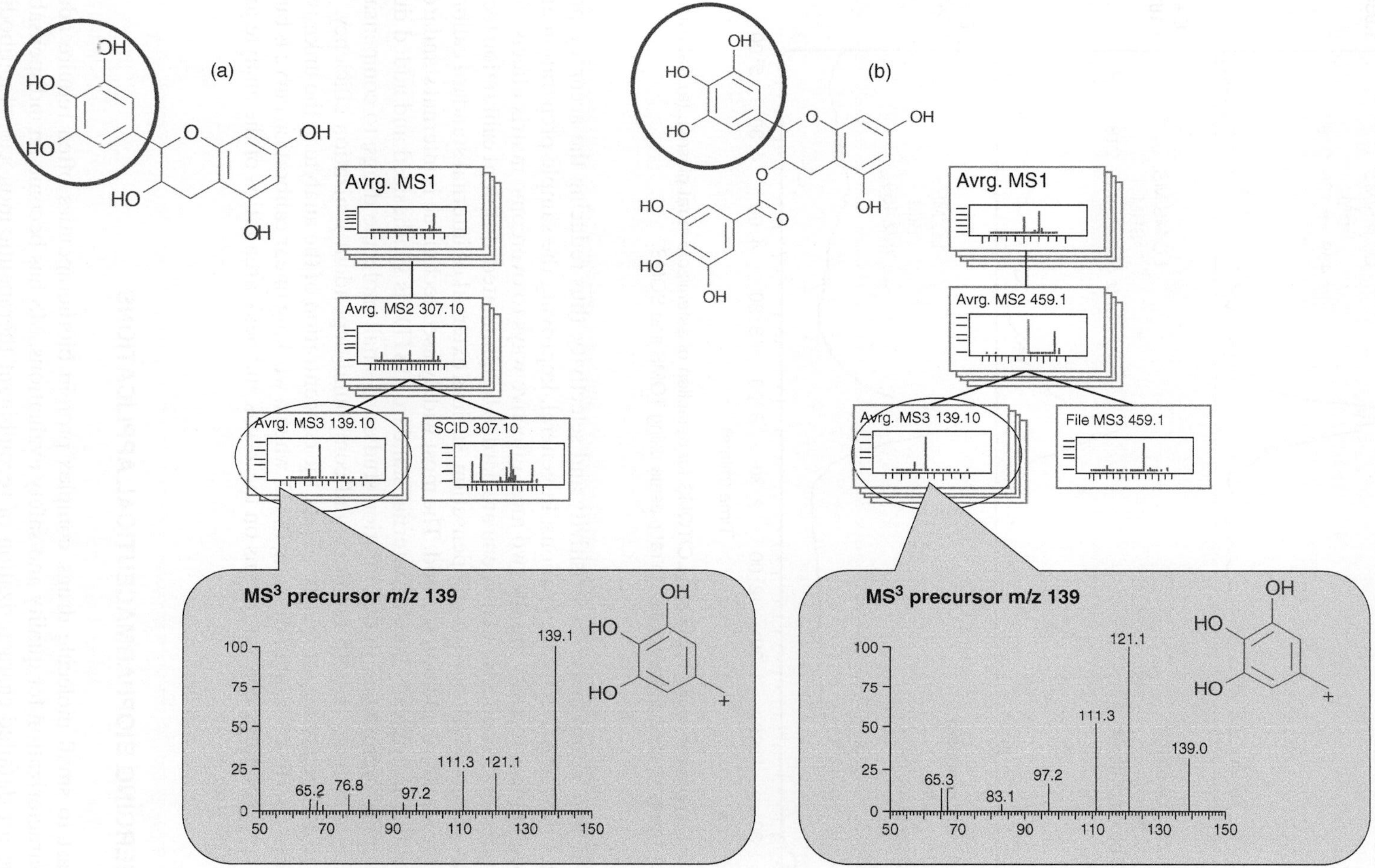


**Figure 6.14.** Spectral trees generated from multiple-stage mass spectra ($MS^n$) for (a) Gallocatechol (GC); (b) Unknown compound – Epigallocatechin gallate (EGCG).

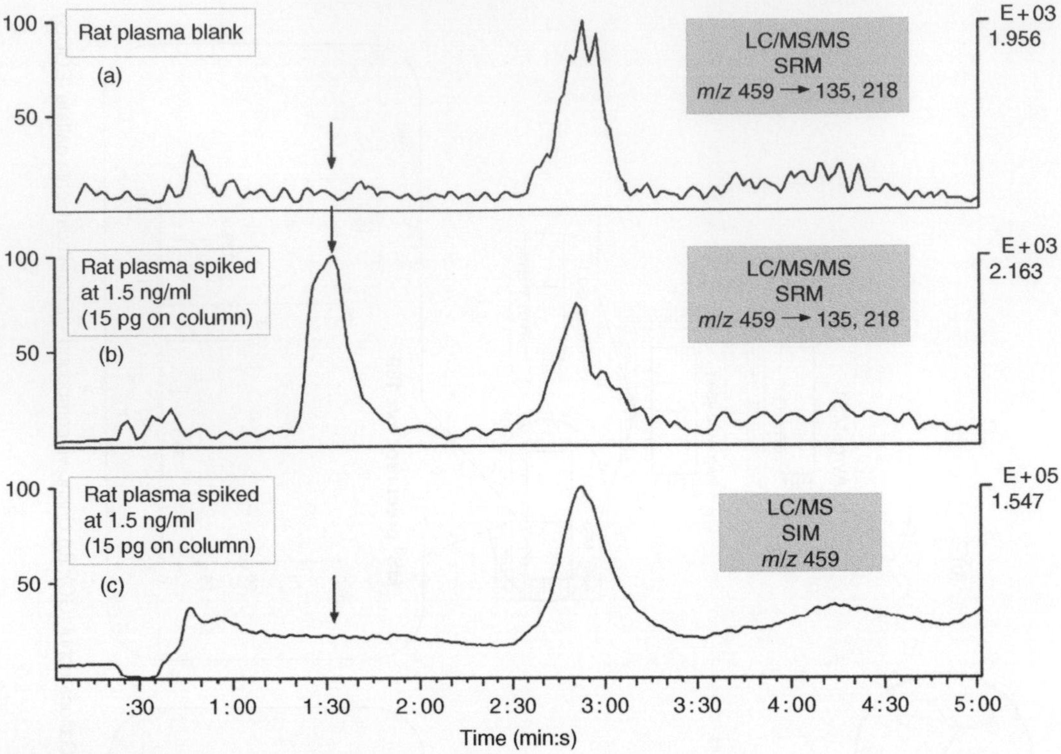


**Figure 6.15.** Comparison of LC/SQMS and LC/TQMS for detection of astemizole in rat plasma. (a) Rat plasma blank; (b–c) 1.5 ng/ml of astemizole spiked in rat plasma using TQMS and SQMS.

samples. It may compromise its sensitivity and selectivity, thus reducing the accuracy, precision, and robustness of its applications. In general, improving the sample preparation and chromatographic selectivity are the two most effective ways to overcome matrix effects.

Sometimes matrix effects phenomena are hard to be eliminated, different calibration techniques have been developed for compensation, such as external or internal standard calibration and the standard addition method. The most widely used technique is internal standards, where a stable isotopically labeled internal standard (ISTD) is synthesized and added during the sample preparation in quantitation studies of small-molecule drugs, to compensate control for variations in both recovery and compound-dependent ionization efficiency. In addition to compensating for matrix effects, the concentration of the analyte in the unknown samples can also be calculated by using ISTD, when a working linear calibration curve is built with the corresponding concentrations on the *x*-axis and peak area ratios of the analyte and ISTD on the *y*-axis.

## 6.4 EMERGING BIOPHARMACEUTICAL APPLICATIONS

In contrast to small molecule drugs, complex protein biotherapeutics often require additional characterization for quality and safety evaluations. MS has become an indispensable technique for detailed characterization of recombinant therapeutic monoclonal antibodies (mAbs) during the product/process development [25, 26].

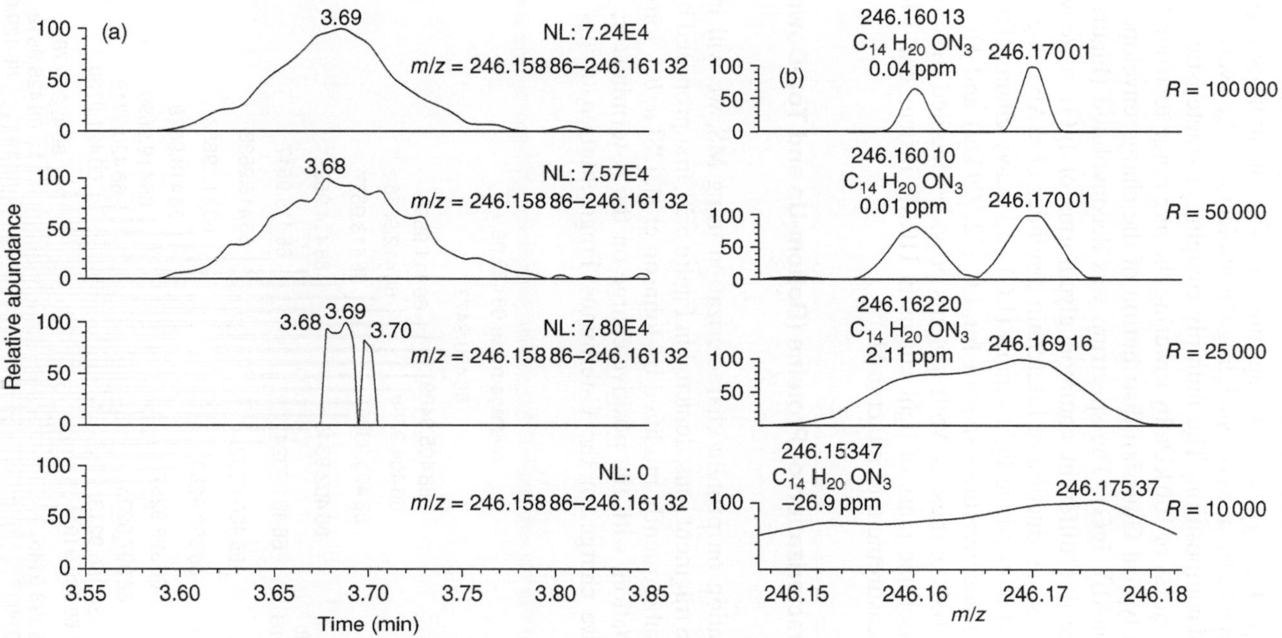


**Figure 6.16.** Example ion chromatograms and mass spectra acquired from LC-HRMS. (a) Extracted ion chromatograms of the analyte at *m/z* 246.1601 with 5 ppm mass error extraction window; (b) Mass spectra of analyte and matrix interference at RT 3.69 minutes, which were acquired at a different resolution. Source: Kellmann et al. 2000 [22]. Copyright 2000. Reprinted with permission of American Chemical Society.

### 6.4.1 Intact Mass Measurement of Proteins

To measure the mass of a small-molecule compound, the monoisotopic ion is usually the most intense peak. This is not the case for larger biomolecules. Figure 6.17 shows the isotopic envelope for a common protein standard, bovine serum albumin (BSA), with elemental composition of $C_{2932}H_{4614}N_{780}O_{898}S_{39}$ and an average mass 66, 432. The monoisotopic mass is not typically used to deconvolute the mass of such large biomolecules. The average mass is used to confirm its elemental composition. The multiply charging characteristic of ESI enables detection of analytes in excess of 100 kDa by lowering the *m/z* range to around 2000–4000.

Figure 6.18a shows a typical Gaussian distribution of the charge envelope for an intact monoclonal antibody (mAb), IgG1. The spectrum was deconvoluted (Figure 6.18b), and the exact masses of the five different common glycoforms of IgG1 were well-resolved (Figure 6.18c). In most cases, reduction of interchain disulfides of mAb using dithiothreitol (DTT) can be performed to produce light chains (LC) and heavy chains (HCs) of mAbs. Figure 6.19a,b show the deconvoluted spectra for LC (22, 929 Da) and HC (49, 475 Da), respectively. LC/MS analysis of these mAb fragments provides detailed structural information, such as the monoisotopic mass of light chains in HRMS instrument, which can't be gained from the mass measurement of intact mAb.

### 6.4.2 Structural Characterization of Proteins (Bottom-Up and Top-Down Approaches)

As background information on protein characterization using MS, we will introduce the nomenclature of peptide fragment ions, depicted in Figure 6.20, first proposed by Roepstorff and Fohlman [27] and subsequently modified by Johnson et al. [28] *a*, *b*, *c* ions are formed by main-chain fragmentation, with the positive charge on the N-terminus. *x*, *y*, *z* ions are formed with the positive charge on the C-terminus. Fragmentation occurs in pairs, for

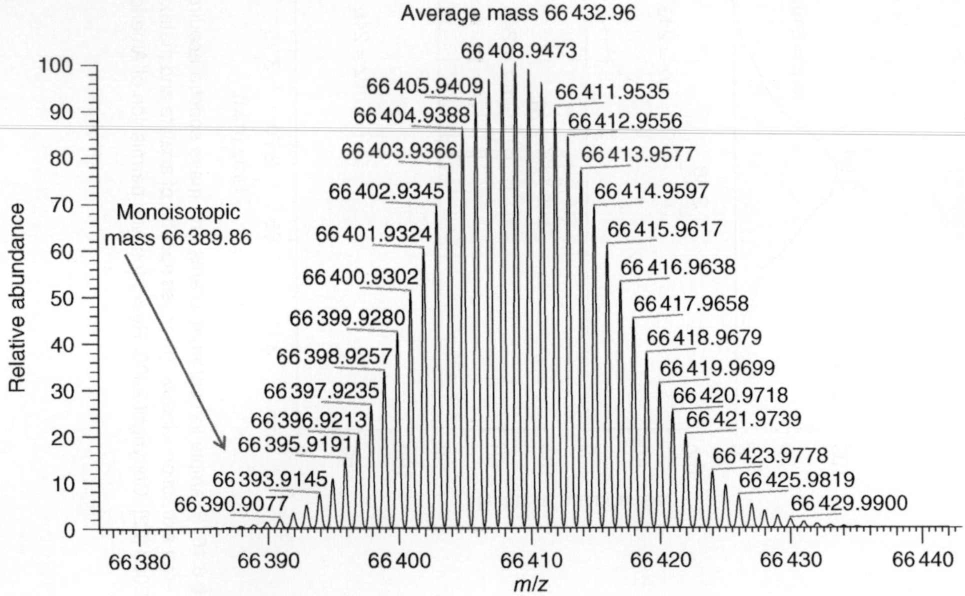


**Figure 6.17.** The simulated mass spectrum of bovine serum albumin (BSA), $C_{2932}H_{4614}N_{780}O_{898}S_{39}$.

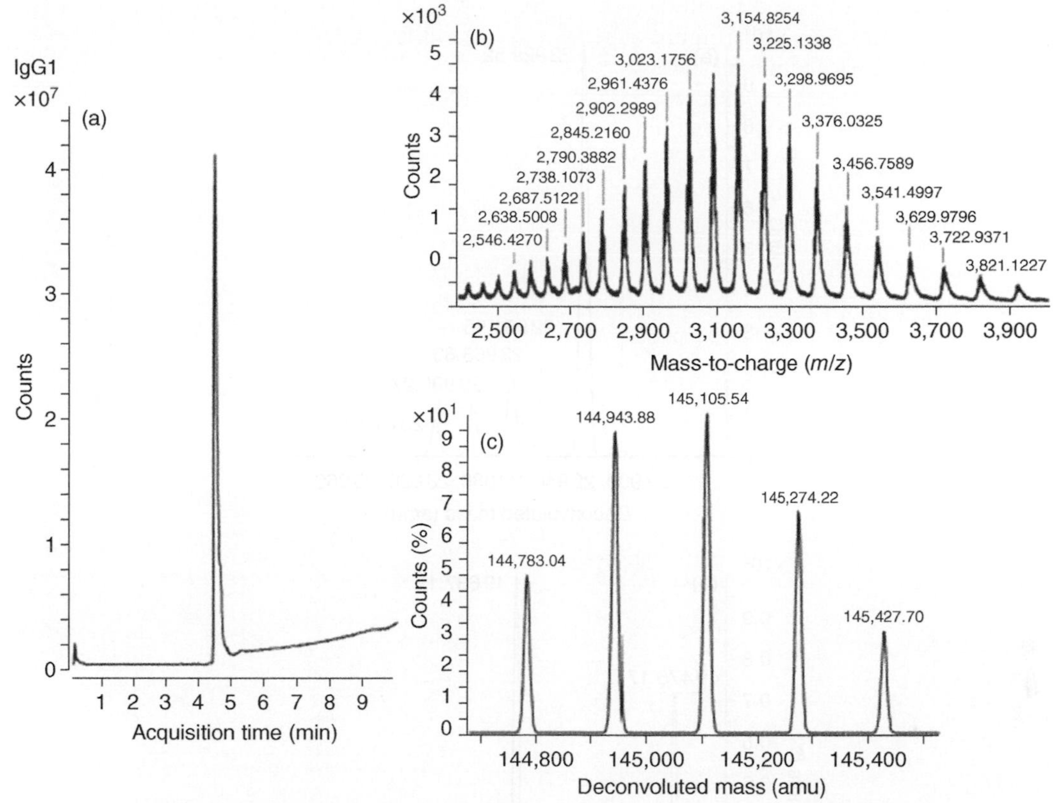


**Figure 6.18.** HPLC/MS analysis of intact IgG under RPC with Q-TOF. (a) RPC/TIC chromatogram; (b) Mass spectrum showing multiply-charged ions; (c) Deconvoluted mass spectra of five major glycoforms of IgG. HPLC conditions: Agilent Poroshell 300SB C8, 5 μm, 75 × 2.1 mm; MPA – water/0.1% formic acid; MPB – 80/20/10 IPA/ACN/water/0.1% formic acid; gradient: 15–20%B in 4 minutes; 20–75%B in 1 minute; 75%–100%B in 5 minutes; flow rate – 1.2 ml/min at 60 °C; MS – Agilent 6530 Q-TOF. Source: Courtesy of Agilent Technologies.

example, *a*/*x*, *b*/*y*, and *c*/*z*, in which *a*, *b*, *y* ions are produced by low-energy CID and *c*/*z* ions are primarily generated by ETD. This characteristic fragmentation of peptide ions in the gas phase provides a useful tool to probe protein structures by the LC/MS/MS methods described as follows.

MS-based structural characterization of protein biotherapeutics may involve bottom-up and/or top-down approaches [25], which provide amino acid sequence verification, localization of disulfide linkages, profiling of N-glycan structures, and elucidation of additional expected and unexpected PTMs. The bottom-up analysis of a protein (also called peptide mapping) on the peptide level by LC/MS/MS is the most versatile and powerful approach to ensure the identity of a protein by verifying the correct amino acid sequence [29]. As illustrated in Figure 6.21, the protein was firstly digested by specific enzymes (e.g. trypsin cleaves proteins at the carboxyl side of lysine and arginine residues, except when either is bound to a C-terminal proline), followed by LC/MS/MS analysis of the digest. The acquired peptide MS/MS spectrum was searched against an *in silico* tryptic digest of the entire protein database, employing different database search engines (e.g. SEQUEST, Mascot, Byonic™). The best overlap between the experimental and theoretical mass spectra identifies the protein.

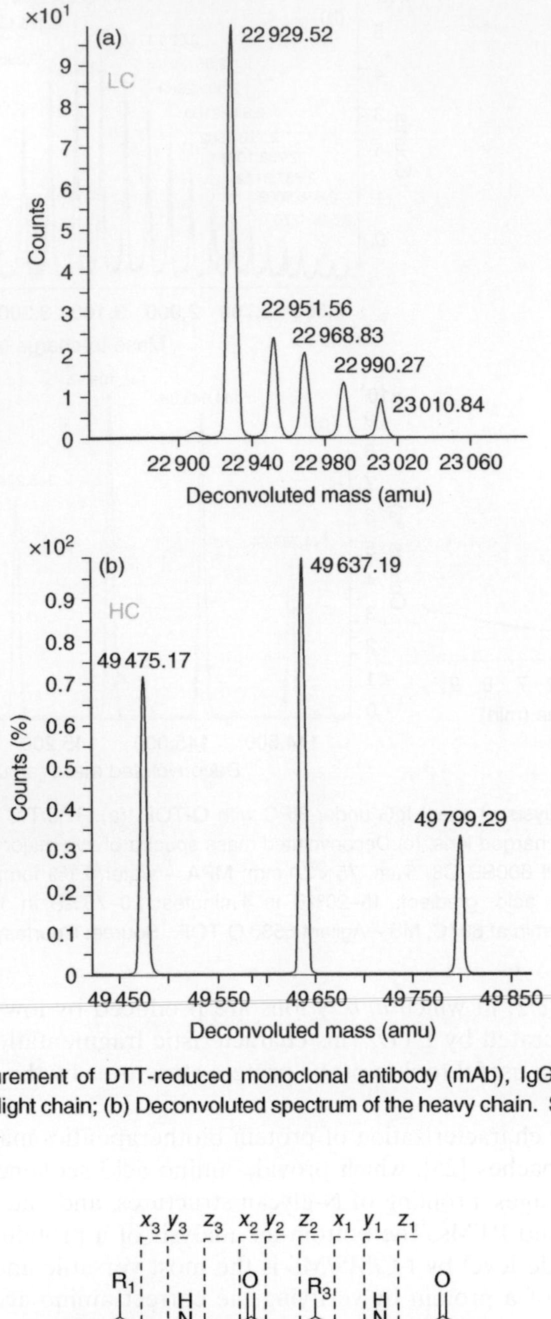


**Figure 6.19.** Mass measurement of DTT-reduced monoclonal antibody (mAb), IgG1, using LC/TOF. (a) The deconvoluted spectrum of light chain; (b) Deconvoluted spectrum of the heavy chain. Source: Courtesy of Agilent Technologies.

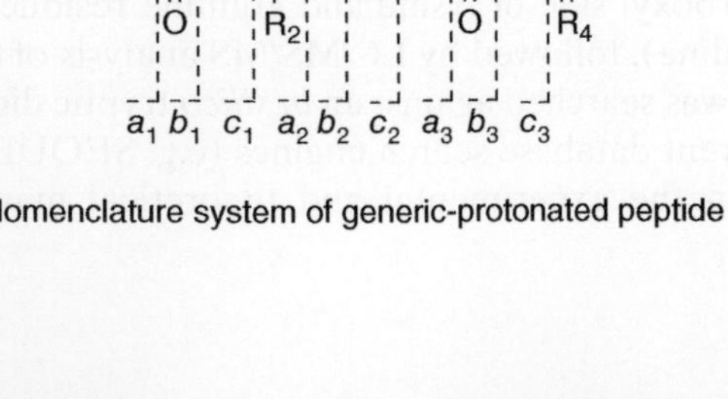

**Figure 6.20.** Nomenclature system of generic-protonated peptide fragmentation.

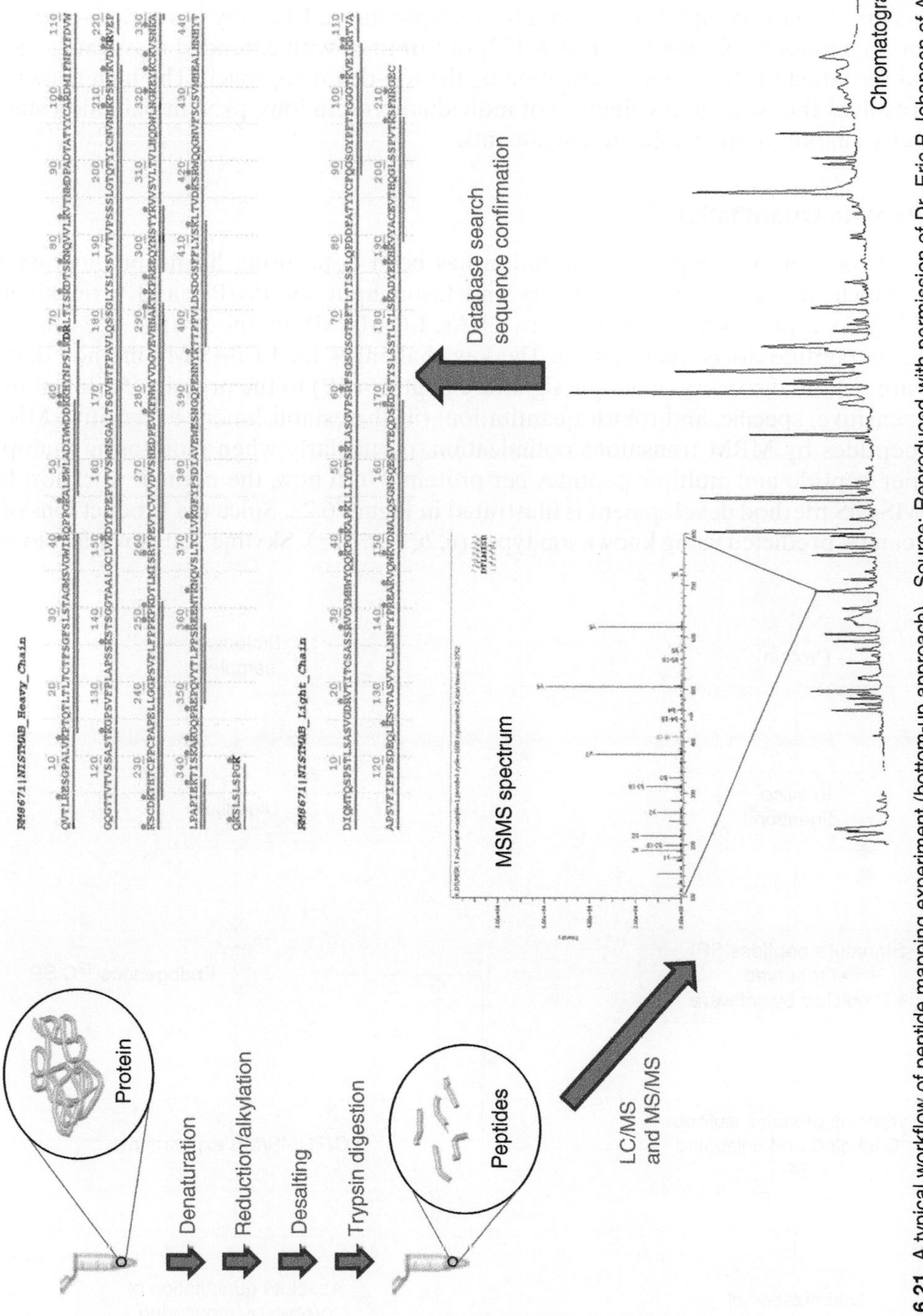


**Figure 6.21.** A typical workflow of peptide mapping experiment (bottom-up approach). Source: Reproduced with permission of Dr. Eric B Johansen of AbbVie Stemcentrx LLC.

Conversely, in the top-down approach, intact proteins are introduced into a mass spectrometer and then subjected to gas-phase fragmentation. It provides a fast and convenient way to obtain the sequence information of protein [30, 31]. For a large protein molecule, however, the top-down analysis has always been challenging, because the highly charged fragment ions produce the complicated spectra for interpretation. Recently it was demonstrated that ultra-high mass resolution by FT-ICR [32] or Orbitrap with extended mass range [8] is beneficial for intact proteins characterization by the top-down approach. The high mass resolution resolves the isotopic distribution of individual protein ions, protein oxidation states and thereby enables more confident assignments.

### 6.4.3 Peptide Quantitation

Historically, protein and peptide quantitation has been done using ligand binding assays (LBAs), which afford sufficient sensitivity and throughput for PK/PD and toxicokinetic studies [26]. As a promising supplement to LBAs, LC/MS/MS methods have been proven successful in peptide drugs quantitation. The key challenges for LC/MS/MS method development are: (i) the discovery of unique signature peptide (SP) to the protein of interest that ensures sensitive, specific, and robust quantitation; (ii) the establishment of sensitive MRM for the peptides by MRM transitions optimization, particularly when monitoring multiple MRMs per peptide and multiple peptides per protein. Until now, the optimal workflow for the LC/MS/MS method development is illustrated in Figure 6.22. Since the product ions of a peptide can be predicted using known ion types (*a*, *b*, *c*, *x*, *y*, *z*). Skyline software, developed

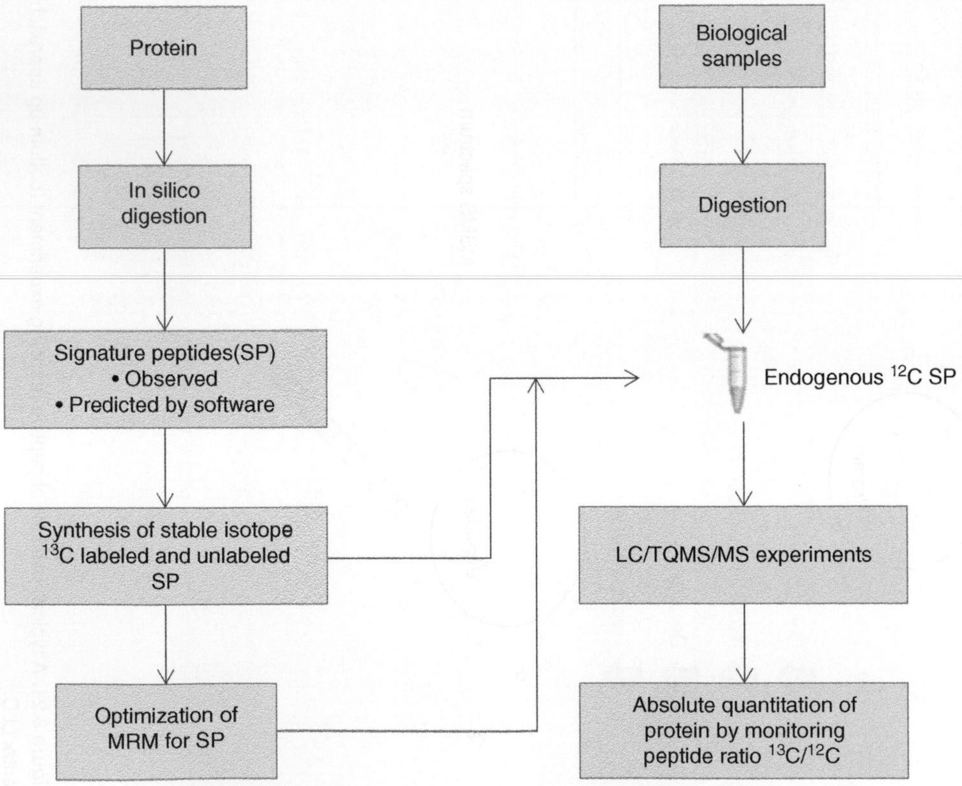


**Figure 6.22.** A typical workflow of peptide quantitation by HPLC/TQMS.

from Dr. MacCoss lab [33], provides an excellent starting point for the development of an LC/MS/MS method, by generating a list of theoretical product ions of a given peptide precursor ion, plus an estimate of the collision energy for these transitions. The next step in the method is to align the experiment data to the predicted data by enzyme-specific proteolytic peptides and product ions in MS/MS. Lastly, in the LC/MS/MS method development, fragmentation energy and cycle time are optimized to achieve maximum sensitivity. Then isotope-labeled SP will be added into digested protein sample in the biological matrix for absolute quantitation.

## 6.5 ENVIRONMENTAL, FOOD SAFETY, CLINICAL, TOXICOLOGY, AND "OMICS" APPLICATIONS

Besides pharmaceutical analysis, LC/MS methods are becoming mainstream in clinical, environmental, and food testing. For example, the routine vitamin D blood test in the clinical laboratories [34] has been dominated by the LC/MS/MS measurement of 25-hydroxy vitamin D2/D3, the metabolites of vitamin D.

Another main application of LC/MS in the field of environmental science is to perform high-throughput screening in the multiresidue testing of targeted and nontargeted organic contaminations. This includes but not limited to pesticides, pharmaceuticals, personal care products, industrial chemicals, hormones, flame-retardants, and plasticizers. These contaminants can enter the environment during its production, consumption, and disposal at ppm or lower levels. Recent improvements in LC/MS capabilities have led to a surge in the use of LC/MS-based techniques for screening, confirmation, and quantitation of these trace level contaminants. In a recent study [35], over 500 pesticides were detected and quantified within one single 12-minutes HPLC method coupled with HRMS. The same LC/MS technology has also been widely applied to the testing of legal and illegal drugs in routine monitoring for clinical and forensic toxicology applications [36, 37].

Remarkable progress and wide application of "omics" technologies [38, 39], including genomics, lipidomics, proteomics, and metabolomics, have occurred during the past two decades. These "omics" research studies are often based on global analysis of biological samples using high-throughput analytical approaches and bioinformatics software. The integrated interdisciplinary omics data provides a better understanding of human diseases, leading to the discovery of new biomarkers for research on novel drug targets and therapies. Currently, LC/MS is the main platform technology as it meets most of the high-performance bar for these assays in sensitivity, selectivity, throughput, robustness, flexibility, and linearity range of quantification of complex biological samples. Figure 6.23 shows the typical MS-based workflow used in metabolomics study [39]. First, by employing LC/MS and LC/MS/MS experiments, the biochemical by-products of cellular metabolites were identified and quantified. Then data interpretation was performed using statistical software, such as principal component analysis (PCA). Different technologies, such as NMR or infrared spectroscopy (IR), can support the MS platforms when needed.

## 6.6 FUTURE PERSPECTIVES

This chapter highlights the power of LC/MS by combining the high resolving power of liquid chromatography and superior mass detection capability of mass spectrometry in qualitative and quantitative analysis. With the continual improvement in ionization methods, HPLC

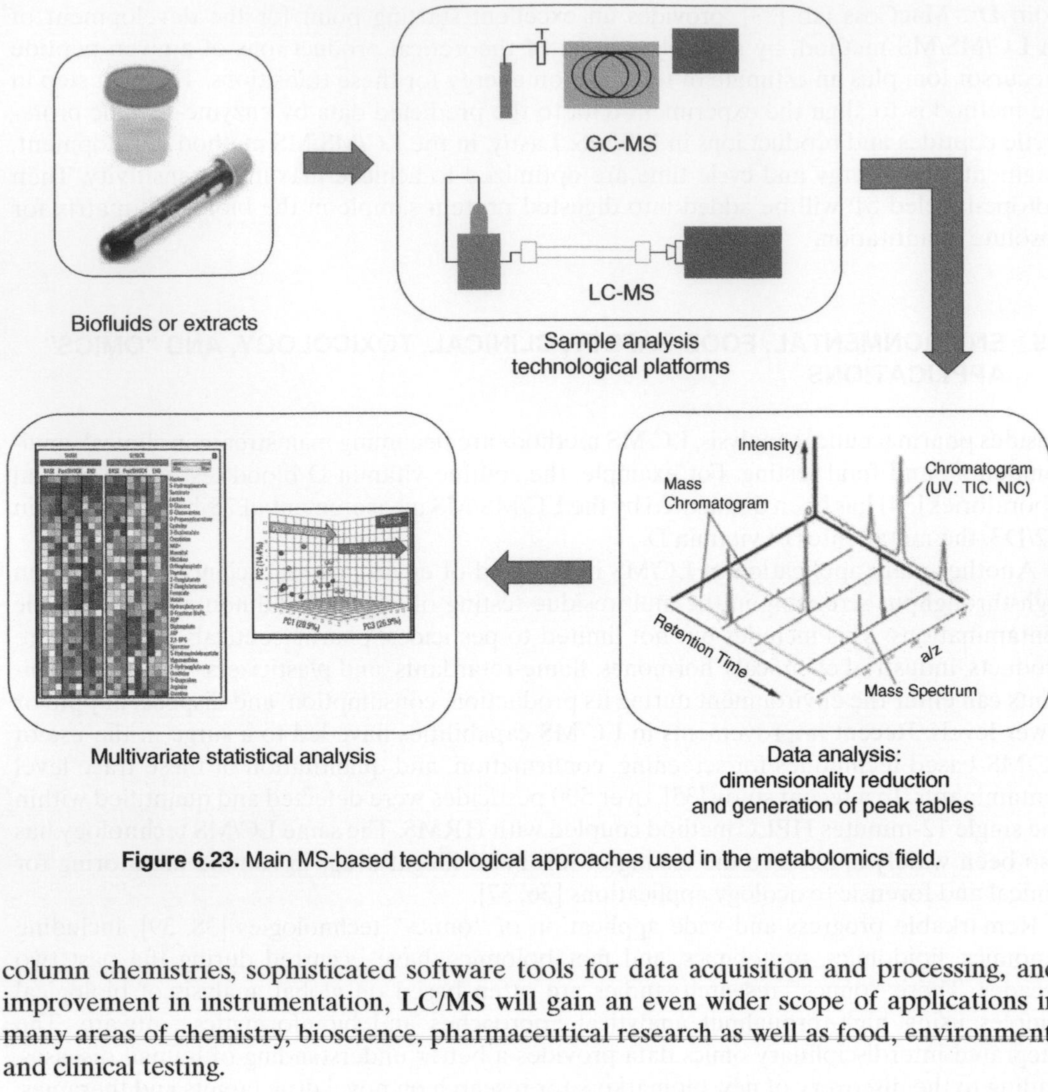


**Figure 6.23.** Main MS-based technological approaches used in the metabolomics field.

column chemistries, sophisticated software tools for data acquisition and processing, and improvement in instrumentation, LC/MS will gain an even wider scope of applications in many areas of chemistry, bioscience, pharmaceutical research as well as food, environment, and clinical testing.

## 6.7 QUIZZES

1. What type of vacuum is required for single-quadrupole MS?
   (a) Atmospheric pressure
   (b) $10^{-5}$ torr
   (c) $10^{-13}$ torr
   (d) With nitrogen only

2. Which MS analyzer is considered to have high resolution?
   (a) SQMS
   (b) Ion trap
   (c) TOF
   (d) TQMS

**3.** **What type of ionization source is most common in LC/MS?**
(a) ESI
(b) EI
(c) APCI
(d) APPI

**4.** **Which MS is most likely used for trace quantitation?**
(a) SQMS
(b) Orbitrap
(c) Ion trap
(d) TQMS

**5.** **What type of MS appears to be most useful for structure elucidation?**
(a) SQMS
(b) Orbitrap
(c) TQMS
(d) FT-ICR

**6.** **Which MS is capable of MS$^n$ fragmentations?**
(a) Ion trap
(b) TOF
(c) SQMS
(d) TQMS

**7.** **Which LC/MS application is used in pharmaceutical development under GMP?**
(a) Structural elucidation
(b) Potency assay
(c) Chiral separation
(d) Cleaning verification

**8.** **Which technique is considered to be the gold standard for structure elucidation in the identification of new chemical entities ?**
(a) SQMS
(b) FTIR
(c) MS/MS
(d) NMR

**9.** **Which mass information is likely used in MS interpretation?**
(a) Average mass
(b) Exact mass
(c) Nominal mass
(d) None of these

**10.** **Which mobile phase additive is not MS-compatible?**
(a) Formic acid
(b) Acetic acid
(c) Phosphoric acid
(d) Ammonia

### 6.7.1 Bonus Quiz

In your own words, describe the reason why LC/MS is becoming a dominant analytical technology platform for bioscience research.

## 6.8 REFERENCES


1. de Hoffmann, E. and Stroobant, V. (2007). *Mass Spectrometry – Principles and Applications*, 3e. Hoboken, NJ: Wiley Interscience.
2. Gross, J.H. (2017). *Mass Spectrometry: A Textbook*, 3e. New York: Springer.
3. Korfmacher, W.A. (2012). *Mass Spectrometry for Drug Discovery and Drug Development*. Hoboken, NJ: Wiley.
4. Dong, M.W. (2006). *Modern HPLC for Practicing Scientists*. Hoboken, NJ: Wiley.
5. Straube, E.A., Dekant, W., and Völkel, W. (2004). Comparison of electrospray ionization, atmospheric pressure chemical ionization, and atmospheric pressure photoionization for the analysis of dinitropyrene and aminonitropyrene LC-MS/MS. *J. Am. Soc. Mass Spec.* 15 (12): 1853–1862.
6. Juraschek, R., Dülcks, T., and Karas, M. (1999). Nanoelectrospray-more than just a minimized-flow electrospray ionization source. *J. Am. Soc. Mass Spec.* 10 (4): 300–308.
7. Cole, R.B. (1997). *Electrospray Ionization Mass Spectrometry: Fundamentals, Instrumentation and Applications*. New York: Wiley.
8. Makarov, A. (2000). Electrostatic axially harmonic orbital trapping: a high-performance technique of mass analysis. *Anal. Chem.* 72 (6): 1156–1162.
9. Hu, Q., Noll, R.J., Li, H. et al. (2005). The Orbitrap: a new mass spectrometer. *J. Mass Spec.* 40 (4): 430–443.
10. Makarov, A., Denisov, E., and Lange, O. (2009). Performance evaluation of a high-field Orbitrap mass analyzer. *J. Am. Soc. Mass Spec.* 20 (8): 1391–1396.
11. Zhang, L.K., Rempel, D., Pramanik, B.N., and Gross, M.L. (2005). Accurate mass measurements by Fourier transform mass spectrometry. *Mass Spec. Rev.* 24 (2): 286–309.
12. Fernandez-Lima, F.A., Becker, C., McKenna, A.M. et al. (2009). Petroleum crude oil characterization by IMS-MS and FTICR MS. *Anal. Chem.* 81 (24): 9941–9947.
13. Wong, M., Murphy, B., Pease, J.H., and Dong, M.W. (2015). Separation science in drug development, Part I: high-throughput purification. *LCGC North Am.* 33 (6): 402–413.
14. Lin, B., Pease, J.H., and Dong, M.W. (2015). Separation science in drug development, Part II: high-throughput characterization. *LCGC North Am.* 33 (8): 534–545.
15. Dong, M.W. (2015). Separation science in drug development, Part III: analytical development. *LCGC North Am.* 33 (10): 764–775.
16. Kou, D., Wigman, L., Yehl, P., and Dong, M.W. (2015). Separation science in drug development, Part 4: quality control. *LCGC North Am.* 33 (12): 900–909.
17. Nagao, T., Yukihira, D., Fujimura, Y. et al. (2014). Power of isotopic fine structure for unambiguous determination of metabolite elemental compositions: in silico evaluation and metabolomic application. *Anal. Chim. Acta* 813: 70–76.
18. Korfmacher, W.A. (2005). Foundation review: principles and applications of LC-MS in new drug discovery. *Drug Discovery Today* 10 (20): 1357–1367.
19. Liu, L. and Pack, B.W. (2007). Cleaning verification assays for highly potent compounds by high-performance liquid chromatography-mass spectrometry: strategy, validation, and long-term performance. *J. Pharm. Biomed. Anal.* 43 (4): 1206–1212.
20. Dong, M.W., Zhao, E.X., Yazzie, D.T. et al. (2012). A generic HPLC/UV platform method for cleaning verification. *Am. Pharm. Rev.* 15 (6): 10–17.
21. Teasdale, A. (2011). *Genotoxic Impurities: Strategies for Identification and Control*. Hoboken, NJ: Wiley.

22. Kellmann, M., Muenster, H., Zomer, P., and Mol, H. (2009). Full scan MS in comprehensive qualitative and quantitative residue analysis in food and feed matrices: how much resolving power is required? *J. Am. Soc. Mass Spec.* 20 (8): 1464–1476.
23. Gu, C. (2012). *Quantitative Analysis of Potential Genotoxic Impurities by High Resolution Mass Spectrometer*, 60th ASMS National Meeting. Vancouver.
24. Keire, D.A., Whitelegge, J.P., Souda, P. et al. (2010). PYY (1-36) is the major form of PYY in rat distal small intestine: quantification using high-resolution mass spectrometry. *Regul. Peptides* 165 (2): 151–157.
25. Zhang, Z., Pan, H., and Chen, X. (2009). Mass spectrometry for structural characterization of therapeutic antibodies. *Mass Spec. Rev.* 28 (1): 147–176.
26. An, B., Zhang, M., and Qu, J. (2014). Toward sensitive and accurate analysis of antibody biotherapeutics by liquid chromatography coupled with mass spectrometry. *Drug Metabol. Disp.* 42 (11): 1858–1866.
27. Roepstorff, P. (1984). Proposal for a nomenclature for sequence ions in mass spectra of peptides. *Biomed. Mass Spec.* 11: 60.
28. Johnson, R.S., Martin, S.A., Biemann, K. et al. (1987). Novel fragmentation process of peptides by collision-induced decomposition in a tandem mass spectrometer: differentiation of leucine and isoleucine. *Anal. Chem.* 59 (21): 2621–2625.
29. Aebersold, R. and Goodlett, D.R. (2001). Mass spectrometry in proteomics. *Chem. Rev.* 101 (2): 269–296.
30. Siuti, N. and Kelleher, N.L. (2007). Decoding protein modifications using top-down mass spectrometry. *Nat. Methods* 4 (10): 817–821.
31. Tran, J.C., Zamdborg, L., Ahlf, D.R. et al. (2011). Mapping intact protein isoforms in discovery mode using top-down proteomics. *Nature* 480 (7376): 254–258.
32. Mao, Y., Valeja, S.G., Rouse, J.C. et al. (2013). Top-down structural analysis of an intact monoclonal antibody by electron capture dissociation-Fourier transform ion cyclotron resonance-mass spectrometry. *Anal. Chem.* 85 (9): 4239–4246.
33. MacLean, B., Tomazela, D.M., Shulman, N. et al. (2010). Skyline: an open source document editor for creating and analyzing targeted proteomics experiments. *Bioinformatics* 26 (7): 966–968.
34. Singh, R.J., Taylor, R.L., Reddy, G.S., and Grebe, S.K. (2006). C-3 epimers can account for a significant proportion of total circulating 25-hydroxyvitamin D in infants, complicating accurate measurement and interpretation of vitamin D status. *J. Clin. Endocrine. Metabol.* 91 (8): 3055–3061.
35. Zhang, A., Chang, J.S., Gu, C., and Sanders, M. (2010). Nontargeted screening and accurate mass confirmation of pesticides using high-resolution LC–orbital trap mass spectrometry. *LCGC North Am.* 8 (3): 40–43.
36. Ojanperä, I., Kolmonen, M., and Pelander, A. (2012). Current use of high-resolution mass spectrometry in drug screening relevant to clinical and forensic toxicology and doping control. *Anal. Bioanal. Chem.* 403 (5): 1203–1220.
37. Wu, A.H., Gerona, R., Armenian, P. et al. (2012). Role of liquid chromatography-high-resolution mass spectrometry (LC-HR/MS) in clinical toxicology. *Clin. Toxicol.* 50 (8): 733–742.
38. Tolstikov, V. (2016). Metabolomics: bridging the gap between pharmaceutical development and population health. *Metabolites* 6 (3): 20.
39. Girolamo, F.D., Lante, I., Muraca, M., and Putignani, L. (2013). The role of mass spectrometry in the "omics" era. *Curr. Org. Chem.* 17 (23): 2891–2905.

# 7

# HPLC/UHPLC OPERATION GUIDE

## 7.1 SCOPE

This chapter describes the operating procedures of common high-performance liquid chromatography (HPLC) modules, chromatography data systems (CDS), columns, samples, mobile phases, and fittings as exemplified by best practices of experienced practitioners (topics shown in Figure 7.1). Concepts in the qualitative and quantitative analysis are discussed together with processes in chromatography data analysis and report generation. Environmental and safety concerns are summarized. Guidelines for increasing HPLC precision and avoiding pitfalls in trace analysis are described. The goal is to provide the laboratory analyst with a concise operating guide for HPLC and ultra-high-pressure liquid chromatography (UHPLC) systems.

Note that method conversion from HPLC to UHPLC conditions and fittings for UHPLC are covered in Chapter 5. The reader is referred to more detailed discussions on this topic in books [1–6], manufacturers' operating manuals, training courses [7, 8], and relevant articles in trade journals [9].

## 7.2 SAFETY AND ENVIRONMENTAL CONCERNS

### 7.2.1 Safety Concerns

The general safety concerns in the HPLC Lab are in-line with most analytical laboratories working with organic solvents and small amounts of biohazardous samples [10]. The high-pressure operation of the HPLC instrument does not pose any significant safety risk to the user since small volumes of incompressible liquids are used, and the modules are designed for these operations. Typical safety risks involve external conditions such as high electric voltages near mobile phases, which can be flammable liquids. The operation and setup of HPLC modules should comply with local, state, and national fire codes such as NFPA 30, NFPA 45, NFPA 70 (National Electrical Codes) and other safety regulations [11]. The toxicity of analytical samples and organic solvents used in mobile phases needs

*HPLC and UHPLC for Practicing Scientists*, Second Edition. Michael W. Dong.

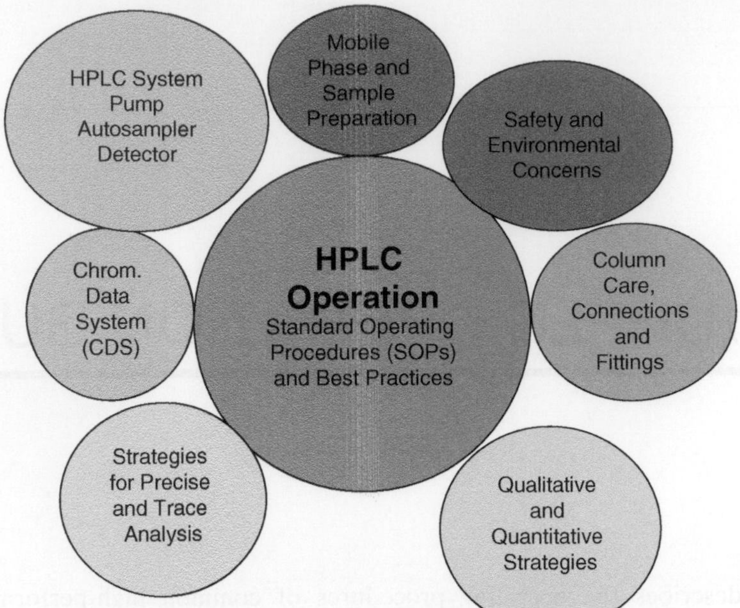


**Figure 7.1.** Schematic diagrams showing the subsections of this chapter.

to be reviewed and integrated into safe laboratory practices and procedures. The material safety data sheets (MSDSs) of each compound or solvent being handled should be consulted before use, preferably with safety and health professionals knowledgeable about workplace chemical exposures and fire safety.

The two common reversed-phase chromatography (RPC) solvents, acetonitrile (ACN) and methanol can be used safely in the laboratory with minimal engineering control and use of personal protective equipment (PPE). Other solvents used in normal-phase chromatography (NPC) or gel-permeation chromatography (GPC) such as tetrahydrofuran (THF), methylene chloride, dimethyl sulfoxide (DMSO), and dimethyl formamide (DMF) need to be handled with a higher level of workplace safety and environmental control. In the case of methylene chloride, OSHA (Occupational Safety and Health Administration) has a specific set of regulations for workplace monitoring and control.

As a general rule, the following practices should be considered when handling chemicals and mobile phase solvents:

- Wear safety glasses and gloves when handling toxic or corrosive chemicals.
- Select PPE that is compatible with the solvents or toxicity of the samples being used.
- Use appropriate respiratory protection when handling substances with acute toxicities to minimize exposure.
- Volatile, flammable, and toxic organic solvents should be handled in a laboratory fume hood and other systems designed specifically for these applications.
- Particular attention should be exercised in weighing solid powder in safety weigh station with recirculating air.
- Toxins and drugs with potential carcinogenicity should be handled on in glove boxes in the powder form. Once enclosed inside a vial or in solution, they can be handled with less occupational risks.

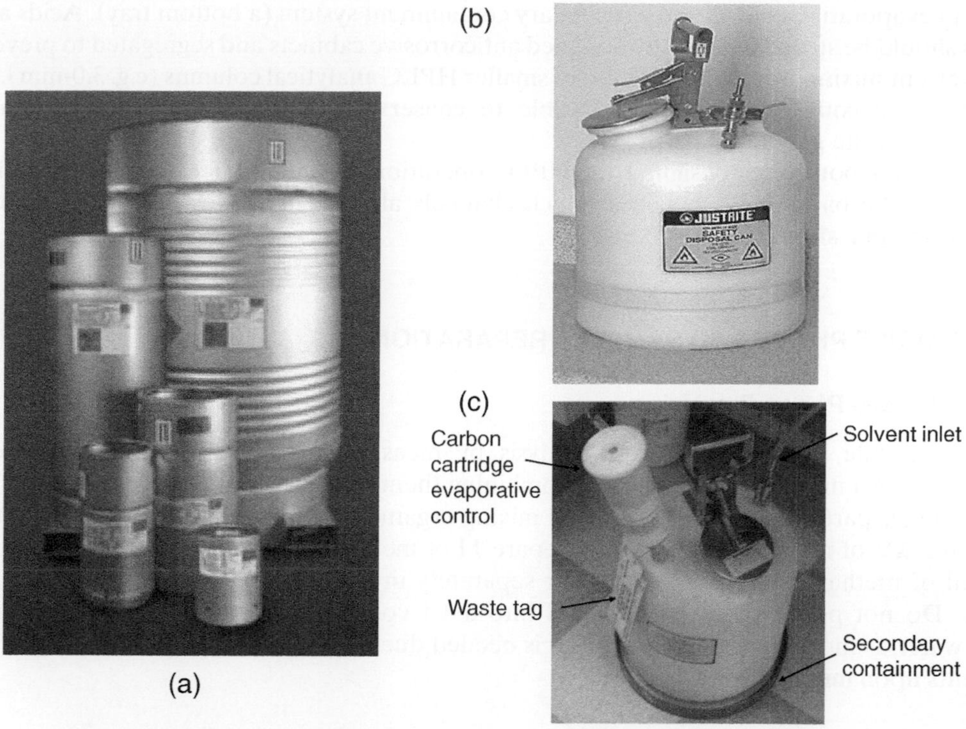


**Figure 7.2.** (a) Solvent cycletainer. (b) Justrite solvent disposal can. (c). A waste disposal can with a waste line connection, an evaporative control cartridge, and a secondary containment tray.

Most HPLC-grade solvents are available in 1-gallon glass bottles and should be transported in appropriate secondary containers such as rubberized carriers. Many common solvents are available in stainless steel "kegs" or cycle-tainers (Figure 7.2a), which are pressurized with nitrogen for convenient dispensing in the laboratory. Electrical grounding of the containers is required in some states. Note that NFPA 45 requires that any dispensing of flammable liquids in a laboratory must use local exhaust ventilation, such as a laboratory fume hood. A common cycle-tainer size is 19-l or 5-gallon. These containers should be transported with safety carts. The potential danger for flask implosion during vacuum filtration and degassing should be noted. This danger can be substantial when vacuum-filtering larger volumes (i.e. 4 l) or if the non-vacuum-grade glassware is mistakenly used. Coated glass equipment and safety shielding should be considered for this operation.

### 7.2.2 Environmental Concerns

The storage and disposal of all chemicals and solvent wastes must follow applicable codes and regulations such as RCRA (Resource Conservation and Recovery Act) [10, 11]. Waste flammable solvents should be stored in rated flammable liquid containers while they are being filled during the HPLC operation. These containers should be properly labeled and transported to a hazardous waste accumulation area within three days after they are filled. Figure 7.2b–c show an example of a solvent disposal safety container (5-gallon), equipped with an evaporative control and a secondary containment system (a bottom tray). Acids and bases should be stored in specially designed anticorrosive cabinets and segregated to prevent inadvertent mixing from spills. The use of smaller HPLC analytical columns (e.g. 3.0-mm i.d.) should be encouraged whenever possible to conserve solvent usage and to minimize hazardous waste generation.

For new laboratories designed for HPLC operation, a stainless steel waste collection system at the back of lab benches, which channels all solvent wastes down to a central collection tank, should be considered.

## 7.3 MOBILE PHASE AND SAMPLE PREPARATION

### 7.3.1 Mobile Phase Premixing

Premix mobile phase for isocratic analysis by measuring the volume of each solvent separately in a measuring cylinder and combining them in the solvent reservoir [6, 12]. This premixing is particularly important when mixing organic solvents with water because of the negative $\Delta V$ of mixing. For example, prepare 1 l of methanol/water (50 : 50) by measuring 500 ml of methanol and 500 ml of water separately in a measuring cylinder and combine them. Do not pour 500 ml of methanol into a 1-l volumetric flask and fill it to volume with water as more than 500 ml of water is needed due to the shrinkage of these imperfect solvents upon mixing.

### 7.3.2 Mobile Phase Additives and Buffers

The use of acidic or basic additives (e.g. 0.1% formic acid, 0.1% ammonia) in mobile phase A (MPA) in RPC is required for any samples containing acidic or basic analytes to control the ionization states of these analytes (see Section 2.3.3). Note that MPA is always the weaker solvent by HPLC conventions to avoid confusion. The use of buffers in MPA is required for critical purity assays where slight changes in mobile phase pH may lead to coelution problems of key analytes. Table 2.3 summarizes common buffers with their respective $pK_a$, UV cutoffs, and compatibility with a mass spectrometer (MS). Since the most effective buffering range is $\pm$1 pH unit of its $pK_a$, picking the right buffer for the desired pH is critical. A buffer concentration of 5–20 mM is sufficient for most applications [2, 5]. pH adjustments should be made in the MPA alone, before mixing it with any organic solvents.

Beware of the possibility of buffer precipitation when mixing it with organic solvents such as ACN. Caution should be exercised when blending ACN with phosphate buffers. The concentration of the phosphate buffer should be kept below 15 mM if possible in MPA or less than 85% ACN in MPB to prevent precipitation during mixing. Buffered mobile phases typically last up to one week pending on the absorption of carbon dioxide or potentials for bacterial growth.

### 7.3.3 Filtration

Filtration through a 0.45-μm membrane filter of all aqueous mobile phases containing buffers or ion-pairing reagents is recommended. Use cellulose acetate membrane filters for aqueous solvents and either polytetrafluoroethylene (PTFE) or nylon filters for organic solvents. The filtration of any HPLC-grade solvents including water from purification systems is not recommended since they are already prefiltered. Use an all-glass filtration apparatus

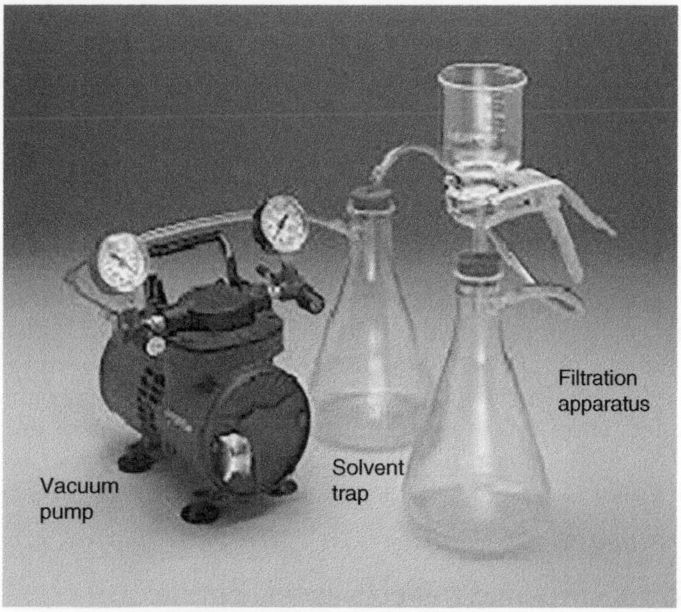


**Figure 7.3.** All-glass solvent filtration flask using 0.45-µm membrane filters and connected to a vacuum pump with a solvent trap.

and 47- or 95-mm membranes with thick-wall vacuum-grade filtration flasks. Examine the flask for chips or cracks before use to prevent chances of implosion. Figure 7.3 shows a typical filtration apparatus used with a vacuum pump equipped with a solvent trap. Note that many manufacturers recommend filtration through 0.2 µm membranes for UHPLC columns. Nevertheless, filtration of buffered mobile phases can be waived if high-purity reagents are used (e.g. >99.995 + % ammonium formate) [13] and normal preventive maintenance is strictly followed with periodic replacements of internal filter elements in the pumps.

### 7.3.4 Degassing

Mobile phase degassing is critical for accurate pump blending and gradient operation [12]. Online vacuum degassers (schematic in Figure 4.23) are standard equipment in most HPLC systems today. Note that vacuum degasser should be turned "ON" all the time and purged when changing solvents. All solvent lines should be filled with solvents in quaternary pumps even though only lines A and B are used. Helium sparging and pressurization is an effective degassing technique particularly for pumping volatile solvents (e.g. pentane or methylene chloride) or labile reagents (e.g. ninhydrin). Degassing by vacuum filtration or sonication is partially effective at best, and such "degassed" solvents will "regas" within a few hours [12].

### 7.3.5 Samples, Diluents, and Sample Preparation

In HPLC analysis, diluted sample solutions (e.g. 0.001–10 mg/ml) are typically placed in 2-ml glass vials into sample trays or microtiter plates for automated injections. Screw caps with thin septa and clear or amber vials are commonly used. Typical fill-volumes are 1–1.5 ml though smaller sample volumes can be accommodated by micro glass inserts or vials with tapered bottoms (e.g. 100–300 µl).

Alternately, the use of solvent weights is advocated for a higher degree of accuracy and convenience, so sample concentrations are expressed as mg/g instead of mg/ml without the use of volumetric flasks. Though this approach appears to have merit, its practice has not caught on in pharmaceutical laboratories.

The best sample diluents are MPA or solvents of equivalent strengths as MPA minus any additives. The default diluent for drug discovery samples is 50% ACN/water. Solvent strengths can be increased to solubilize low-solubility samples though injection volumes of samples containing stronger diluents should be lowered to avoid peak splitting or other chromatographic anomalies.

The ideal sample preparation technique for quantitative analysis is dilute-and-shoot [14, 15]. Drug substance sample solutions are injected without filtration while oral drug product extracts (tablets or capsules) would require a filtration step with a syringe filter (e.g. 13–25-mm i.d., 0.2–0.45 μm membrane filters). More complex samples may require substantial extraction, clean-up, analyte enrichment, or other sample preparation procedures often with the aid of internal standards for accurate quantitation. These preparation procedures are discussed elsewhere in books [15, 16] and other sources [17].

## 7.4 BEST PRACTICES IN HPLC/UHPLC SYSTEM OPERATION

This section summarizes best practices and standard operating procedures (SOPs) used by experienced practitioners in HPLC system operation. They are categorized by each module and the column. Guidelines for enhancing the precision of retention time and peak areas are included. Many operating procedures are based on a popular HPLC system (e.g. Waters Alliance) with updates for modern UHPLC systems as needed.

### 7.4.1 Pump Operation

- Place solvent line sinkers (10-μm filters) into intermediary solvent reservoirs when switching solvents to prevent cross-contamination or buffer precipitation.
- Cap the solvent reservoirs to minimize atmospheric contaminations or evaporation.
- Turn on the online vacuum degasser.
- Set an upper-pressure limit (e.g. 4,000 psi for HPLC or 3000 psi below system limits for UHPLC); set lower-pressure limit if available (typically at 100 psi to trigger pump shutdown if solvent runs out).
- Perform "dry" primes by opening the prime/purge valve and draw out 10 ml of each solvent line (needed if solvent lines are "dry" to prime the pump head). Perform wet prime for three to five minutes daily to purge solvent lines and when changing mobile phases in the reservoirs.
- Program the pump to purge out the column with strong solvents (e.g. ACN for RPC) and shut down the pump after the sample sequence is complete.

Cautions when using buffers or high-pH mobile phases:

- Do not let buffers sit in the HPLC system due to the danger of precipitation. Flush buffers from column and system with 10% MeOH in water.
- Use the piston seal wash feature to wash the back of the piston to prolong seal life (if available and when using high-salt buffer). Keep the peristaltic seal-wash pump primed and the reservoir filled. A typical seal wash solvent is 10% MeOH in water.

- Titanium-based biocompatible systems are preferable for application with high-salt or high-pH mobile phases. High-pH compatible pump and injector rotor seals are recommended for mobile phases pH >10.

### 7.4.2 HPLC Column Use, Precaution, Connection, and Maintenance

The following operating guides are recommended for maintaining RPC columns. Consult other references [2, 3] or the vendor's column instructions for other column types (size-exclusion chromatography [SEC], normal-phase chromatography [NPC], ion-exchange chromatography [IEC], chiral) with respect to special precautions, mobile phase compatibility, and column regeneration procedures.

#### *7.4.2.1 Column Use*

- Store RPC columns in ACN or MeOH or a mixture of water and organic solvents.
- Cap unused columns with "closed" fittings to prevent columns from drying out.
- Always flush the column with a strong solvent (ACN or MeOH) before use to eliminate any highly retained analytes.
- Use guard columns or in-line filters if "dirty" samples are injected. (see Section 3.6).

#### *7.4.2.2 Column Precautions*

- Do not exceed the pH range and upper temperature/pressure limits of the column (typically pH of 2–8, 60–80 °C, and 6000–18 000 psi for silica-based RPC columns).
- *Note:* Many modern HPLC silica-based columns can be used at a wider pH range of 1.5–10 (e.g. hybrids and bonded phases using polyfunctional silanes. see Figure 3.11).
- Never let buffers sit immobile inside the column.

***7.4.2.3 Column Connections*** Stainless steel fittings and ferrules (e.g. Swagelok) are used in high-pressure fluidic connections inside an HPLC or UHPLC system. For HPLC column connections, it is more convenient to use finger-tight reusable PEEK fittings (e.g. Upchurch [Idex], <5000 psi). UHPLC fittings used for column inlet connections are described in Section 5.5.2. Note that regular HPLC PEEK fittings are sufficient for UHPLC column outlet connections to detectors.

For those who elect to use stainless fittings and ferrules for column connections, here are some advice and guidelines.

- Use 1/16″ compression nuts and ferrules from Parker, Swagelok, or Rheodyne. These are compatible with most columns and interchangeable with each other. Use Waters fittings for Waters columns. Make sure that the tubing "bottoms out" inside the fitting when attaching a new ferrule. See Figure 7.4 for ferrule, shapes, and correct seating depths of the various fittings with the tubing bottoming out.
- Use short lengths of 1/16″ o.d. and 0.007″ i.d. stainless or PEEK tubing for general applications and 0.003–0.005″ tubing for narrow-bore and UHPLC columns. Stainless steel tubing can be square-cut with a cutting wheel though precut tubing with small i.d. (<0.007″) is usually purchased.
- Note Agilent systems use 1/32″ o.d. capillaries with a short length of 1/16″ tube braced at the ends. The inner diameters are 0.08 mm (black), 0.12 mm (red), or 0.17 mm (green) (0.003″–0.007″).
- Finger-tight HPLC fittings and PEEK tubing (<5000 psi) are convenient, inexpensive, and easy to use. Note that PEEK is not compatible with DMSO, THF, and methylene chloride often used in GPC.

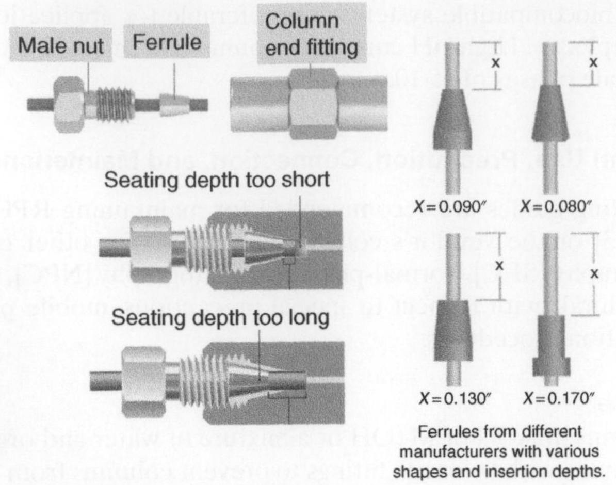


**Figure 7.4.** Schematic diagrams of column connection hardware and the importance of having the correct seating depths of the stainless steel ferrule. If the seating depth is too low, an additional void volume is created. If the seating depth is too long, leaks might occur. Note that various manufacturers offer fittings of different seating depths and ferrule shapes, which might not be interchangeable. Source: Adapted from Fittings Primer (IDEX) and other resources. https://www.idex-hs.com/literature-tools/educational-materials/fittings-primer/.

#### *7.4.2.4 Column Maintenance and Regeneration*

- Typical column lifetime is 3–24 months or 1000–3000 injections, depending on the type of mobile phase and samples injected.
- Column performance (efficiency) decreases with time as signified by increased back-pressure and peak widths.
- Monitor the column back-pressure and efficiency performance and take corrective action immediately; don't wait until the column is plugged.
- If the pressure is abnormally high, try back flushing the column as soon as possible with a strong solvent. The inlet frits of some HPLC columns can be replaced if plugged.
- Regenerating the column by flushing with a series of strong solvents may restore performance of a contaminated column. For RPC columns, use a sequence of water, MeOH, $CH_2Cl_2$, and MeOH. Consult vendor's instructions whenever possible.
- If column "voiding" occurs in an HPLC column inlet, some performance might be restored by filling the inlet void with a similar packing.
- In general, UHPLC columns packed with smaller particles are difficult to repair due to their high efficiency and smaller column volumes.

### 7.4.3 Autosampler Operation

The following guidelines are recommended for autosamplers: [12]

- *Flush solvents* (required for some *XYZ*-type autosamplers): Use a methanol or ACN and water mixture without buffer (e.g. 50%) as the flush solvent. Degas the flush solvent to eliminate any bubbles. (Note that integrated-loop type autosamplers use the mobile phase to flush the loop and do not require a flush or carrier solvent).
- Some UHPLC autosamplers (e.g. Acquity UPLC Classic) require two wash solvents: A strong wash (e.g. 90% ACN/water) and a weak wash (10% ACN/water). Typical volumes

are 200 µl (strong) and 600 µl (weak wash) per injection in the partial loop injection mode (where the sample loop is only partially filled).

- *Purging:* Purge the injector daily and before sample analysis to remove bubbles in the sampling syringe. This step is critical for sampling precision for autosamplers such as Waters Acquity UPLC Classic or Alliance. If the bubble cannot be dislodged by automated purging, manual purging may be needed by disconnecting the knurl-nut attached to the plunger and quickly moving the syringe plunger up and down manually.
- *Needle wash*: The outside tip of the sampling needle can be washed automatically in several ways depending on the device by dipping it either in a designated wash vial or into a rinse station fed by fresh solvent.
- *Filling sample vials*: Fill each vial with enough sample solutions for all injections (e.g. 1–1.5 ml in a 2-ml vial). Some autosamplers use a side port needle that might require at least a 0.5-ml sample volume in a 2-ml vial (e.g. Waters Alliance). Use screw caps with thin septa and clear or amber vials. Some UHPLC autosamplers use preslit septa (Waters UPLC Classic).
- *Injection volumes*: Typical injection volume range is 5–50 µl for HPLC or 1–20 µl for UHPLC autosamplers. Injection sizes of <5 µl might lead to poorer precision for conventional *XYZ*-type HPLC autosamplers (see Figure 7.5 for case study). UHPLC autosamplers can inject precisely down to 1 µl.
- For large-volume injections, a larger sampling syringe and sampling loop are required.

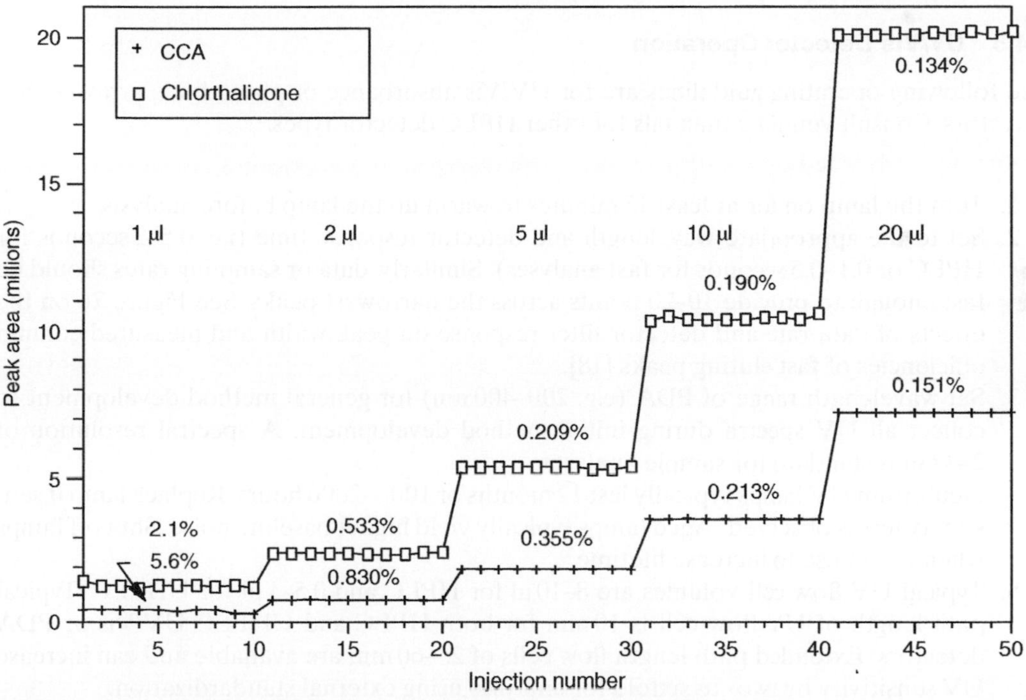


**Figure 7.5.** Diagram illustrating the precision of an XYZ HPLC autosampler vs. injection volumes. Note that peak area precision decreases (higher RSD) with small injection volumes. The phenomenon is caused by the finite sampling volume precision of the autosampler sampling syringe and its stepper motor, which was about 0.01 µl for this particular autosampler. Source: From M.W. Dong 2000 [12].

- *Temperature control*: Temperature controllers for autosampler trays are useful options, particularly for sample solutions of limited thermal stability.
- *Carryovers*: If excessive carryover is encountered, explore the use of pre- or postinjection flush procedures or an alternate material for the rotor seal (e.g. using PEEK vs. Vespel polyimide to reduce carryover of highly basic compounds or proteins).
- *Injector rotor seal*: The use of high-pH mobile phases with pH > 10 would require a change of the injector rotor seal materials from Vespel to Tefzel or PEEK.
- *Injector volumes*: Typical volumes of sample loops and sampling syringes are 100 and 250 µl for HPLC autosamplers, respectively; 10 or 50 µl for UHPLC autosamplers. Put a warning label outside the autosampler if a nonstandard sampling syringe or loop is installed.

### 7.4.4 Column Oven and Switching Valve

There are two types of column ovens: a forced air-circulatory and a still-air aluminum block-type. Typical temperature range is ambient plus 5–80 °C or 100 °C. A Peltier device can be added to extend the lower temperature limit to 5 °C. The use of a column oven is mandatory since retention time stability in RPC is highly sensitive to column temperatures. Most column ovens are equipped with low-volume heat-exchangers to preheat the mobile phase before entering the column. Optional column switching valves (2-, 4-, or 6-column) can be installed to increase the versatility of using multiple columns in assays or methods development.

### 7.4.5 UV/Vis Detector Operation

The following operating guidelines are for UV/Vis absorbance or photodiode array (PDA) detectors. Consult vendor's manuals for other HPLC detector types.

1. Turn the lamp on for at least 15 minutes to warm up the lamp before analysis.
2. Set to the appropriate wavelength and detector response time (i.e. 0.5–2 seconds for HPLC or 0.1–0.5 seconds for fast analyses). Similarly, data or sampling rates should be fast enough to provide 10–20 points across the narrowest peaks. See Figure 7.6 on the effects of data rate and detector filter response on peak width and measured column efficiencies of fast eluting peaks [18].
3. Set wavelength range of PDA (e.g. 200–400 nm) for general method development to collect all UV spectra during initial method development. A spectral resolution of 2–4 nm is standard for sample analyses.
4. Deuterium UV lamps typically last 12 months or 1000–2000 hours. Replace lamp if sensitivity loss is observed. Aged lamps typically yield higher baseline noise. Shut off lamps when not in use to increase lifetime.
5. Typical UV flow cell volumes are 8–10 µl for HPLC and 0.5–1 µl for UHPLC. Typical path length of UV flow cell is 10 mm for both HPLC and UHPLC UV/Vis or PDA detectors. Extended path length flow cells of 25–60 mm are available and can increase UV sensitivity by two- to sixfold for analyses using external standardization.

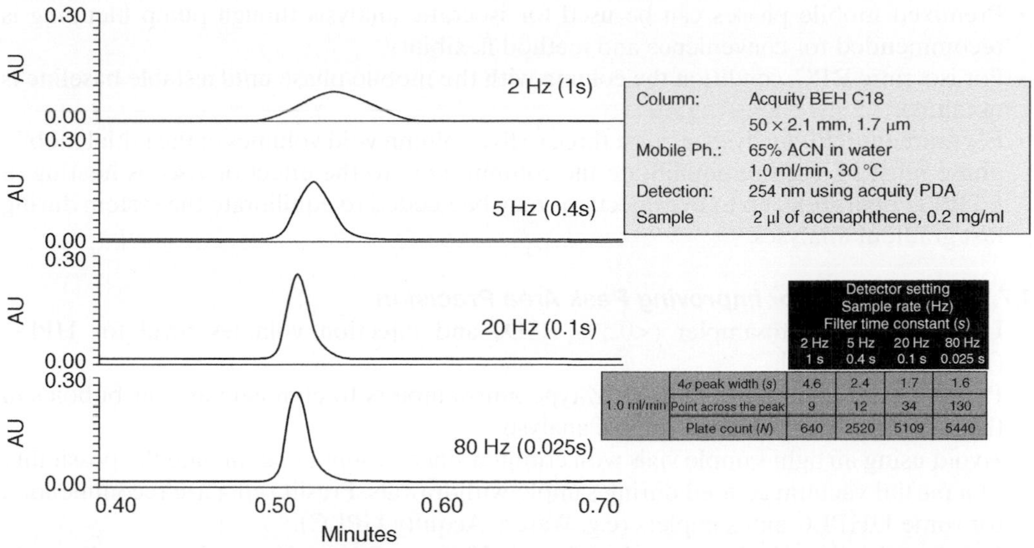


**Figure 7.6.** Comparative chromatograms illustrating the effect of detector response (filter time constant) and sample rate on peak width and plate counts on a fast eluting peak. Source: Fountain et al. 2009 [18]. Reprinted with permission of Elsevier.

### 7.4.6 HPLC System Shutdown

The following procedures are recommended for HPLC system shutdown:

- Turn off column oven first and let the column cool down before turning off the pump flow. Flush any buffered mobile phases out of the system with 10% MeOH in water and stop the flow.
- Turn off the UV lamp to preserve source life, though the power of the detector should be left in the "on" position.
- For high-sensitivity gradient analysis, it is a good practice to inject a blank at the end of a sample sequence to clean out the system and injector before system shutdowns.

### 7.4.7 Guidelines for Increasing HPLC Precision

Guidelines for increasing HPLC precision for retention time and peak area are presented as follows. The reader is referred to Ref. [12] for details of a case study.

#### *7.4.7.1 Guidelines for Improving Retention Time Precision*

- The use of precise pumps and column ovens is mandatory for automated and precise analysis [12, 14].
- Most column ovens are operative from ambient +5 °C upward. Therefore, the standard default column temperature is 30 or 35 °C. Column temperatures at 25 °C down to ~5 °C can be set in column ovens equipped with Peltier cooling capability.

- Premixed mobile phases can be used for isocratic analysis though pump blending is recommended for convenience and method flexibility.
- For isocratic RPC, condition the column with the mobile phase until a stable baseline is obtained.
- For gradient RPC analysis, at least three to five column void volumes of the initial mobile phase must be used to equilibrate the column. Due to the effect of viscous heating in UHPLC operation, up to five injections may be needed to equilibrate the system during fast gradient analyses.

#### 7.4.7.2 *Guidelines for Improving Peak Area Precision*

- Use a precise autosampler (<0.5% RSD) and injection volumes >5 μl for HPLC (Figure 7.5) [12, 14].
- Purge the sampling syringe in *XYZ*-type autosamplers to eliminate any air bubbles in the sampling syringe before sample analysis.
- Avoid using airtight sample vials with crimped silicone septa to eliminate the possibility of a partial vacuum created during sample withdrawals. Preslit septa are recommended for some UHPLC autosamplers (e.g. Waters Acquity UPLC).
- Improve the signal/noise ratio of peaks to >50 if possible and use a data sampling rate of at least 10–20 pt/pk (see case study in Figure 7.7).

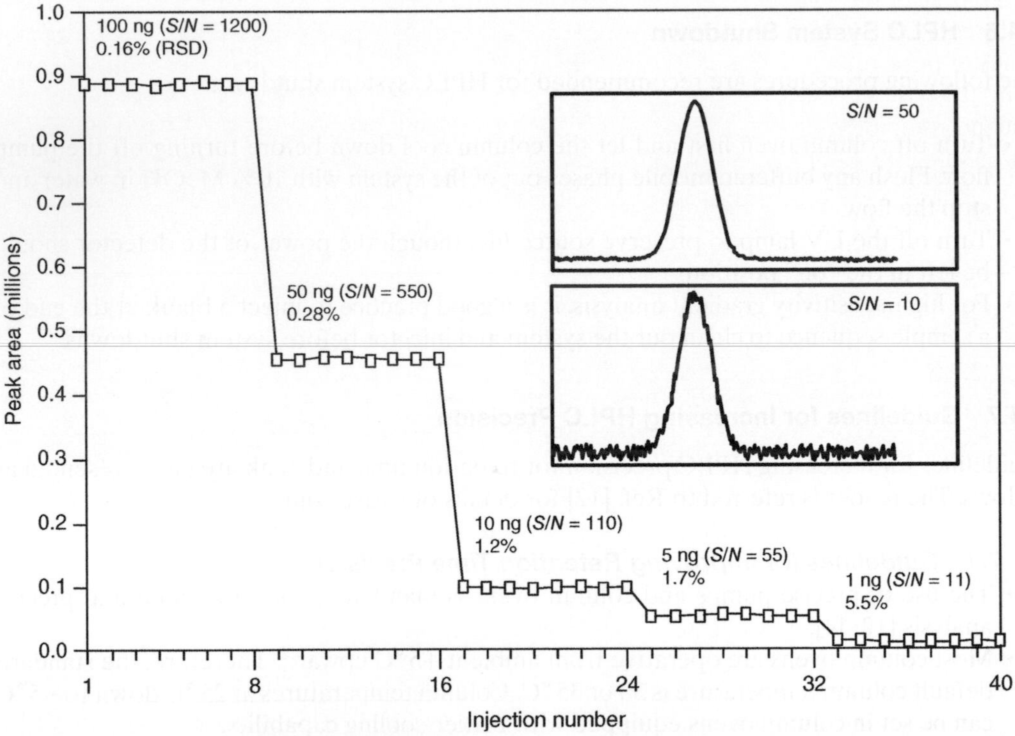


**Figure 7.7.** Diagram illustrating typical autosampler precision vs. peak signal/noise ratio (keeping injection volume constant at 10 μl). Note that the peak area precision worsens (increasing RSD) because precision was limited by the statistical variation of the integration of noisy peaks when the signal-to-noise ratios (*S/N*) are less than 100. Source: From Dong 2000 [12].

- Plot baseline with chromatograms at expanded scales in reports and watch out for integration problems. Use more advanced integration algorithms (e.g. ApexTrack) if possible for complex chromatograms with sloping baselines [19].

A comprehensive discussion on chromatographic peak integration can be found elsewhere [19, 20].

## 7.5 FROM CHROMATOGRAMS TO REPORTS

While the goal of the quantitative analysis is to obtain a chromatogram with adequate separation of key analytes, several processing steps are needed to convert the raw chromatographic data into useful quantitative information or reports. In the past three decades, CDSs have made tremendous improvements in performance and refinements. Chapter 4 gives a brief history of CDS from strip chart recorders to the client–server data networks. The reader is referred elsewhere for more details of CDS [19]. In today's HPLC laboratories, particularly those in regulated industries, analysts tend to spend as much time on a CDS as those in front of a chromatograph.

While the functional details of individual CDS vary greatly, Figure 4.25 shows the typical key processes or functions (integration, calibration, and quantitation) to convert sample chromatograms (raw data from the detector) into useful reports. These processes are controlled by a set of user-specified methods residing in the data system as summarized in Table 7.1 (Examples and terminologies from Waters Empower CDS).

"Integration" is the process of converting digital chromatography raw data into peak data (series of peak retention times and associated peak areas). In a traditional integration algorithm, the baseline and component peaks are established by monitoring the slope of the raw data and comparing them with a "threshold" to determine the "peak start." A series of integration events in the processing method can be used to customize this integration process. A newer integration algorithm using the second derivatives of the raw data for peak detection is available in some CDS and appears to be superior for the automated integration of complex chromatograms with sloping baselines (Figure 7.8) [19].

**Table 7.1. Data System Method Types and Functions (Waters Empower CDS)**

| Data system method type | Primary functions |
|---|---|
| Instrument method | Controls and documents parameters of pump and detector<br>– Pump: flow, gradient conditions, degasser, (oven)<br>– Detector: wavelength, bandwidth, filter response, sampling rate |
| Sample set or sequence method | Controls and documents parameters of autosampler and sample/standard information<br>– Injection sequence: vial #, inject volume, # injections, run time<br>– Functions: inject samples, equilibrate, calibrate, quantitate<br>– Standard/sample info: name, amount, sample weight, label claim, level |
| Processing method | Controls and documents integration parameters, component names, calibration and quantitation information<br>– Integration: threshold, peak width, minimum peak area, integration events<br>– Component table: component names, retention times, response factors |
| Report method | Performs automated custom calculations, formats, and prints reports |

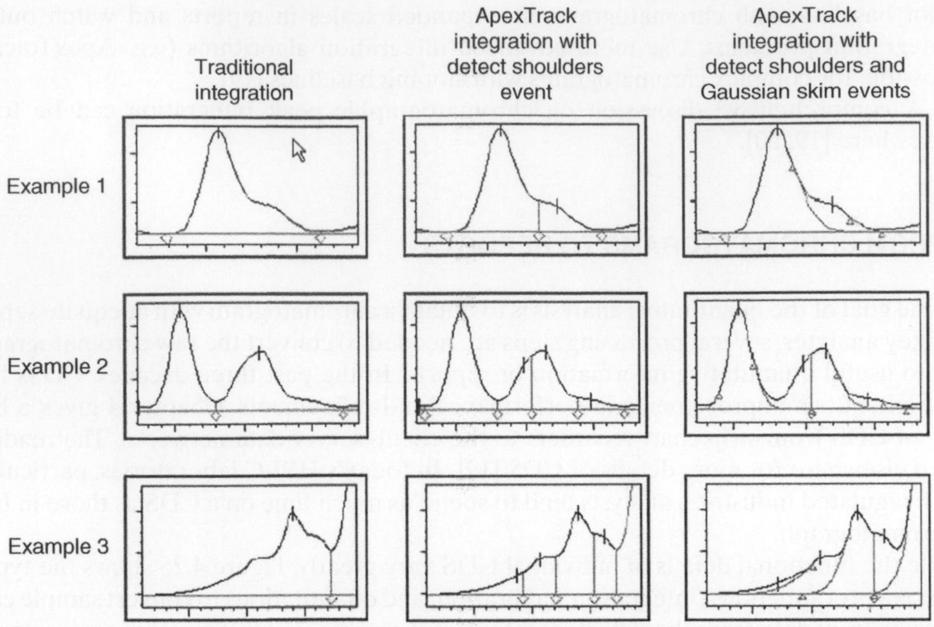


**Figure 7.8.** Diagrams illustrating the use of traditional integration and "ApexTrack" integration algorithm (Waters Empower) based on second derivative peak detection. Note that for complex chromatograms, the traditional integration algorithm may require some manipulations while the ApexTrack appears to integrate difficult peaks in an automated fashion. Source: Courtesy of Waters Corporation.

"Calibration" is the process of establishing a calibration curve of the specified analyte from a set of injected calibration standard solutions. Figure 7.9 shows a typical calibration curve plotting the peak area of the analyte against the amount injected. The response factor ($R_f$) can be calculated from the slope of the curve or by dividing the peak area with the amount according to the equation below. In "quantitation," peaks in the unknown samples are identified by comparing them with retention times of the component list in the processing method. The amount of the sample (either concentration or weight) can then be calculated by dividing the peak area by its respective response factor.

$$\text{Response Factor } (R_f) = \frac{\text{Area}_{\text{std}}}{\text{Amount}_{\text{std}}} = \frac{\text{Area}_{\text{sample}}}{\text{Amount}_{\text{sample}}}$$

While single point calibration is typically for UV detectors, in MS, ELSD (evaporative light scattering detector), or CAD (charged aerosol detector) detection, it is customary to establish a multiple-level calibration standard curve [14].

"Reporting" is the process of generating a formatted report from the result files using a reporting method. There are two types of reports – sample and summary reports (in Waters Empower's terminology). Examples are shown in Figures 7.10 and 7.11. The sample report documents the entire sample and method information, the sample peak result table, and the chromatogram, and may include additional data such as spectral info from the PDA or the MS (Figures 4.26 and 7.10). The summary report summarizes data of specific sample results and might include sums, averages, and relative standard deviation (RSD) of specific analytes. For instance, summary reports are used for summarizing system suitability testing (Figure 7.11), content uniformity assays, and dissolution testing. Customized calculations can be used in both report types. These reports can also be exported to the Laboratory Information Management System (LIMS) or other archival systems.

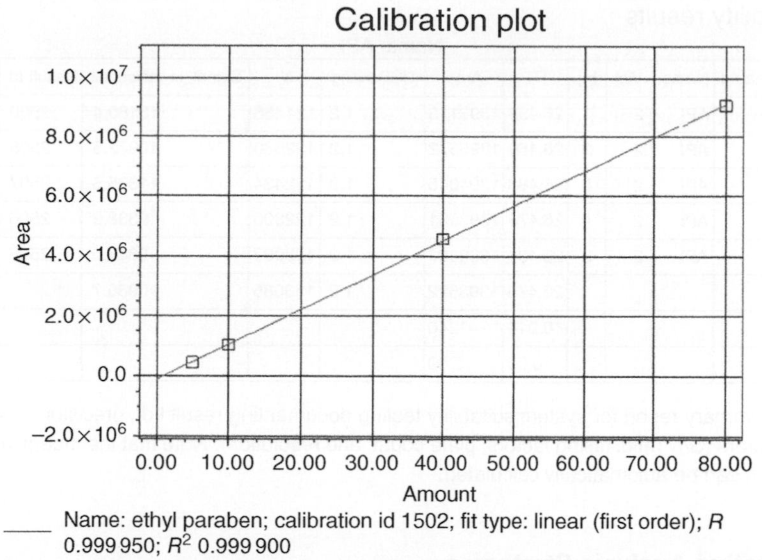


**Figure 7.9.** A typical calibration curve using external standardization (linear calibration through the origin). Note that the slope of the curve is equal to the response factor of the analyte.

**Sample information**

| | | | |
|---|---|---|---|
| Sample name: | sst-nxm | Acquired by: | DonqMi |
| Sample type: | Unknown | Date acquired: | 12/2/2004 6:30:02 PM |
| Vial: | 2 | Acq. method set: | ch56_DG_uv_pda |
| Injection #. | 1 | Date processed: | 12/9/2004 8:07:54 AM |
| Injection volume: | 30.00 ul | Processing method: | mwd_api_pda1 |
| Run time: | 55.0 minutes | Channel name: | Wv In Ch1 |
| Sample set name: | | Proc. chnL descr.: | PDA 280.0 nm |

**Auto-scaled chromatogram**

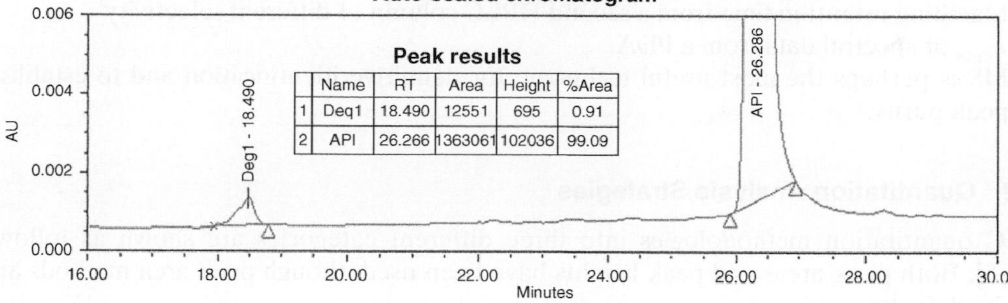


**Peak results**

| | Name | RT | Area | Height | %Area |
|---|---|---|---|---|---|
| 1 | Deq1 | 18.490 | 12551 | 695 | 0.91 |
| 2 | API | 26.266 | 1363061 | 102036 | 99.09 |

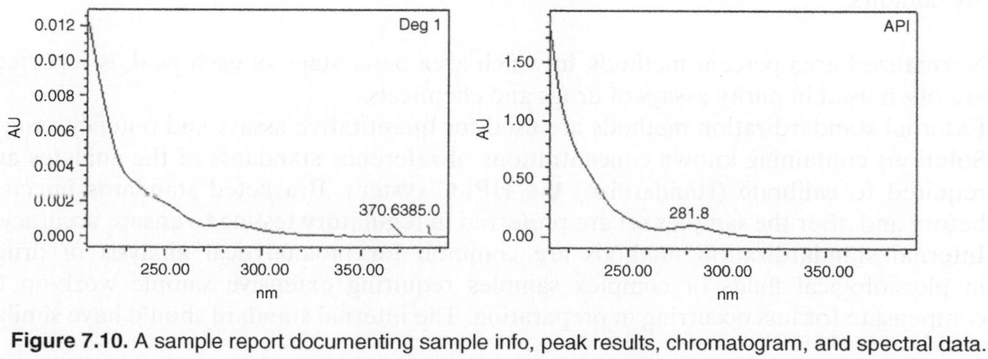

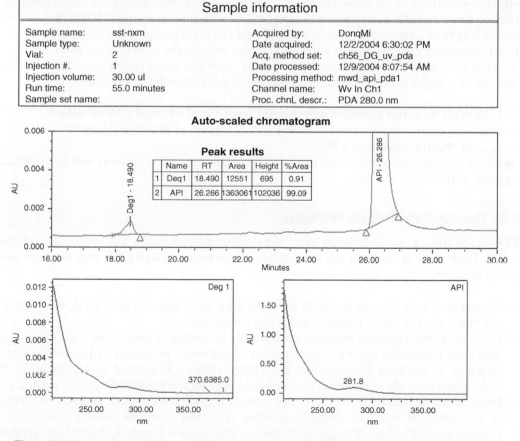

**Figure 7.10.** A sample report documenting sample info, peak results, chromatogram, and spectral data.

System suitability results

Name: API

| | Sample name | Name | Vial | Inj | RT | Area | USP tailing | *N* | Signal_to_noise_all | Result Id | USP resolution |
|---|---|---|---|---|---|---|---|---|---|---|---|
| 1 | sst-nxm | API | 2 | 1 | 26.458 | 1393113 | 1.2 | 124666 | 13180.4 | 2509 | 23.2 |
| 2 | sst-nxm | API | 2 | 2 | 26.466 | 1395572 | 1.2 | 122530 | 10922.5 | 2508 | 23.0 |
| 3 | sst-nxm | API | 2 | 3 | 26.488 | 1394045 | 1.2 | 122434 | 11395.5 | 2507 | 23.0 |
| 4 | sst-nxm | API | 2 | 4 | 26.471 | 1392961 | 1.2 | 122800 | 9338.3 | 2506 | 22.8 |
| 5 | sst-nxm | API | 2 | 5 | 26.491 | 1392397 | 1.2 | 122993 | 9847.0 | 2505 | 22.8 |
| Mean | | | | | 26.475 | 1393622 | 1.2 | 123085 | 10936.7 | | 22.9 |
| S. D. | | | | | 0.014 | 1240 | | | | | |
| %RSD | | | | | 0.05 | 0.09 | | | | | |

**Figure 7.11.** A summary report for system suitability testing documenting result i.d., precision, and other system suitability parameters (*S/N* ratio, tailing factors, plate count, and resolution). Note that the means and precision of each table column can be automatically calculated.

### 7.5.1 Qualitative Analysis Strategies

In qualitative analysis, the goal is to establish the identity of unknown components in the sample [2, 4, 6]. In HPLC with UV detection, peaks are identified by its retention time as compared with those of the known analytes in the reference standard solution. Since absolute retention times are affected by many parameters, relative retention times (RRTs) (retention time ratio against a reference component, typically the main component) are often used for related substances analysis in more complex samples. Spiking the sample with known analytes is another technique to ascertain peak identity by retention times (e.g. a minor peak riding at the tail of a major peak). Other techniques to aid peak identification are as follows:

- Peak ratio from two detection wavelengths from a dual-channel UV/Vis detector.
- Matching retention time from a second HPLC column of different selectivity.
- $\lambda_{max}$ or spectral data from a PDA.
- MS is perhaps the most useful technique for definitive identification and to establish peak purity.

### 7.5.2 Quantitation Analysis Strategies

HPLC quantitation methodologies into three different categories are shown as follows [2, 4–6]. Both peak areas and peak heights have been used, though peak area methods are used by defaults.

- Normalized area percent methods, in which area percentage of each peak is reported, are often used in purity assays of drugs and chemicals.
- External standardization methods are used for quantitative assays and potency assays. Solutions containing known concentrations of reference standards of the analytes are required to calibrate (standardize) the HPLC system. Bracketed standards injected before and after the samples set are preferred in regulatory testing to ensure accuracy.
- Internal standardization methods are common for bioanalytical analysis of drugs in physiological fluids or complex samples requiring extensive sample work-up to compensate for loss occurring in preparation. The internal standard should have similar

structures to the analytes and is added before sample work-up. For UV detection, the internal standard must be resolved from any potential sample components. In most bioanalytical LC/MS assays, isotopically labeled analytes (e.g. deuterated analytes) are commonly used to compensate for isolation recovery and nebulization/ionization efficiency.

## 7.6 SUMMARY OF HPLC OPERATION

The following is a summary of important procedures, a typical sequence of events, for HPLC operation:

- Filter and degas mobile phases.
- Prime pump, rinse column with strong solvents, and equilibrate column with the mobile phase until a steady UV baseline is obtained.
- For *XYZ*-type autosamplers, purge injector and make sure there are no bubbles in the sampling syringe.
- Perform system suitability testing (for regulatory testing, see Section 11.3.7).
- Analyze samples.
- Process and report data.
- Rinse the column and shut down the pump and lamp.

Additional guidelines for using UHPLC equipment and high-pH mobile phase operation are available elsewhere [21, 22].

## 7.7 GUIDES ON PERFORMING TRACE ANALYSIS

Trace analysis is difficult and requires rigorous procedures to eliminate potential contaminations and interferences. These assays are particularly challenging for gradient HPLC separations, where any trace contaminants in the weaker mobile phase (MPA) are concentrated on the head of the column during column equilibration and emerge as ghost peaks. The reader is referred to books [5, 14] and research articles [13] on this topic. This section summarizes guidelines to minimize some of the difficulties and avoid potential pitfalls:

- Use high-purity reagents particularly for those additives used in MPA. Rinse all glassware including the solvent reservoirs.
- Run a procedure blank to ensure that the blank chromatogram does not contain any interfering peaks (see Figure 7.12 for comparison between a good and a bad blank chromatogram in gradient HPLC). A more comprehensive discussion of ghost peaks can be found elsewhere [23].
- The purity of MPA is critical since several milliliters of MPA are used to equilibrate the column for gradient analysis. One successful strategy is to use a high-purity buffer (>99.995%) and eliminate the buffer filtration step to minimize the potential for contamination [13].
- The pH adjustment step of MPA may pose potential problems. The pH calibration buffers often contain preservatives that may contaminate MPA if the pH electrode is dipped during pH adjustments (see Figure 7.13) [22]. This problem can be eliminated by pouring out small aliquots into separate containers to check pH.

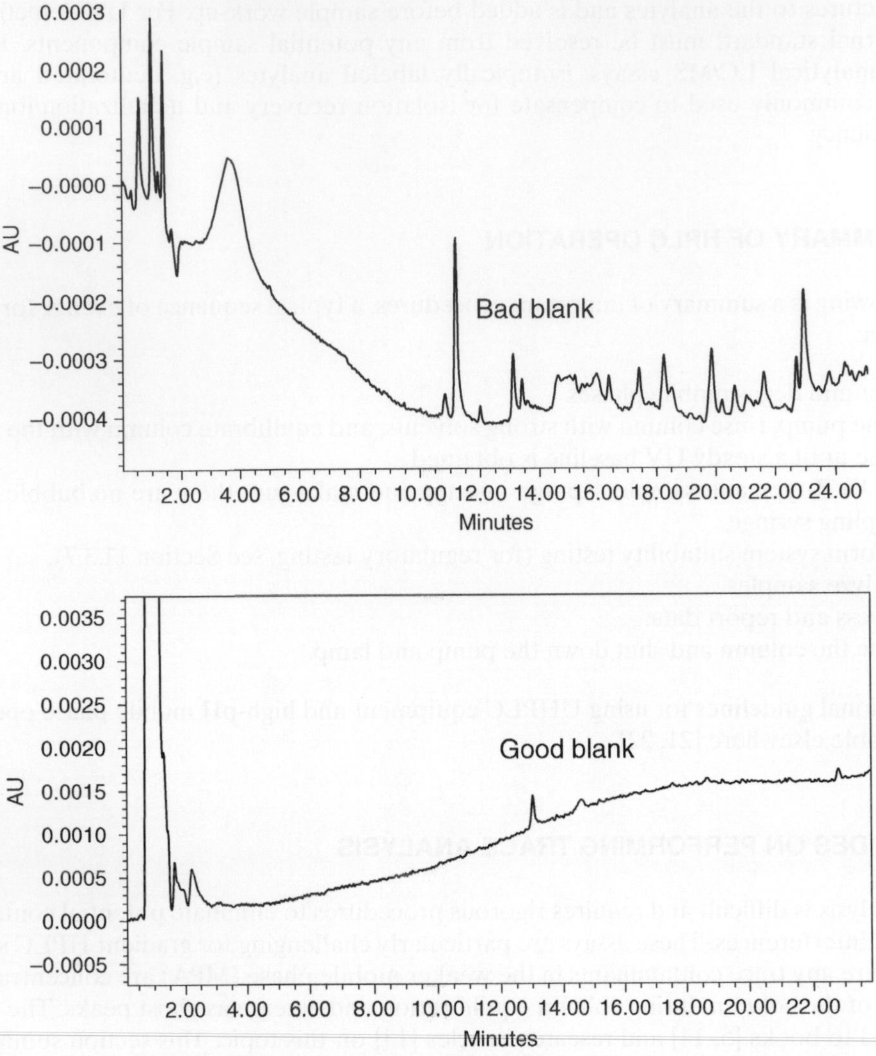


**Figure 7.12.** A good vs. a bad blank chromatogram from a gradient trace analysis for impurity testing of pharmaceuticals. The "ghost" peaks from the blank injection are derived mostly from the trace contaminants in the weaker mobile phase, which are concentrated during column equilibration.

- Enhance the sensitivity of the HPLC method by:
  - Selecting a detection wavelength at or near the $\lambda_{max}$ of the analytes or at far UV (210–230 nm) if the molar absorptivity of the analyte at $\lambda_{max}$ is too low.
  - Increasing the injection volume or the concentration of the sample.
  - Using a detector with higher sensitivity (e.g. a UV/Vis detector typically has higher sensitivity than a PDA detector) and a flow cell with an extended path length if warranted.
- Minimize gradient shifts:
  - Substantial gradient shifts due to absorbance or refractive index changes of the mobile phase during broad gradients are particularly serious for running

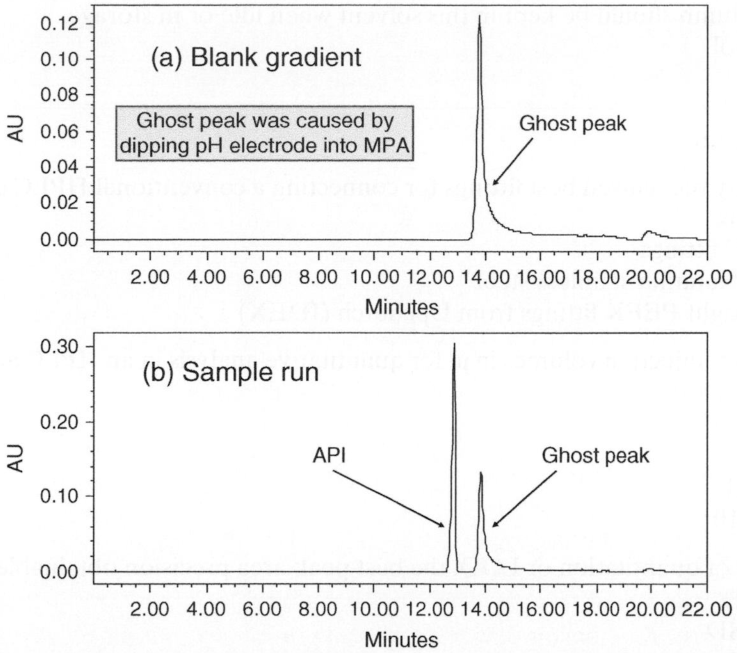


**Figure 7.13.** (a) A ghost peak stemming from contamination of mobile phase A with the pH calibration buffer preservative (sodium *o*-phenylphenate) during pH adjustment procedure caused by dipping the pH electrode into mobile phase A. (b) Note that this ghost peak is substantial and elutes close to the main component in a pharmaceutical assay. This ghost peak is eliminated by not dipping the pH electrode into the mobile phase A during preparation. Source: Dong et al. 2003 [22]. Reprinted with permission of Elsevier.

MS-compatible trifluoroacetic acid (TFA) or formic acid gradients with low UV detection.

— Gradient shifts can be reduced by lowering the concentrations of the mobile phase additives (i.e. from 0.5–0.1% TFA to 0.03–0.05%) and by balancing absorbance of MPA and MPB. See example in Figure 10.7.

## 7.8 SUMMARY

In summary, more effective and successful HPLC assays can be accomplished by following some of the "best practices" in mobile phase preparation, strategies in qualitative, quantitative, and trace analysis, and standard operating procedures of the HPLC system and CDS.

## 7.9 QUIZZES

**1.** Filtration of this mobile phase is needed.
- (a) HPLC-grade solvents
- (b) Water from purification system
- (c) Mobile phase with ion-pairing reagents
- (d) 0.05% formic acid in water

2. An RPC column should be kept in this solvent when idle or in storage.
   (a) Methanol
   (b) Water
   (c) Buffer
   (d) Chloroform

3. The generally recognized best fittings for connecting a conventional HPLC column is
   (a) Swagelok
   (b) UHPLC fittings
   (c) Parker-Hannifin stainless steel
   (d) Finger-tight PEEK fittings from Upchurch (IDEX)

4. The minimum injection volumes in µl for quantitative analysis in an HPLC and a UHPLC autosamplers are
   (a) 50 and 10
   (b) 5 and 1
   (c) 5 and 0.1
   (d) 20 and 10

5. At the limit of quantitation or LOQ, the best peak area precision obtainable is about
   (a) 5% RSD
   (b) 20% RSD
   (c) 1% RSD
   (d) 50% RSD

### 7.9.1 Bonus Quiz

In your own words, describe in what circumstances when the filtration of the mobile phase can be eliminated and what benefits can one derive from skipping this filtration step.

In your own words, describe the common steps in an HPLC operation for quantitative assays.

## 7.10 REFERENCES


1. Snyder, L.R. and Kirkland, J.J. (2010). *Introduction to Modern Liquid Chromatography*, 3e. Hoboken, NJ: Wiley. Chapters 5, 11 and 17.
2. Snyder, L.R., Kirkland, J.J., and Glajch, J.L. (1997). *Practical HPLC Method Development*, 2e. New York: Wiley-Interscience.
3. Neue, U.D. (1997). *HPLC Columns: Theory, Technology, and Practice*. New York: Wiley-VCH.
4. Katz, E. (1987). *Quantitative Analysis Using Chromatographic Techniques*. Chichester, United Kingdom: Wiley.
5. Meyer, V.R. (2010). *Practical HPLC*, 5e. Chichester, United Kingdom: Wiley.
6. Ahuja, S. and Dong, M.W. (eds.) (2005). *Handbook of Pharmaceutical Analysis by HPLC*. Amsterdam, the Netherlands: Elsevier/Academic Press.
7. HPLC training Short Courses at National Meetings such as Pittcon, Amer. Chem. Soc., and Eastern Analytical Symposium.
8. Separation Science. Provides Online Training, Webinars, and Conferences in Separation Science, United Kingdom. http://www.sepscience.com/Techniques/LC (accessed 01 February 2019).
9. Stoll, D. "HPLC Troubleshooting" – Columns in LC.GC North Amer.
10. (1995). *Prudent Practices in the Laboratory: Handling and Disposal of Chemicals*. Washington, DC: National Research Council.

11. Furr, A.K. (2000). *CRC Handbook of Laboratory Safety*, 5e. Philadelphia: CRC Press.
12. Dong, M.W. (2000). *Today's Chemist at Work* 9 (8): 28.
13. Dong, M.W. (2014). *LCGC North Am.* 32 (8): 552.
14. Harris, D.C. (2015). *Quantitative Chemical Analysis*, 9e. New York, NY: W. H. Freeman.
15. Choi, C. and Dong, M.W. (2005). *Handbook of Pharmaceutical Analysis by HPLC* (ed. S. Ahuja and M.W. Dong). Amsterdam, the Netherlands: Elsevier. Chapter 5.
16. Mitra, S. (ed.) (2003). *Sample Preparation Techniques in Analytical Chemistry*. Hoboken, NJ: Wiley-Interscience.
17. Rayne, D. "Sample Preparation" – Columns in LC.GC North Amer.
18. Fountain, K.J., Neue, U.D., Grumbach, E.S., and Diehl, D.M. (2009). *J. Chromatogr. A* 1216 (32): 5979.
19. Mazzarese, R.P. (2005). *Handbook of Pharmaceutical Analysis by HPLC* (ed. S. Ahuja and M.W. Dong). Amsterdam, the Netherlands: Elsevier. Chapter 21.
20. Dyson, N. (1998). *Chromatographic Integration Methods*, RSC chromatography monographs, 2e. Cambridge, United Kingdom: Royal Society of Chemistry.
21. Guillarme, D. and Veuthey, J.-L. *Guideline for the use of UHPLC Instruments*. University of Geneva, Switzerland: LCAP.
22. Dong, M.W., Miller, G., and Paul, R. (2003). *J. Chromatogr. A* 987 (1–2): 283.
23. Williams, S. (2004). *J. Chromatogr. A* 1052 (1–2): 1.

# 8

# HPLC/UHPLC MAINTENANCE AND TROUBLESHOOTING

## 8.1 SCOPE

This chapter provides an overview of high-performance liquid chromatography (HPLC) system maintenance practices and summarizes strategies and guidelines for HPLC troubleshooting. It describes common maintenance procedures that can be performed by the user for both HPLC and ultra-high-pressure liquid chromatography (UHPLC) systems. Frequently encountered troubleshooting problems are classified into four categories (pressure, baseline, peak, and data). Each problem is described with symptoms and possible solutions for mitigation. Several troubleshooting case studies are used to illustrate the problem diagnosis and resolution process. For a more detailed discussion of the subject, the reader is referred to textbooks [1–4], magazine columns [5], training software [6], manufacturers' operating and service manuals [7, 8], and other resources [9–13].

## 8.2 HPLC SYSTEM MAINTENANCE

Common HPLC maintenance procedures that can be performed by the user are described, including procedures such as replacing check valves, filters, detector lamps, flow cells, and autosampler sampling syringes. Other more elaborate maintenance tasks such as replacing pump pistons/seals, high-pressure needle seals of autosamplers, or wavelength calibration of UV detectors are typically handled by the service specialists or internal metrology staff. Increasingly in UHPLC pumps, entire pump heads in are replaced with factory-refurbished units or exchange of the full modules during annual calibration. Most laboratories that work in a regulated environment have an annual preventive maintenance program for their HPLC systems when most of the wearable items are replaced. System calibration and performance verification procedures typically follow preventive maintenance (PM) (see Section 11.3.7). Note that maintenance and troubleshooting procedures for individual HPLC

*HPLC and UHPLC for Practicing Scientists*, Second Edition. Michael W. Dong.

modules can vary significantly with different manufacturers or service providers. These procedures are often available digitally today with data systems. Other online resources or videos (free or subscription basis) are available at various websites (e.g. sepscience.com, chromatographyonline.com, chromacademy.com, academysavant.com, lcresources.com, chromatographyforum.org).

### 8.2.1 HPLC Pump

Common maintenance tasks that can be easily performed by the user are the replacement of solvent filters (sinkers), in-line filters, check valves, and piston seals. Most modern HPLC pumps are designed for easy maintenance, with front panel access to many internal components. Figure 8.1 is a diagram of a slide-out HPLC pump unit (Waters Alliance) showing the two pump heads with check valves, the purge valve, the in-line filter, and other components. Procedures for replacing some consumables items are summarized as follows:

- *Solvent line filter (Sinker)*: The solvent line sinker, typically a 10-μm sintered stainless-steel or glass filter, can be replaced by a direct connection to the Teflon solvent line (Figure 8.2a). The sinkers should be replaced annually. A partially plugged solvent sinker can restrict solvent flow, which can result in poor retention time precision.
- *In-line filter*: The in-line filter element (typically 0.5 μm) can be replaced by first disconnecting the connection tubing, opening up the filter body, and replacing the used filter element with a new unit (Figure 8.2b). A partially plugged in-line filter can cause higher than expected system back pressure.
- *Check valve*: A check valve (ball and seat valve, which works by gravity) can be replaced by first disconnecting the solvent tube, unscrewing the check valve housing, and replacing the inside cartridge with a new unit. The cartridge typically contains a ruby ball and sapphire seat and must be installed with the flow arrow pointing upward (Figure 8.2c). A used check valve cartridge can often be cleaned by sonication in water, isopropanol,

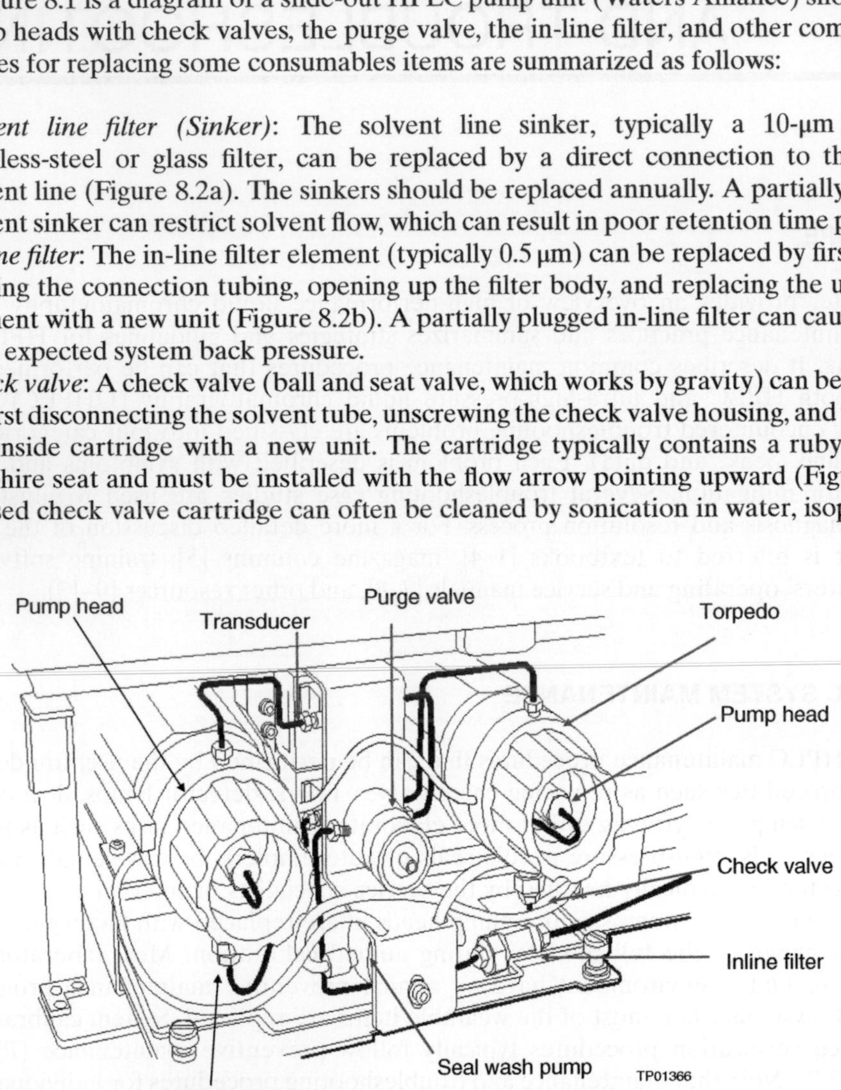


**Figure 8.1.** Diagram the inside of the Waters Alliance 2690 pump module after opening the front panel, showing the two pump heads and various components. Source: Courtesy of Waters Corporation.

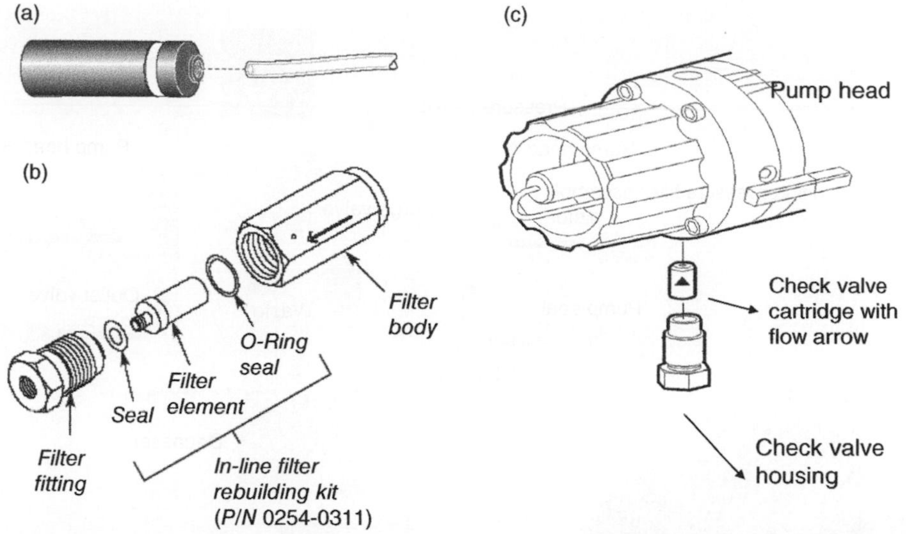


**Figure 8.2.** (a) The solvent sinker. (b) The in-line filter showing the inside filter element. (c). The pump head with the check valve housing and the check valve cartridge (Waters Alliance). Source: Courtesy of Waters Corporation.

or in 6N nitric acid if heavily contaminated. Higher than expected pressure pulsation is the symptom of a bad or contaminated check valve (see Figure 8.3a,b). Note that the inlet check valve is typically more problematic than the outlet check valve. Similarly, the inlet check valve of the "A" solvent line containing MPA (aqueous mobile phase) is most problematic for trapping air bubbles due to the high surface tension of water.

- *Piston seal*: Leaks from piston seals can be detected under the pump head manually by placing the finger under the pump head or leak sensors. The piston seal replacement procedure can be more elaborate and is highly dependent on the particular pump model. A piston seal service kit from the manufacturer should be purchased, and the procedure from the service manual followed closely for this operation. Typically, the piston should

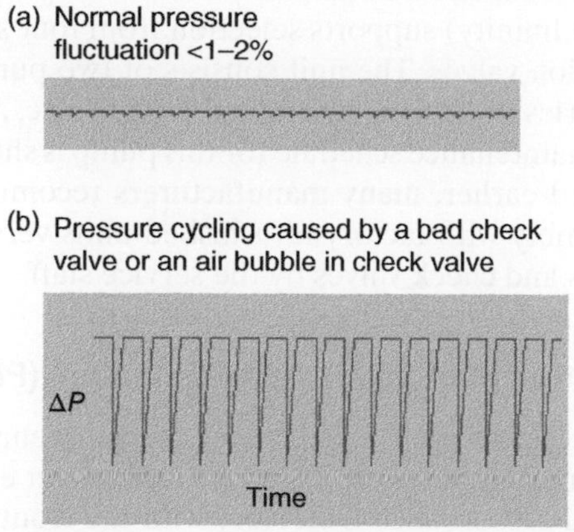


**Figure 8.3.** (a) Pressure profile of a normal HPLC pump showing pressure fluctuation <2%. (b) Pressure profile showing significant fluctuations of a malfunctioning pump due to a bad check valve or a trapped bubble in the pump head (airlock of the inlet check valve). Source: Adapted with permission from Academy Savant.

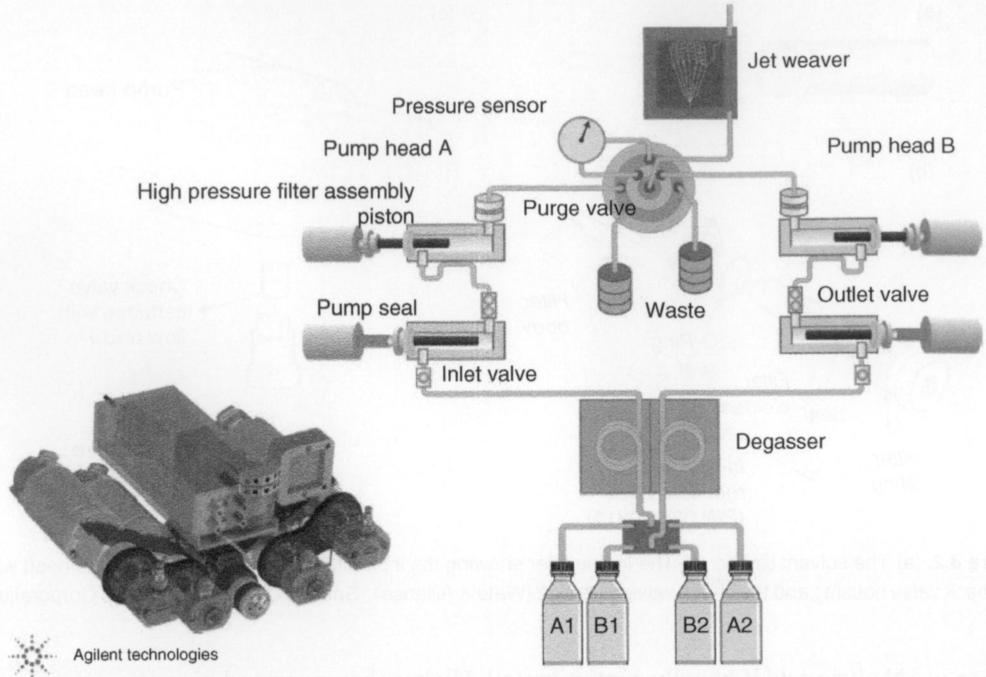


**Figure 8.4.** A drawing and a schematic diagram of the Agilent Infinity Binary Pump showing its various internal components. Source: Adapted with permission from Agilent Technologies.

be fully retracted before dismantling the pump head and replacing the seal. A lower than expected system pressure and a leak behind the pump head are indications for the need for the replacement of the seal. A higher than expected retention time can also serve as a diagnostic indicator.

A schematic diagram of a UHPLC pump and the recommended maintenance schedule are shown in Figures 8.4 and 8.5 from one manufacturer as an example. The binary high-pressure mixing pump (Agilent 1290 Infinity) supports selection from four solvents (A1, A2, and B1, B2) via low-pressure selection valves. The unit consists of two pumps, each with two pump heads in a dual-piston in-series design, an automated purge valve, and a Jet Weaver external mixer. The recommended maintenance schedule for this pump is shown in the second column of Figure 8.5. As mentioned earlier, many manufacturers recommend the replacement of the entire pump head-assembly with factory refurbished units versus the traditional way of on-site replacement of seals and check valves by the service staff.

### 8.2.2 UV/Vis Absorbance or Photodiode Array Detectors (PDA)

Modern UV absorbance detectors are designed for easy maintenance and often have front panel access to the lamp and the flow cell. Figure 8.6 shows an example of a modern UV detector (Waters 2487 Dual Absorbance Detector) with the front panel removed, showing the location of the deuterium lamp and the flow cell cartridge. Both units are self-aligning and do not require any user adjustment upon replacement. Procedures for replacing these items are summarized as follows:

**Solvent inlet**

*Clean or replace:*

- Solvent inlet filter

**Column Compartment**

- Replace the rotor seal (Column valve)

*Check:*

- Leak sensor
- Drain tube

**Pump**

*Replace:*

- PTFE frit or high pressure filter
- PUMP seals or pump heads
- Outlet ball valve
- AIV cartridge
- Passive inlet valve
- Wash seals

*Clean:*

- Pistons
- Support ring

*Check:* Leak sensor

- Drain tube
- Piston springs

**Autosampler**

*Replace:*

- *Needle
- *Needle seat
- Rotor seal

*Check:*

- Leak Sensor
- Drain tube

* ... if necessary

**Detector**

*Replace:*

- *Lamp
- *Cell window
- *Cartridge

*Check:*

- Flow Cell or cartridge

*Check:*

- Leak sensor
- Drain tube

Note there are variations from instrument model to instrument model.

Agilent Technologies

**Figure 8.5.** Maintenance recommendations for Agilent Infinity UHPLC systems. Source: Adapted with permission from Agilent Technologies.

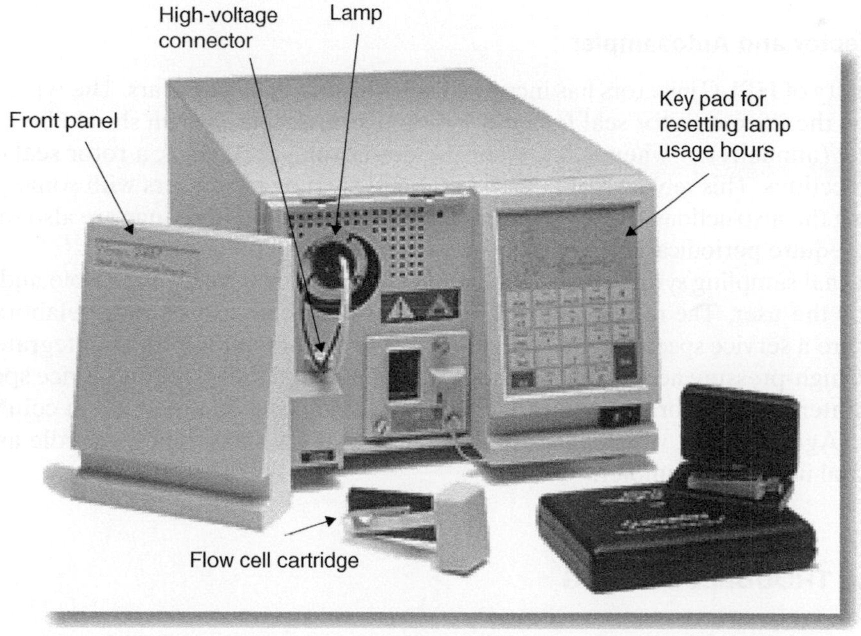


**Figure 8.6.** A picture of a Waters 2487 UV/Vis absorbance detector with the front panel removed showing the position of the UV lamp (source) and the flow cell cartridge. Source: Courtesy of Waters Corporation.

- *UV lamp*: Turn off the detector and unplug the power cord as a safety precaution. Let the lamp cool for five minutes. Disconnect the high-voltage power connector to the old lamp and loosen to remove the securing screws. Replace with the new lamp and reattach the high-voltage connector. Reset the lamp usage hour setting in the detector. Gloves should be worn for this operation to avoid touching the lamp surfaces with bare fingers. Note that many sources may contain mercury switches, which should be cut out for disposal as hazardous waste.
- *Flow cell*: Disconnect the inlet and outlet solvent tubes, and then loosen to remove the securing screws to the flow cell. Remove the flow cell assembly from the instrument and inspect the flow cell windows for dirt, particles, contaminants, or window cracks by viewing it against a bright light source. Most flow cells can be disassembled for cleaning or window replacement.
- Alternately, the damaged flow cell may be returned to the manufacturer for repair or reconditioning. Some of the optical components (e.g. windows, lens, and mirrors) inside the detector might require cleaning or replacement after several years of use. Indicators for the need to service these optical items are low source energy or low sensitivity performance even after a new lamp has been installed. Occasionally, the monochromator might need adjustment to restore wavelength accuracy. These procedures are best performed by a factory-trained specialist.
- UHPLC UV detectors have designs similar to those of HPLC detectors except their flow cells are significantly smaller (e.g. 0.5–1 μl) although their standard path lengths are identical (i.e. 10 mm).

### 8.2.3 Injector and Autosampler

The reliability of HPLC injectors has increased significantly in recent years. The typical wearable item is the injector rotor seal for an *XYZ*-type autosampler, which should be replaced periodically (annually) or when leaks occur. Under normal conditions, a rotor seal can last >30 000 injections. This replacement can be accomplished by most users with some practice by following the instructions. The sampling needle and the sampling syringe are also wearable items that require periodical replacement to restore sampling precision.

An external sampling syringe of an *XYZ*-type autosampler is easily accessible and can be replaced by the user. The replacement of the sampling needle can be more elaborate and might require a service specialist on some models. For autosamplers with an integrated-loop design, the high-pressure needle seal also requires annual replacement by a service specialist.

The maintenance list for a common UHPLC autosampler is summarized in column 3 of Figure 8.5 (Agilent 1290), which includes the replacement of the sampling needle and seat, the rotor seal in the autosampler.

## 8.3 HPLC TROUBLESHOOTING

*An ounce of prevention is worth a pound of cure.*

*Benjamin Franklin*

The best troubleshooting strategy is to prevent problems from occurring by exercising best practices in daily HPLC operation (see Chapter 7) and by performing periodic preventive maintenance. Nevertheless, HPLC problems do occur and often at the most inconvenient times. This section summarizes common HPLC troubleshooting and diagnostic strategies. Discussion focuses on common problems, their associated symptoms, and typical remedial actions. Several troubleshooting case studies are used to illustrate these strategies. Note that in addition to built-in diagnostic firmware often embedded with each HPLC modules, most UHPLC manufacturers often include maintenance/diagnostics/system monitoring software with their system or chromatography data system (CDS) to aid users or service personnel on maintenance feedback and troubleshooting (e.g. Agilent Lab Advisor, Waters Empower UPLC [ultra performance liquid chromatography] Console).

There are also numerous free or subscription-based resources available from various training organizations (Chromacademy.com, sepscience.com) and manufacturers of HPLC instruments and columns.

### 8.3.1 General Problem Diagnostic and Troubleshooting Guide

The following outline is a high-level practical guide for problem diagnosis and troubleshooting. The user is referred to more detailed procedures from additional resources elsewhere [3–9].

- *Verify that a problem exists* by repeating the experiment. A problem that occurs only once may not be a real problem.
- *Go back to a documented reference application* such as the one shown in Figure 8.7, which documents the chromatogram, operating conditions, expected retention time and pressure, peak heights, baseline noise, and plate count. Running this reference application with known good columns, samples, and mobile phases can often lead to a quick diagnosis. For instance, pump problems are indicated if pressure is lower than expected. A system blockage is indicated if the back pressure is higher than normal. The detector or autosampler might be faulty if peaks are smaller than expected. In this regard, it is

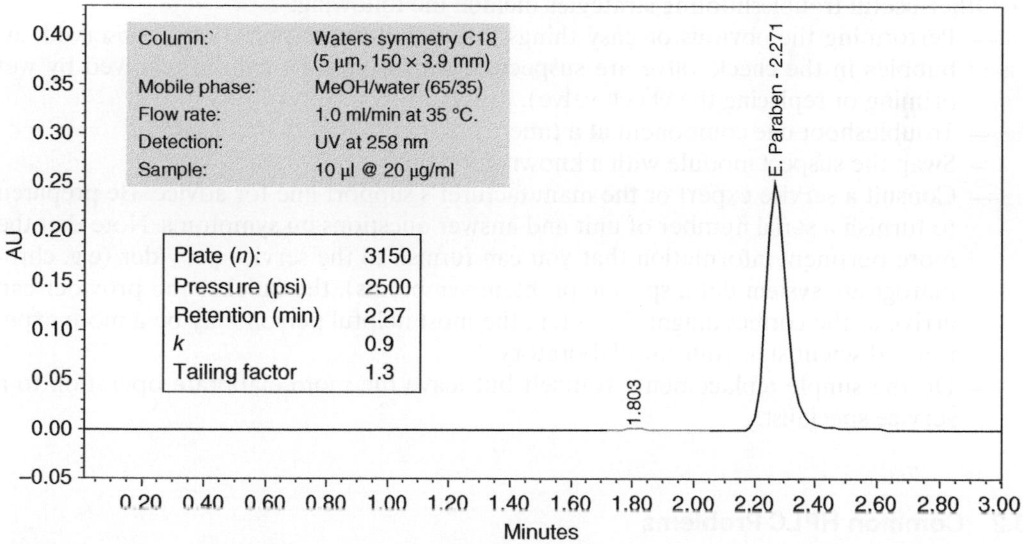


**Figure 8.7.** HPLC chromatogram of ethylparaben used as a reference application for problem diagnostics. The inset shows the HPLC conditions and the typical performance parameters such as plate count, pressure, and retention time.

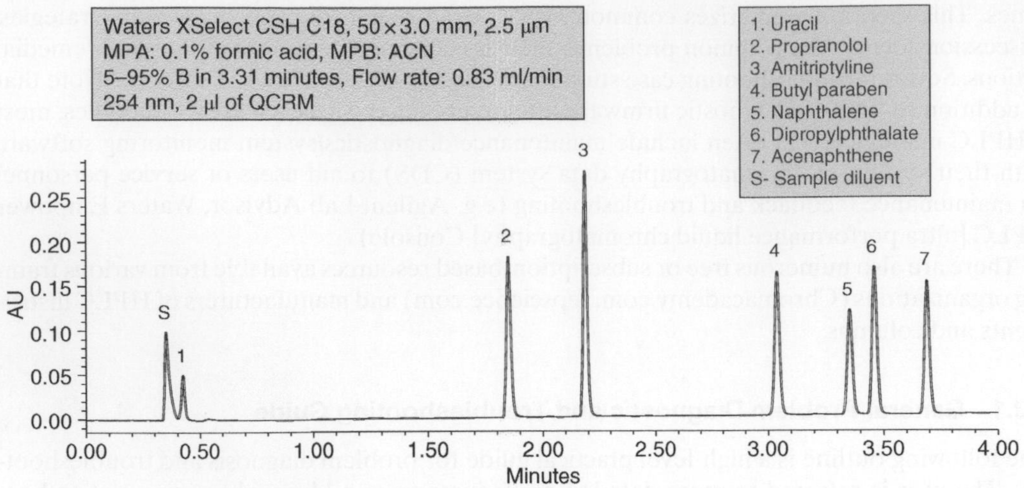


**Figure 8.8.** A quality control reference material (QCRM) useful for troubleshooting gradient systems, columns, and detectors (UV and MS). Source: Courtesy of Waters Corporation.

often useful to document the operating parameters and chromatograms in an analytical method (i.e. pressure, chromatograms of blank, retention marker, and sensitivity solutions). A multicomponent test mix available from some column or instrument vendor can be used to troubleshooting problems associated with gradient analysis (e.g. proportioning valve), column or MS detection (see example in Figure 8.8).

- *Isolate problem areas by visual inspection* of equipment for leaks and loose cables or tubing connections. Use the built-in diagnostics of each module or in the CDS to locate the problem area (e.g. lamp energy reading for the detector or compression test for the autosampler and the pump).
- Other useful troubleshooting strategies include the following:
  - — Performing the obvious or easy things first (i.e. if pump pulsation is observed, air bubbles in the check valve are suspected, which typically can be resolved by wet priming or replacing the check valve).
  - — Troubleshoot one component at a time.
  - — Swap the suspect module with a known good one.
  - — Consult a service expert or the manufacturer's support line for advice. Be prepared to furnish a serial number of unit and answer questions on symptoms. Note that the more pertinent information that you can furnish to the service provider (e.g. chromatogram, system data, specific problem symptoms), the quicker the provider can arrive at the correct diagnosis. Often, the most helpful person may be a more experienced scientist in your own laboratory.
  - — Do the simple replacements yourself but leave the more elaborate operation to a service specialist.

### 8.3.2 Common HPLC Problems

Common HPLC problems are caused by component malfunctions (pump, degasser, injector, detector, data system, column) and faulty preparation of the mobile phase or sample preparation. Problems can be categorized into several areas:

- Pressure problems
- Baseline problems (chromatogram)
- Peak problems (chromatogram)
- Data performance problems

Each area is discussed with typical symptoms and suggestions for mitigation.

***8.3.2.1 Pressure Problems and Causes*** *Pressure too high*: Higher than expected system pressure is caused by partial blockages in system components such as filter, guard column, column, connection tubing, injector, or detector tubing. The solution is to isolate the source for the partial blockage by disconnecting one component at a time. For instance, if higher than expected pressure is experienced after disconnecting the column (or the guard column), the blockage is occurring upstream in the injector or the pump. If the pressure is still high after disconnecting tubing to the injector, the problem then lies in the pump. The most likely location of blockage is the column that is packed with small particles. High column back pressure can be remedied by back-flushing or replacing the inlet frit of the column. A high-pressure problem should be investigated early before total blockage occurs.

*Pressure too low*: Lower than expected system pressure is caused by leaks (piston seal, column connections, injector), pump malfunctions (lost prime, air bubbles trapped in check valve, vapor lock, faulty check valves, broken piston), or inadequate solvent supply (empty solvent reservoir, plugged solvent sinker, bent solvent lines, or wrong solvent mixture). Problem diagnosis can be made by visual inspections for leaks and by monitoring the pressure reading of the pump to look for pulsations and other symptoms.

*Pressure cycling*: Pressure cycling is caused by air bubbles trapped in check valves or malfunctioning of check valves. A typical pressure profile of an HPLC pump is shown in Figure 8.3 with a pressure fluctuations specification of $<\pm 2\%$ of the nominal pressure in HPLC. Note that most modern UHPLC pumps including UHPLC pumps have much lower pulsation specifications (e.g. 30–50 psi). A malfunctioning check valve is most likely the result of a trapped air bubble or contamination of the ball and seat mechanism in the inlet check valve causing incomplete closures. It usually manifests itself by significantly increased pressure fluctuations that are synchronous to the pump strokes (Figure 8.3). An air bubble can be dislodged by degassing all solvents and by wet priming the pump at high flow rates. If that is not successful, change the solvent to acetonitrile, methanol, or isopropanol that wets the check valves more effectively. Malfunctioning check valves should be dismantled and cleaned or replaced by a new unit.

***8.3.2.2 Baseline Problems (Chromatogram)*** Common baseline problems are illustrated with several diagnostic examples shown in Figures 8.9 and 8.10. The description of each symptom and its suggested remedial actions are as follows.

*Noisy baseline*: Short-term detector noise can be estimated by measuring the peak-to-peak signal fluctuation of the baseline at an expanded scale. The noise of a modern UV detector should be close to the noise specification or $\pm 1 \times 10^{-5}$ AU. Noisy baseline such as the one shown in Figure 8.9a is typically caused by the low energy of an aging UV lamp, which should be replaced. Low light energy can also be caused by a contaminated detector flow cell or high UV absorbance of the mobile phase (Figure 8.9d). Detector noise can be caused by a large air bubble trapped in the flow cell or pressure fluctuations due to a small leak in the flow cell.

A noisy baseline in a refractive index (RI) detector can be caused by insufficient mobile phase degassing or inadequate temperature thermostatting.

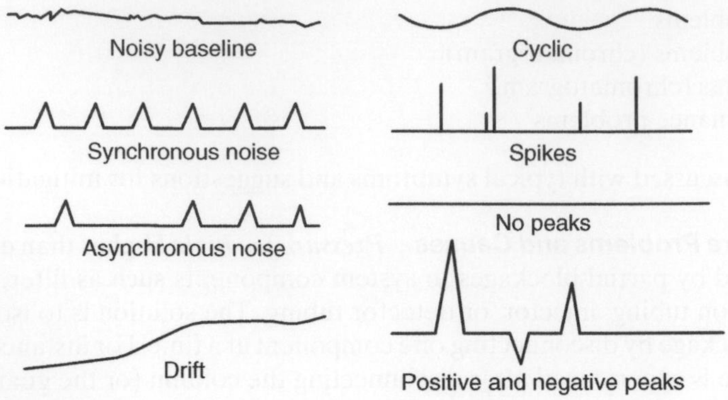


**Figure 8.9.** Examples of various baseline problems. Source: Courtesy of Waters Corporation.

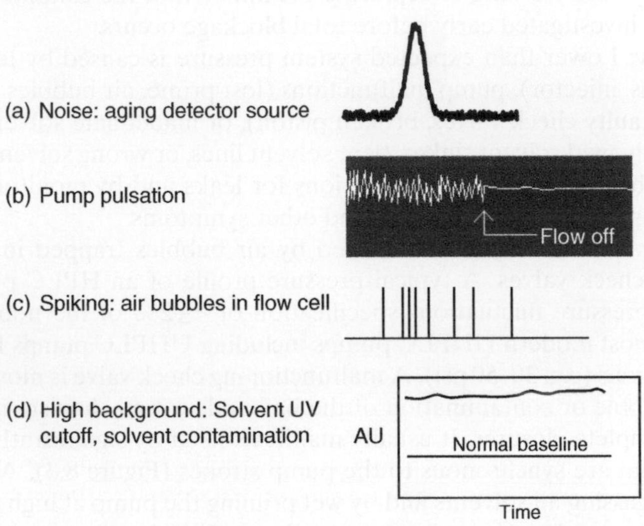


**Figure 8.10.** Examples of various baseline problems and possible causes.

*Synchronous and asynchronous noise*: Cyclical baseline noise in sync with pump strokes is likely pump-related (Figure 8.3b). For instance, a flow rate of 1 ml/min delivered by a pump with a piston volume of 100 μl would operate at 10 cycles per minute. One definitive test for pump-related noise is to check if the noise is eliminated after the pump is turned off (Figure 8.9b). This synchronous noise can be caused by pump malfunctions (air bubbles in the pump head, check valve problems, or broken plunger). Inadequate mixing of the mobile phases is a frequent cause if solvent blending at the pump is used, particularly for high-pressure mixing binary pumps with a small external mixer. This cyclical noise can be remedied by using a larger mixer or switching to a premixed mobile phase.

Asynchronous detector noise (which is not in sync with the pump and occurs at lower frequency shown in Figure 8.8) can be caused by electrical sources such as ground loop problems, power/voltage fluctuations, or electrical shielding problems. These situations might be difficult to diagnose and remedy. Asynchronous noise can also be caused by the detector, such as leaks, loose detector connections, or electric problems.

*Baseline drift*: Baseline drift in a UV absorbance detector often reflects the change of the energy output of the UV lamp with time. Baseline drift is reduced significantly in dual-beam absorbance detectors because the inherent fluctuation of the lamp energy is compensated. A specification of $<1 \times 10^{-4}$ AU/h is typical. Baseline shifts associated with gradient analysis are quite normal. They are caused by the difference of absorbance and refractive index of the initial and final mobile phases (see case study in Figure 8.11). These gradient shifts can be minimized by balancing the absorbance of the two mobile phases (e.g. MPA = 0.05% TFA in water and MPB = 0.03% TFA in acetonitrile) and by using a flow cell designed to minimize the effects of refractive index changes (e.g. a tapered flow cell). Other causes are contaminants in the solvents or reagents in the mobile phase.

Nonspecific drifts can be caused by strongly retained peaks slowly bleeding off a contaminated column. These problems can be remedied by purging the column with a strong solvent until the baseline is stable. Drift can also be caused by temperature fluctuations or small leaks in the flow cell. Drifts caused by thermal effects are particularly prominent in refractive index and conductivity detectors. Cyclic baseline drift can be caused by the cyclical thermal effects such as periodic cooling from a ventilation source or cyclical voltage fluctuations such as those caused by the periodic current draws from a high-wattage compressor on the same circuit as the HPLC detector.

*Spikes on the baseline*: Spikes on the baseline (Figure 8.9c) are likely caused by air bubbles out-gassing in the detector flow cell. They can be eliminated by mobile phase degassing and by placing a pressure restrictor in the detector outlet (e.g. with 50-psi back pressure device). Spikes can also be caused by poor signal wire connections (loose or damaged wiring) or malfunction of the detector or the data system.

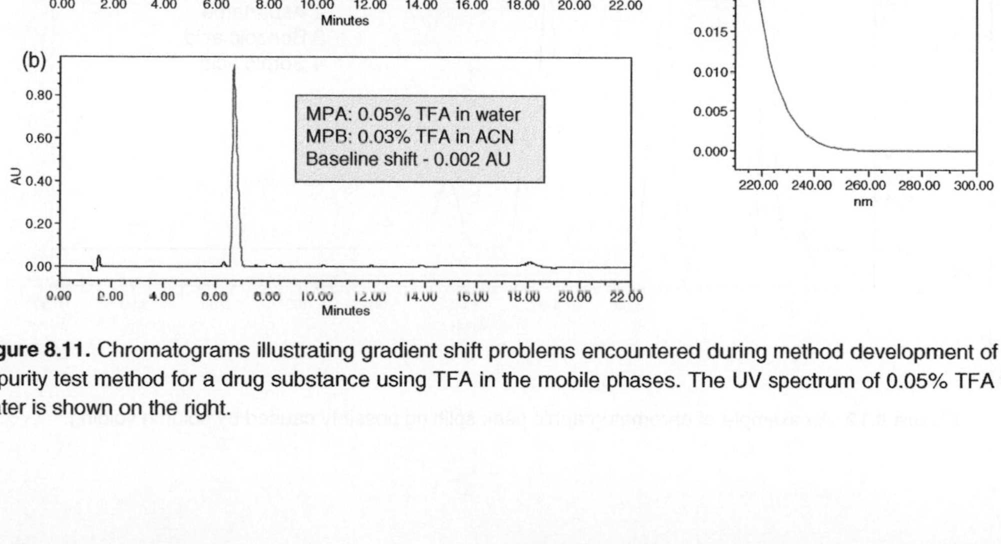

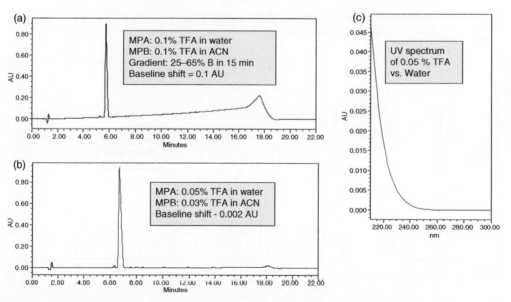


**Figure 8.11.** Chromatograms illustrating gradient shift problems encountered during method development of an impurity test method for a drug substance using TFA in the mobile phases. The UV spectrum of 0.05% TFA vs. water is shown on the right.

***8.3.2.3 Peak Problems (Chromatogram)*** Peak problems such as poor peak resolution, broad peaks, split peaks, tailing or fronting peaks, and extra peaks are most often caused by the column [13] and its interaction with the mobile phase and the sample/diluent.

*No peaks in the chromatogram* could be due to a number of reasons including the injector not making an injection, a pump not delivering flow, big leaks, a dead detector, a miswired data system, incorrect mobile phase, a particularly retentive or adsorptive column, or a bad/incorrect sample. The best procedure to diagnose such problems is to go back to reference test conditions such as those shown in Figures 8.7 and 8.8. If the reference conditions can be duplicated successfully, thereby inferring that the system is functioning properly, then the troubleshooting effort can be directed to the specific application conditions (column, mobile phase, sample, etc.).

*Broad peaks and split peaks*: Abnormally broad peaks and split peaks are indications of degraded column performance caused by sample contamination, partially blocked inlet frit, or column voiding (Figures 8.12 and 8.13). They are remedied by backflushing the column, refilling the column inlet with packing, or, more often nowadays, by replacing the column. Note that broader than expected peaks for early eluting peaks are often caused by extra-column band broadening from the HPLC system (Figure 8.14). These extra-column band broadening effects are particularly deleterious for small-diameter UHPLC columns.

Anomalous peak shapes can also be caused by injecting samples dissolved in diluent stronger than the mobile phase (Figure 8.15). If possible, the diluent for dissolving the sample should be of the same or a weaker strength than MPA or the initial mobile phase in gradient analysis. If stronger solvents must be used as the final sample diluent, the volume should be kept small (e.g. <2–5 μl in HPLC) to prevent peak broadening or peak splitting of the early eluting peaks. Peak splitting can also result when the sample pH is radically different from the pH of the mobile phase.

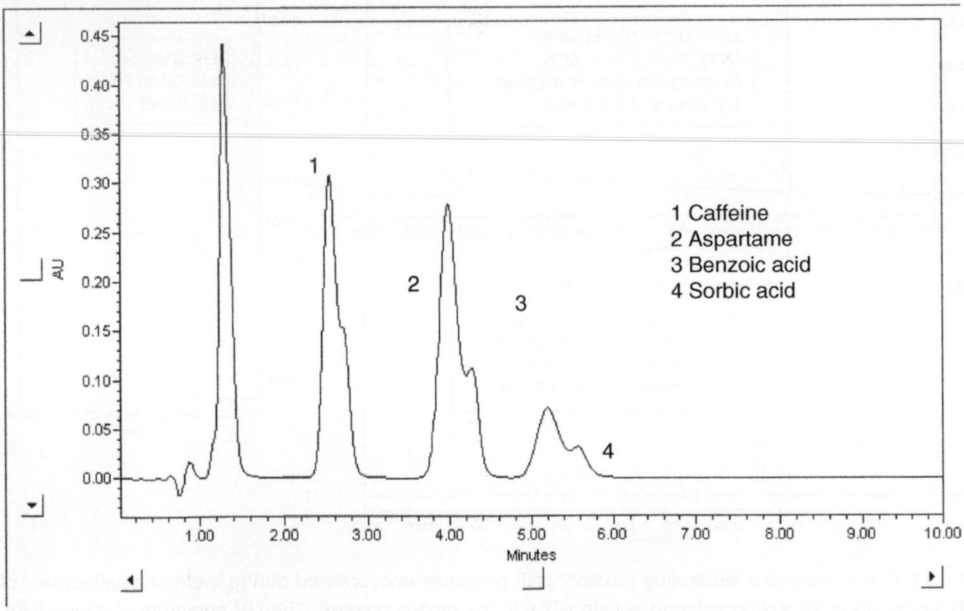


**Figure 8.12.** An example of chromatographic peak splitting possibly caused by column voiding.

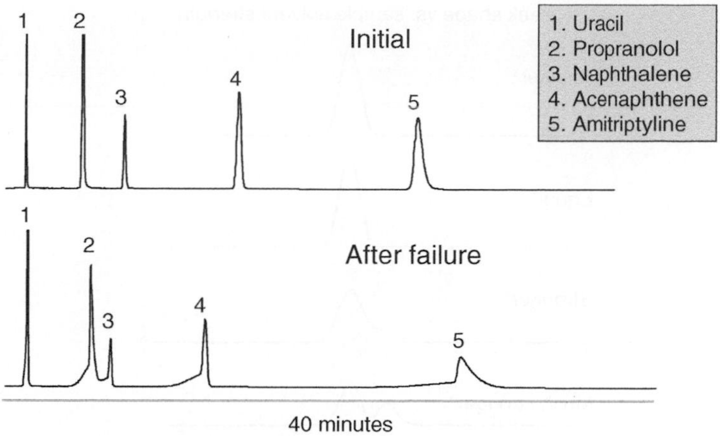


**Figure 8.13.** Comparative Chromatograms of a good and failed C18-silica column due to exposure to high pH mobile phase. Source: Courtesy of Waters Corporation.

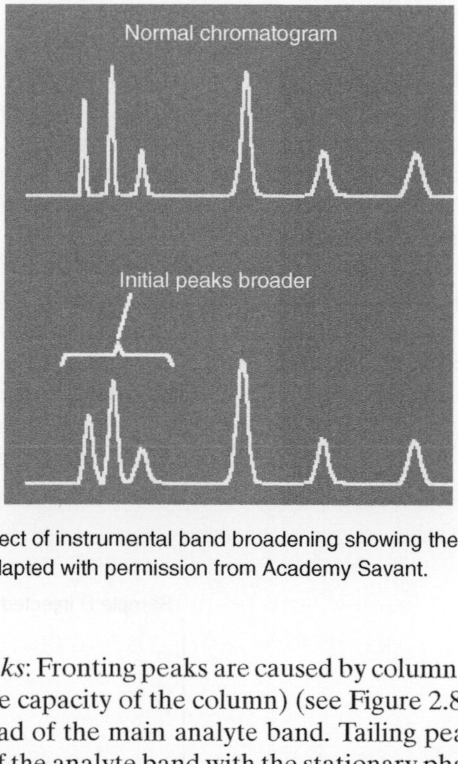


**Figure 8.14.** Example of the effect of instrumental band broadening showing the more pronounced effect on the early eluting peaks. Source: Adapted with permission from Academy Savant.

*Fronting and tailing peaks*: Fronting peaks are caused by column overload (sample amount exceeding the sample capacity of the column) (see Figure 2.8), resulting in some of the analytes eluting ahead of the main analyte band. Tailing peaks are caused by the secondary interaction of the analyte band with the stationary phase. A common example is the tailing peaks of basic analytes caused by strong interactions with the acidic residual silanol groups of the silica-based bonded phases. Peak tailing of a basic analyte can be reduced by adding an amine modifier (0.1% triethylamine) to the mobile phase or by replacing the column with lower silanophilic activity (see Chapter 3). Asymmetric peak shapes can also be caused by chemical reactions or isomerization of the analyte during chromatography [1].

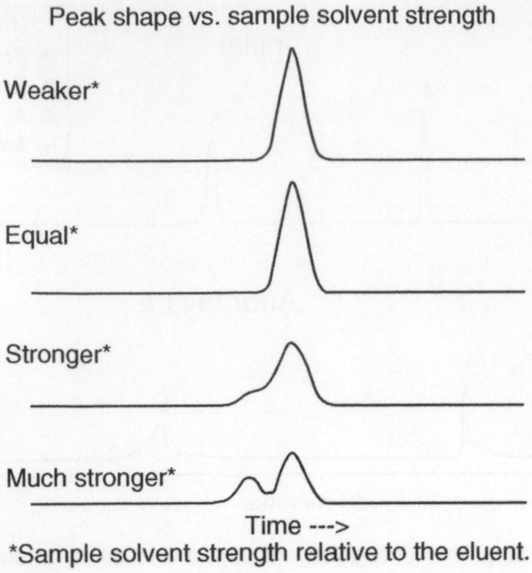


**Figure 8.15.** Examples of the peak anomaly by injecting the same sample solution with different diluents weaker and stronger than the mobile phase strength. Source: Adapted with permission from Academy Savant.

*Negative and positive peaks*: Negative peaks or dips in the chromatogram are common near the solvent front. They are caused by refractive index changes or by sample solvents with less absorbance than the mobile phase. These baseline perturbations are usually ignored by starting data integration after the solvent front. If all peaks are negative, it might be an indication of the wrong polarity in the detector wire connection unless an indirect UV detection technique is used. Negative peaks are more commonly encountered when using an RI detector.

*Ghost peaks and extra peaks*: Ghost peaks associated with a blank gradient (running a gradient by injecting just the sample solvent) are usually caused by trace contaminants in the weaker mobile phase (MPA). This situation can be minimized by using purified reagents and by exercising stringent mobile phase preparation precautions (see Chapter 7.7 in trace analysis). Unexpected extra peaks, particularly those that are unusually broad, are typically caused by late eluting peaks from prior injections during isocratic analysis (Figure 8.16). Some of these peaks can be so broad that they are often mistaken for baseline drifts.

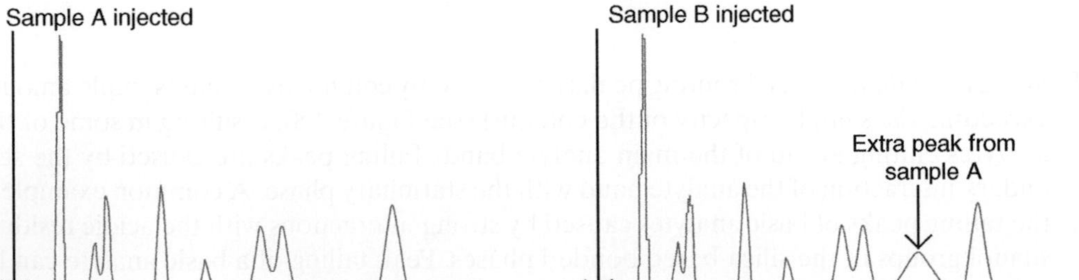


**Figure 8.16.** Example of an unexpected broad peak caused by the elution of a highly retained peak from the previous injection. Source: Adapted with permission from Academy Savant.

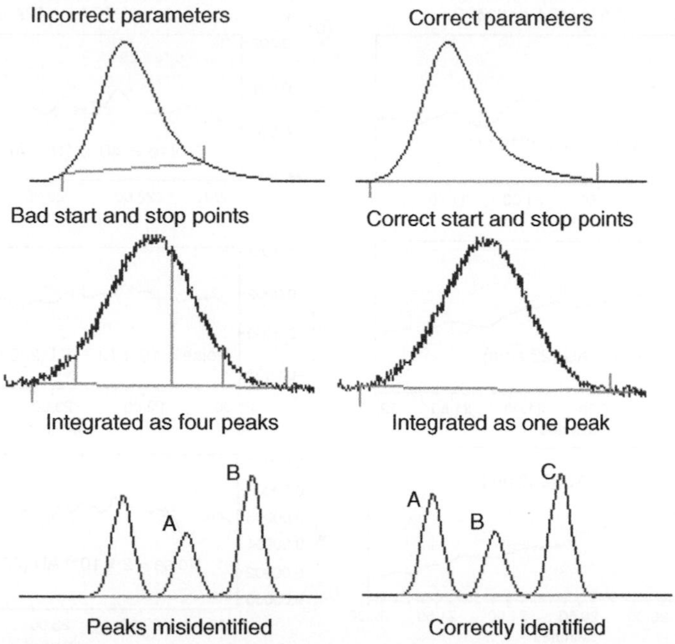


**Figure 8.17.** Example of various peak integration problems. Source: Courtesy of Waters Corporation.

***8.3.2.4 Data Performance Problems*** Data performance problems such as poor precision and inaccurate results might be difficult to diagnose. Data problems can stem from system malfunctions or problems relating to the mobile phase, method calibration, or sample preparation. Poor system precision of retention time and peak area and their possible causes and remedies are covered in Section 7.4.7 and several published articles [10, 11]. Poor method accuracy can be caused by peak coelution (peak purity), sample stability, sample preparation/recovery problems, matrix interferences, equipment calibration, or integration problems (Figure 8.17). One excellent guideline to follow is to always print reports with chromatograms that show the peak baselines to check for proper peak area integration. Data problem situations typically require the insights of an experienced practitioner for the correct diagnosis and remedy. They should be investigated further by eliminating each potential cause as illustrated in some of the following case studies.

## 8.4 TROUBLESHOOTING CASE STUDIES

Several case studies concerning baseline and data performance problems, as well as troubleshooting examples, are selected here to illustrate problem diagnosis and resolution.

### 8.4.1 Case Study 1: Reducing Baseline Shift and Noise for Gradient Analysis

Figure 8.11 illustrates a gradient baseline shift problem encountered during the method development for a method to analyze impurities of a drug substance using a mobile phase containing 0.1% trifluoroacetic acid (TFA). Due to the absorbance of TFA in the far UV region

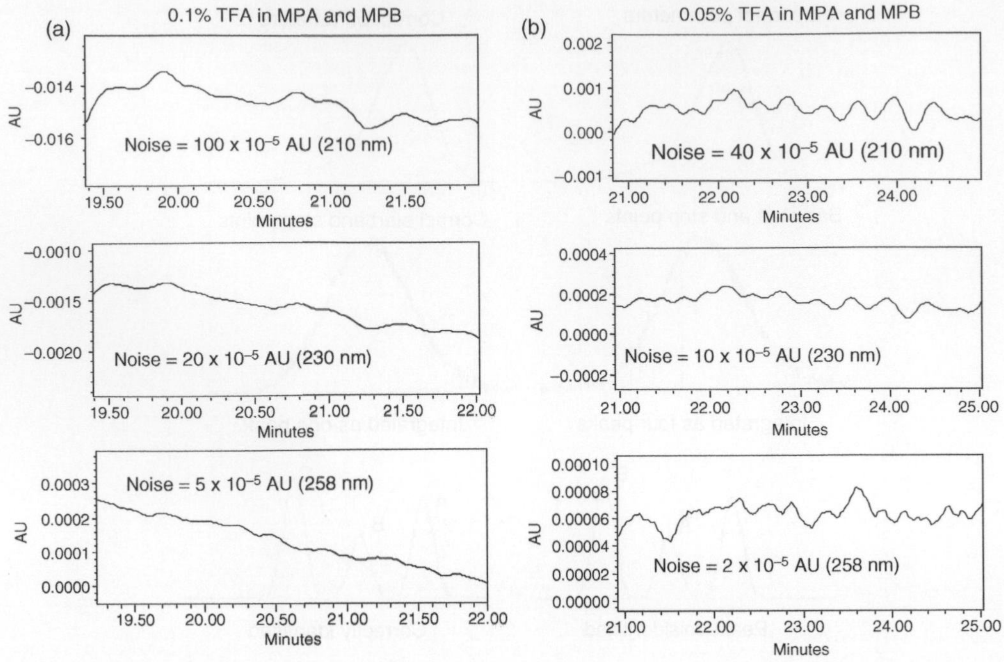


**Figure 8.18.** Baseline fluctuations from blending 0.1% TFA in water and 0.1% TFA in ACN vs. that of 0.05% TFA in water and ACN. The magnitude of baseline noise is dependent on the monitoring wavelength and is caused by pump blending issues and the absorbance of TFA.

(200–230 nm) (Figure 8.11c), considerable baseline shift (0.1 AU at 230 nm) (Figure 8.11a) and substantial baseline fluctuations (Figure 8.18a) were observed. These baseline problems were substantially reduced by lowering the concentration of TFA in both the starting mobile phase (MPA) and final mobile phase B (MPB) from 0.1% to 0.05% (Figures 8.10b and 8.18b). Further reduction of the gradient shift and baseline fluctuation can be achieved by adjusting the concentration of TFA in MPB until it matches the absorbance of MPA of 0.05% TFA in water (i.e. by balancing the absorbance of MPB with MPA). Note that MPB used in Figure 8.11b was adjusted to contain only 0.03% TFA in acetonitrile. The spectral wavelength (chromophoric) shift of TFA in acetonitrile, a more nonpolar solvent environment than water, may also contribute to this baseline shift issue. Further reduction of the baseline noise at low UV detection may be achieved by increasing the mixing volumes of the system.

### 8.4.2 Case Study 2: Poor Peak Area Precision

Two real-world examples are used to illustrate how to resolve this common problem encountered during the calibration of an autosampler (Chapter 4). In Table 8.1, the precision of peak area was found to be >1% RSD (relative standard deviation), which was above the acceptance criterion of 0.5% RSD expected in the operational qualification procedure. Repeating the experiment showed similar precision results. Since a worn sampling syringe of the autosampler is the most common cause of poor precision, it was replaced. The peak area precision was found to be <0.31% RSD after this replacement.

In the second scenario shown by the data in Table 8.2, poor precision was encountered in both the retention time and peak area of an analyte. Scrutiny of the data showed that both

**Table 8.1. Peak Area Precision Data Before and After Replacing the Sampling Syringe of the Autosampler**

| Run | Peak area (before) | Peak area (after) Replacing syringe |
|---|---|---|
| 1 | 1 394 782 | 1 419 308 |
| 2 | 1 432 841 | 1 421 143 |
| 3 | 1 445 966 | 1 409 706 |
| 4 | 1 422 240 | 1 410 025 |
| 5 | 1 422 555 | 1 410 880 |
| 6 | 1 410 891 | 1 408 446 |
| 7 | 1 418 816 | 1 409 981 |
| 8 | 1 404 683 | 1 412 092 |
| 9 | 1 425 581 | 1 410 867 |
| 10 | 1 428 552 | 1 414 600 |
| Average | 1420691 | 1412705 |
| RSD (%) | **1.02** | **0.31** |

**Table 8.2. Peak Area and Retention Time Precision Data Before and After Replacing the HPLC Column**

| Run | $t_R$ (min) | Peak area | Run | $t_R$ (min) | Peak area |
|---|---|---|---|---|---|
| 1 | 2.350 | 998 451 | 10 | 2.454 | 1 159 680 |
| 2 | 2.371 | 1 014 657 | 11 | 2.458 | 1 181 497 |
| 3 | 2.386 | 1 023 041 | 12 | 2.465 | 1 200 950 |
| 4 | 2.401 | 1 036 924 | 13 | 2.464 | 1 219 282 |
| 5 | 2.414 | 1 05 5259 | 14 | 2.465 | 1 233 082 |
| 6 | 2.427 | 1 073 619 | 15 | 2.47 | 1 248 339 |
| 7 | 2.437 | 1 097 006 | 16 | 2.47 | 1 260 375 |
| 8 | 2.444 | 1 116 404 | 17 | 2.471 | 1 273 832 |
| 9 | 2.448 | 1 139 731 | 18 | 2.476 | 1 284 628 |
| | | | Average | 2.437 | 1 145 375 |
| | | | RSD (%) | 1.56 | **8.45** |
| | *After replacing column*: | | Average | 2.483 | 1 365 489 |
| | | | RSD (%) | 0.20 | **0.12** |

retention time and peak area tended to trend upward. Troubleshooting the autosampler by replacing the sampling syringe yielded no improvements. The poor precision of retention time hinted at possible column problems. The column was therefore replaced, and acceptable precision was restored. While column should not impact peak area precision in general, problems are often fixed pragmatically without a complete understanding of the cause.

### 8.4.3 Case Study 3: Poor Assay Accuracy Data

Table 8.3 shows a set of data generated during the assay of five pharmaceutical suspension samples (in duplicate preparation) prepared for a release testing of this clinical trial material. Data showed at least two suspected out-of-specification (OOS) results because they were beyond the specification range of 90–110% of label claim. Data review indicated that the

**Table 8.3. Assay Data of Five Separate Samples (in Duplicate Preparation) of a Drug Suspension Containing a Low Solubility Drug Substance**

| Sample | Preparation | % Label claim |
|---|---|---|
| 1 | a | 99.5 |
| | b | 100.2 |
| 2 | a | 99.1 |
| | b | 99.6 |
| 3 | a | 100.8 |
| | b | **75.7** |
| 4 | a | 100.0 |
| | b | 99.5 |
| 5 | a | **84.3** |
| | b | **91.7** |

HPLC system was functioning properly and passed system suitability testing. Subsequent laboratory investigation indicated that incomplete extraction of drug substance from the suspension was likely the root cause due to the relatively low solubility of the drug in the extraction solvent. In the subsequent laboratory investigation, the two retained suspect OOS sample extracts were resonicated and analyzed and found to have assay values of over 98%. The extraction procedure of the analytical method was thus revised to include an extra vortexing step and a longer ultrasonication extraction time in the sample preparation.

### 8.4.4 Case Study 4: Equipment Malfunctioning and Problems with Blank

Case study 4 shows several examples of problems caused by equipment malfunctions and other troubleshooting case studies. The first one involved a situation of poor retention time reproducibility of a gradient assay. It involved the analysis of a complex natural product, using a narrow bore column (2.1 mm i.d.) at 0.5 ml/min. System suitability test showed retention times to be erratic and could vary by one to two minutes without any obvious trends. Flow rate accuracy was found to be acceptable. However, the compositional accuracy test failed (see Section 11.3.6 on HPLC calibration). The tentative diagnosis was that of a malfunctioning of the proportioning valve. After its replacement, the retention time precision performance was re-established.

The second is an example of operator error encountered during isolation of an impurity of a drug substance using a preparative HPLC system. The symptom was a classic example of detector baseline "spiking" (Figure 8.10c) caused by bubble formation in the flow cell. Pump blending was used since gradient elution was needed for this isolation. Since the system was only equipped with helium degassing, helium was turned on to degas the solvent, and a back-pressure restrictor (50 psi) was placed at the detector outlet. Surprisingly, detector spiking became worse. It was later found that a tank of argon was used for degassing by a new analyst who replaced an empty helium tank. Though argon, like helium, is also an inert gas, it is not effective for solvent degassing due to its relatively high solubility in the mobile phase. The spiking ceased after actually degassing the mobile phase with helium.

The last example shown in Figure 8.19 involves ghost peak problems in blank injections, which can be time-consuming to diagnose. After eliminating the common reagent purity problems in MPA or MPB (by substituting mobile phase additives and solvents)

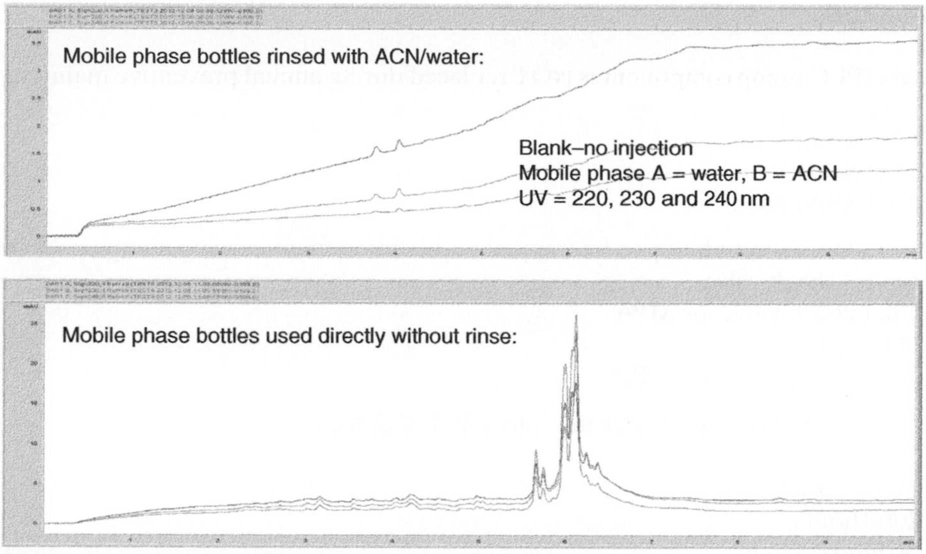


**Figure 8.19.** Chromatogram of a blank injection showing high levels of ghost peaks in the blank gradient caused by contaminants in the solvent reservoir bottles. Problem resolved after the bottles were rinsed with acetonitrile and water.

and potential system contamination (by cleaning the HPLC system with a mixture of water/MeOH/ACN/isopropanol [1 : 1 : 1 : 1] with 0.2% formic, a recipe from the equipment manufacturer), the ghost peak problems persisted. Later on, the root cause was found to be residues left in the solvent reservoir bottles due to a change of the bottle washing procedures from a central glassware service. Simple rinsing the bottles with acetonitrile and water eliminated the ghost peaks in the blank injection.

## 8.5 SUMMARY AND CONCLUSION

Table 8.4 summarizes the symptoms of common HPLC problems and the most probable causes. Developing HPLC troubleshooting skills often takes many years of operating experience and a working understanding of the principles of the instrument as well as considerable patience to eliminate all the typical causative factors.

**Table 8.4. Summary of Symptoms of Common HPLC Problems and Probable Causes**

| | Symptom | Most Probable Cause |
|---|---|---|
| Pressure | Too high | Column, filter |
| | Too low | Leaks from column, pump |
| Baseline | Noisy | Old lamp, mobile phase |
| | Wandering | Degassing, mobile phase, contaminated column |
| Peak | No peak | Detector, injector |
| | Bad peaks | Column, mobile phase, connections |
| Data performance | Retention time | Degassing, column, temperature changes, insufficient equilibration time |
| | Peak area | Autosampler, column, integration, lamp |
| | Accuracy | Sample prep, autosampler |

## 8.6 QUIZZES

**1.** Which HPLC pump component is NOT replaced during annual preventive maintenance?
(a) Filter
(b) Check valve
(c) Piston seal
(d) Microprocessor

**2.** An air bubble is most likely to be trapped in this component causing pump problems.
(a) Outlet check valve
(b) Inlet check valve for MPA
(c) Degasser
(d) Inlet check valve for MPB

**3.** The lifetime of a typical Deuterium source in UV detector is
(a) 300 hours
(b) 20 000 hours
(c) 2000 hours
(d) 50 hours

**4.** Synchronous baseline noise is typically caused by issues in
(a) pump
(b) mobile phase
(c) lamp
(d) column

**5.** Sharp spikes in UV detector signals are typically caused by
(a) dirty mobile phases
(b) dirty cell windows
(c) air bubbles
(d) pump issues

**6.** UV baseline shift during gradient operation is usually caused by
(a) dirty mobile phases
(b) RI and absorbance effects
(c) air bubbles
(d) detector problems

**7.** Peak shape problems in the chromatogram are generally caused by
(a) column
(b) detector
(c) pump
(d) autosampler

**8.** Extra-column band broadening effect under isocratic conditions is more problematic for
(a) late-eluting peaks
(b) gradient analysis
(c) larger columns
(d) early eluting peaks

**9.** If possible, the strength of the sample diluent should be
(a) equal to 50% ACN/water
(b) same as MPB

(c) stronger than MPA
(d) same or weaker than MPA

**10.** Poor precision in peak area is typically an indication of a problem in
(a) column
(b) detector
(c) pump
(d) autosampler

### 8.6.1 Bonus Quiz

In your own words, describe the most common reasons for having a very high pressure pulsation in HPLC pumps and the best procedure for mitigation.

## 8.7 REFERENCES


1. Snyder, L.R., Kirkland, J.J., and Dolan, J. (2010). *Introduction to Modern Liquid Chromatography*. Chapter 17, 3e. Hoboken, NJ: Wiley.
2. Snyder, L.R., Kirkland, J.J., and Glajch, J.L. (1997). *Practical HPLC Method Development*, 2e. New York: Wiley-Interscience.
3. Dolan, J. and Snyder, L.R. (1989). *Troubleshooting LC Systems*. Totowa, NJ: Humana Press.
4. Sadek, P.C. (2000). *Troubleshooting HPLC Systems: A Bench Manual*. New York: Wiley.
5. Dolan, J. (2016). "HPLC Troubleshooting," Columns in LC.GC North Amer. 1983–2016. http://www.lcresources.com/training/tsbible.html (available as a searchable database at LC Resources website with full articles in LCGC from 1983 to 2016).
6. *Troubleshooting in HPLC, CLC-70*, (CD-based Instruction), and WLC-4043p (computer-based online instruction), Academy Savant, Fullerton, California. http://www.academysavant.com/clc-40.htm.
7. (2014). *Acquity UPLC System, Operator's Guide*, 71500082502/Revision F. Milford, MA: Waters Corporation.
8. Otero, M. (2013). *Introduction to Agilent 1290 Infinity Maintenance*. Agilent Technologies.
9. Neue, U. (2002). *HPLC Troubleshooting Guide*, Waters Applications Notes, 720000181EN. Milford, MA: Waters Corporation.
10. Dong, M.W. (2000). *Today's Chemist at Work* 9 (8): 28.
11. Grushka, E. and Zamir, I. (1989). *Chemical Analysis*, vol. 98 (ed. P. Brown and R. Hartwick), 529. New York: Wiley Interscience.
12. LinkedIn Discussion Groups, *Analytical Chemistry-Method Development and International HPLC user group*.
13. Neue, U.D. (1997). *HPLC Columns: Theory, Technology, and Practice*. Chapter 17. New York, NY: Wiley-VCH.

# 9

# PHARMACEUTICAL ANALYSIS

## 9.1 INTRODUCTION

### 9.1.1 Scope

This chapter provides an overview of high-pressure liquid chromatography (HPLC) in the pharmaceutical analysis in drug development – from drug discovery to quality control of commercial drug products. The focus is on HPLC methodologies of drug substances (DS) and drug products (DP) of small molecule drugs for identity, potency, and purity assessments. LC/MS analysis in pharmaceutical analysis and bioanalytical studies is discussed in Chapter 6. Characterization and quality control of recombinant biologics are discussed in Chapter 12, and regulatory aspects of pharmaceutical testing are covered in Chapter 11. HPLC method development, validation, and transfer are discussed in Chapters 10 and 11. Additional details on the use of HPLC in drug development and production are available in reference books [1, 2] and articles [3–8].

### 9.1.2 Glossary and Abbreviations

A listing of the common terms or abbreviations (acronyms) in pharmaceutical development is shown here as a quick reference. The reader is referred elsewhere for further clarification and explanations [9].

| | |
|---|---|
| Target | A protein (receptor) related to a disease state |
| Hit or lead | A compound that binds to the target and possibly mitigates the disease |
| Drug candidate | A lead with optimized structure motifs to be developed as a new drug |
| CRO, CMO | Contract Research Organization, Contract Manufacturing Organization |

*HPLC and UHPLC for Practicing Scientists*, Second Edition. Michael W. Dong.

| | |
|---|---|
| API, NCE | Active pharmaceutical ingredient, new chemical entity |
| DS, DP, CTM | Drug substance, drug product, clinical trial materials |
| FDA, EMA | U.S. Food and Drug Administration, European Medicines Agency |
| IND, IMPD, CTA | Investigational new drug, Investigational Medicinal Product Dossier, Clinical Trial Applications (documents to the agency to conduct clinical trials in human) |
| NDA, MAA | New Drug Application, Market Authorization Application |
| BLA, MAA | Biologic License Application, Market Authorization Application |
| ICH | International Conference on Harmonization, a consortium organization offering technical guidelines on pharmaceutical development well accepted by the industry and regulatory agencies (https://www.ich.org/) |
| USP, EP | United States Pharmacopeial Convention (www.usp.org), European Pharmacopoeia (https://www.edqm.eu/en/european-pharmacopoeia-9th-edition) |
| GLP, GMP | Good laboratory practice, Good Manufacturing Practice, regulations for conducting nonclinical drug development and manufacturing. |
| CMC | Chemistry, Manufacturing and Controls |
| QA, QC | Quality assurance, Quality control (quality unit) |
| Release testing | QC testing conducted before DS/DP can be released for official use |
| QbD, DoE | Quality by Design, Design of Experiments |

## 9.2 OVERVIEW OF DRUG DEVELOPMENT PROCESS

Drug development is a highly complex, expensive, and multidisciplinary process [9, 10]. The pharmaceutical industry is a major employer of analytical chemists skilled in HPLC since it is the dominant analytical technique for characterization and quality assessments in drug development and production (see Table 9.1) [1, 2]. The modern molecular approach to drug development often starts with an understanding of the pathophysiology of a disease and a molecular target(s) (in the patient or pathogen), followed by the synthesis of new chemical entities (NCEs) that bind to the molecular target (hits and leads). These NCEs with acceptable physicochemical properties and preliminary safety/efficacy profiles are subsequently optimized by medicinal chemists by varying structural motifs of a lead into a development candidate(s). The drug candidate is then synthesized at kilogram levels by process chemists, characterized by analytical chemists, and formulated into appropriate dosage forms by pharmaceutics scientists (formulators). If safety and efficacy profiles of the drug candidate are acceptable, it may eventually become a new commercial drug product after a stringent regulatory approval process including clinical trials to ensure safety/efficacy in patients.

Clinical trial materials (CTMs), both DS and DP, used in the clinics must be produced under Good Manufacturing Practice (GMP) regulations. Quality specifications and quality control (QC) testing are required to release the CTM product for use, with increasing levels of regulatory oversight during the progression of the clinical trials [11, 12]. A schematic diagram of an organizational structure for technical development in a pharmaceutical company

**Table 9.1. Primary Characterization and Quality Control Tests for DS and DP**

| Tests for drug substance | Tests for drug product (oral solid dosage form) |
|---|---|
| • Appearance: visual<br>• Identity: IR, HPLC, (MS, NMR)<br>• Assay: HPLC<br>• Impurity (chemical): HPLC, (LC/MS)<br>• Impurity (chiral): HPLC, (SFC)<br>• Moisture: Karl Fischer<br>• Particle size distribution,<br>• Solid state characterization: XRPD, (DSC/TGA), (dynamic vapor sorption)<br>• Salt counterions: IC, titration<br>• Residual solvents: GC<br>• Inorganics, metals: ROI, IC, ICP-OES, ICP-MS | • Appearance: visual<br>• Identity: HPLC, UV<br>• Potency/content: HPLC<br>• Impurity (chemical): HPLC<br>• Uniformity of dosage units: by HPLC or by weight variation<br>• Moisture: Karl Fischer<br>• Dissolution: HPLC, UV, disintegration<br>• Hardness,<br>• Solid state characterization: XRPD<br>• Microbial testing<br>• Functional excipients: HPLC |

IR = infrared spectroscopy, NMR = nuclear magnetic resonance, XRPD = X-ray powder diffraction, DSC/TGA = differential scanning calorimetry/thermogravimetric analysis, ROI = residue on ignition, ICP = inductively coupled plasma spectrometry, OES = optical emission spectrometry, N/A = generally not applicable or required for QC, techniques in brackets are characterization test not used generally in QC testing.

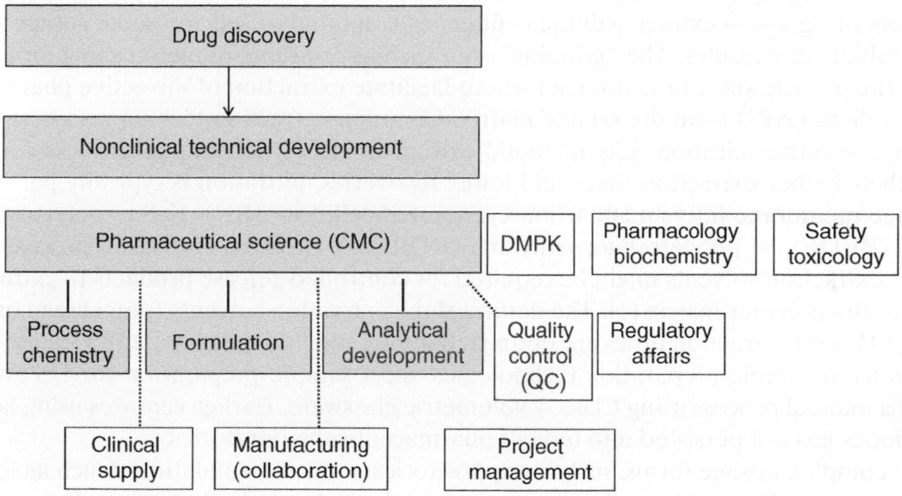


**Figure 9.1.** Schematic diagram showing the organization chart of a nonclinical technical development operation or CMC organization in a pharmaceutical company.

is shown in Figure 9.1. The pharmaceutical science unit is responsible for chemistry, manufacturing, and control (CMC) functions and has three typical subgroups [7].

1. A process chemistry group to scale-up the synthetic route of the drug substances.
2. A pharmaceutics group for formulation development and production of drug products for the clinic.
3. An analytical development group to develop methodologies to characterize and assess the quality of DS/DP lots and to support process development.

The pharmaceutical science unit works collaboratively with others such as drug metabolism and pharmacokinetics (DMPK), safety/toxicology, pharmacology/biochemistry, clinical quality control, regulatory affairs, project management, and clinical supplies. Its responsibilities are the technical and process development of all CTM and the documentation of the CMC sections in regulatory filings before the New Drug Application (NDA) registration. A brief synopsis of these respective functions are: *DMPK* works to understand the pharmacokinetics and transformation of the NCE and how to improve on bioavailability; *Pharmacology studies* the mode of action, efficacy in biological systems and the clinics; *Safety/Toxicology* conducts toxicological evaluations of the CTM and impurities in animal studies; *QC* is the laboratory branch of quality department responsible for method validation, establishing product specifications, release testing, stability studies, and preparation of technical documents for regulatory filing; *Regulatory Affairs* deals with regulatory agencies and their required filings; *Project Management* coordinates timelines and activities; *Manufacturing Collaboration* lines up and manages external suppliers and CMOs; *Clinical Supplies* predicts usage and ensures adequate supply of CTM for clinical trials.

## 9.3 SAMPLE PREPARATION PERSPECTIVES

In general, the "dilute and shoot" approach can be used for most DS and parenteral products since they are highly purified materials with filtration steps in their final processing [1, 13]. A process of "grind → extract → dilute → filter" is commonly used for solid dosage forms such as tablets or capsules. The "grinding" step such as grinding, rough crushing, or milling reduces the particle size of the dosage form to facilitate extraction of the active pharmaceutical ingredient (API) from the sample matrix. Common extraction techniques are shaking, vortexing, or ultrasonication. Liquid–liquid extraction (LLE) techniques are less common due to their higher extraction times and lower recoveries. Filtration is typically performed on syringe membrane filters made from regenerated cellulose (RC), Nylon, polyvinylidene fluoride (PVDF), or polytetrafluoroethylene (PTFE). A two-step extraction process using different extraction solvents might be required for controlled-release products to extract the API from the polymer matrix [1]. The nature of the extraction solvents (percentage organic solvent, pH) and extraction times are optimized during method development. Despite many innovations in sample preparation technologies, most sample preparation for DS and DP remains a manual process using Class A volumetric glassware. Earlier ventures using laboratory robotics has not persisted into today's pharmaceutical laboratories.

More complex dosage forms, such as suppositories, topical formulations (such as lotions, gels, creams, and oil-based formulations), and physiological samples (serum or plasma) might require more elaborate sample clean-up and extraction such as LLE or solid-phase extraction (SPE). The final sample extract is placed in an HPLC vial or a microtiter plate for HPLC or LC/MS analysis. Further details on sample preparation can be found elsewhere [1, 13, 14].

## 9.4 HPLC, SFC, AND HPLC/MS IN DRUG DISCOVERY

Modern drug discovery typically starts in basic research to understand the molecular target(s) (e.g. a receptor, or an enzyme) associated to the disease state of the body and the design of NCEs for intervention (inhibition or activation). In small-molecule drug discovery, medicinal chemists synthesize many novel compounds (NCEs) and optimize their therapeutic indices (ratio of therapeutic effects vs. toxicities) by changing their molecular structural motifs. In a

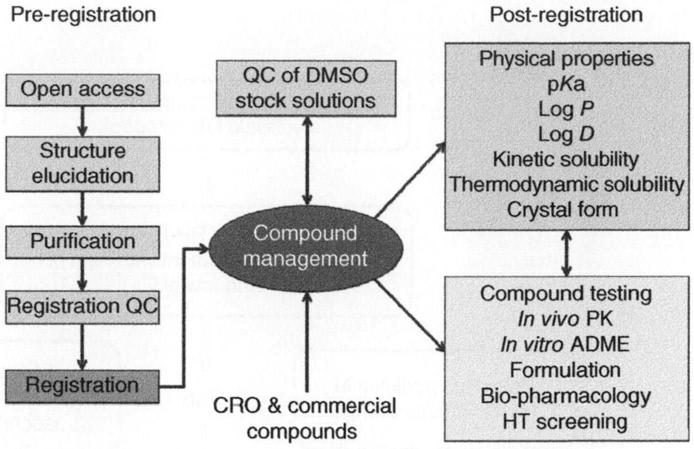


**Figure 9.2.** Schematic diagram of the sample workflow of a central analytical laboratory supporting high-throughput purification and characterization of NCEs in small-molecule drug discovery. Source: Reprinted with permission of Wong et al. 2015 [5].

pharmaceutical company, a large number of NCEs are synthesized at the milligram or gram scale, followed by purification and characterization by a central analytical laboratory, before archival (as solids and prealiquoted stock solutions in dimethylsulfoxide [DMSO]) in a Compound Management organization for further biological assays [2, 5, 6].

Figures 9.2 and 9.3 show the sample workflow and its purification strategy used in a central support laboratory supporting the small-molecule discovery programs of a pharmaceutical company [5]. The high-throughput purification process includes initial sample screening using RP-UHPLC/UV/MS with both low-pH and high-pH mobile phases followed by reversed-phase chromatography (RPC) or supercritical fluid chromatography (SFC) purification and sample recovery using Genevac or lyophilization to generate solid samples in a powder state for further characterization. SFC (both achiral and chiral) has emerged as a preferred purification technique in drug discovery due to the ease of compound recovery [5, 15]. Normal-phase chromatography (NPC) purification is rarely used since it often generates oils rather than free-flowing solids, which are difficult to manage.

The same central analytical laboratory in the case study in the reference generates Certificates of Testings (COTs) for compound registration reports by the use of many high-throughput characterization techniques for testing – such as, for identity (MS, NMR, FT-IR), purity (achiral and chiral using RPC/UV/CAD or SFC/UV), optical rotation (polarimetry), water content (Karl Fisher), kinetic solubility (RPC/UV/CLND), $pK_a$ (pH-UV titration), $\log D$ (RPC/UV or RPC/MS/MS), and quantitation of DMSO stock solutions (RPC with CAD or CLND detection with absolute calibration) [6].

## 9.5 HPLC TESTING METHODOLOGIES FOR DS AND DP

After obtaining an NCE with acceptable drug like properties and safety/efficacy profiles, a more detailed analytical characterization is required before its nomination as a clinical development candidate. This milestone triggers a pivotal toxicological evaluation of the NCE called Good Laboratory Practice (GLP) toxicology (tox) study, comprising of studies to ensure safety in two animal species (rodent and nonrodent), which represents a substantial financial commitment from the sponsor. A CMC team is now assembled to synthesize and

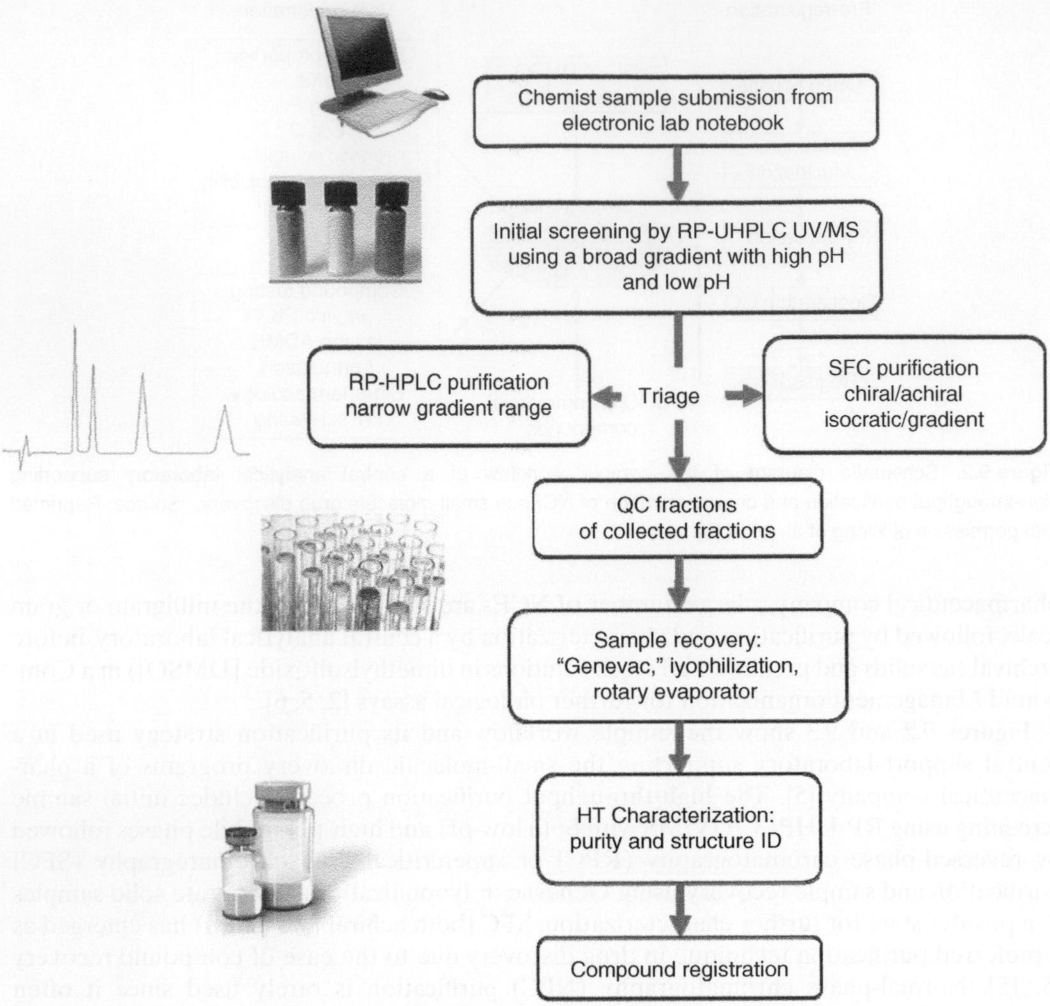


**Figure 9.3.** Schematic diagram of the sample workflow for high-throughput purification in a central analytical laboratory. Source: Reprinted with permission of Wong et al. 2015 [5].

release this GLP tox lot (e.g. 0.5–1 kg DS). The first task for the analytical chemist is to develop a stability-indicating HPLC method to investigate the purity and stability of the DS [2, 7]. Planning for regulatory filing to conduct clinical trials in humans then begins in earnest. Initial quality specifications and QC methodologies are established since CTM are produced under GMP regulations.

The roles and responsibilities of the analytical chemist in analytical development or quality control functions are shown in Table 9.2. These roles are typically separate functions in major companies but can be merged in smaller organizations. The goals are to gain a solid scientific understanding of the physicochemical properties of the DS or DP lots to safeguard the quality and consistency of the CTM during drug development. HPLC and other analytical methodologies used in the characterization and QC are described in this section.

**Table 9.2. Roles and Responsibilities of the Analytical Chemist in Drug Development**

| Analytical development | Quality control |
|---|---|
| • Characterization of physicochemical properties<br>• Method development<br>• Elucidation of the structure of key impurities and degradation products<br>• Analytical support for chemical process and formulation development<br>• Performing initial stability studies of DS and DP<br>• Collaboration with multidisciplinary technical development team(s), and oversee CMO activities to resolve technical issues | • Release testing of RM, SM, DS intermediates, DS, and DP, and provide COA<br>• Method validation (and method transfer to CMO)<br>• Establish phase-appropriate specifications for QC of clinical trial materials<br>• Perform stability studies to establish shelf life, packaging, shipping, and storage conditions |

CMO = contract manufacturing organization, RM = raw materials, SM = starting materials, COA = Certificate of Analysis, CMC = Chemistry, Manufacturing and Controls, IND = investigational new drug, CTA = clinical trial applications, IMPD = investigational medicinal product dossier, NDA = new drug application.

### 9.5.1 Identification Test (DS, DP)

Identification testing is used to confirm the presence of the API in samples of DS or DP lots. Typical tests include an HPLC retention time match and a spectroscopic confirmation (e.g. IR) against a qualified reference standard (see Section 9.5.3.3) plus visual observations of the DS or DP (e.g. appearance, size, form, color). Identification testing is performed for shipping verification, release testing, and stability studies. More substantial identification tests such as MS (single stage or tandem-MS) and NMR (Hydrogen-1 and Carbon-13 including 2-D NMR) are used for structure elucidation as characterization tests (test results are archived for each lot though they are usually not as QC tests reported in the CoAs).

### 9.5.2 ASSAY (Rough Potency and Performance Testing, DP)

With few exceptions, RPC/UV is the primary methodology used in potency testing (main component only) and purity evaluation (determinations of process impurities and degradation products). For the rough potency testing of the dosing solutions during initial toxicology evaluations and performance testing of DP (uniformity of dose and dissolution testing), a simple, nonstability-indicating isocratic RPC/UV method using Fast LC is typically used (see Figure 9.4) [16]. Note that this rough potency testing is not used during release testing of CTM, which must use a stability-indicating HPLC method.

***9.5.2.1 Testing for Uniformity of Dosage Units*** According to the USP<905> [17], a content uniformity test is required to ensure processing consistency (e.g. 10 single-tablet assays). A test of weight variation can be substituted for a content uniformity assay if the product contains >50 mg of an API and comprises >50% by weight of the dosage form unit. Typical acceptance criteria proposed for uniformity of dosage units are that each of the dosage units must lie between 85.0% and 115.0% of the label claim with an RSD (relative standard deviation) NMT 9.0%. If these criteria are not met, testing of additional tablets or capsules is required according to USP<905>. An example of a content uniformity report is shown in Table 9.3.

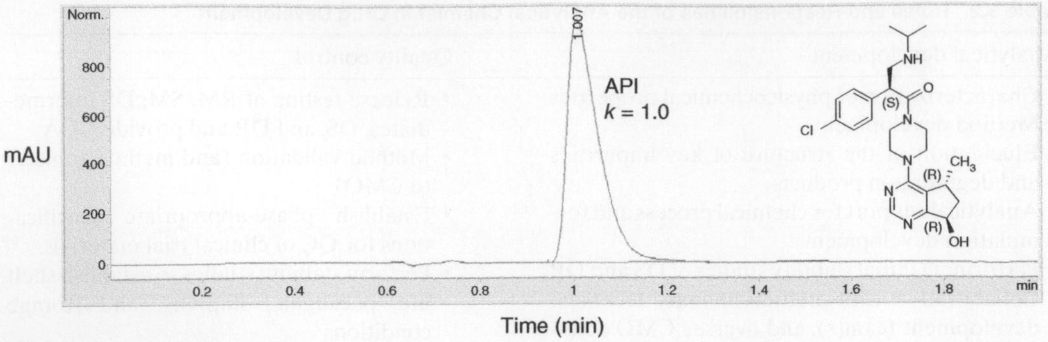


**Figure 9.4.** Example of a fast LC method for potency assay for content uniformity assay or dissolution testing. Structure of the API shown in the inset. HPLC conditions: Waters X-Bridge C18 (50 × 4.6 mm, 3 μm), mobile phase – 30% ACN in 0.05% trifluoroacetic acid, flow rate – 1.0 ml/min at 30 °C; detection – 280 nm; sample – 10 μl of API at 0.1 mg/ml in 0.1 N HCl.

**Table 9.3. An Example of a Uniformity of Dosage Unit Report**

| | Sample name | Result ID | % API |
|---|---|---|---|
| 1 | CAPSULE 1 | 1507 | 99.5 |
| 2 | CAPSULE 2 | 1508 | 101.0 |
| 3 | CAPSULE 3 | 1509 | 97.3 |
| 4 | CAPSULE 4 | 1510 | 102.9 |
| 5 | CAPSULE 5 | 1511 | 102.4 |
| 6 | CAPSULE 6 | 1512 | 98.6 |
| 7 | CAPSULE 7 | 1513 | 100.2 |
| 8 | CAPSULE 8 | 1514 | 105.5 |
| 9 | CAPSULE 9 | 1515 | 98.6 |
| 10 | CAPSULE 10 | 1516 | 101.4 |
| Mean | | | **100.7** |
| Minimum | | | **97.3** |
| Maximum | | | **105.5** |
| %RSD | | | **2.41** |

***9.5.2.2 Dissolution Testing*** Dissolution testing is used to measure the release of the API under standardized conditions (USP<711>) using the Apparatus Type I (basket method) or Type II (paddle method) or other apparatus types [1, 17]. This *in-vitro* evaluation is performed to ensure product lot-to-lot consistency of release profile and may be compared or correlated with the *in-vivo* bioavailability of the solid dosage formulation from clinical studies to establish a predictive relationship called an *in-vitro in-vivo* correlation (IVIVC). Dissolution testing is performed to check product consistency during formulation development, stability studies, and product releases. Table 9.4 lists typical HPLC dissolution method attributes. Figure 9.5 shows examples of the dissolution profiles of a controlled release product from a formulation development study.

**Table 9.4. Typical Method Attributes for HPLC Methods for Dissolution Testing**

| Method | Isocratic HPLC with UV detection |
|---|---|
| Sample preparation | None, except filtration of the samples withdrawn from the dissolution apparatus |
| Quantitation | External standardization |
| Typical specification limits | >Q + 5% in 30–60 minutes for immediate release products (Stage 1)<br>±10–15% of release specification at specified time points for controlled release products |
| Method validation acceptance criteria | Accuracy: Three levels in triplicate at specification levels; <±5% of true values<br>Precision: 12 sample units NMT 10–20% RSD<br>Specificity: API resolved from major impurities/degradants and product components<br>Linearity: Linearity established between 10% and 150% of analyte concentration, $r > 0.99$ with % y-intercept NMT 5.0% |

*Q* is a value of amounts of API dissolved specified in USP <711>, February, 2011. NMT = not more than.

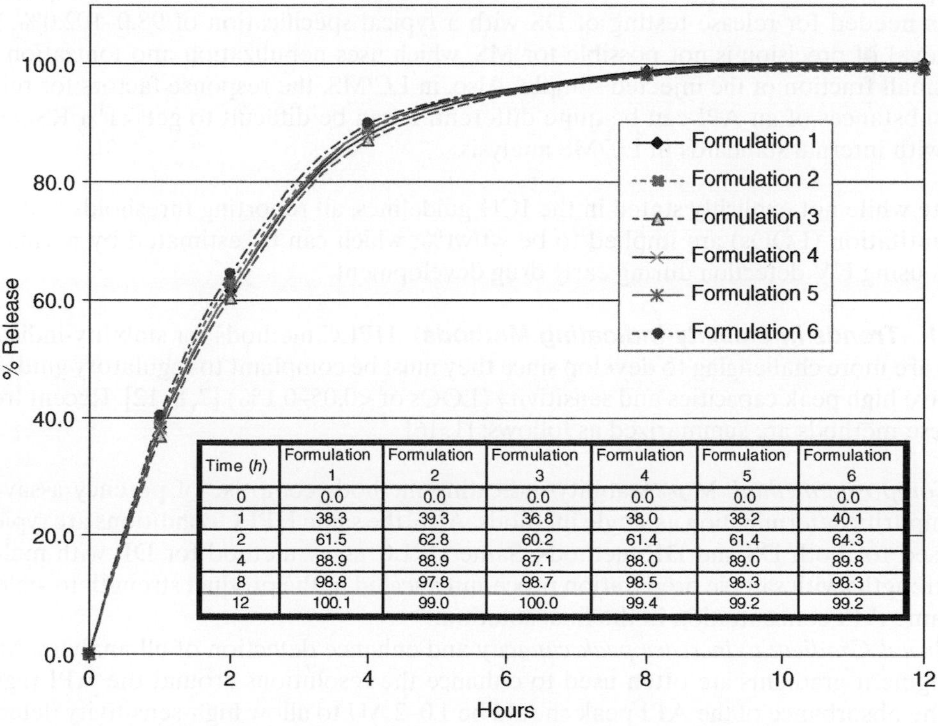


| Time (*h*) | Formulation 1 | Formulation 2 | Formulation 3 | Formulation 4 | Formulation 5 | Formulation 6 |
|---|---|---|---|---|---|---|
| 0 | 0.0 | 0.0 | 0.0 | 0.0 | 0.0 | 0.0 |
| 1 | 38.3 | 39.3 | 36.8 | 38.0 | 38.2 | 40.1 |
| 2 | 61.5 | 62.8 | 60.2 | 61.4 | 61.4 | 64.3 |
| 4 | 88.9 | 88.9 | 87.1 | 88.0 | 89.0 | 89.8 |
| 8 | 98.8 | 97.8 | 98.7 | 98.5 | 98.3 | 98.3 |
| 12 | 100.1 | 99.0 | 100.0 | 99.4 | 99.2 | 99.2 |

**Figure 9.5.** Dissolution profiles of six samples of an experimental controlled release product, manufactured under different processing conditions. The inset shows the actual data obtained.

### 9.5.3 Stability-Indicating Assay (Potency and Purity Testing of DS and DP)

The accurate quantitation of the API, its impurities, and degradants is important to ensure drug potency, purity, and stability for drug quality assurance. These assessments typically utilize stability-indicating HPLC methods, which separate the API from all process impurities, degradation products, and excipients (in case of DP) [2, 12]. Requirements and validation parameters of these methods are published by the International Conference on Harmonization (ICH) [12]. These stability-indicating assays are typically gradient RPC-UV methods to yield in a single chromatographic run quantitative data for potency (API levels) and impurities levels of DS and DP during release testing and stability studies [1, 16]. Invariably, broad gradient RPC/UV methods are used for this application for the following reasons.

*Mass balance*: The retention mechanism of RPC is a weak dispersive force (hydrophobic interaction), so all components in the drug sample should elute from the column under high mobile phase strength conditions (e.g. 100% MeOH or ACN). The assurance of total elution of all impurities in the sample is less certain for hydrophilic interaction chromatography (HILIC), NPC, ion-exchange chromatography (IEC), or other separation modes with stronger interactions [4].

*UV detection is standard*: Nearly all drugs have UV chromophores. UV detection is highly quantitative and capable of peak area precision of <0.2% RSD. This high precision is needed for release testing of DS with a typical specification of 98.0–102.0%. This level of precision is not possible for MS, which uses nebulization and ionization of a small fraction of the injected sample. Also, in LC/MS, the response factors for related substances of an API can be quite different. It can be difficult to get <1% RSD even with internal standards in LC/MS analysis.

Note while not explicitly stated in the ICH guidelines, all reporting thresholds and limits of quantitation (LOQs) are implied to be wt/wt%, which can be estimated by normalized area% using UV detection during early drug development.

**9.5.3.1 *Trends in Stability-Indicating Methods*** HPLC methods for stability-indicating assays are more challenging to develop since they must be compliant to regulatory guidelines and have high peak capacities and sensitivity (LOQs of <0.05–0.1%) [7, 8, 12]. Recent trends for these methods are summarized as follows: [1, 16]

- *Composite method*: Most stability-indicating methods comprise of potency assay and impurity determination as single methods. Also, the same HPLC conditions are typically used for both DS and DP methods. Same HPLC assay method for DP with multiple strengths with sample preparation procedures scaled to the product strength to yield the same API concentration in the final solution.
- *Broad Gradient to increase peak capacity* and enhance detection of all analytes. Multi-segment gradients are often used to enhance the resolutions around the API regions. The absorbance of the API peak should be 1.0–2 AU to allow high-sensitivity detection of impurities, but <2.0 AU to prevent detector signal saturation. In some cases, multiple detection wavelengths can be used for getting the proper sensitivity for impurities without saturation of the API.
- *Only degradants are reported in DP methods*: In DP methods, only degradants are reported according to regulations since process impurities are supposedly controlled in the DS method. However, process impurities are typically monitored and tracked in DP methods to make sure that they are not coeluting with degradants or the API.

- *MS-compatible*: MS-compatible methods are preferred to facilitate peak tracking and identification of unknowns for early-phase methods.
- *UHPLC methods*: UHPLC methods are preferred for faster speed and higher resolution. Nevertheless, since UHPLC instruments may not be available in all laboratories, UHPLC method is often converted back to HPLC conditions, which are adopted to be the primary regulatory methods for global manufacturing.

***9.5.3.2 Potency Determination (DS, DP)*** The potency assay is a quantitative determination of the API in DS and DP samples [1, 2]. A highly purified lot of the API is qualified as a reference standard first, using a mass-balance approach to determine its purity factor on an "as-is" basis (see Section 9.5.3.3). This reference standard is then used as a calibration standard for all subsequent potency assays. Specification limits are typically stringent for DS (98.0–102.0%) and 90.0–110.0% of the label claims for DP (or 95.0–105.0% for EP). For tablets or capsules, a composite assay of 10–20 units is typically used to minimize unit-to-unit variation. Alternately, a portion of the composite ground powder equal to the average tablet weight (ATW) may be extracted and assayed. Quantitative extraction of the API from the solid formulations is critical. Table 9.5 summarizes the typical stability-indicating HPLC assay and method validation acceptance criteria for potency and impurity determinations.

Equations for potency calculations (purity as is) for DS and % Label Claim (LC) for DP are shown as follows.

$$\%\ \text{Purity "as is"(DS)} = \%\ \text{Potency(DS)} = \frac{A_{\text{std}}}{W_{\text{std}}} \times \frac{W_{\text{std}} \times P_{\text{std as is}}}{A_{\text{std}}} \times 100\%$$

$$\%\ \text{LC} = \frac{A_{\text{samp}}}{A_{\text{std}}} \times \text{Std conc.} \times \frac{V_{\text{samp}}}{N_{\text{tab}}} \times \frac{100}{\text{Label Claim}}$$

$$\text{Where: Std. conc.} = \frac{W_{\text{std}}(\text{mg})}{V_{\text{Stock}}(\text{ml})} \times \text{Purity} \times \text{CF}$$

| | |
|---|---|
| % Purity "as is" | = purity factor of a DS as assayed against an API reference standard of the same form |
| $P_{\text{std}}$ "as is" | = the purity factor assigned to the reference standard after correction for water, residue on ignition (ROI), and residual solvent. It can be reported as a free base (after salt correction) or as a salt (ee Section 9.5.3.4). |
| % LC | = percentage label claim of API per tablet or capsule |
| $A_{\text{samp}}$ | = area of the API peak in the sample solution |
| $W_{\text{samp}}$ | = weight of the sample (DS) |
| $A_{\text{std}}$ | = area of the API peak in the reference standard solution |
| $W_{\text{std}}$ | = weight of the reference standard |
| CF | = weight conversion factor if the API is a different salt form than that of the reference standard |
| Purity | = purity of reference standard after correction for moisture, residual solvents, and purity of the compound |
| $V_{\text{samp}}$ | = volume of the sample solution |
| $N_{\text{tab}}$ | = number of tablets or capsules tested, |
| 100 | = conversion factor to % |

**Table 9.5. Typical Method Attributes for Stability-Indicating Assays of DS and DP**

| Method | HPLC with UV detection (frequently PDA) |
|---|---|
| Sample preparation | Dissolve drug substances in volumetric flasks at 0.1–5 mg/ml (DS)<br>"Grind → extract → dilute to volume → filter" of 10–20 dosage units (DP) |
| Quantitation | External standardization using a well-qualified reference standard (potency)<br>Area normalization against the API peak using RRF (impurities) |
| Typical specification limits (potency) | 98–102% purity on a dried basis<br>90–110% of label claim (USP), 95–105% of label claim (EP) |
| Method validation acceptance criteria for potency assay[a] | Accuracy: Three levels in triplicate (80%, 100%, and 120% of nominal assay concentration); < ±2% from true value<br>Precision: Six determinations at 100% level; < ±2.0 RSD<br>Specificity: API resolved from impurities and degradation products<br>Linearity: Five levels between 80% and 120% ($r > 0.999$)<br>Robustness: Method accuracy unaffected by slight perturbations |
| Typical specification limits (impurities) and thresholds for reporting, identification, and qualification | ICH guidelines for late-stage and commercial products or IQ guidelines in early development as supported by toxicology/clinical/impurity qualification data |
| Method validation acceptance criteria for impurity assay[a] | Accuracy: Three levels in triplicate (LOQ – 1%), < ±10–20% of spikes values<br>Precision: Six replicates at LOQ, <10–20% RSD<br>Specificity: Resolve key specified impurities and degradants<br>Linearity: Five levels between LOQ – 2% ($r > 0.98$)<br>Sensitivity: LOQ of at least 0.05% for DS and 0.1% for DP |

$r$ = coefficient of linear correlation, NMT = not more than, LOQ = limit of quantitation

[a]Typical values as examples.

**9.5.3.3 *Qualification of Reference Standard for the API*** A qualified reference standard of the API is used in potency assays. During early development, a highly purified reference standard is prepared (e.g. 50–100 g for the first reference standard batch) to serve as calibration reference standard for identity and potency assays. The reference standard undergoes substantial characterization and testing to establish a purity factor and is requalified periodically. An example of the Certificate of Analysis (COA) of such a reference standard for an NCE is shown in Table 9.6 with the purity factor calculation shown below using a mass balance approach to correct for water, residual solvents, and ROI. A correction for the counter ion is typically needed for pharmaceutical salts as drug potency is typically expressed as a free base since the counter ion is not efficacious.

$$\text{Purity factor calculation} = \%\text{purity by HPLC} \times [1 - 0.01(\%\text{water} + \%\text{residual solvents} + \%\text{ROI} + \%\text{salt})]$$

**Table 9.6. Certificate of Analysis (COA) of a Reference Standard of an API**

| | | |
|---|---|---|
| Lot number: XYZ | Date of manufacturing: 12 April 2016 | Retest date: March 2017 |
| Formula: $C_{24}H_{33}Cl_2N_5O_2$ | Molecular weight: 494.5 g/mol | Storage conditions: Refrigerated with desiccant |
| Test | Test name | Result |
| Visual | Appearance | White solid |
| USP<197> | Identity by FTIR | Consistent with structure |
| USP<761> | Identity by NMR | Consistent with structure |
| USP<941> | Identity by XRPD | Consistent with qualification spectrum |
| M1234 | Purity by HPLC | 99.5% |
| M1235 | Impurity by HPLC | Total = 0.44%, RRT 0.75 = 0.08%, RRT 0.81 = 0.16%, RRT 0.97 = <0.05%, RRT 1.05 = 0.14%, |
| M1236 | Impurities by Chiral HPLC | Not detected |
| USP<921> | Water content | 1.4% |
| USP<231> | Heavy metals | ≤20 ppm |
| M2234 | Metals by ICP-AES | Pd ≤ 2.6 ppm, Ru ≤ 2.6 ppm |
| M2235 | Residual solvents | Total 0.03%, ethyl acetate = 0.03% |
| USP<281> | Residual on ignition (ROI) | 0.0% |
| Combustion | Elemental analysis | C = 56.9%, H = 6.9%, N = 13.7% |
| M2236 | Counter ion content by ion chromatography (IC) | 7.5% chloride |
| DSC | Glass transition temperature | $T_g$ = 142.5 °C |

The purity factor or potency as a free base assigned to this material is 90.6%.
Purity factor calculation = % purity by HPLC × [1–0.01(%water + %residual solvents + %ROI + %salt)]
Testing performed by organizations A, B, C, and D followed by signatories by analyst and QC manager with the date.

**9.5.3.4 *Quantitation of Impurities and Degradation Products (DS, DP)*** The stability-indicating HPLC method must separate the API from all key process impurities and degradation products (degradants) with UV detection for accurate quantitation [12]. For DP samples, the excipients should also not interfere with quantitation of the API or impurities. This is usually achieved by chromatographic separation or occasionally, by a proper choice of the detection wavelength. Method development can be challenging for complex DS or DP as described in Chapter 10.

Figure 9.6 shows an example of a UHPLC stability-indicating assay of a DP. The top chromatogram is the separation of a retention marker solution containing the API spiked with key impurities and degradants. The bottom chromatogram is from an extract of a tablet during an accelerated stability study placed at 50 °C/75%RH for three months [4]. The levels of degradants (or impurities) are calculated according to the equation below indexed to the peak area of its parent API. In early drug development, when impurities or degradants are not available as reference material, normalized area% analysis is often used with the relative response factor (RRF) assumed to be 1.

$$\text{Degradant}(\%) = \frac{\text{Area of degradant}}{\text{Area of parent API}} \times \frac{100\%}{\text{RRF of degradant}}$$

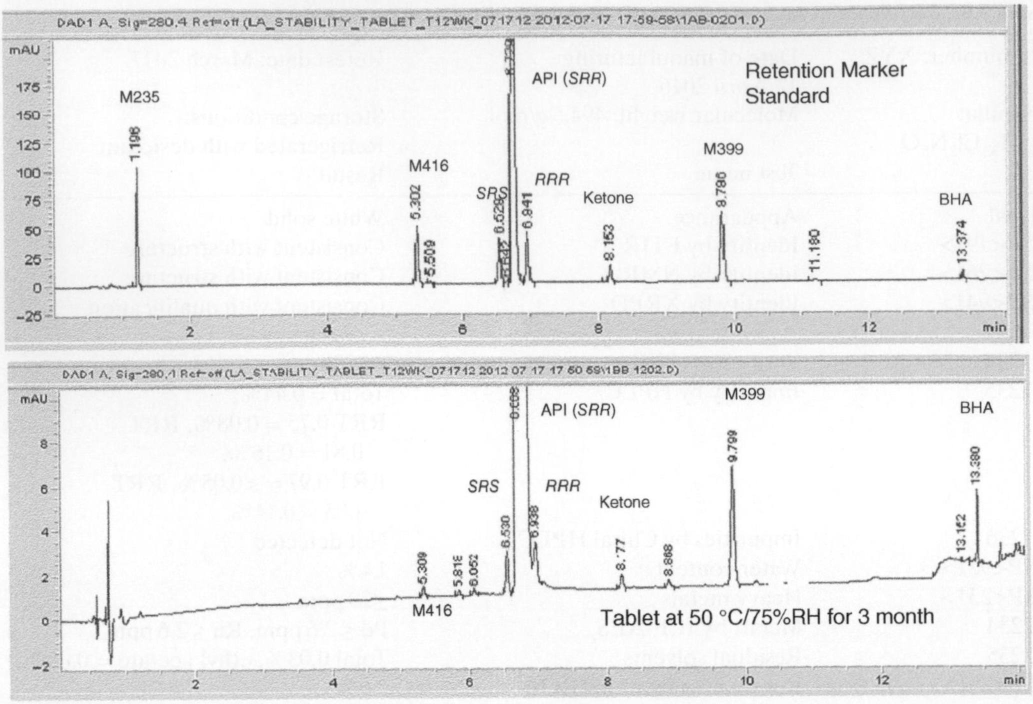


**Figure 9.6.** UHPLC chromatogram of a retention marker solution and a three-month stability sample (extract of a tablet kept in a stability chamber at 50 °C/75%RH). This is an example of a stability-indicating assay used extensively in the pharmaceutical industry to establish shelf life. HPLC conditions were as follows: column – ACE Excel 2 C18, 100 mm × 3.0 mm, 2 μm; mobile phase A – 20 mM ammonium formate pH 3.7; mobile phase B – 0.05% formic acid in acetonitrile (ACN); flow – 0.8 ml/min, 40 °C (typical initial backpressure 450 bar); gradient – 5–15%B in 2 minutes, 15–40%B in 10 minutes, 40–90%B in 1 minute; UV detection @ 280 nm; Sample – tablet extract in 20% ACN in 0.1 N HCl, 3 μl injection, Structure of the API shown in Figure 9.4. Source: Reprinted with permission of Dong 2013 [4].

RRF is relative response factor of the degradant against API at the monitoring wavelength.

***9.5.3.5 Case Study of an HPLC Method of an API with Multiple Chiral Centers*** Currently, a major fraction of small-molecule drugs in development have one or more chiral centers [9, 18, 19]. Since stereoisomers can have different pharmacological or toxicological properties, it is a regulatory expectation to use single enantiomers for clinical development of NCEs. HPLC/UV with achiral or chiral stationary phases remains the primary analytical methodology for the assessment of overall chemical purity and the determination of stereoisomeric or enantiomeric impurities. Achiral RPC/UV is the method of choice for the determination of the overall chemical purity of DS. For APIs with multiple chiral centers, the content of diastereomers is a more important quality attribute since enantiomers are unlikely to be present in a well-controlled asymmetric synthesis process. Ideally, diastereomers can be quantified simultaneously in a single stability-indicating method as shown in this case study.

Figure 9.6a shows a UHPLC stability-indicating method of a DP from an API with three chiral centers (structure shown in Figure 9.4). There are a total of $2^3$ or 8 possible stereoisomers for this molecule or four pairs of diastereomers. Note that diastereomers can be separated by achiral RPC while enantiomers by chiral HPLC only. An achiral RPC/UV stability-indicating assay capable of separating all four pairs of diastereomer was developed as discussed in Section 10.3.4 using a multisegment gradient approach to increase the resolution of the region around the API [16].

Figure 9.6a shows a chromatogram of a retention marker solution from a UHPLC method developed to support DP development (not a regulatory method). Peak designations shown in the chromatograms are: API (*SRR*, absolute configuration for the drug molecule with three chiral centers); *SRS* and *RRR* (process impurities diastereomers. Note that *SSR* is not found in any DS lots); M235, M416, and M399 (degradants designated by their parent ions in MS); ketone (an oxidative degradant); BHA (butyl hydroxyanisole, an antioxidant additive) [4]. Figure 9.6b shows the chromatogram of an extract of a tablet formulation kept in a stability chamber at 50 °C/75%RH for three months, indicating increased levels of degradants for M416, *SRS*, *RRR*, ketone, and M399. Note that due to the chemistry of each chiral center, some stereoisomers are degradants, while others aren't (i.e. the chiral center is "stable" and does not undergo isomerization reactions). The data summary of this accelerated stability study is shown in Table 9.7.

**Table 9.7. Results of a Three-Month Accelerated Stability study in a New Drug Product Formulation**

| Peak ID | | M235 (area%) | M416 (area%) | SRS (area%) | API (area%) | RRR (area%) | Ketone (area%) | M399 (area%) |
|---|---|---|---|---|---|---|---|---|
| RT (Min) | | 1.20 | 5.30 | 6.53 | 6.73 | 6.94 | 8.15 | 9.80 |
| Temperature/RH% | RRT | 0.18 | 0.79 | 0.97 | 1.00 | 1.03 | 1.21 | 1.46 |
| 5 °C | Coated-yellow tablet, 100 mg | | | 0.01 | 99.9 | | | |
| 25 °C/60 | Coated-yellow tablet, 100 mg | | 0.01 | 0.01 | 99.9 | | 0.01 | 0.01 |
| | Coated-yellow tablet, 100 mg, 1 des | | 0.01 | 0.01 | 99.9 | | 0.01 | 0.01 |
| | Coated-yellow tablet, 100 mg, open dish | | 0.01 | 0.01 | 99.8 | | 0.03 | 0.01 |
| 30 °C/65 | Coated-yellow tablet, 100 mg | | 0.01 | 0.01 | 99.9 | | 0.02 | 0.02 |
| 40 °C/75 | Coated-yellow tablet, 100 mg | | 0.02 | 0.02 | 99.8 | | 0.04 | 0.06 |
| | Coated-yellow tablet, 100 mg, 1 des | | 0.03 | 0.02 | 99.7 | | 0.05 | 0.06 |
| | Coated-yellow tablet, 100 mg, open dish | | 0.02 | 0.03 | 99.7 | 0.01 | 0.03 | 0.10 |
| 50 °C/75 | Coated-yellow tablet, 100 mg | 0.01 | 0.02 | 0.09 | 99.3 | 0.06 | 0.04 | 0.35 |

RRT = relative retention time; des = desiccant.
Source: Reprinted with permission of Dong 2013 [4].

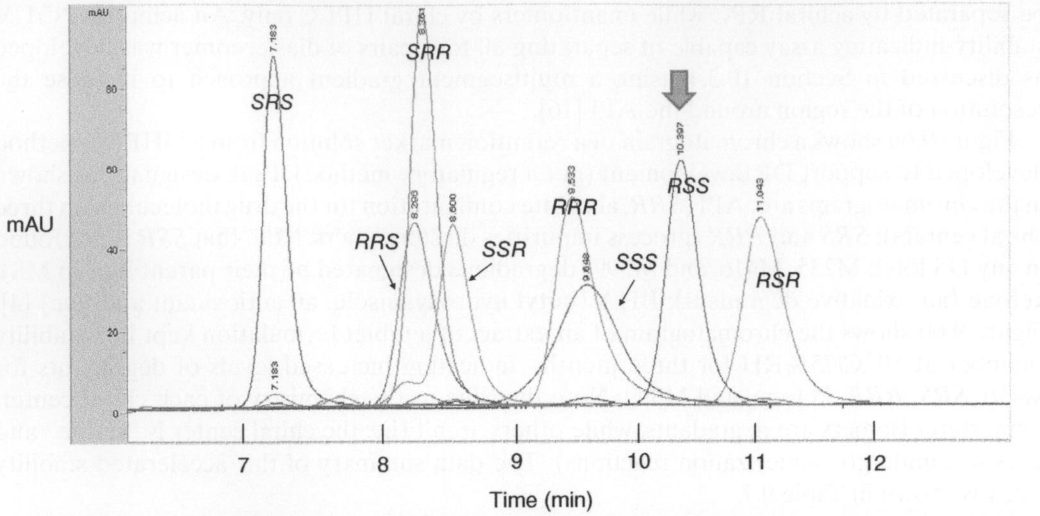


**Figure 9.7.** Chiral HPLC method for the API showing overlaid chromatograms of all eight stereoisomers of Ipatasertib (an investigational AKT inhibitor). HPLC method conditions: column: Chiralpak IA, 150 × 4.6 mm, 5 μm; MPA: hexanes/ethanol/ethanolamine (100 : 10 : 0.05 v/v); MPB: ethanol/ethanolamine (100 : 0.05 v/v); Flow rate: 1.0 ml/min at 30 °C; Gradient: 0–10%B in 15 minutes, 10–50%B in 5 minutes, 50–0%B in 5 minutes; Detection: 280 nm; Injection volume = 5 μl of API at 2 mg/ml in ethanol, Structure of the API shown in Figure 9.4.

Determination of enantiomers is commonly performed by chiral HPLC using chirospecific phases (CSP) [1, 3]. While SFC is becoming the dominant technique for preparative and column screening purpose, chiral HPLC is likely preferred for QC assays. Today, the use of polysaccharide CSPs predominates in chiral separations because of their versatility and wide applicability (particularly the immobilized CSPs). They are used primarily under NPC conditions with an additive to keep the analytes in a nonionized state. RPC is less useful than NPC since solutes are separated in hydrated states often with less selectivity differences between these isomers.

Figure 9.7 shows the overlay of eight chromatograms generated using a chiral NPC method developed for QC analysis of the DS, which resolves the *RSS* enantiomer from the API (*SRR*) and all the other six stereoisomers [19]. This method was used to quantitate the enantiomer (*RSS*) only as diastereomers (*RRR*, *SRS*, and *SSR*) are quantitated by the primary regulatory RPC method (Section 10.3.4). Note that a basic additive is needed to keep the analytes in a nonionized state to lessen peak tailing. This method had a LOQ of ~0.2% and was qualified and used for release testing and stability studies. This method uses an immobilized polysaccharide column under NPC gradient condition with ethanolamine as a mobile phase additive. There was some initial discussion on whether the development of a chiral HPLC for the enantiomer was truly needed with sufficient QC on the chiral purity of the starting materials of the API. Analytical data of many subsequent batches verified that no quantifiable level of the *RSS* enantiomer was found.

#### 9.5.3.6 *Control of Chemical and Chiral Purities of Starting Materials for Multichiral APIs*

In the case study described earlier, the first GMP synthesis of the API lot was achieved by coupling of the two regulatory starting materials (SMs): $\beta^2$-amino acid (SM1 with one chiral center) and *trans* cyclopentyl pyrimidine core (SM2 with two chiral centers) [20]. The chemical and chiral purity of both SM has stringent quality specifications. Figure 9.8

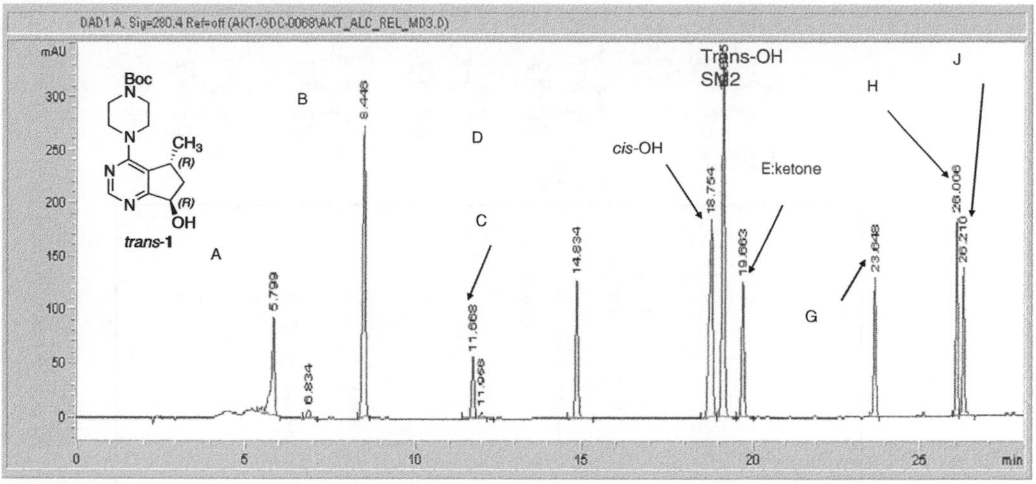


**Figure 9.8.** The achiral HPLC chromatogram of the retention time marker solution for the purity assays of RSM2, illustrating the use of a multi-segment gradient method. This method separates RSM2 from its *cis*-isomers, its immediate precursors (ketone), and other available synthetic precursors (labeled peaks A to J). HPLC conditions: column – ACE 3C18, 150 × 4.6 mm, 3 μm: MPA – 20 mM ammonium formate pH 3.7, MPB – methanol; Flow rate – 0.8 ml/min at 30 °C; Gradient program – 5–30%B in 5 minutes, 30–80%B in 15 minutes, 80–100%B in 6 minutes; Detection – 280 nm; Injection volume = 10 μl of RSM2 at 0.5 mg/ml. Structures of RSM2 is shown in the inset, and the *cis*-isomers and its precursor ketone are shown in Figure 9.9.

shows the RPC/UV QC methods used for the chemical purity of SM2, which separates the cis and trans isomers, its immediate precursors (ketone), and all other synthetic precursors A to J. A multisegment gradient using methanol as the strong solvent was used for this achiral RPC method for this complex SM2. Figure 9.9 shows the chiral HPLC method used for the determination of all the four stereoisomers of SM2 in addition to the two enantiomers of its immediate precursor ketone. This method uses an immobilized polysaccharide column under NPC gradient condition with diethylamine as a mobile phase additive.

***9.5.3.7 DP with Multiple APIs or Natural Products*** Another category of complex pharmaceutical samples that can benefit from the higher peak capacities of UHPLC is DP with multiple APIs (examples shown in Figure 5.3) and from natural products (Figures 5.4 and 5.5). Accurate analysis of these types of complex sample was hitherto an unmet need in pharmaceutical analysis, and UHPLC can now provide a more satisfactory solution [18].

***9.5.3.8 Stability Studies*** Stability studies are conducted during drug development to establish storage conditions, shelf life, and packaging conditions for both DS and DP [12, 21]. Data from these studies are part of the CMC sections in regulatory submissions to the regulatory authorities. These stability studies consume major fractions of analytical resources within a company. Table 9.7 is a stability summary report for a three-month stability study of an oral tablet formulation under accelerated and stressed conditions, indicating increased levels of degradants (particularly for hydrolytic degradant M399) under high heat and humidity conditions. Note that such HPLC data are highly reproducible by different labs for many time points during a stability study, which can last for two to three years [4].

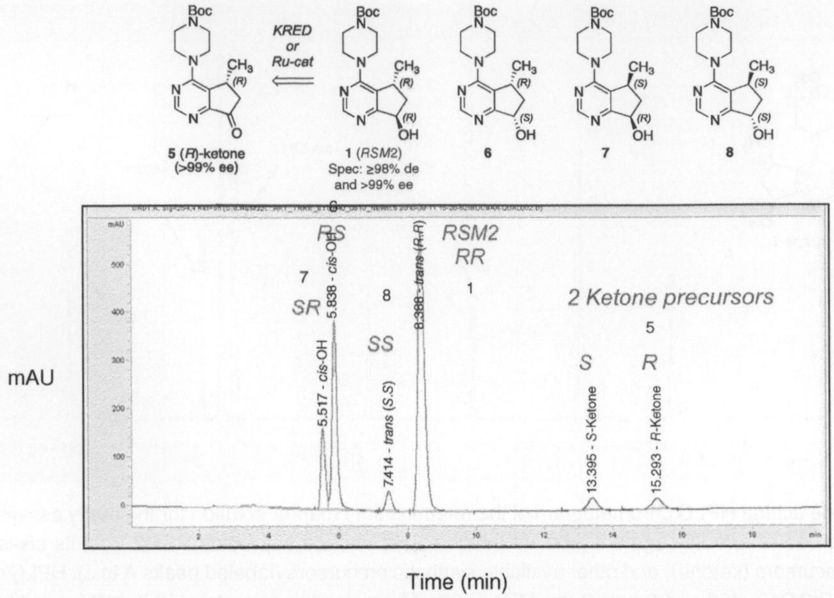


**Figure 9.9.** Chiral LC chromatogram of a retention time marker solution of RSM2 spiked with its *S,S*-enantiomer, two *cis* diastereomers (*SR* and *RS*) and its two ketone immediate precursors (*S* and *R*). Chiral HPLC method conditions: Column – Chiralpak IC (150 × 4.6 mm, 5 μm); MPA – hexane; MPB: 0.1% diethylamine in ethanol; Flow rate – 0.8 ml/min at 40 °C; Gradient – 40–100%B in 15 minutes; Detection – 254 nm; Injection volume = 5 μl of RSM2 at 1.5 mg/ml in ethanol. Structure of the various isomers of RSM2 and its precursor ketone are shown in the inset.

### 9.5.4 Assay of Preservatives

A preservative is a substance that extends the shelf-life of drug products by preventing oxidation or inhibiting microbial growth [1]. Preservatives must be monitored in the products since they are considered to be functional excipients. Preservative levels are usually low in DP because the regulators permit preservatives to be added in products only to prevent microbial proliferation arising under in-use conditions, not to cover poor microbiological manufacturing conditions, thus sensitive analytical methods are required. Common preservatives are antioxidants for solid dosage forms such as butylated hydroxytoluene (BHT), and antimicrobials for liquid formulations such as parabens, sodium benzoate, or sorbic acid. For these additive components, typical assay specifications are 85–115% of the label claims. Assay of preservatives can be developed separately for DP or can be part of the primary stability-indicating assay.

### 9.5.5 Assay of Pharmaceutical Counterions

Accurate determinations of counterions are mandatory for the release testing and quality control (QC) of all pharmaceutical salts to confirm the identity of the salt form and mass balance of the DS. In addition, the residues of some counterions used during the synthesis of API can also be relevant for safety consideration (e.g. residual iodides in a chloride salt can boost sensitization reactions and interact with thyroid metabolism). Ion chromatography (IC) is the standard analytical technique used in most pharmaceutical laboratories capable of excellent accuracy, specificity, and sensitivity for both cations and anions [1, 22]. Figure 9.10

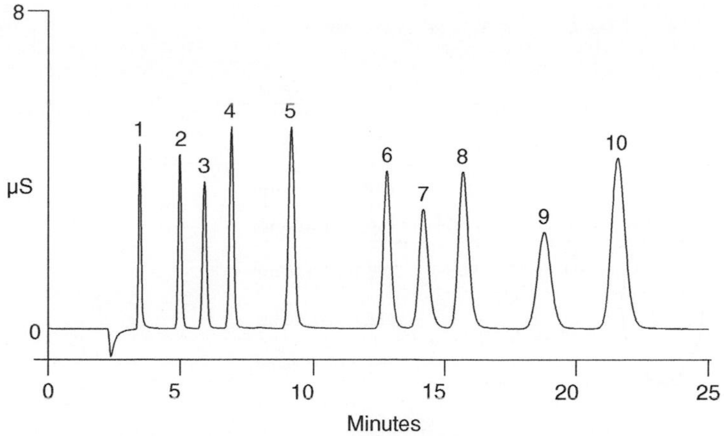


**Figure 9.10.** Separation of the common anions and disinfectant byproduct anions on an IonPac AS9-HC and an AG9-HC column. Columns: 4 × 250 mm IonPac AS9-HC and 4 × 50 mm AG9-HC. Eluent: 9 mM sodium carbonate. Flow rate: 1 ml/min. Injection volume: 25 μl. Detection: suppressed conductivity utilizing the anion self-regenerating suppressor (4 mm), recycle mode. Ions: 1. Fluoride (3 mg/l); 2. Chlorite (10 mg/l); 3. Bromate (20 mg/l); 4. Chloride (6 mg/l); 5. Nitrite (15 mg/l); 6. Bromide (25 mg/l); 7. Chlorate (25 mg/l); 8. Nitrate (25 mg/l); 9. Phosphate (40 mg/l); 10. Sulfate (30 mg/l). Source: Ahuja and Dong 2005 [1]. Copyright 2005. Reprinted with permission of Elsevier, Chapter 8.

shows an application of a rapid separation of common anions by IC. Recent publications have shown the potential application of a mixed-mode chromatography (MMC)/CAD method for pharmaceutical anions (see Figure 3.30) [23]. Since chloride is the most common salt form, the use of microtitration using $AgNO_3$ titrants with coulometric detection was found to be simple, accurate, and quite amenable for release testing purposes [22].

### 9.5.6 Assay of Potential Genotoxic Impurities (PGI)

Of increasing regulatory concerns is the presence of potential genotoxic impurities (PGI) in pharmaceuticals [24–26]. PGI can be residual reagents or side products, which can be reactive, and have diversified structures. These PGI may need to be controlled at parts-per-million levels. Thus, very sensitive and selective methodologies are required. The development of analytical procedures for PGI can be extremely challenging and resource-intensive. While generic methods using LC/UV, LC/MS, or LC/MS/MS may be effective at times, additional tactics with specific sample extraction and derivatization are often needed in practice. An analytical strategy for PGI has been proposed [24]. Most laboratories develop methods for PGI to demonstrate the impurity-purging capability of the chemical scale-up processes during development so that QC release testing would not be required for the final product at the commercial stage.

## 9.6 CLEANING VERIFICATION

Cleaning verification is a GMP requirement in the pharmaceutical manufacturing of drug substances and drug products. Cleaning verification must be performed to demonstrate the "cleanliness" of the production equipment at the completion of each manufacturing step by

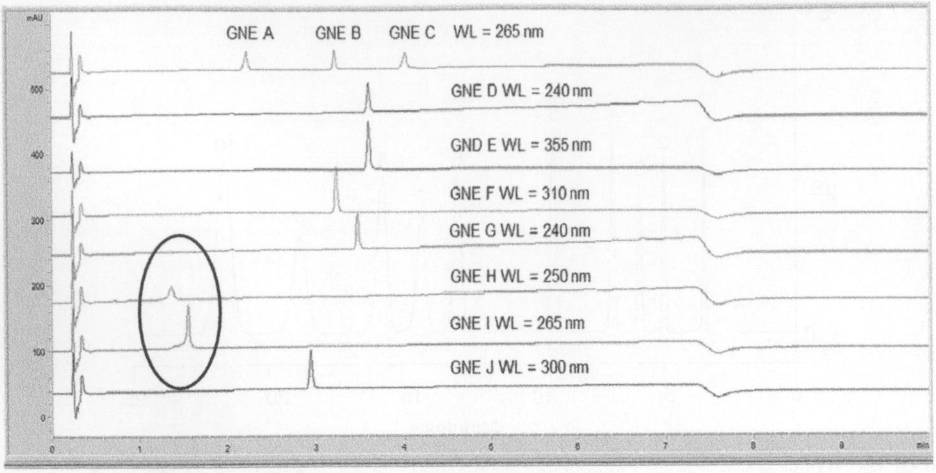


**Figure 9.11.** Stacked HPLC chromatograms of 10 new chemical entities (NCEs) (GNE A–J) HPLC conditions adapted for multiple APIs are shown in [27]. This standard method has been qualified for GMP cleaning verification use for GNE A, B, and C. Detection is by UV at 265 nm. Note that two of the less hydrophobic NCEs (circled) show peak fronting due to the injection of a relatively large injection volume of 20 μl. Source: Reprinted with permission of Dong et al. 2012 [27].

visual inspection and the analysis of cleaning swabs or rinse solutions to confirm that API and other cleaning agents have been adequately removed to pre-established acceptance limits [27]. Cleaning validation is particularly important when manufacturing equipment is not dedicated to a single DS or DP.

The analytical procedures can be nonspecific (e.g. gravimetric [residue on evaporation or ROE] or total organic carbon [TOC]) or product-specific techniques such as ion mobility spectrometry (IMS) or HPLC using UV, evaporative light scattering detector (ELSD), CAD, or MS detection. Figure 9.11 shows overlaid chromatograms of 10 NCEs in a 10-minutes generic HPLC/UV platform method. This generic method is amenable to cleaning verification of diverse API rinse solutions at a concentration of 0.2–10 μg/ml. Lower detection limits can be accommodated by sample concentration, the use of a long-path-length flow UV flow cells, or MS detection. Using UHPLC, similar assays can be achieved in about two to three minutes [27].

## 9.7 BIOANALYTICAL TESTING

Bioanalytical testing is the analysis of drugs and their metabolites in physiological fluids (plasma, serum, tissue extracts), typically from experimental animals or human subjects during preclinical or clinical studies [14, 28, 29]. In the last decade, conventional HPLC methods using UV, fluorescence, or electrochemical detection after elaborate sample clean-up are now replaced by LC/MS/MS methodologies. Extremely challenging concentration levels (such as pg/ml) in complex analytical matrixes may be required for highly active APIs. For this reason, detection methods with enhanced specificity and sensitivity are required. Sample preparation needed to simplify the sample (please note that a plasma sample contains thousands of components, while a DS/DP sample a few), may include off-line procedures such as protein precipitation, LLE, or SPE. The 96-well microplate is becoming the standard format

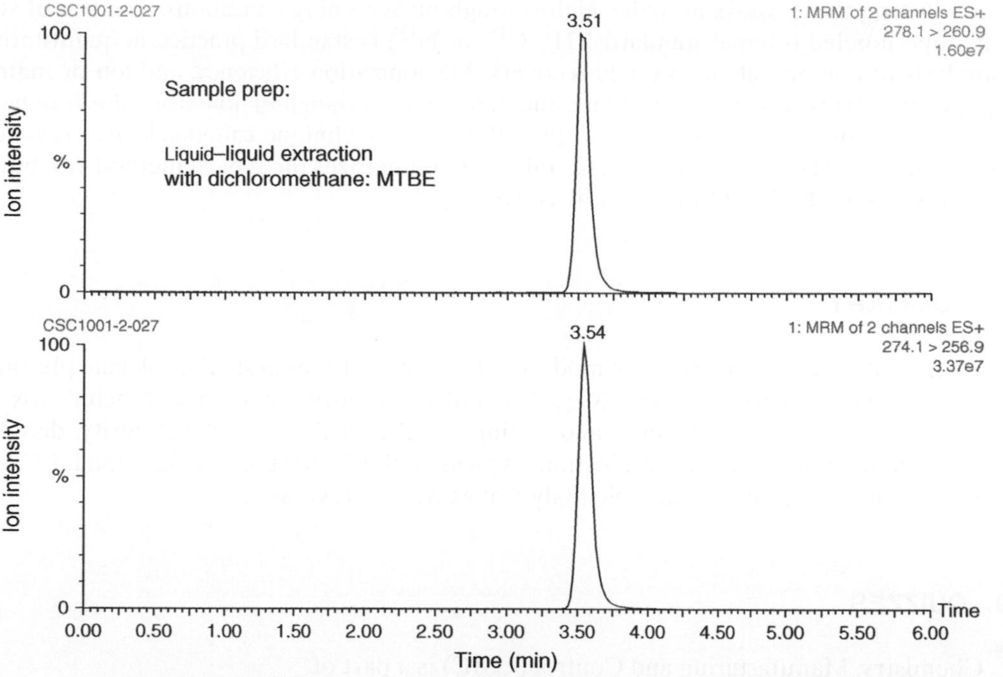


**Figure 9.12.** An example of bioanalytical analysis using LC/MS/MS analysis for the quantitation of clinical bioavailability samples. Sample preparation used liquid–liquid extraction of dichloromethane and *t*-butylmethylether. A tetra-deuterated analog of the API was used as an internal standard (upper trace) for the quantitation of the actual API of the subject sample (lower trace). Note that a fragmented daughter ion ($M/z$ = M -17, 257) was used for the actual quantitation of the API with excellent sensitivity (low ng/ml) and selectivity.

**Table 9.8. Typical Method Attributes for LC/MS/MS Methods for Bioanalytical Testing**

| Method | Isocratic or gradient HPLC with MS/MS detection |
|---|---|
| Sample preparation | SPE, protein precipitation, liquid–liquid extraction or on-line clean up; typically using the 96-well microplate format |
| Quantitation | Internal standardization based on a stable isotope labeled analog of the analyte is preferred |
| Method validation acceptance criteria | Accuracy: Three levels, five determinations each; <15–20% of the spiked values |
| | Precision: Three levels, five determinations each; NMT 15–20% RSD |
| | Specificity: Analytes resolved from other metabolites or endogenous components; check blank samples from six subjects |
| | Linearity: Regression coefficient of calibration curve >0.99 based on six to eight levels covering the entire dynamic range |
| | Stability: Room temperature, frozen storage, freeze/thaw, in biological matrices |

SPE = solid-phase extraction.

for conducting these assays in under high-throughput screening conditions. The use of stable isotope-labeled internal standards ($H^2$, $C^{13}$, or $N^{15}$) is standard practice in quantitative bioanalysis to compensate for sample recovery, MS ionization efficiency, and ion or matrix suppression effects. For high sensitivity and selectivity, a daughter ion from the fragmentation of the parent analyte is used for quantitation – a technique called selective reaction monitoring (SRM) shown in Figure 9.12. Table 9.8 summarizes the typical method attributes of LC/MS/MS methods for bioanalytical assays.

## 9.8 SUMMARY

This chapter provides an overview of modern HPLC applications in small-molecule pharmaceutical analysis from drug discovery to quality control. Applications discussed include assays for potency and content uniformity and testing for chemical purity, chiral purity, dissolution performance, and cleaning verification. Typical method attributes are described. LC/MS applications in drug discovery and bioanalytical assays are reviewed.

## 9.9 QUIZZES

1. Chemistry, Manufacturing and Control (CMC) is a part of
   (a) clinical development
   (b) nonclinical drug development
   (c) drug discovery
   (d) preclinical research
2. Analytical chemists does NOT typically work in this department
   (a) Quality control
   (b) DMPK
   (c) Pharmacology
   (d) Analytical development
3. Which separation mode or technique is generally not used in high-throughput purification in drug discovery?
   (a) NPC
   (b) RPC/UV
   (c) SFC
   (d) IEC
4. Efficient assays for dissolution testing DO NOT generally have this feature
   (a) Does not use reference standard
   (b) Stability-indicating
   (c) Used for drug products
   (d) Analysis time <2 minutes
5. A stability-indicating HPLC method does NOT have this feature
   (a) Used in stability studies
   (b) Must report process impurities for DP
   (c) Separates both impurities and degradation products
   (d) Can also measure levels of APIs

6. Dissolution testing IS NOT used for
   (a) indication for bioavailability
   (b) release of API from DP
   (c) DP degradation
   (d) DP performance
7. Chiral HPLC generally uses this column
   (a) C18
   (b) Derivatized polysaccharides
   (c) Amino
   (d) Achiral column
8. The primary stability-indicating method for DS with multiple chiral centers is typically
   (a) an isocratic method
   (b) an NPC chiral HPLC method
   (c) a chiral HPLC method that separates enantiomers
   (d) an achiral RPC method that separates diastereomers
9. Which technique is NOT used for determination of pharmaceutical counter ions in QC?
   (a) IC
   (b) LC/MS
   (c) MMC/CAD
   (d) Microtitration
10. Which technique is NOT used for cleaning verification?
   (a) RPC/UV
   (b) C/MS
   (c) NPC/UV
   (d) RPC/CAD

### 9.9.1 Bonus Quiz

1. In your own words, describe why gradient RPC/UV is typically used for stability-indicating assays of DS and DP. Is MS considered to be a viable detector?
2. In your own words, explain why the purity factor of a DS requires corrections for water, residual solvents, and ROI besides HPLC purity. Why is purity factor or potency often expressed as a free base with salt correction?
3. In your own words, explain why regulatory agencies are concerned with stability data.

## 9.10 REFERENCES


1. Ahuja, S. and Dong, M.W. (eds.) (2005). *Handbook of Pharmaceutical Analysis by HPLC*. Amsterdam, the Netherlands: Elsevier. Chapters 5, 6, 8, 13, 17, and 18.
2. Kazakevich, Y.V. and LoBrutto, R. (eds.) (2007). *HPLC for Pharmaceutical Scientists*. Hoboken, NJ: Wiley. Chapters 11, 13, and 14.
3. Guillarme, D. and Dong, M.W. (2013). *Am. Pharm. Rev.* 16 (4): 36.
4. Dong, M.W. (2013). *LCGC North Am.* 31 (6): 472.
5. Wong, M., Murphy, B., Pease, J.H., and Dong, M.W. (2015). *LCGC North Am.* 33 (6): 402.
6. Lin, B., Pease, J.H., and Dong, M.W. (2015). *LCGC North Am* 23 (8): 534.
7. Dong, M.W. (2015). *LCGC North Am.* 33 (10): 764.

8. Kou, D., Wigman, L., Yehl, P., and Dong, M.W. (2015). *LCGC North Am.* 33 (12): 900.
9. Hill, R.G. and Rang, H.P. (eds.) (2012). *Drug Discovery and Development: Technology in Transition*, 2e. Edinburgh, Scotland: Churchill Livingston, Elsevier.
10. Dong, M.W. (2016). Drug development process. Short Course presented at Pittcon, Atlanta (March).
11. FDA (2018). *Code of Federal Regulations*, Title 21, parts 210 and 211, Government Publishing Office, Washington, D.C.
12. ICH Harmonized Tripartite Guideline (2003). *Stability testing of new drug substances and products*, Q1A (R2); *Validation of analytical procedures*, Q2(R1), 2015; *Impurities in new drug substances*, Q3A(R2), 2006; *Impurities in new drug products*, Q3B(R2), 2006; *Specifications: test procedures and acceptance criteria for new drug substances and new drug products*, Q6A, 1999. International Conference on Harmonisation, Geneva, Switzerland.
13. Bishop, E.J., Kou, D., Manius, G., and Chokshi, H.P. (2011). *Sample Preparation of Pharmaceutical Dosage Forms* (ed. B. Nickerson). New York, NY: Springer Science. Chapter 10.
14. Wilson, I.D. (ed.) (2003). *Bioanalytical Separations*, Handbook of Analytical Separations, vol. 4. Amsterdam, the Netherlands: Elsevier Science.
15. Webster, G.K. (ed.) (2014). *Supercritical Fluid Chromatography: Advances and Applications in Pharmaceutical Analysis*. Singapore: Pan Stanford.
16. Dong, M.W. (2013). *LCGC North Am.* 31 (8): 612.
17. (2017). *United States Pharmacopeia and National Formulary, USP 40-NF 35*. Rockville, MD: United States Pharmacopoeial Convention, Inc.
18. Dong, M.W., Guillarme, D., Fekete, S. et al. (2014). *LCGC North Am.* 32 (11): 868.
19. Dong, M.W., Al-Sayah, M., Goel, M., and Remarchuk, T. (in preparation).
20. Remarchuk, T., St-Jean, F., Carrera, D. et al. (2014). *Org. Process Res. Dev.* 18 (12): 1652.
21. Huynh-Ba, K. (ed.) (2009). *Handbook of Stability Testing in Pharmaceutical Development*. New York, NY: Springer.
22. Dong, M.W. and Woods, R.M. (2016). *LCGC North Am.* 34 (10): 792.
23. Zhang, K., Dai, L., and Chetwyn, N. (2010). *J. Chromatogr. A* 1217 (37): 5776.
24. Venkatramani, C.J. and Al-Sayah, M.A. (2014). *Am. Pharm. Rev.* 17 (5): 64.
25. ICH Guideline M7 (2014). Assessment and control of DNA reactive (mutagenic) impurities in pharmaceuticals to limit potential carcinogenic risk. *International Conference on Harmonisation*, Geneva, Switzerland.
26. Teasdale, A. (ed.) (2011). *Genotoxic Impurities: Strategies for Identification and Control*. Hoboken, NJ: Wiley.
27. Dong, M.W., Zhao, E.X., Yazzie, D.T. et al. (2012). *Am. Pharm. Rev.* 15 (6): 10.
28. Guidance for Industry. Bioanalytical method validation. U.S. Department of Health and Human Services Food and Drug Administration Center for Drug Evaluation and Research (CDER) Center for Veterinary Medicine (CVM). https://www.fda.gov/downloads/drugs/guidances/ucm368107.pdf.
29. European Medicines Agency. Guideline on bioanalytical method validation. http://www.ema.europa.eu/docs/en_GB/document_library/Scientific_guideline/2011/08/WC500109686.pdf (accessed 07 January 2019).

# 10

# HPLC METHOD DEVELOPMENT

## 10.1 INTRODUCTION

### 10.1.1 Scope

This chapter provides an overview of high-pressure liquid chromatography (HPLC) method development focusing on reversed-phase stability-indicating methods for small-molecule drugs. The chapter describes the traditional approach to method development, starting with the initial selection of a column, detector, and mobile phase, followed by parameter optimization to improve the resolution of all key analytes. Modern method development trends including the use of automated screening systems and software products are reviewed. Other method development approaches described here include a three-pronged template approach and a universal generic gradient method development strategy. Case studies are presented to illustrate the various approaches and parameter selection processes. The reader is referred to other sources such as textbooks [1–4], and review articles [5, 6], for further discussions and approaches for method development. Sample preparation perspectives, method validation, and method transfer processes are described in Sections 9.3, 11.4, and 11.5. HPLC methods for large biomolecules and recombinant biotherapeutics are covered in Chapter 12.

### 10.1.2 Considerations Before Method Development

Development and validation of new analytical methods are costly and time-consuming. Before starting this process, a thorough literature search should be conducted for existing methods and approaches for the intended analytes or similar compounds. This search should include a review of relevant sources such as chemical abstracts, compendial monographs (e.g. *USP, EP*), journal articles (e.g. *Anal. Chem., J. Chrom. A & B., J. Sep. Sci., Analyst, JAOAC, etc.*), manufacturers' literature and applications notes, and the Internet (Google, Google Scholar, etc.). Although these types of searches may not uncover a directly usable method, it often provides a good starting point or other useful information. New analytical methods are typically needed for the following reasons:

*HPLC and UHPLC for Practicing Scientists*, Second Edition. Michael W. Dong.

- Existing methods are not available (e.g. New Chemical Entity (NCE) development program)
- Existing methods are not reliable, accurate, sensitive, robust, or regulatory compliant
- New instrumentation or technique has higher performance (e.g. ultra-high-pressure liquid chromatography [UHPLC], hydrophilic interaction chromatography [HILIC], liquid chromatography–mass spectrometry [LC/MS]).

### 10.1.3 HPLC Method Development Trends in Pharmaceutical Analysis

Best practices and emerging trends in HPLC method development are described here and discussed further in case studies. This includes examples that demonstrate:

- *Instrument developments*: Use of multisolvent pumps, photodiode array detectors (PDA), and MS in method development systems. The increased use of UHPLC, automated column/mobile phase screening systems, and software prediction tools are used to facilitate the method development process.
- *Detection developments*: The trend toward MS-compatible, gradient methods for stability-indicating (purity) methods for active pharmaceutical ingredients (APIs) and drug products (DP).
- *Orthogonal method approach*: The use of a secondary and orthogonal method(s) to ensure the separation of all impurities and the absence of coeluting or hidden impurities under the main peak.
- *LC/MS or LC/MS/MS*: A standard platform technology for trace quantitative analysis (e.g. bioanalytical assays, genotoxic impurities, and metabolomics. These are discussed in Chapters 6, 9, and 13).

## 10.2 A FIVE-STEP STRATEGY FOR TRADITIONAL HPLC METHOD DEVELOPMENT

A five-step strategy for traditional HPLC method development (initially described by Snyder et al.) [1] is summarized as follows:

1. Define method and separation goals
2. Gather sample and analyte information
3. Initial method development: "scouting" runs and getting the first chromatogram(s)
4. Method fine-tuning and optimization (the most time-consuming step)
5. Method prequalification (for regulatory methods)

Steps 1–5 are discussed herein while method validation is discussed in Section 11.4.

### 10.2.1 STEP 1: Defining Method Types and Goals

There are three types of methods: quantitative, qualitative, and preparative methods.

*Quantitative methods:* A *quantitative method* is the most common HPLC method type used today and generates information on the concentration or amount of the analyte(s) in the sample. A very common quantitative method is a purity assessment of process impurities and degradation products in pharmaceutics. This type of method is called a related substance or "stability-indicating" method in the pharmaceutical industry. A stability-indicating method is also a potency method as it can measures the active ingredient(s) in a drug substance or drug product sample without interference from process-related impurities, degradation products, and excipients.

For accurate quantitation in HPLC, it is necessary to use reference standards to calibrate the system before the sample analysis. Pharmaceutical potency assays typically use an external standard approach that relies on a single nominal concentration that covers the sample concentration range (or the formulation dosage strength range). An internal standard approach is popular among inhaled product assay methods that feature a broad analyte concentration range to detect very low analyte levels in samples. Multiple external calibration standards used to generate linear regressions are common in LC/MS clinical assay methods. A quantitative method is also adopted as a qualitative method to serve in analyte confirmation (as described below). A quantitative method used in regulatory testing is more difficult to develop and requires extensive efforts for method validation.

*Qualitative methods*: Assay or potency methods may also be routinely used as an identification test to confirm the presence (or absence) of analyte(s) in a sample by matching retention time and/or UV spectral data (from a PDA detector) with the same data from a reference standard analyzed in the same run. The generated data is often used as a secondary confirmation in pharmaceutical analysis. This type of methods can also be used as a limit test where a standard of the analyte is prepared equivalent to the maximum permitted quantity of the analyte in an unknown sample. The unknowns and standards are analyzed by the assay method, and all samples that show an analyte peak larger than the limit test standard fail to qualify, while those below the limit test are acceptable. Some examples of limit tests include cleaning verification or the remaining levels of the starting materials in an in-process control (IPC) test.

*Preparative methods*: Preparation methods are used to isolate purer components of interest from a crude sample for further work, for example, identification/structure elucidation or to generate reference standards. A preparative method requires a scale-up of the analytical method (flow rates, sample size, etc.) and involves the use of large-diameter columns. Method validation is not required since the goal is to generate purified components or enriched fractions for analysis. An analytical HPLC system is often used to isolate a small amount of purified material(s) (low milligram quantities) by loading up a larger sample amount and collecting the eluent containing the component to be isolated. However, the use of equipment used for regular analytical testing is not recommended for this type of semipreparative work due to the high risk of equipment contamination. It is always preferable to use dedicated equipment or specific preparative-scale LC instruments for such purification applications.

In drug discovery, supercritical fluid chromatography (SFC) is becoming the preferred preparative purification technique because it provides excellent selectivity, high resolving power as well as easy sample recovery for achiral and chiral compounds. The use of $CO_2$ as the eluent in SFC offers reduced costs and easy eluent evaporation over traditional liquid chromatography approaches.

***10.2.1.1 Method Goals and Acceptance Criteria for Quantitative Assays*** During method validation of quantitative assays, a series of studies are performed to demonstrate adequate method performance by passing typical acceptance criteria for peak resolution, accuracy, precision, linearity, specificity, and sensitivity. For instance, pharmaceutical potency methods typically require a resolution of >1.5 of the API from the closest eluting component;

**Table 10.1. Pertinent Sample and Analyte Information**

| Sample/analyte | Information |
|---|---|
| Sample | Number of components; concentration range of analytes |
| Sample matrix | API; formulation (excipients) |
| Analyte(s) | Chemical structure; molecular weight; purity |
| | Solubility; Log *P*; Log *D*; acidic/neutral/basic functional groups; the number of ionizable groups; N atoms, p$K_a$ values |
| | Chromophores; $\lambda_{max}$; the number of chiral centers; possibility of isomeric forms (e.g. stereoisomers, regioisomers, tautomers, rotamers) |
| | Stability/reactivity; storage conditions; toxicity |
| Others | Availability of reference standard materials |

a precision of retention time and peak area of <1% RSD (relative standard deviation); and linearity in the range of 70–130% of the label claim or the target value. Other desirable characteristics include analysis time approximately 5–30 minutes and minimal sample work-ups. Method validation is discussed further in Section 11.4.

### 10.2.2 STEP 2: Gathering Sample and Analyte Information

After defining method goals, the next step is to gather sample and analyte(s) information such as those listed in Table 10.1. These physicochemical data of the analytes are useful for the selection of the appropriate chromatographic separation and detection method. If some of these data are unavailable (e.g. p$K_a$, solubility, molecular weight, purity, Log *P*, Log *D*), separate studies can be initiated, or data obtained during initial method development. For NCEs, chemical structures and molecular formulae are typically well established. Particular attention should be directed to acidic, basic, aromatic, or reactive functional groups. Estimates of p$K_a$, solubility, chromophoric, or stability data can be inferred and/or calculated by software using the structure (e.g. chemicalize.com or acdlabs.com) if needed. Such data can help with the selection of the appropriate mobile phase pH and should always be considered.

Toxicity data and Material Safety Data Sheets (MSDSs) should be gathered to develop guidelines on safe handling procedures. Certificates of Analysis (COAs) from material suppliers and the technical packages from the API manufacturers are invaluable sources of pertinent data, which may include reference chromatograms, purity assessments, and spectral data (MS, NMR, IR, and UV).

### 10.2.3 STEP 3: Initial HPLC Method Development

During initial HPLC method development, a set of conditions (detector, column, mobile phases) is selected to obtain the first "scouting" chromatograms of the sample. In most cases, these are reversed-phase chromatography (RPC) on a C18 or a C8 column with UV detection. For stability-indicating methods or purity assays, which separate the API from all expected impurities and degradation products, gradient analysis is the obvious choice because of its higher peak capacities and the ability to analyze both early and late eluting components with better resolution and sensitivity.

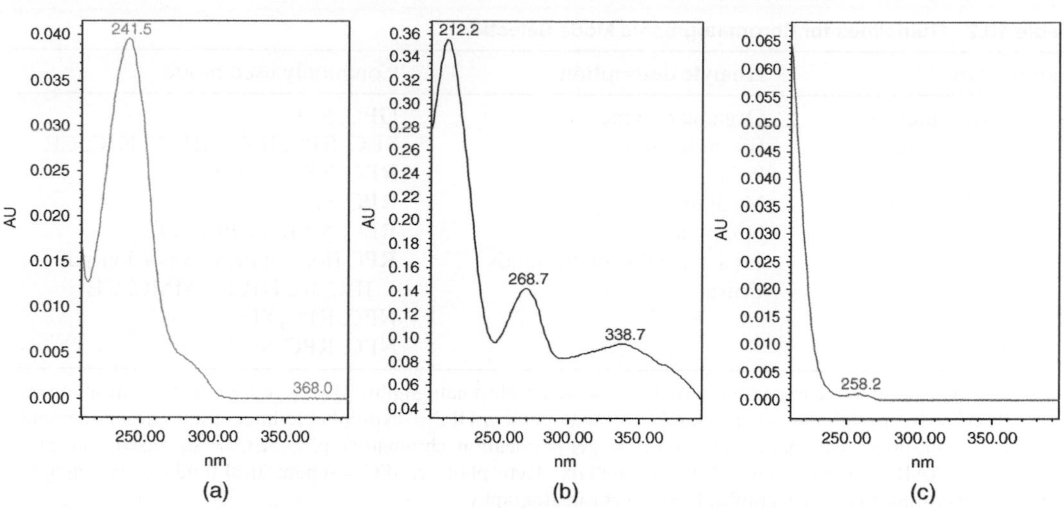


**Figure 10.1.** UV absorption spectra of three analytes illustrating the decision process in setting HPLC detector wavelength. (a) The $\lambda_{max}$ at 241 nm is a clear choice. (b) The three $\lambda_{max}$ at 212 nm, 269 nm, and 339 nm give rise to different possibilities: 269 nm is the obvious choice though 339 nm may be selected for higher selectivity if interference from other matrix components is a problem. (c) The selection of $\lambda_{max}$ at 258 nm is an issue due to low sensitivity. Far UV (210–230 nm) offers better sensitivity and a more universal detection.

***10.2.3.1 Initial Detector Selection*** For chromophoric analytes having reasonable UV absorbance (e.g. most drugs), the UV/Vis or PDA detector is a clear choice. UV/V is detectors are robust, reliable, relatively sensitive, easy-to-use, and very precise since the entire injected sample passes through the UV flow cell. The detection wavelength is usually set at the maximum absorbance wavelength ($\lambda_{max}$) of the main component or at a far UV wavelength (i.e. 200–230 nm) to improve sensitivity (Figure 10.1a–c). If possible, it is best to choose a wavelength ≥220 nm to minimize background absorbance from the mobile phase components (buffer salts, solvents, etc.). For related substances/purity methods, the detector is usually set to the $\lambda_{max}$ value for the parent compound. Many impurities in the sample would be related to the parent structure and are likely to absorb well at the parent $\lambda_{max}$ value. Late-phase related substance/purity methods typically require impurities to have their relative response factors (RRF) determined. Hence accurate %w/w values can be determined. This process ensures that impurities that are structurally different from the parent compound (and therefore have a very different UV absorption spectrum, e.g. the chromophoric group is degraded), can still be accurately quantified.

For nonchromophoric analytes (with low or no UV/Vis absorbance), derivatization can be considered, but this tends to be challenging and often less quantitative. Typically other detection modes are selected, which may include refractive index (RI) detectors, evaporative light scattering detectors (ELSDs), or charged aerosol detectors (CADs). Mass spectrometry (MS) is a possible choice for "ionizable" analytes. Tandem MS (or MS/MS) is the standard detection for bioanalytical assays, biomarker analysis, drug discovery screening, and genotoxic impurity assays due to its specificity and sensitivity. However, its use for routine quality control (QC) assays of pharmaceuticals is still limited due to the cost and complexity of operation (relative to LC/UV methods). Other detection options include fluorescence, conductivity detection for ionic species, and electrochemical detection (ECD) for neuroactive species in biochemical research.

**Table 10.2. Guidelines for Chromatographic Mode Selection**

| Sample type | Analyte description | Commonly used mode |
|---|---|---|
| Macromolecules | Organic polymers | GPC, RPC |
| ($M_w > 2000$ amu) | Biomolecules | SEC, RPC, IEC, HILIC, HIC, CE |
| Organics | Polar | RPC, NPC, HILIC |
| ($M_w < 2000$ amu) | Medium polarity | RPC, SFC |
| | Nonpolar | RPC, NARP, NPC, SFC |
| | Ions, ionizable compounds | RPC (ion suppression or ion-pair), IEC, IC, HILIC, MMC, CE |
| | Chiral | NPC, RPC, SFC |
| Preparative | | NPC, RPC, SFC |

RPC = Reversed-phase chromatography, IP = ion-pair chromatography, IEC = ion-exchange chromatography, HILIC = hydrophilic interaction liquid chromatography, HIC = hydrophobic interaction chromatography, NPC = normal-phase chromatography, GPC = gel-permeation chromatography, SEC = size-exclusion chromatography, NARP = nonaqueous RP, CE = capillary electrophorese, SFC = supercritical fluid chromatography, MMC = mixed-mode chromatography, IC = ion chromatography.

***10.2.3.2 Selection of Chromatographic Mode*** Table 10.2 lists guidelines for the selection of chromatographic modes based on the analyte properties that include molecular weight, polarity, and analysis type. The retention mechanism and the attributes of each mode are described briefly in Chapter 2. It is clear that RPC is applicable to a wide range of analyte properties, classes, and types of analysis. The dominance of RPC is primarily due to its versatility and robustness. RPC offers high resolution, reproducibility, and predictability. Additionally, RPC offers the assurance that all components in the sample are likely to be eluted from the RPC column because of the weak dispersive interactions between the stationary phase and most analytes [7]. With the relatively stronger interactions and retention mechanisms of other chromatographic modes (e.g. IEC, NPC, and HILIC), it is rather difficult to confirm the elution of all components in the sample. For this reason, it is a good practice during the development of API stability-indicating RPC methods to extend the gradient step to 100%B or 100% of the strong solvent and hold for at least 10 minutes to elute highly retained components in pivotal API lots. Finally, RPC with volatile mobile phase additives or buffers is MS-compatible and offers the option for quick peak tracking and identification.

***10.2.3.3 Selecting Samples*** For the development of stability-indicating related substances or purity methods, the ideal sample would be a test mix containing the API spiked with all the expected process impurities and degradation products. Unfortunately, reference materials of NCE impurities are generally unavailable in early-phase drug development. In most cases, synthetic precursors and intermediate compounds or the mother liquor from the last crystallization step may be available and can be helpful.

The analytical chemist usually generates degradation products from forced degradation studies to challenge methods (see Section 11.4.1.1). This is an important activity and a significant piece of work that involves the exposure of samples to a variety of conditions (also described/eventually required by regulators) that include light, heat, oxidative stress, acid, base, and so on. The purpose of this work is to ensure that the degradation pathways of the active ingredient are well understood and that the analytical procedures can detect principle degradants to ensure the quality and safety of drugs. Reference standards are more likely to be available in late-stage drug development programs. Similarly for generic drugs, reference standards are typically available and can often be purchased from a variety of commercial chemical companies or standards agencies (if a monograph exists).

***10.2.3.4 Initial Selection of HPLC Column and Mobile Phase*** A silica-based C18 or C8 column remains a good starting point because of its robustness, high efficiency, reproducibility, and retention of a range of mid-polar to hydrophobic compounds. Additional guidelines are as follows:

Begin method development with a newly purchased column, so the work is developed on a representative column from the manufacturer. Extensively used or old columns (>1-year-old) may not have the same retention and selectivity properties when compared to a newly manufactured column and should be avoided in method development as they can cause significant problems when attempting to reproduce the developed method on new columns. The analyst is encouraged to select a new column packed with 3- or sub-3-μm high-purity silica-bonded phases from a reputable column manufacturer. Three-micrometer or sub-3-μm packing is preferable for faster analysis and works well on either HPLC or UHPLC equipment [8]. Columns packed with sub-2-μm packing are best reserved for high-throughput screening (HTS) applications or fast methods for IPC methods on UHPLC equipment [9]. Columns packed with superficially porous particles (SPPs) are excellent choices [10]. For various applications, the following column dimensions are suggested:

- 50–100 mm × 3.0-mm i.d. for simple samples (e.g. assay methods of the main component)
- 100–150 mm × 3.0-mm i.d. for stability-indicating assays or multicomponent testing method of complex samples
- 30–50 mm × 2.1-mm i.d. for LC/MS analysis and HTS

While 4.6-mm i.d. remains the standard column i.d. for most laboratories, the optimum flow rate for 4.6-mm i.d. columns packed with 3-μm particles is ~2 ml/min, which is too high for routine testing [8]. It should be noted that 2.1-mm i.d. columns have stringent requirements for system dispersion and are only compatible with UHPLC or highly optimized HPLC equipment. Issues that lead to band broadening, poor peak shape, and variable method performance (especially for analytes with low retention) are typically seen when analysts attempt to use 2.1-mm i.d. columns on suboptimal equipment. For QC applications, 3.0-mm i.d. columns are preferred with optimum flow rates at ~1 ml/min and are compatible with both HPLC and UHPLC equipment.

Guidelines on mobile phase selection are discussed in Section 2.3 and are illustrated in the following case studies. After evaluation of the initial chromatograms, further fine-tuning may be needed until all key analytes are baseline resolved. Other bonded phases or column chemistries can be selected to enhance the resolution of critical pairs (two components that are coeluting or partially resolved) or to fulfill other method performance criteria.

***10.2.3.5 Generating the First Chromatogram*** After selecting the appropriate column and mobile phases, the method development is initiated by making the first injection of the sample into the HPLC. As explained earlier, the first chromatogram is typically performed with a C18 or C8 column using a broad gradient (e.g. 5–100% acetonitrile) with an acidic mobile phase A (MPA) at pH 2–4 as illustrated in the following case study.

***10.2.3.6 Case Study: Initial Purity Method Development of an NCE Using a Broad Gradient*** In this case study, a broad scouting gradient was used to obtain the first sample chromatogram for a drug substance method. The method goal was to develop a

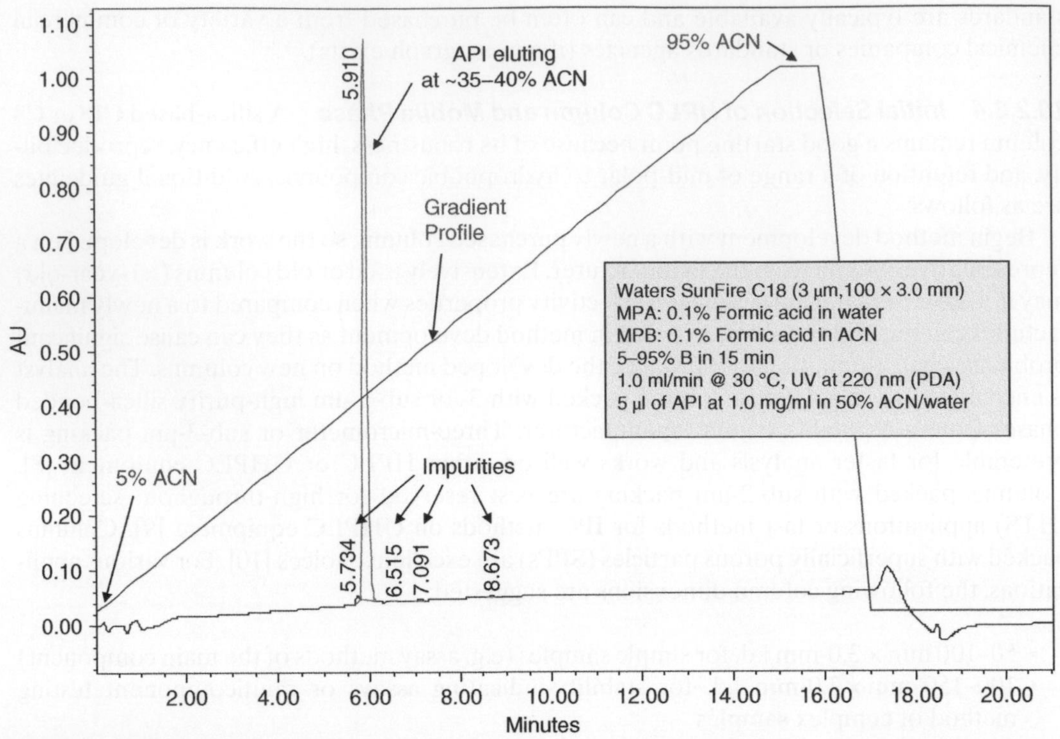


**Figure 10.2.** The first HPLC chromatogram of a drug substance sample obtained with a broad generic gradient during initial method development. The API elutes at %B of 35–40% of acetonitrile, showing the presence of several impurities between five and nine minutes.

MS-compatible stability-indicating method for an NCE. The API is a basic salt with a p$K_a$ of ~10. It has good solubility in water and has a UV spectrum shown in Figure 10.1c. The first chromatogram was generated by a high-purity silica-based C18 column (100 × 3.0 mm, 3 µm) with a broad gradient of 5–95%B in 15 minutes (MPA = 0.1% formic acid in water; MPB = 0.1% formic acid in acetonitrile), as shown in Figure 10.2. An acidic mobile phase of pH 2–4 was chosen as it typically yields better peak shapes for basic analytes because residual silanols are not ionized at this low pH. While most basic drugs are ionized at acidic pH, these drug molecules typically have sufficient hydrophobicity for adequate retention with C18 stationary phases.

The API sample was dissolved in a 50/50 acetonitrile/water diluent (as a default diluent) to 1.0 mg/ml. A PDA was used to collect data between 210 and 400 nm, allowing chromatograms to be plotted at various wavelengths of 260, 230, 220, and 210 nm, as shown in Figure 10.3. As expected from the UV spectrum, far UV detection at 210–230 nm generated greater sensitivity, though significant baseline gradient shifts were observed. A detection wavelength of 220 nm produced a strong signal for the API at ~1 AU with an acceptable gradient shift of the baseline. Figure 10.2 shows that all sample-related peaks appeared to elute between five and nine minutes (or eluting at a mobile phase strength of ~30–50% acetonitrile [ACN]).

These initial gradient runs were used to establish the purity of the sample, estimated hydrophobicity and UV spectral data of the API and its related substances, and the appropriate detection wavelengths. Next, the initial scouting gradient was reduced to a narrower range of 20–60% ACN yielding a chromatogram with enhanced resolution of the impurities

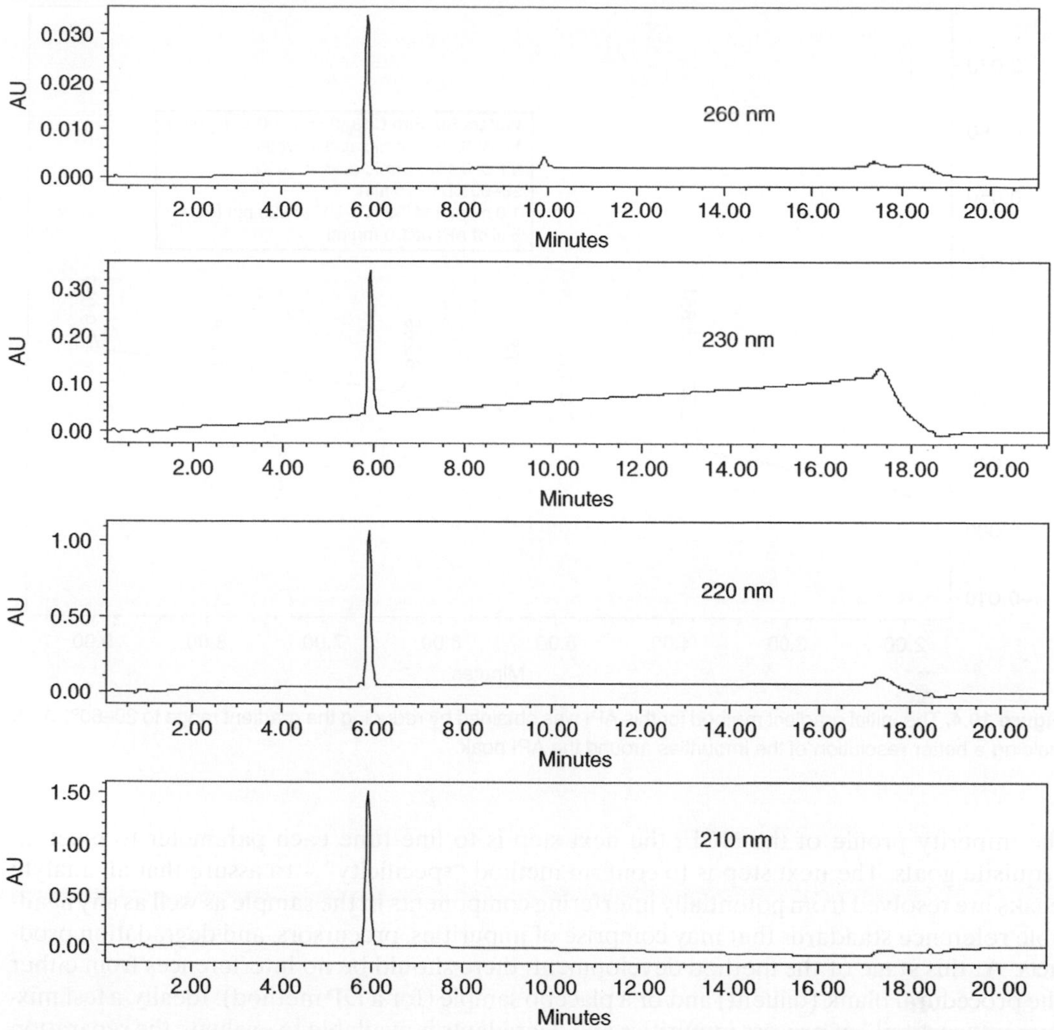


**Figure 10.3.** The PDA data collected from Figure 10.2 were plotted at 260, 230, 220, and 210 nm. As expected from the UV spectrum of the compound shown in Figure 10.1c, far UV detection at 210–230 nm yielded more sensitive detection, though some baseline gradient shifts were discernable. An excellent detection wavelength at 220 nm appears to yield an absorbance of the API at ~1 AU accompanied by a relatively flat baseline.

around the API peak (Figure 10.4). The optimized method conditions are depicted in the final chromatogram in Figure 10.7 and discussed in Section 10.3.1.

### 10.2.4 STEP 4: Method Fine-Tuning and Optimization

The goal of most HPLC analysis is to achieve adequate resolution of all key analytes with sufficient precision and sensitivity in a reasonable time. For ICH (International Conference on Harmonization) compliant stability-indicating methods [11], all expected impurities and degradation products must be adequately separated from the API and each other to ensure accurate UV detection and quantitation. After obtaining the initial chromatograms depicting

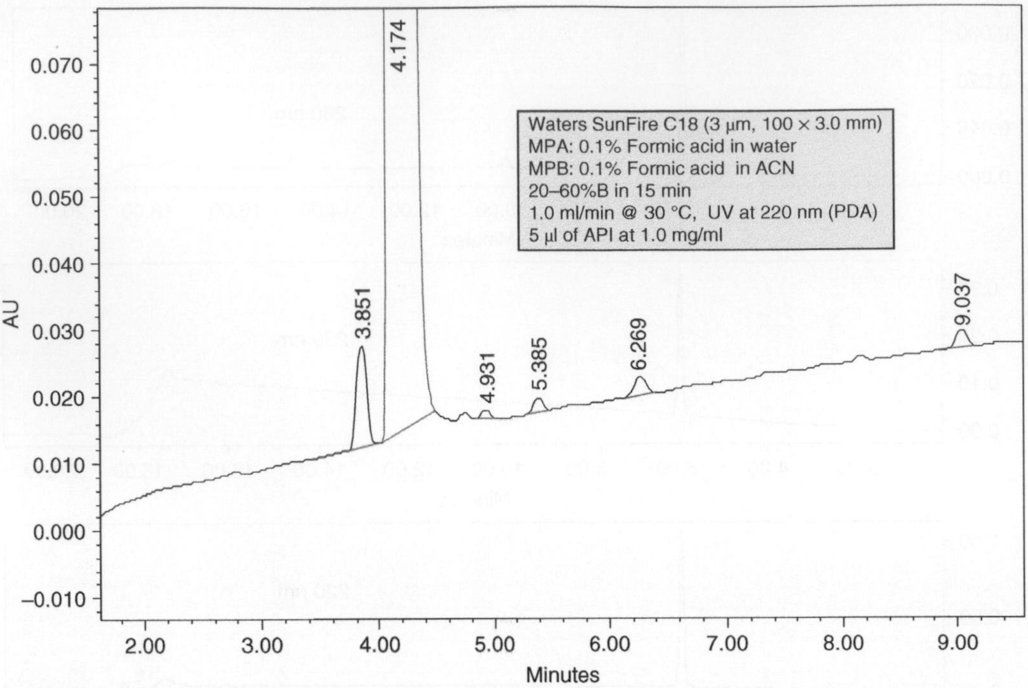


**Figure 10.4.** The initial gradient method for this API was obtained by reducing the gradient range to 20–60% ACN yielding a better resolution of the impurities around the API peak.

the impurity profile of the NCE, the next step is to fine-tune each parameter to meet all requisite goals. The next step is to confirm method "specificity" – to assure that all analyte peaks are resolved from potentially interfering components in the sample as well as any available reference standards that may comprise of impurities, precursors, and degradation products. At this stage of the method development, there should be no interferences from either the procedural blank (diluent) and/or a placebo sample (for a DP method). Ideally, a test mixture or "cocktail" of process impurities and degradants is available to evaluate the separation capability of the initial method. If these reference standards are not available, a "stressed" or forcedly degraded sample can be used to challenge the method (see Section 11.4.1.1). Sufficient testing should be performed to ensure that impurities are not "hidden" under the main API peak. This study is usually done by evaluating peak purity by PDA and MS and by developing an "orthogonal" HPLC method with different selectivity (see Section 2.7).

At this stage, any coelution of the key analytes is minimized by fine-tuning the separation parameters and by adjusting selectivity ($\alpha$) (see Section 2.2.4). This process is iterative and may be challenging for complex API or drug products. Afterward, considerable efforts are focused on improving other method performance parameters such as sensitivity, peak shape, robustness, and analysis time. During drug development, it is not uncommon for the method development process to be reoptimized due to the appearance of new impurities and degradation products from the manufacturing process changes of the API or drug product. Parameters typically optimized are as follows:

- *Mobile phase parameters*: Percentage organic solvent (%B), buffer type and concentration, pH, solvent type (e.g. acetonitrile or MeOH)

- *HPLC operating parameters*: Flow rate ($F$), temperature ($T$), gradient range ($\Delta\Phi$), gradient time ($t_G$), number of gradient segments, and their respective starting and ending %B
- *Column parameters*: Bonded phase type (e.g. C18, C8), length ($L$), column diameter ($d_c$), particle size ($d_p$)
- *Others*: Detector setting (wavelength), sample amount (injection volume and sample concentration and diluent of the final sample)

***10.2.4.1 Mobile Phase Parameters (%B, pH, Buffer, Solvent Type)*** In HPLC, the mobile phase controls the retention and selectivity of the separation for a given stationary phase and can be conveniently and continuously adjusted during method improvement. Lowering the solvent strength (%B of the strong solvent) increases retention and typically increases resolution in isocratic separations (see Figure 2.15). The effect of pH, buffer concentration, and organic solvent type for analyte retention on selectivity is described in Section 2.3.6. Furthermore, the influence of these parameters on fine-tuning selectivity is illustrated in the following case studies.

***10.2.4.2 Operating Parameters (F, T, $\Delta\Phi$, $t_G$, Gradient Segment)*** Flow rate ($F$) does not affect analyte retention ($k$) or selectivity ($\alpha$) in an isocratic analysis, but it can be an important factor affecting both average retention ($k^*$) and selectivity ($\alpha$) within gradient elution (Section 2.6.2). Increasing column temperatures ($T$) reduces retention and solvent viscosity in RPC and can have significant effects on selectivity. The gradient range ($\Delta\Phi$) of the strong solvent (typically %B) is defined as "final %B – initial %B" and is usually determined from the initial work using a broad gradient scouting analysis.

A narrower gradient range is useful to increase the resolution of analytes around the main component. Gradient time ($t_G$) in conjunction with $\Delta\Phi$ defines the gradient slope ($\Delta\Phi/t_G$). An increase in $t_G$ often increases the overall resolution of complex samples, as the peak capacity is roughly proportional to the square root of $t_G$ for broad linear gradients. Both $T$ and $t_G$ can be continuously varied, and their combined change offers an exceptional opportunity for fine-tuning separations using retention modeling software programs (e.g. DryLab, LC Simulator, or ChromSword). For purity analysis of complex samples, a multiple-segment gradient approach is an effective strategy as shown in Section 10.4.3.

***10.2.4.3 Column Parameters (Bonded Phase Type, L, $d_p$, $d_c$)*** While the C18 stationary phase is traditionally used, the selectivity of different bonded phases is often exploited for the separation of closely eluting pairs of key analytes in critical analyses. Column screening with a variety of stationary phases that offers different retention mechanisms can occur during either initial method development or the fine-tuning steps but may be dependent on the availability of suitable samples. Automated column/mobile phase screening systems equipped with multiport selection valves (Figure 10.5) are typically used with an "orthogonal" column set (e.g. C18 (C8), C18 AQ, phenyl, cyano, polar-embedded, and pentafluorophenyl).

Column dimensions ($L$, $d_p$, and $d_c$) are matched to the intended application and can be optimized to enhance efficiency, speed, or sensitivity. One useful strategy is to use short UHPLC columns (50 × 3.0 mm, sub-3-μm) for rapid column/mobile phase screening, followed by optimization with longer columns (to maximize resolution) using the selected bonded phase [5]. SPPs are preferred in RPC because of the higher efficiency as compared to similar totally porous particles (TPPs) [10].

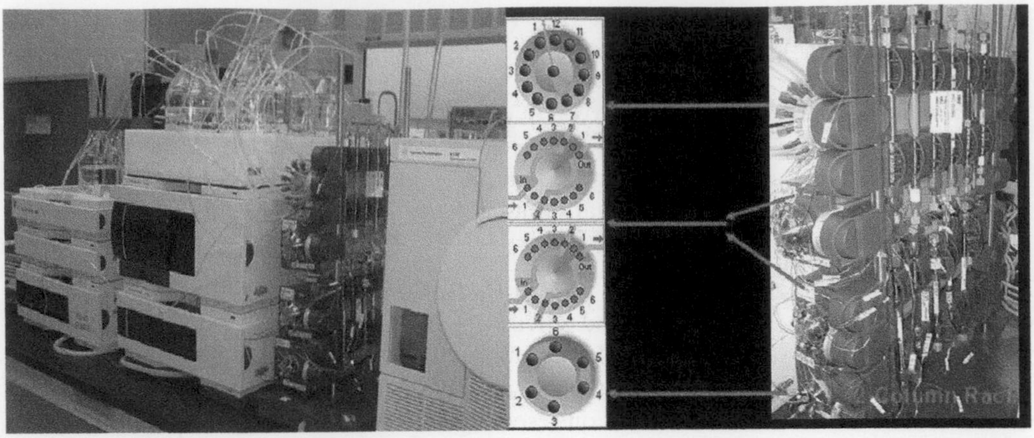

**Figure 10.5.** An illustration of an automated column/mobile phase screening HPLC/UV/MS system with added-on selection valves (12 each for column and mobile phase selection).

***10.2.4.4 Detector Setting and Sample Amount*** Finally, detector setting (detection wavelength, response time, and sampling rate) and sample loading parameters are optimized.

There are two choices for detection wavelength setting: $\lambda_{max}$ of the API for maximum sensitivity of the main component or a non-$\lambda_{max}$ setting (e.g. 220 nm) for detection of related substances that can provide an adequate sensitivity of the main component. Ideally, RRFs of the impurities would have been determined to ensure accurate %w/w impurity data. If RRFs of all impurities have not been determined, all impurities are assumed or assigned an RRF value of 1.0.

During the optimization of UV detection sensitivity optimization, the goal is to maximize the signal-to-noise ratio (S/N), while maintaining method linearity and good peak shape(s). Ideally, the peak height of the main component should be 1–1.5 absorbance units. This enhancement is often achieved by increasing the sample concentration and/or the injection volume. For large-volume injections (i.e. >20 μl), the sample solvent strength should be equal to or weaker than the starting mobile phase (MPA or initial %B) to prevent chromatographic anomalies (Figure 8.14). Typical concentrations of APIs range from 0.1 to 5 mg/ml. While a weaker diluent is preferred for chromatographic considerations, the selected diluent is ultimately dependent on its solubilization power for the API and associated impurities. This consideration is of utmost importance particularly for those APIs and impurities that have limited aqueous solubility.

### 10.2.5 STEP 5: Method Prequalification

For regulatory methods, the final step in the method development process is likely a method prequalification step to ensure that the method can be validated without issues. This prequalification involves several experiments on method specificity, sensitivity, linearity, and precision and should be completed in several hours. Method validation is a GMP (good manufacturing practice) activity, which requires written protocols and preestablished acceptance criteria before execution. Therefore, it makes sense to conduct test experiments to ensure that the newly developed method would pass the acceptance criteria before conducting method validation.

**Table 10.3. Summary of Factors Optimized During RPC Gradient Method Development**

| Stage/approaches | Factor | Suggestions and comments |
|---|---|---|
| *Initial development* | | |
| Scouting gradients | Column | Modern columns packed with 3- or sub-3-μm C18 from high-purity SPP or TTP silica support |
| | Detection wavelength | Use PDA to evaluate $\lambda_{max}$ of all analytes. Select $\lambda_{max}$ of main component or far-UV wavelength (200–230 nm) |
| | Mobile phase | MPA: Water acidified at pH 2–4 for acidic or basic analytes<br>MPB: MeOH or ACN |
| | Operating conditions | Use broad gradients (5–100%B) for initial evaluation<br>Evaluate the first chromatogram to define gradient range $\Delta\phi$ for next evaluation using a narrower gradient range<br>Set $T = 30$–$35\,°C$, $F \sim 1$ ml/min for 3.0 mm i.d. columns |
| *Method optimization* | | |
| Fine-tuning and finalization | Mobile phase and operating conditions | Evaluate method with impurity "cocktail," precursors, and formulation placebo<br>Modify pH, solvent type, $T$, and $t_G$ to improve resolution |
| | Column | Evaluate method with different bonded phases to maximize selectivity and resolution<br>Further evaluation with longer columns to increase resolution for the selected phase |
| | Detector and sample | Finalize detection wavelength and sample conc. Alternately, injection volume can be optimized for peak shape and method sensitivity |

### 10.2.6 Summary of Method Development Steps

Table 10.3 summarizes the steps in the HPLC method development process for stability-indicating gradient methods.

### 10.2.7 Phase-Appropriate Method Development and Validation

Phase-appropriate method development is a proactive approach advocated by H. Rasmussen et al. to closely align the method development and regulatory requirements in the rapidly changing manufacturing processes during drug development [2]. Early-phase methods tend to be broad-gradients and MS-compatible methods developed for release testing of clinical trial material (CTM). This approach features a primary method that is backed up by a secondary (orthogonal) method used to evaluate pivotal CTM lots. Examples of primary and secondary methods are shown in Figure 10.6. The secondary method ensures no peak coelution amongst impurities and that no hidden peaks are observed underneath the API peak. In addition, the secondary backup method can be quickly validated if the primary method is found to be inadequate. Finally, Phase 3 methods are developed when synthetic routes, formulation, processing conditions, and specifications are "locked." This approach has both

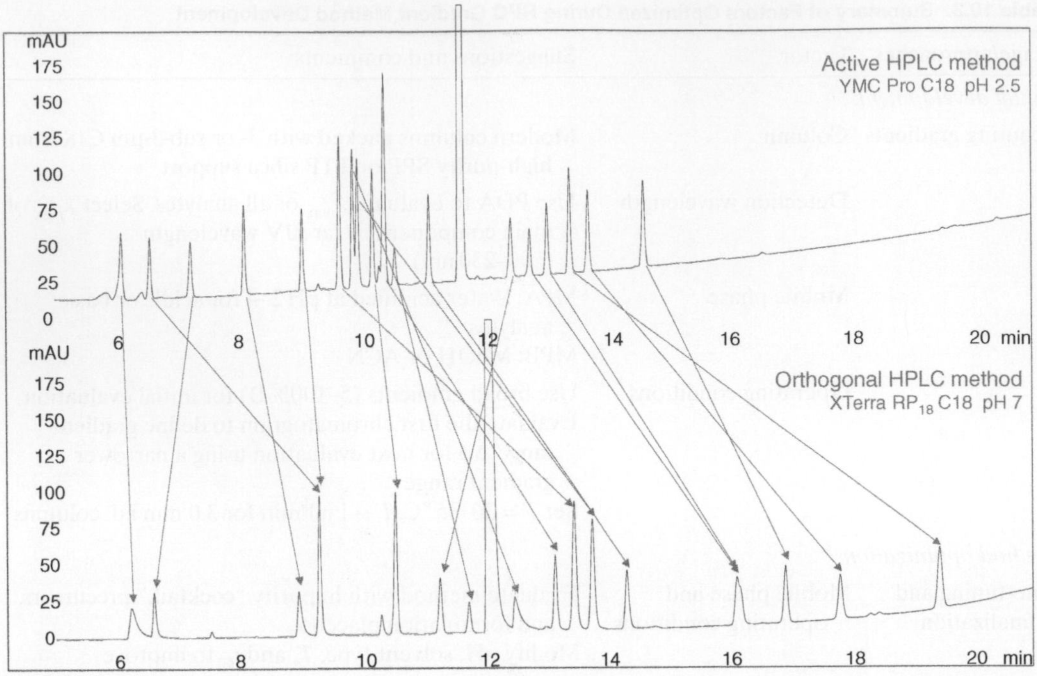


**Figure 10.6.** Examples of a primary (top) and a secondary (orthogonal, bottom) phase 1 methods used in phase-appropriate method development and validation. Source: Reprinted with permission of Rasmussen et al. 2005 [2]. Copyright 2005, Elsevier.

the scientific vigor and flexibility to ensure the best possible methods that enable prompt responses to address process changes. Submissions of regulatory filings for conducting clinical trials and commercial product include all critical analytical methods and their validation data (see Chapter 11).

### 10.2.8 Method Development Software Tools

Table 10.4 is a summary of automated systems and software platforms for HPLC method development. Several manufacturers have preconfigured method scouting systems for column/mobile phase screening. Many software platforms are available for off-line predictive modeling with inputs from limited experiments (e.g. DryLab). Other platforms support direct instrument control and on-line optimization (e.g. ChromSword Auto and ACD Labs AutoChrom) and statistical analysis of experimental results (e.g. Fusion QbD). The reader is referred to key references [12–16] and the manufacturers' websites for further details. Examples of method development using DryLab simulation and S-Matrix's Fusion QbD software are shown in case studies (Sections 10.3 and 10.4).

## 10.3 CASE STUDIES

Four cases studies are shown here to illustrate the method development process including the use of software platforms. These are as follows:

**Table 10.4. Summary of Automated Systems and Software Platforms for HPLC Method Development**

| Vendor, system | Description |
|---|---|
| *Method scouting systems* | |
| Agilent 1290 Infinity II method development solution | An automated system based on Agilent 1260/1290 UHPLC systems with column and mobile phase selection valves operated under OpenLAB CDS for column/mobile screening, method scouting, and gradient/temperature optimization |
| Shimadzu Nexera Method scouting system | A method scouting system based on the Nexera X2 UHPLC system with column/mobile phase selection valves for automated screening and gradient optimization |
| Hitachi Chromaster Ultra with ChromSwordAuto | A UHPLC instrument with six ACE columns controlled and interfaced with ChromSwordAuto software for unattended and automated method development, robustness assessment, and column screening |
| *Software platform and automated method development systems* | |
| Molnar Institute DryLab Ver. 4 | A popular computer modeling software (offline) based on linear solvent strength model in gradient elution. Relatively simple and easy-to-use predictive program for optimizing isocratic and gradient methods based on results from 2 to 12 actual runs |
| ChromSwordAuto (Dr. Galushko Software) | A software platform provides direct control of Agilent, Thermo, and Hitachi HPLC systems for automated method scouting and optimization of solvent, buffer, gradient, column, flow rate, and temperature |
| ACD AutoChrom | Combines instrument control for LC/MS and LC/UV system with software for logical method development, which includes predictive (from chemical structures) and modeling algorithms |
| S-Matrix Fusion QbD | An automated method development software that controls most major HPLC/UHPLC (Waters, Agilent, Thermo) systems using their respective CDS to provide column/mobile phase screening. In addition, this software provides multivariate optimization based on fractional factorial Design of Experiments with statistical analysis of results imported back from CDS |

1. A phase-0 drug substance method to illustrate the selection of bonded phases and mobile phases to improve resolution and sensitivity
2. The development of a drug substance method for an NCE illustrating the use of DryLab modeling software for gradient and column optimization
3. A stability-indicating method for a drug product with two APIs illustrating the use and optimization of high-pH mobile phases.
4. The use of S-Matrix Fusion QbD in the development of a UHPLC method for an NCE with multiple chiral centers illustrating the Design of Experiments (DoEs) approach.

### 10.3.1 A Phase-0 Drug Substance Method for an NCE

This case study is a follow-up of the initial method development example shown in Figures 10.2–10.4. The HPLC conditions used during the initial development shown in Figure 10.4 were found to be inadequate due to the coelution of the API with the immediate synthetic precursor. Varying mobile phases and LC gradient conditions were ineffective

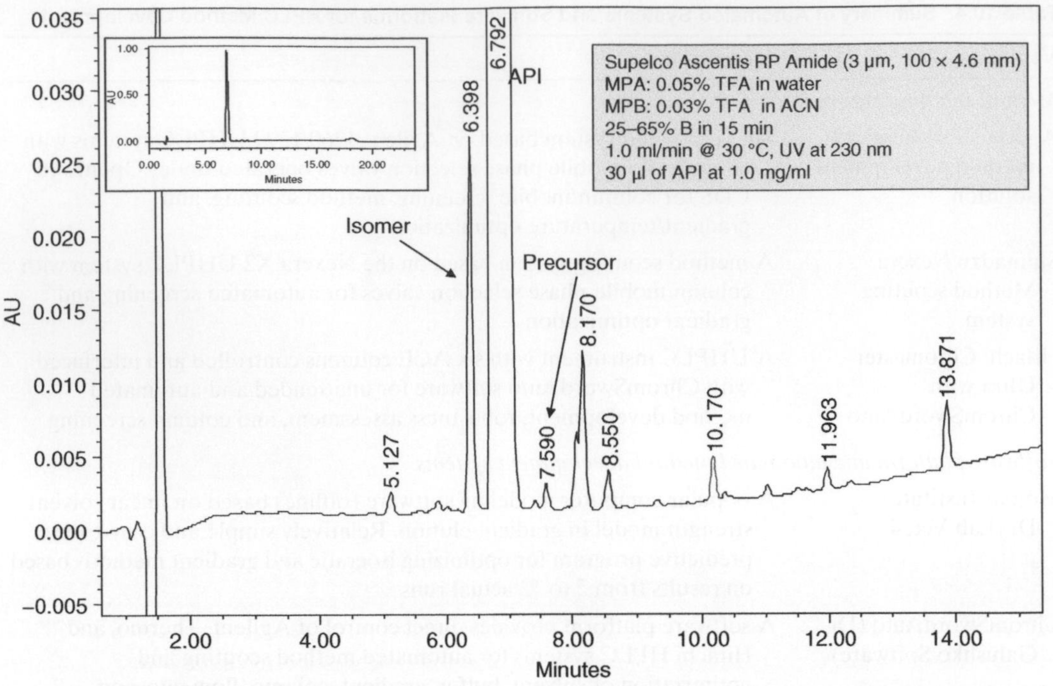


**Figure 10.7.** HPLC analysis of the final phase-0 method of the NCE sample in Figure 10.2.

in separating this critical pair; however, switching the mobile phase acidifier from 0.1% formic acid to 0.1% trifluoroacetic acid (TFA) yielded higher retention and provided a partial separation. At this stage, two polar-embedded columns (Waters XTerra RP18 and Supelco Discovery RP-amide) were evaluated, leading to a successful separation with the latter column. Next, the Supelco Ascentis RP-amide column (now Sigma-Aldrich) with 0.1% TFA was used, though a high gradient baseline shift was observed at 230 nm. Optimal chromatographic conditions (Figure 10.7) with adequate sensitivity (limits of quantitation, LOQ = 0.05%) were found by reducing the level of TFA in MPB and matching the absorbance of MPA (0.05% TFA in water) and MPB (0.03% TFA in acetonitrile). A set of optimum conditions with adequate sensitivity (LOQ = 0.05%) was found (shown in Figure 10.7). The validation of this method is described in Section 11.4.2. Under these conditions, coelution of two impurity peaks was observed near RT = 8.17. However, the chosen method was deemed adequate and fit-for-purpose as a phase-0 method for the release testing of this NCE lot in a toxicological study.

### 10.3.2 Stability-Indicating Method Development for an NCE Using DryLab

This case study started with an initial isocratic stability-indicating method of an NCE as shown in Figure 10.8 developed by a company that synthesized this new drug candidate for a pain indication. In this scenario, mobile phase modifiers were not used since the NCE is a neutral compound (an unusual scenario for a drug). This isocratic method was judged inadequate for many reasons. Firstly, there was a significant coelution of the two important impurities (impurities 4 and 4a). Secondly, a light degradant (compound x, a geometric isomer of the API) was found to be partially resolved and eluted at the front of the API peak.

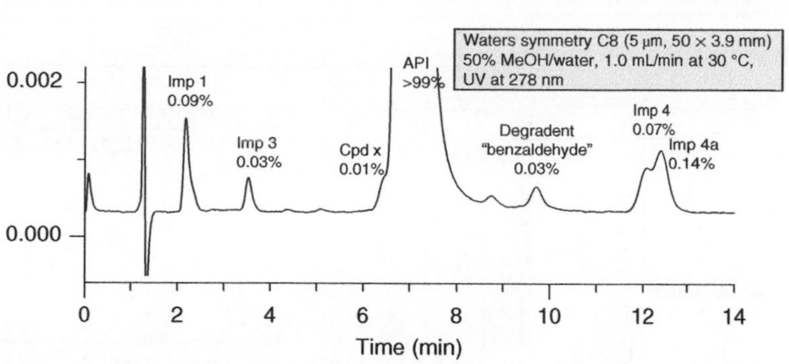


**Figure 10.8.** HPLC analysis of the original isocratic stability-indicating method for a neutral NCE showing coelution with two critical pairs: compound X, API, impurity 4 and impurity 4a.

Finally, isocratic methods often miss late elutors, and in this specific example, an initial gradient assessment revealed the presence of a late-eluting impurity (impurity 5 – not shown or eluting in the Figure 10.8 under isocratic conditions).

After an initial scouting gradient run, DryLab 2000 was used to explore the optimum temperature ($T$) and gradient conditions. Four gradient runs were performed with a Waters Symmetry C8, 5 μm, 150 × 3.9 mm column, used under a broad linear gradient range (35–100% MeOH) at $T = 23$ and 50 °C at two separate gradient run times, $t_G = 30$ and 90 minutes. Retention data of all peaks were entered manually (Figure 10.9) into the DryLab software to enable

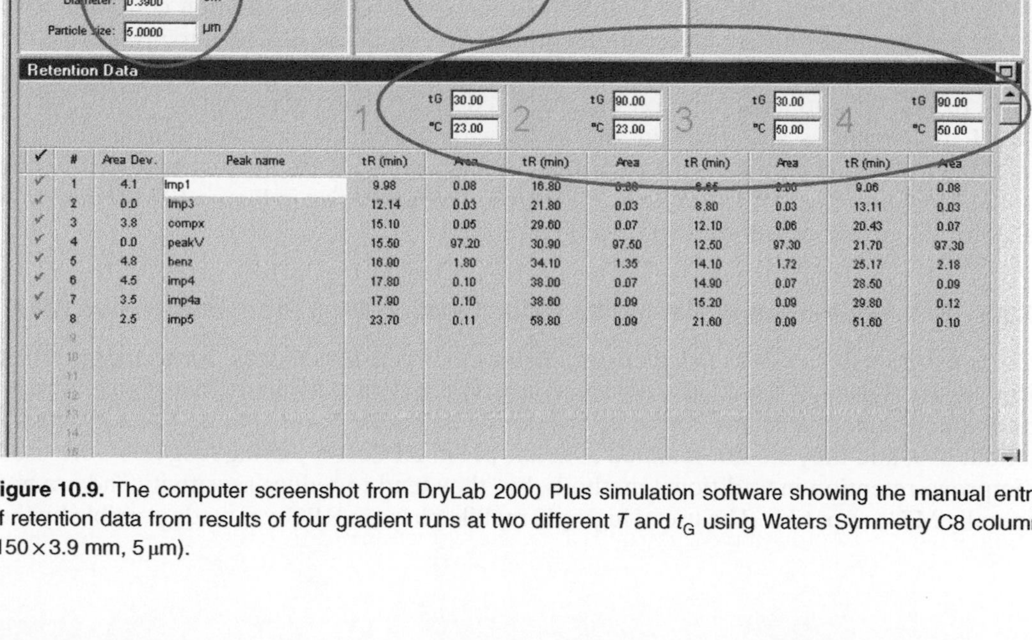

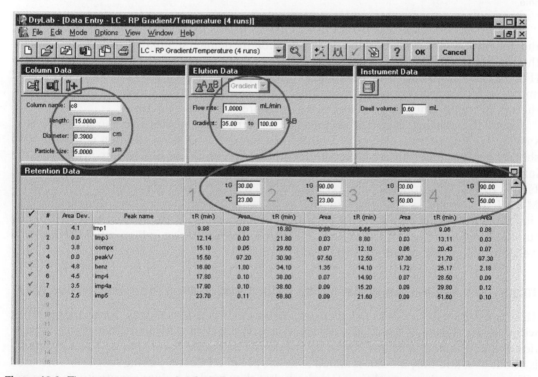

**Figure 10.9.** The computer screenshot from DryLab 2000 Plus simulation software showing the manual entry of retention data from results of four gradient runs at two different $T$ and $t_G$ using Waters Symmetry C8 column (150 × 3.9 mm, 5 μm).

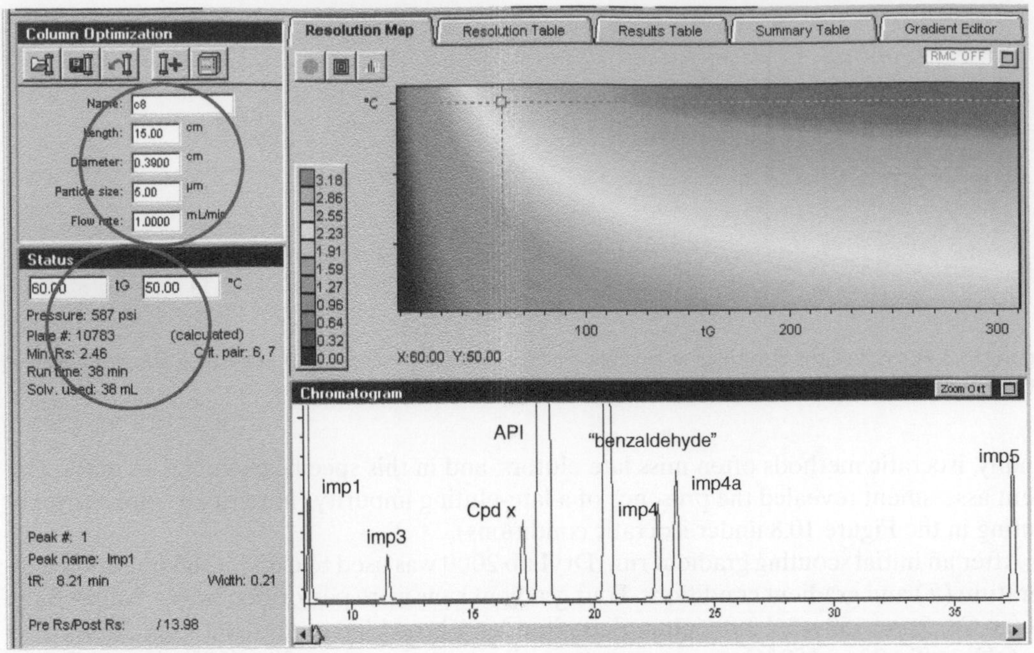


**Figure 10.10.** DryLab color-coded resolution map simulated from data from the experimental results Figure 10.9. The simulated chromatogram at $T = 50\,°C$, and $t_G = 60$ minutes showed excellent resolution but long analysis time.

the generation of a resolution map and retention modeling (simulated chromatograms) for the recommended optimized conditions. The resultant color-coded critical resolution map (Figure 10.10) with one of the optimum chromatograms was simulated at conditions of 50 °C and $t_G = 60$ minutes. Overall, the resolution was excellent in the simulated chromatogram though further method improvements were explored to reduce run time.

Using the existing data, the column parameters were modified in the software (e.g. varying length, internal diameter, and particle size) to evaluate separation condition with shorter analysis time. An *in silico* column optimization study was designed to investigate the impact of a shorter column packed with smaller particles (e.g. 75 × 4.6 mm, 3.5 μm shown in Figure 10.11). An optimal separation was predicted via DryLab under conditions of 35–80% MeOH in 18 minutes at 2.0 ml/min and 40 °C with the smaller column dimensions. The simulated optimal conditions with a run time of approximately 22 minutes were verified by running a confirmatory HPLC experiment (chromatogram shown in Figure 10.12) compared well with the predicted chromatogram in Figure 10.11.

### 10.3.3 Stability-Indicating Method for a Combination Drug Product with Two APIs

This case study illustrates the challenging method optimization process for a drug product with two APIs using a "cocktail" test mix consisting of both APIs and all available impurities and potential degradants. The $pK_a$ of both APIs, highly water-soluble bases (opioids), is about 8–9, and they are not retained at acidic pH in RPC. Ion-pairing reagents were used to improve retention in traditional methods, but they tend to reduce sensitivity and are not generally MS-compatible. The use of high-pH mobile phase in RPC was made possible by the

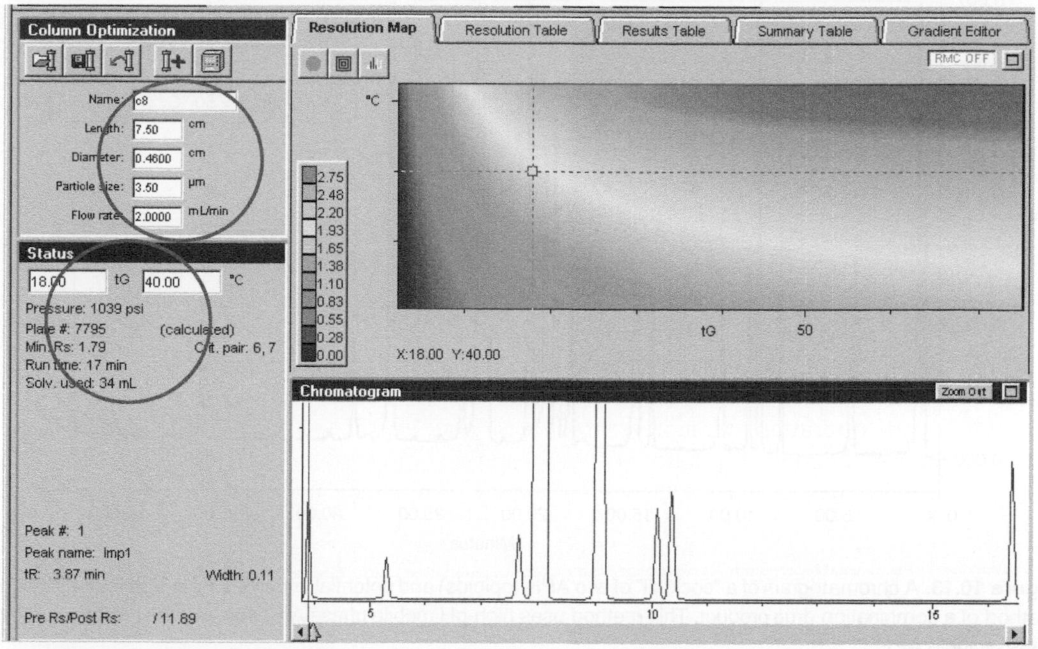


**Figure 10.11.** Screenshot of DryLab color-coded resolution map simulated from data from Figure 10.9 for column optimization. The simulated chromatogram at $T = 40\,°C$, and $t_G = 18$ minutes was predicted for a Waters Symmetry C8 column (75 × 4.6 mm, 3.5 μm). It showed excellent resolution and a run time of 22 minutes.

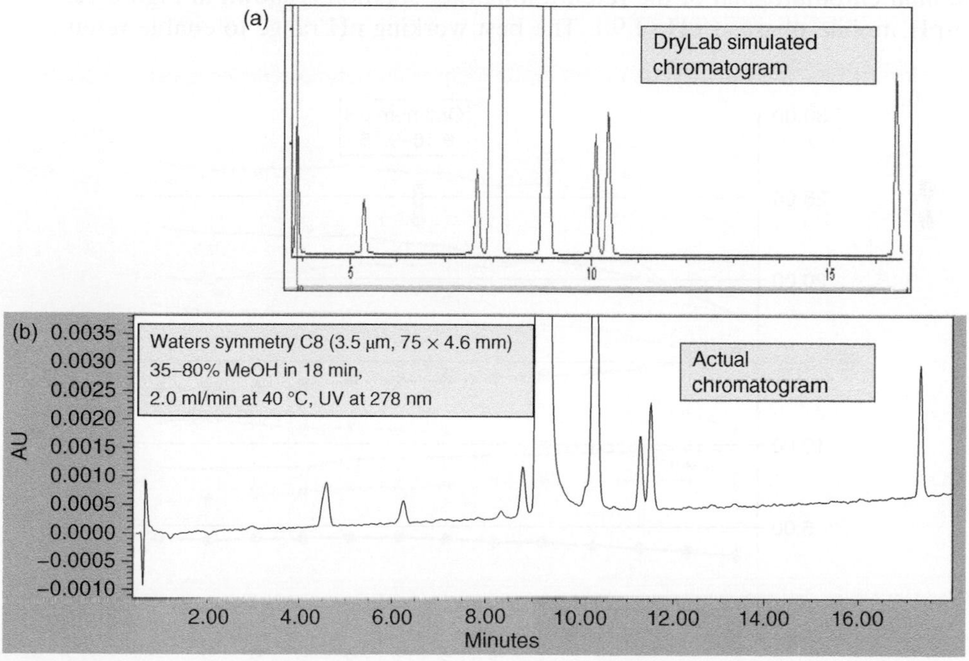


**Figure 10.12.** A comparison between the simulated and the actual (bottom) chromatogram obtained (bottom) by running the new column under the predicted optimum conditions. Excellent correlations were obtained.

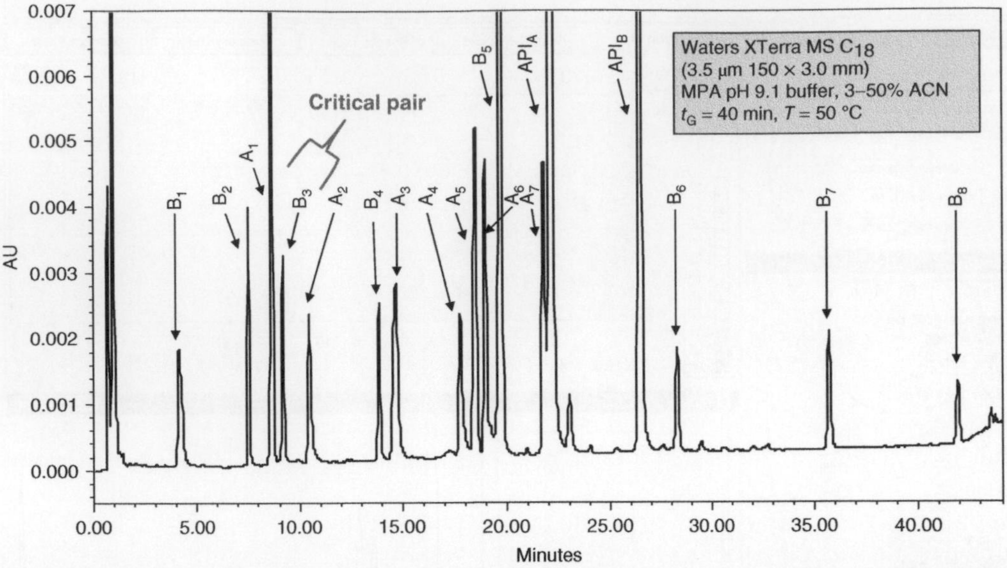


**Figure 10.13.** A chromatogram of a "cocktail" of two APIs (opioids) and potential impurities for a stability-indicating method of a combination drug product. This method uses high-pH mobile phase A to increase the retention of the water-soluble APIs.

availability of high-pH compatible silica-based bonded phases and was applied successfully for this application [17].

The final chromatogram of the retention marker solution is shown in Figure 10.13 using a high-pH mobile phase at pH of 9.1. The best working pH range to enable retention and

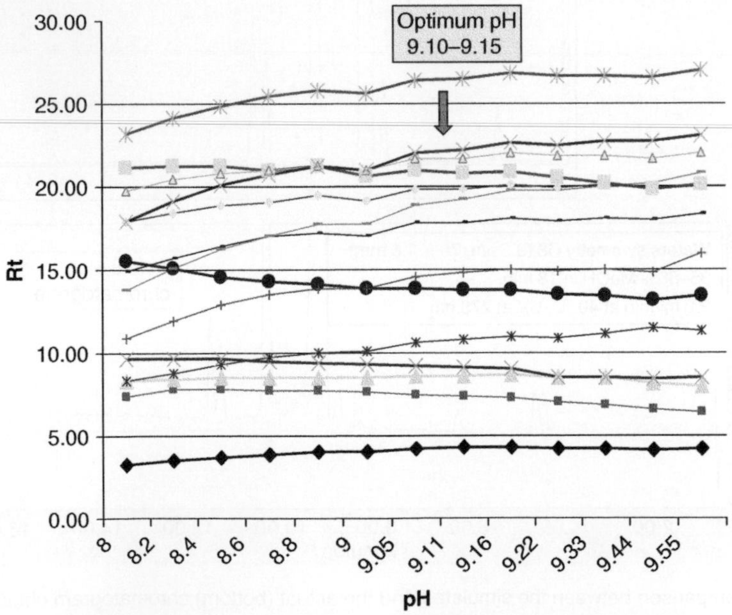


**Figure 10.14.** A plot of retention data vs. pH of the MPA showing the significant impact of pH on the separation. Optimum pH was found to be at pH 9.1.

separation for this particular analysis was investigated with the results plotted in Figure 10.14. The mobile phase pH was found to be a useful but sensitive parameter, controlling the resolution of several critical pairs as their ionization values ($pK_a$ values) are close to the pH of the mobile phase. For adequate resolution of all key analytes, the pH of MPA must be maintained between 9.10 and 9.15. Under these conditions, the peak shapes of all components were excellent, and there was no indication of split peaks [17]. Nevertheless, the mobile phase pH must be tightly controlled, and method robustness could be a potential issue.

### 10.3.4 Automated Method Development System Employing Fusion QbD Software

In this case study, a UHPLC system (Waters ACQUITY UPLC equipped with a PDA, a column oven with a four-column selection valve, and an add-on six-port solvent selector connected to port A1) controlled by Waters Empower chromatography data system (CDS) in conjunction with S-Matrix Fusion QbD software was employed. This software uses a Quality by Design (QbD) and DoE approaches to systematically explore the effect of multiple variables on responses as defined by chromatographic parameters such as peak resolution [16]. Equipment and summary of the software functions (shown in Figure 10.15) were used in the development of a stability-indicating method of an NCE with three chiral centers described in an earlier publication [7, 18] and Figure 5.13. The focus was on a UHPLC method yielding baseline resolution of all four diastereomers of the multichiral NCE.

The results of the mobile phase screening study at six different pH's (2.0, 2.8, 3.7, 5.0, 7.0, and 10) are highlighted in Figure 10.16 leading to the selection of 20 mM ammonium formate at 3.7 as the best pH for the separation of the four diastereomers. The experimental setup

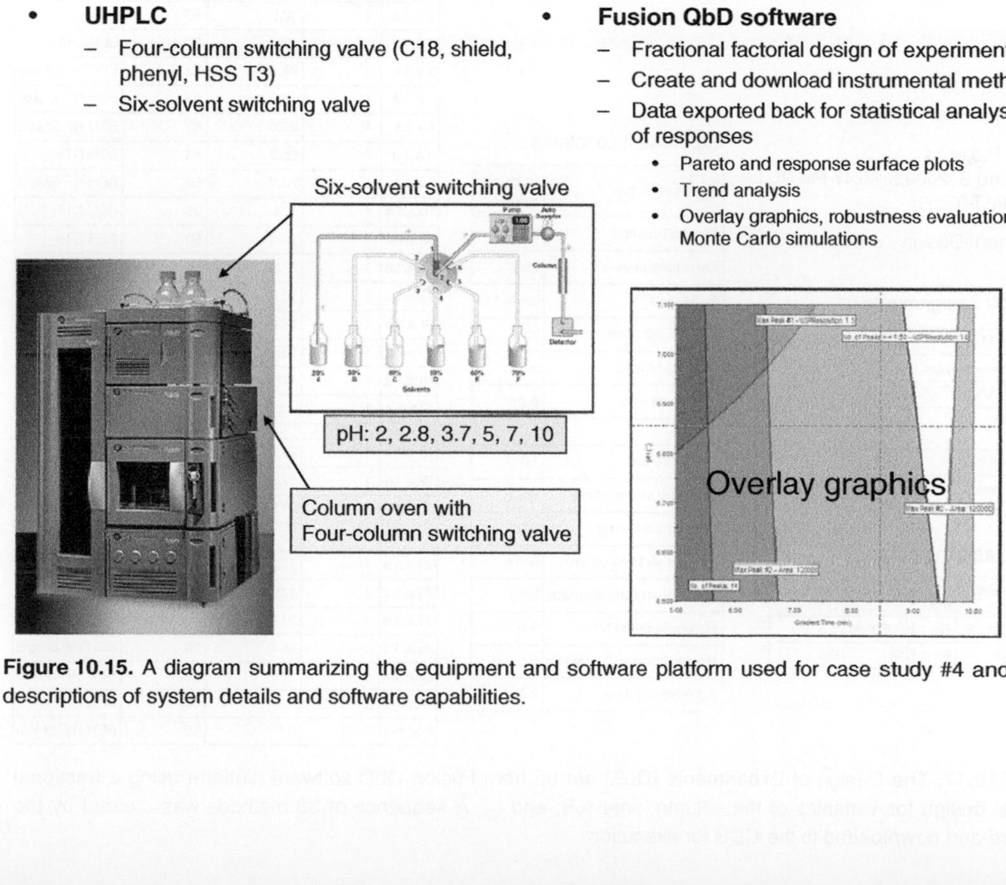
• UHPLC
– Four-column switching valve (C18, shield, phenyl, HSS T3)
– Six-solvent switching valve
• Fusion QbD software
– Fractional factorial design of experiments
– Create and download instrumental methods
– Data exported back for statistical analysis of responses
• Pareto and response surface plots
• Trend analysis
• Overlay graphics, robustness evaluation via Monte Carlo simulations
Six-solvent switching valve
pH: 2, 2.8, 3.7, 5, 7, 10
Column oven with
Four-column switching valve
Overlay graphics

**Figure 10.15.** A diagram summarizing the equipment and software platform used for case study #4 and brief descriptions of system details and software capabilities.

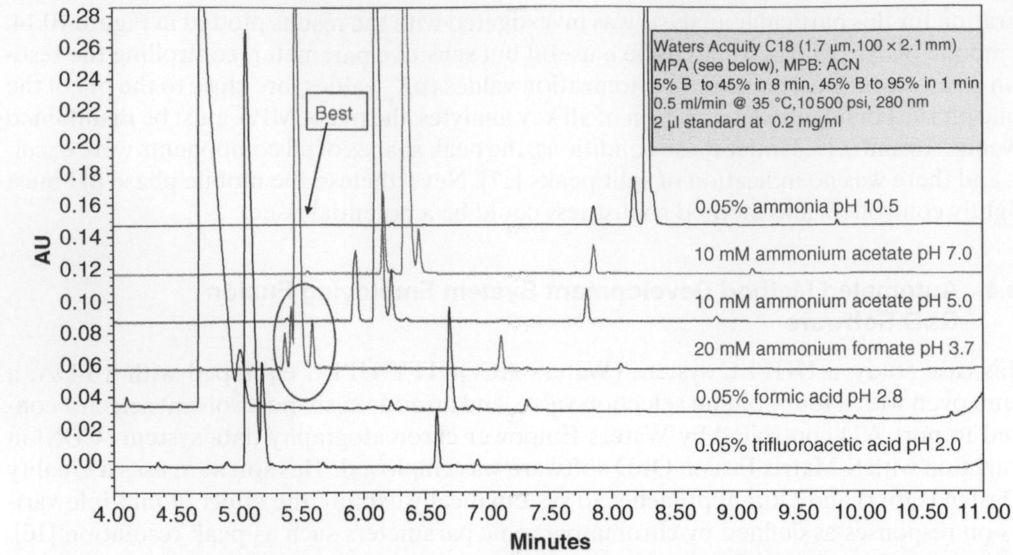


**Figure 10.16.** Overlaid chromatograms of mobile phase screening studies at pH 2.0, 2.8, 3.7, 5.0, 7.0, and 10. The best resolution of the four diastereomers (SRR [API], SRS, RRR, and SSR) was achieved at pH 3.7. Note that severe peak tailing was observed at pH 2.0 and 2.8.

Project: Project 1
Date: June 8, 2009 6:35:47 PM PDT [GMT-07:00]

Experiment Design - experiment 1

Reservoir assignments

| Reservoir A1 | Level |
|---|---|
| Mobile phase solvent 1 | – |
| Reservoir A2-1 | Level |
| Buffer type | Level 1 |
| Reservoir B1 | Level |
| Mobile phase solvent 2 | – |

Column assignments

| Column valve position | Column level | Pump flow rate |
|---|---|---|
| ValvePostior_2 | Be-1 RP Shield | 0.5 |
| ValvePostior_4 | Be-1 C18 | 0.5 |

Experiment constants

| Constant name | Constant value |
|---|---|
| Injection volume | 2.0 |
| Oven temperature | 35.0 |
| Buffer type | Level 1 |
| Initial % aqueous | 95 |
| Initial % organic | 5 |
| Initial hold time | 0.1 |
| Final hold time | 1.0 |
| Ramp up to wash time | 0.1 |
| Column wash time | 2.0 |
| Column wash % organic | 95.0 |
| Ramp down from wash time | 0.1 |
| Re-equilibration time | 3.0 |
| Re-equilibration % organic | 5.0 |
| Equilibration time | 2.0 |

Experiment design matrix

| Run no. | Sample set no. | Gradient time (min) | Final % organic (%) | Column type (*) |
|---|---|---|---|---|
| Wash-1 | 1 | 0.1 | 43 | BEH C18 |
| Wash-2 | 1 | 0.1 | 43 | BEH RP Shield |
| 1.a.1.a | 1 | 10.0 | 43 | BEH C18 |
| 2.a.1.a | 1 | 10.0 | 43 | BEH RP Shield |
| 3.a.1.a | 1 | 10.0 | 50 | BEH RP Shield |
| 4.a.1.a | 1 | 10.0 | 50 | BEH C18 |
| 5.a.1.a | 1 | 25.0 | 35 | BEH RP Shield |
| 6.a.1.a | 1 | 17.5 | 43 | BEH RP Shield |
| 7.a.1.a | 1 | 25.0 | 43 | BEH RP Shield |
| 8.a.1.a | 1 | 25.0 | 43 | BEH C18 |
| 9.a.1.a | 1 | 25.0 | 50 | BEH RP Shield |
| 10.a.1.a | 1 | 17.5 | 35 | BEH C18 |
| 11.a.1.a | 1 | 10.0 | 50 | BEH C18 |
| 12.a.1.a | 1 | 17.5 | 35 | BEH RP Shield |
| 13.a.1.a | 1 | 10.0 | 35 | BEH C18 |
| 14.a.1.a | 1 | 13.8 | 46 | BEH RP Shield |
| 15.a.1.a | 1 | 13.8 | 39 | BEH C18 |
| 16.a.1.a | 1 | 17.5 | 43 | BEH RP Shield |
| 17.a.1.a | 1 | 10.0 | 35 | BEH RP Shield |
| 18.a.1.a | 1 | 25.0 | 35 | BEH C18 |
| 19.a.1.a | 1 | 17.5 | 43 | BEH C18 |
| 20.a.1.a | 1 | 21.3 | 39 | BEH RP Shield |
| 21.a.1.a | 1 | 25.0 | 50 | BEH C18 |
| 22.a.1.a | 1 | 17.5 | 43 | BEH C18 |
| 23.a.1.a | 1 | 17.5 | 50 | BEH C18 |
| 24.a.1.a | 1 | 21.3 | 46 | BEH C18 |
| 25.a.1.a | 1 | 25.0 | 50 | BEH RP Shield |
| 26.a.1.a | 1 | 17.5 | 50 | BEH RP Shield |
| Wash-3 | 1 | 0.1 | 50 | BEH C18 |
| Wash-4 | 1 | 0.1 | 50 | BEH RP Shield |

**Figure 10.17.** The Design of Experiments (DoE) set up from Fusion QbD software platform using a fractional factorial design for variables of the column, final %B, and $t_G$. A sequence of 30 methods was created by the software and downloaded to the CDS for execution.

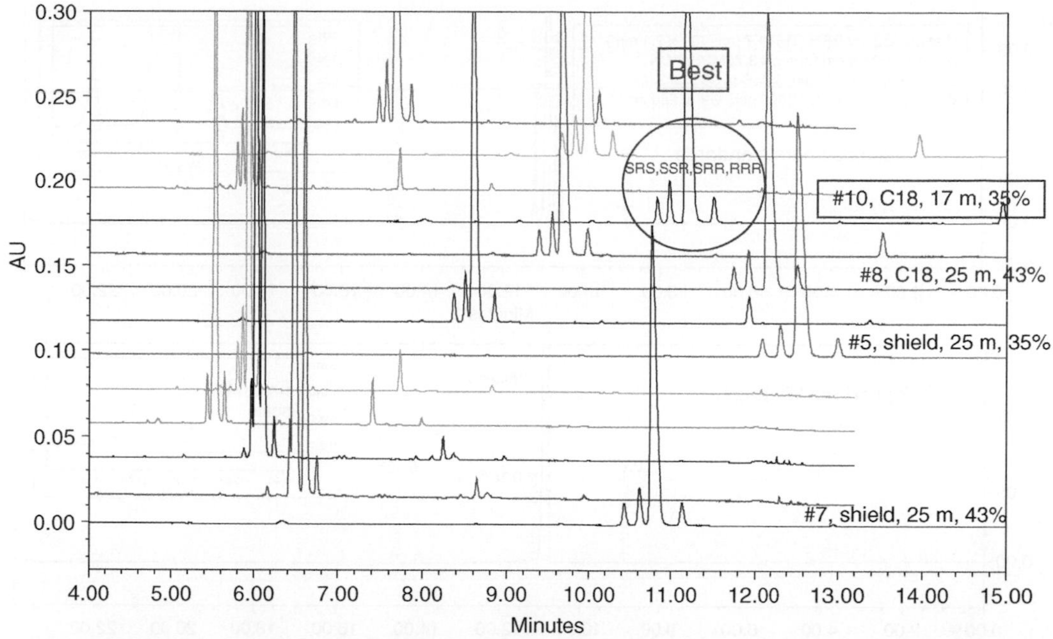


**Figure 10.18.** Chromatograms generated via the DoE study set up in Figure 10.17. Note that experiment #17 was found to yield the best resolution of the diastereomers in minimum run time.

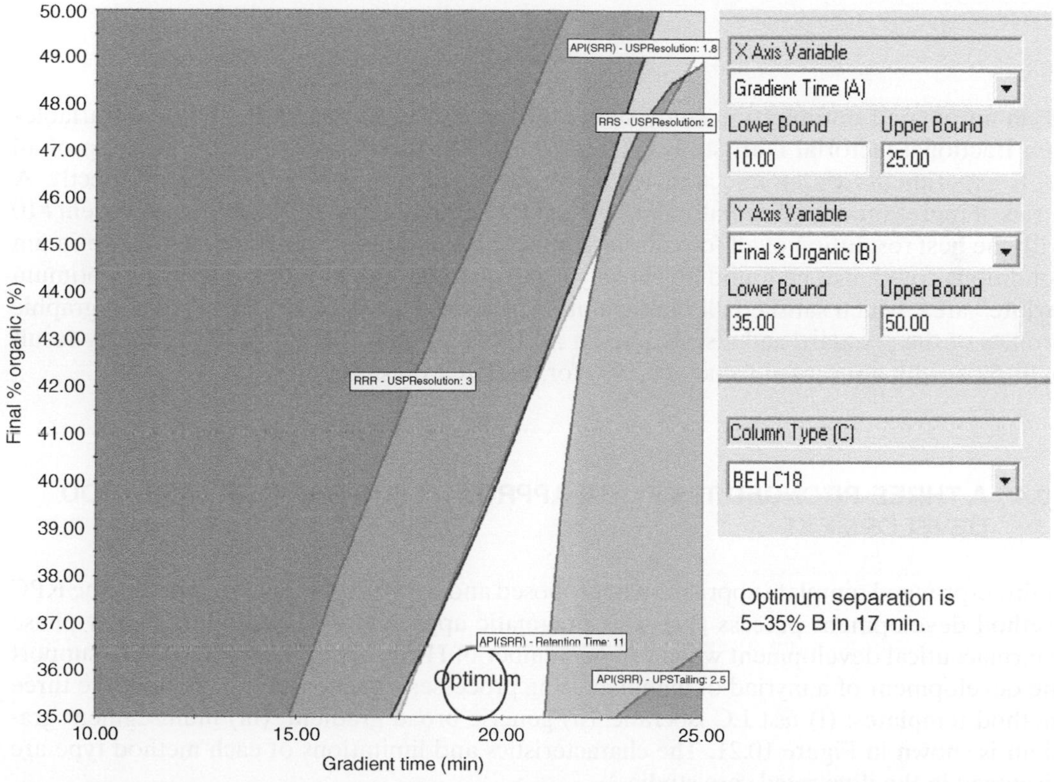


**Figure 10.19.** A graphic overlay displaying the design space (white) showing the acceptance criteria (resolution) with the variable range of $t_G$ (10–25 minutes) and %B (35–50%). The best-operating conditions were found to be 5–35%B in 17 minutes.

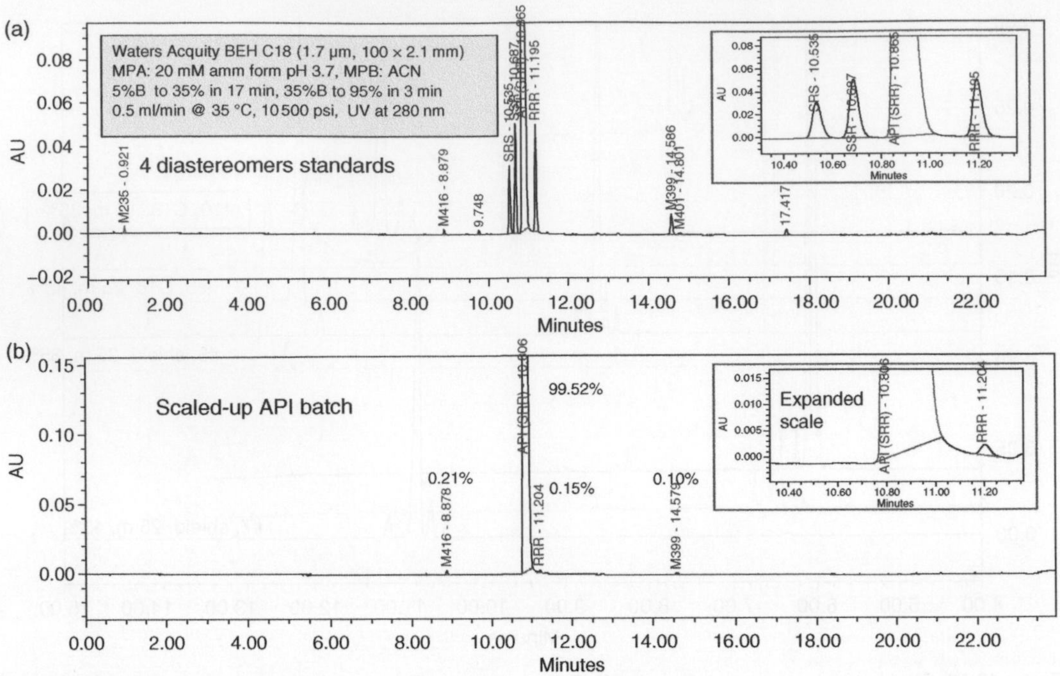


**Figure 10.20.** (a) Comparative UHPLC chromatograms of the standard solution and (b) an API scaled-up batch showing a sample purity of 99.52% with 0.15% of RRR diastereomer.

of an automated optimization study shows the column type, final %B, and $t_G$ as variables in a fractional factorial DoE study (Figure 10.17). The software automates the setup of all DoE experiments via an injection sequence downloaded to the HPLC system directly. A series of representative chromatograms of the DoE study (Figure 10.18) show experiment #10 with the best resolution of all four diastereomers with minimum analysis time. The optimum conditions could also be found objectively using overlaid graphics that display an optimum "white" area, which satisfied all the separation criteria (Figure 10.19). The chromatographic profiles of the standard and a scaled-up an API batch (Figure 10.20) yielded a purity result of 99.52% with a reported value of 0.15% for the RRR diastereomer.

## 10.4 A THREE-PRONGED TEMPLATE APPROACH FOR RAPID HPLC METHOD DEVELOPMENT

A three-pronged template approach was proposed and published in 2013 to expedite the RPC method development process [19]. This pragmatic approach is most useful in early-phase pharmaceutical development when a large number of HPLC methods are needed to support the development of a myriad of manufacturing processes. A schematic diagram of the three method templates: (i) fast LC isocratic, (ii) generic broad gradient, (iii) multisegment gradient is shown in Figure 10.21. The characteristics and limitations of each method type are discussed in the illustrated case studies.

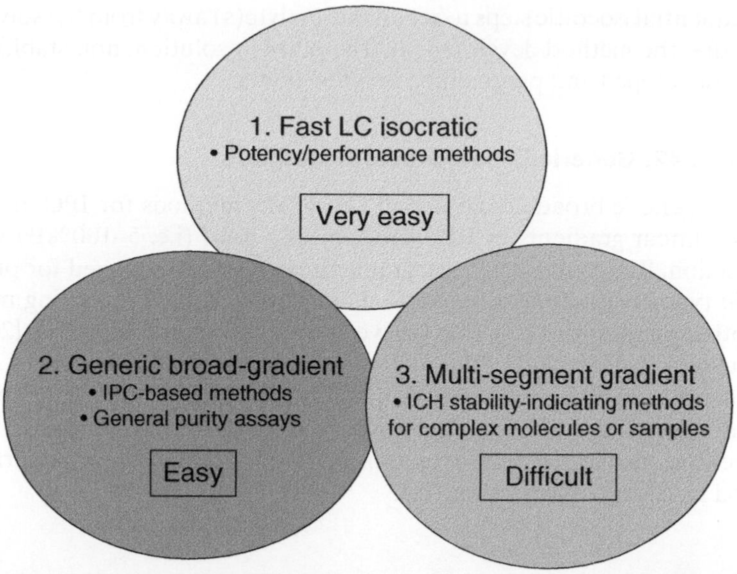


**Figure 10.21.** Diagram depicting the three method types in the three-pronged template approach and their relative ease of implementation. ICH = International Conference on Harmonization. IPC = In-process control. Source: Reprinted with permission of Dong 2013 [19].

### 10.4.1 Template #1: Fast LC Isocratic Potency or Performance Methods

For potency testing (i.e. quantitative determination of the main component only), a fast LC isocratic method is recommended for simplicity and analysis speed. An example of a fast LC isocratic method (Figure 10.22) includes the use of a short column (e.g. 50 × 4.6 mm, 3.5 μm, C18) with UV detection at 280 nm ($\lambda_{max}$ of API). In most cases, a retention factor ($k$ value) of 1–2 with an analysis time <2 minutes appears to be feasible. Real applications where such methods are employed include dosing solution analysis, content uniformity to support drug product manufacturing activities (quantity of API per dose and consistency), as well as dissolution testing (release of API with time) of drug products.

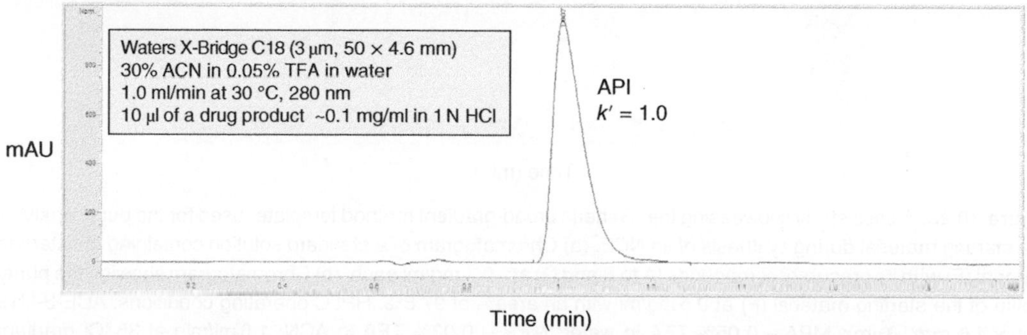


**Figure 10.22.** An example of the "Fast LC isocratic potency/performance method template" used for content uniformity and dissolution (performance) testing of a drug product. Source: Reprinted with permission of Dong 2013 [19].

The use of sequential isocratic steps to retain the analyte(s) away from the solvent front can be used to expedite the method development. These low-resolution, non-stability-indicating methods can be developed and prequalified in a few hours.

### 10.4.2 Template #2: Generic Broad Gradient Methods

Today, the use of generic broad-gradient HPLC/UV/MS methods for IPC or HTS is quite common. Broad, linear gradients in RPC are typically used (i.e. 5–100%B) with UV–Vis and/or MS detection. The use of a generic gradient can also be deployed for purity analysis of many sample types (e.g. simple drug substances/products, raw or starting materials, and reagents) and other applications (e.g. cleaning verification for multiple NCEs) [20]. Examples are shown in Figures 10.23 and 9.11 [20].

Generic gradient method(s) can often be used with little or no modifications and can be qualified and transferred quickly, leading to productivity gains and substantial time-savings from generic method standardization. An extension of this approach to a modernized generic gradient method is described in Section 10.5.

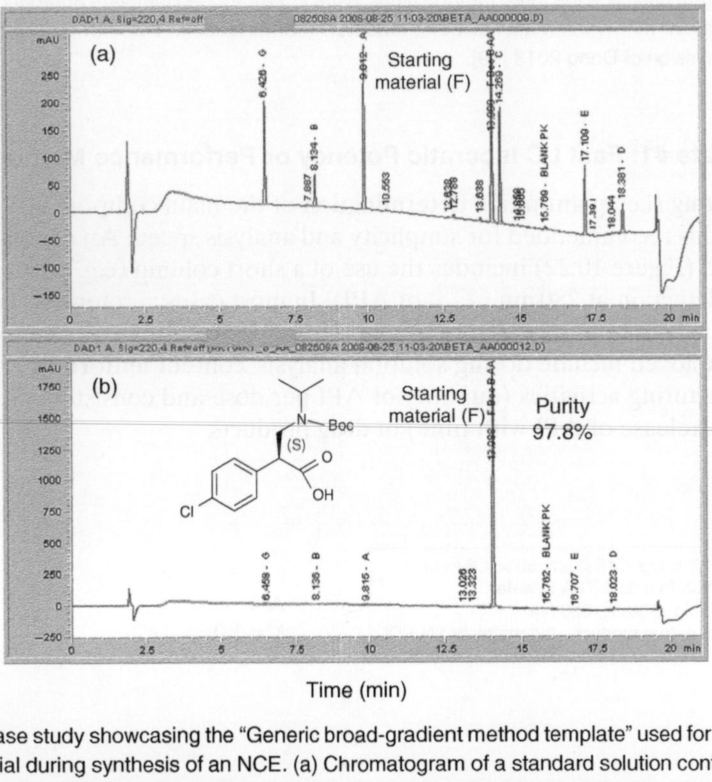


**Figure 10.23.** A case study showcasing the "Generic broad-gradient method template" used for the purity analysis of a starting material during synthesis of an NCE. (a) Chromatogram of a standard solution containing the starting material (F) with its precursor compounds (A to E and G) at ~0.1 mg/ml each. (b) Chromatogram showing the purity profile of the starting material (F) at 0.5 mg/ml with an area% of 97.8%. HPLC operating conditions: ACE-3-C18 (150 × 4.6 mm, 3 μm); MPA – 0.05% TFA in water; MPB – 0.03% TFA in ACN; 1.0 ml/min at 35 °C; gradient program – 5–95%B in 15 minutes, 95% for 2 minutes, UV at 220 nm. Source: Reprinted with permission of Dong 2013 [19].

### 10.4.3 Temple #3 Multisegment Gradient Methods for NCEs

A survey of stability-indicating methods for pharmaceuticals or complex samples indicated that many final methods follow a multisegment gradient pattern shown in Figure 10.24. Rationales for the multiple gradient segments are as follows:

- Isomers, impurities, and degradants of the API usually have structures similar to the parent molecule and often have similar hydrophobicity and retention.
- A shallow gradient with the API eluting toward the end of a shallow gradient segment would maximize the resolution around the elution time of the API. The final %B of this segment is defined by the hydrophobicity of the API.
- A steep gradient segment, followed by a purging step, is needed to elute highly retained components (e.g. dimers)
- An initial low-strength gradient segment may be added to retain polar impurities (e.g. hydrolytic degradants).

An HPLC analysis of a complex molecule with multiple chiral centers in a retention marker solution spiked with potential impurities and a drug product extract was performed using a three-segment gradient (Figure 5.13) [18]. The middle segment is a shallow gradient (15–40%B in 25 minutes) used to separate the API (with an absolute configuration of

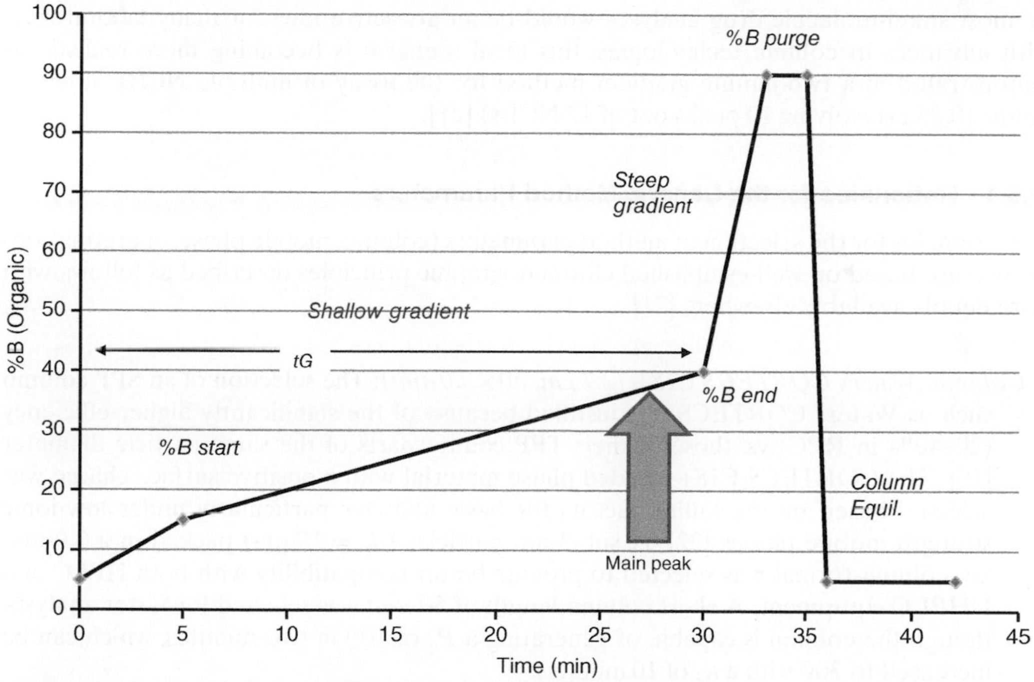


**Figure 10.24.** An example of a gradient profile of the "Multisegment gradient method template" used in Figure 10.27a, showing a shallow middle segment (15–40%B in 25 minutes) with the main component eluting toward the end. This segment is preceded by an early segment (5–15%B in 5 minutes) and followed by a steep segment (40–90%B in 3 minutes) with a 2-minute purging step at 90%B. Source: Reprinted with permission of Dong 2013 [19].

SRR) from its diastereomers (SRS and RRR) and other impurities (M416, M456, and M399 – designated their parent MS ions). In the first gradient segment, a low-strength initial gradient allows for the retention of a hydrolytic degradant (M235), while the final segment depicts a steep gradient that elutes any dimers or other late-eluting species. Traditionally, a single, broad, linear gradient is used during initial method development and column screening, but it is unlikely to provide sufficient resolution around the API to resolve closely eluting impurities. This understanding of the rationale for the multisegment gradient program offers helpful insights into the fine-tuning method development steps.

### 10.4.4 Summary of the Three-Pronged Approach

The three-pronged template approach offers a practical starting point for HPLC method development by focusing on the goals and attributes of the methods. It provides easy templates for method development of simpler samples and offers a targeted pathway during method fine-tuning for complex samples. This approach is an efficient method development strategy in early-phase drug development.

## 10.5 A UNIVERSAL GENERIC METHOD FOR PHARMACEUTICAL ANALYSIS

A single, universal HPLC or UHPLC method with a primary column(s) that works well for most small-molecule drug analyses would be an attractive idea for many laboratories. With advances in column technologies, this ideal scenario is becoming more realistic as demonstrated in a two-minute gradient method for the assay of multiple NCEs shown in Figure 10.25a (resolving 10 peaks out of 12 NCEs) [21].

### 10.5.1 Rationales for the Generic Method Parameters

The rationales for the selection of method parameters (column, mobile phase, operating conditions) are based on well-established chromatographic principles described as follows with more details available elsewhere [21].

*Column: Waters CORTECS C18+ (2.7 μm, 50 × 3.0 mm)*: The selection of an SPP column such as Waters CORTECS was justified because of the significantly higher efficiency (20–40% in RPC) vs. those of their TPP counterparts of the same particle diameter [10]. The CORTECS C18+ bonded phase material with a positive surface charge was selected based on the tailing factors for basic analytes particularly under low-ionic strength mobile phases [22]. A sub-3-μm particles ($d_p$ = 2.7 μm) packed in a 3.0-mm i.d. column format was selected to provide better compatibility with both HPLC and UHPLC equipment. A short column length of 50 mm was selected for faster analysis, though the column is capable of generating a $P_c$ of 100 in two minutes, which can be increased to 300 with a $t_G$ of 10 minutes.

*Mobile phase A (MPA): 0.05% formic acid in water; Mobile phase B (MPB): acetonitrile*: A simple MPA of 0.05% formic acid in water was selected initially to permit easy mobile phase preparation in addition to its ability to support excellent MS ionization efficiency. An alternate MPA of 20 mM ammonium formate with a higher ionic strength buffered at pH 3.7 would likely yield better peak shapes for most NCEs on most bonded phase columns and is the preferred choice for improved method robustness.

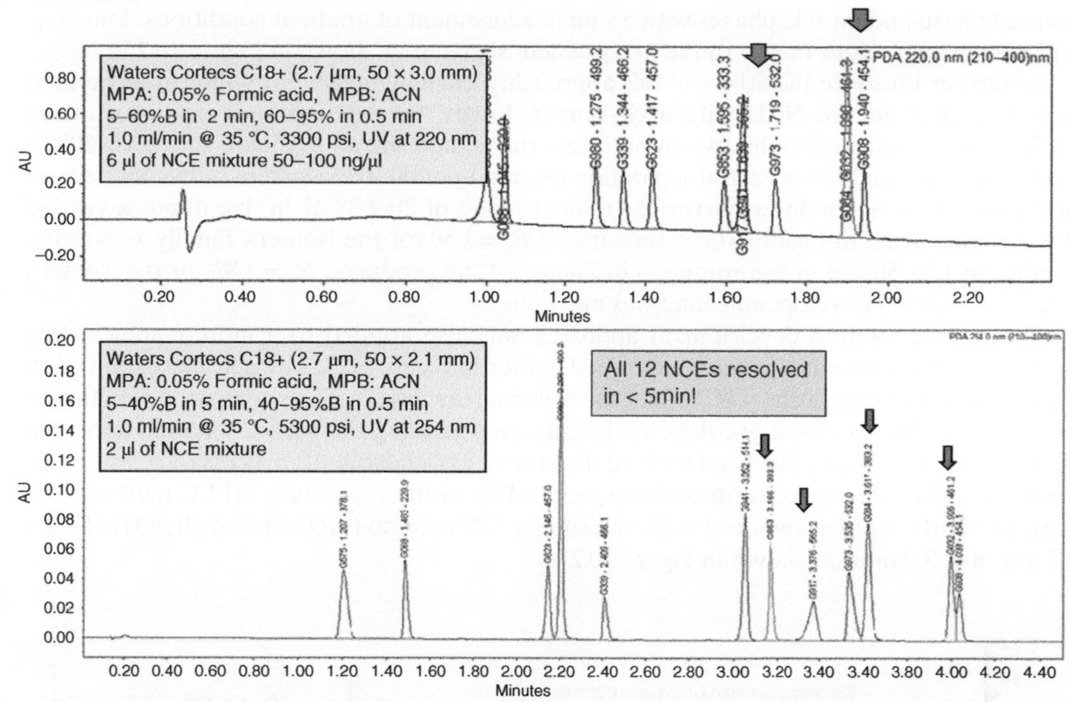


**Figure 10.25.** (a) The proposed universal generic gradient HPLC method using a sub-3-µm SPP column in the analysis of multiple NCEs. This 2-min method resolves 10 peaks in the 12-NCE mixture. (b) A chromatogram depicting the achievement of near-baseline resolution of all 12 NCEs in the test mixture by adjustments of gradient conditions using the same bonded phase in a 2.1 mm column format. Each peak is designated by its code name, retention time in minutes, and M+1 parent ion. Source: Reprinted with permission of Dong 2016 [21].

*Gradient Conditions: 5–60%B in two minutes, 60–95%B in 0.5 minutes, 1.0 ml/min at 40 °C*: A two-segment gradient program of 5–60%B and 60–95%B (rather than a single broad gradient of 5–95% acetonitrile) offers a better chromatographic resolution for multiple NCEs [9, 20]. A quick purging gradient is needed to elute hydrophobic impurities and other later elutors. For this preferred column format (2.7 µm, 50 × 3.0 mm), a flow rate of 1.0 ml/min is optimal with a column temperature of 30–40 °C. An initial operating pressure of 3300 psi is observed for this universal method rendering it compatible with both HPLC and UHPLC equipment.

*Detection: UV at 220 or 254 nm and MS ESI+*: A UV detection wavelength of 220 or 254 nm was used for this method and could be modified to match the $\lambda_{max}$ of the NCEs to maximize UV detection sensitivity. A typical MS scanning range of 150–1000 amu using electrospray ionization in the positive mode (ESI+) was found to work well for most small-molecule NCEs.

### 10.5.2 Adjustment of the Generic Method for Stability-Indicating Assays

The two-minute generic method was deployed as a starting point for stability-indicating assays for many NCEs. It is important to note that all 12 NCEs were resolved via the same bonded phase and mobile phases with a simple adjustment of gradient conditions (longer $t_G$ and narrower gradient range) in the first gradient segment as shown in Figure 10.25b.

To further illustrate the utility of this approach, a challenging stability-indicating method was developed for an NCE with a cis-isomer. Firstly, the NCE sample spiked with the cis-isomer was injected to the two-minute generic method shown in Figure 10.25a (5–60%B in two minutes) yielding a partial separation ($R_s = 0.8$) of the two isomers shown in the inset of Figure 10.26. Secondly, a narrower gradient range of 20–60%B in five minutes yielded the chromatogram in Figure 10.26a showing an $R_s = 1.50$ for the isomers. Finally, separation conditions (30–50%B in five minutes) in Figure 10.26b produced $R_s = 1.82$ for the isomers. The total method development time was one hour.

This simple method development approach was also applied to a more complex drug molecule with three chiral centers described earlier in Figure 5.13. The chromatogram of an existing 42-minute regulatory HPLC method shown in Figure 10.27a is compared with that of a five-minute two-segment gradient method developed using the generic method approach shown in Figure 10.27b. The total method development (adjustment) process took about one hour using the primary column with the same MPA in the regulatory HPLC method. The same method conditions worked well for a similar SPP column (Agilent Poroshell HPH-C18, 2.7 μm, 50 × 3.0 mm) as shown in Figure 10.27c.

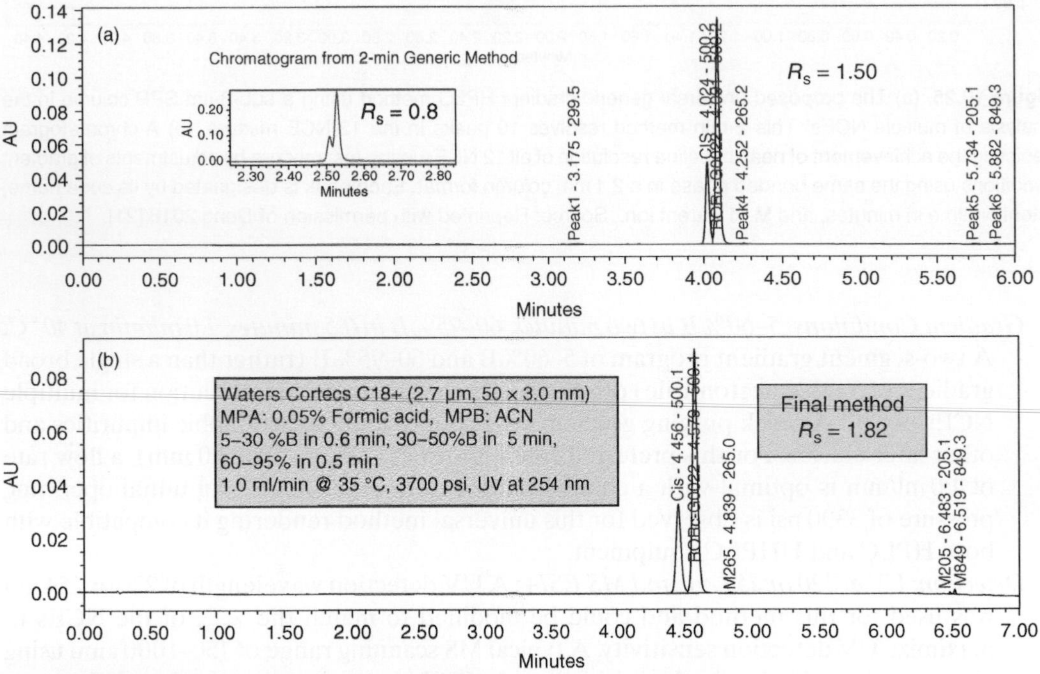


**Figure 10.26.** A case study on the development of a stability-indicating method of an NCE with a cis-isomer using the generic method approach. (a) Chromatogram of the NCE sample spiked with the cis-isomer with a revised gradient program of 20–60%B in 5 minutes. The $R_s$ between the cis- and trans-isomers is 1.50. The chromatogram in the inset shows the partial resolution ($R_s$ = 0.8) of the two isomers using the 2-minute generic method in Figure 10.26a. (b) Chromatogram of the final stability-indicating method with a total method development time of only one hour. $R_s$ is 1.82 for the cis- and trans-isomers. Source: Reprinted with permission of Dong 2016 [21].

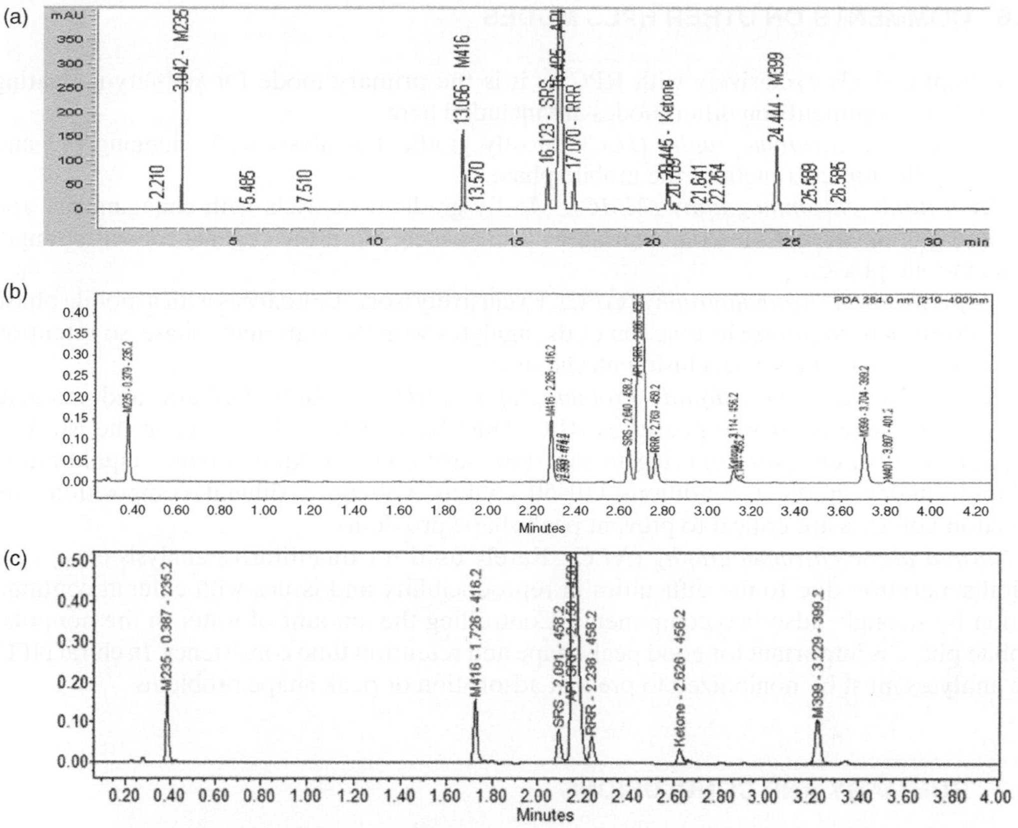

**Figure 10.27.** (a) Chromatogram of the existing 42-minute regulatory HPLC method for a complex NCE molecule with three chiral centers. HPLC operating conditions: ACE C18, 3 μm, 150 mm × 4.6 mm; MPA – 20 mM ammonium formate pH 3.7, MPB – 0.05% formic acid in ACN; 5–15%B over 5 minutes, 15–40%B over 25 minutes, 40–90%B in 10 minutes, 90%B in 3 minutes; 1.0 ml/min at 30 °C; 3500 psi; UV at 280 nm, 10 μl of ~0.5 mg/ml API spiked with impurities. (b) Chromatogram of the 6-minutes, two-gradient segment method using the generic method approach for the same sample. HPLC conditions: Cortecs C18+ (2.7 μm, 50 × 3.0 mm); mobile phases and detection wavelength are the same as in Figure 10.27a; 5–40%B in 4 minutes, 40–95%B in 1 minute; 1.0 ml/min @ 35 °C, 3500 psi, 2 μl of marker solution. (c) Chromatogram using the same conditions on an Agilent Poroshell HPH-C18 (2.7 μm, 50 × 3.0 mm); 4000 psi. Source: Reprinted with permission of Dong 2016 [21].

### 10.5.3 Summary of the Universal Generic Method Approach

A two-minute universal HPLC/UV/MS method for pharmaceutical analysis using modern SPP columns under optimized operating conditions has been proposed. The advantages of this generic method are fast analysis, reasonable peak capacity (~100), and excellent peak shapes for many NCEs using simple mobile phases. This method may function as a general assay method for multiple NCEs and a standard method for cleaning verification. Additionally, the generic method appears to be a good starting point for the development of stability-indicating assays of many drug molecules using a multigradient segment approach with a longer gradient time.

## 10.6 COMMENTS ON OTHER HPLC MODES

This chapter deals exclusively with RPC as it is the primary mode for stability-indicating assays. Brief comments on other modes are included here.

*Ion-exchange chromatography (IEC)*: Mostly gradient analysis with changing pH and increasing the ionic strength of the mobile phase.

*Mixed-mode chromatography (MMC)*: Mostly gradient analysis with changing pH and increasing ionic strength in combination with increasing % of the organic solvent strength in the mobile phase.

*Gel-permeation chromatography (GPC)*: Exclusively isocratic analysis with a mobile phase with additives to minimize interaction of the analytes with the stationary phase, so retention is purely based on the size-exclusion mechanism.

*Hydrophilic-interaction liquid chromatography (HILIC)*: Both isocratic and gradient analysis with similar mobile phases as RPC. Must have at least 3% water in the MPA to maintain the water layer adhering to the stationary phase. Requires larger equilibration volumes under gradient conditions (10–50 column volumes), diluent composition and injection volumes are critical to prevent peak shape problems.

*Normal-phase chromatography (NPC)*: Rarely used for quantitative analysis except for chiral separation due to the difficulties in reproducibility and issues with column contamination by strongly adsorbed components. Controlling the amount of water in the nonpolar mobile phase is important for good peak shape and retention time consistency. In chiral NPC, the analytes must be nonionized to prevent adsorption or peak shape problems.

## 10.7 SUMMARY AND CONCLUSIONS

HPLC method development is a challenging process typically conducted by more experienced scientists in the organization. It can be a time-consuming and iterative process. With all the software and automated systems available today, method development is still mostly a trial-and-error process [13], expedited by a logical sequence of scouting runs and fine-tuning steps to achieve the desired resolution and method performance. This chapter describes a traditional strategy of first defining method development goals and gathering sample/analyte information, followed by initial method scouting and subsequent method fine-tuning. A popular approach is the use of a broad generic gradient using an acidic mobile phase with a C18 column for overall sample assessment to define the initial separation conditions. It is often preferable to develop MS-compatible methods to facilitate the method troubleshooting and identification of unknowns. Method fine-tuning typically involves the adjustment of mobile phase factors (%B, buffer, pH, solvent type) or operating parameters ($F$, $T$, $\Delta\Phi$, $t_G$), which affect selectivity. Modeling software or automated systems are useful for the optimization of mobile phase and operating parameters, particularly for complex separations. Several case studies are also included to illustrate the logical sequence and decision-making process in method development.

Two additional approaches are discussed here: First, a three-pronged template approach to expedite the method development process in early drug development; and secondly, a modernized universal generic gradient method approach which works well for the assaying of multiple NCEs and can be quickly adjusted for stability-indicating assays.

## 10.8 QUIZZES

1. A UV detector is the standard detector for pharmaceutical analysis by HPLC because of these reasons. Which one is not true?
   (a) Most drugs are chromophoric
   (b) UV detector is precise
   (c) Has less coelution issues
   (d) It is easy-to-use and sensitive

2. Which statement is NOT true for the initial development of stability-indicating assays?
   (a) Uses C18 column with an acidic mobile phase and a broad gradient
   (b) Uses a quat pump with PDA
   (c) Uses MS compatible mobile phases
   (d) Uses IEC rather than RPC

3. The next step after initial method development with a broad gradient is often this step:
   (a) Use a narrower gradient range to increase resolution around the API peak
   (b) Change detection to CAD or MS
   (c) Explore selectivity with different bonded phase
   (d) Use automated method development systems or software platform

4. Column/mobile phase screening systems and sophisticated software are needed for:
   (a) everyone
   (b) labs performing method development frequently
   (c) drug discovery labs
   (d) QC labs

5. What kind of software is DryLab considered to be?
   (a) Simulation modeling
   (b) Design of experiments
   (c) Quality by Design
   (d) Automated method development

6. HPLC methods using high-pH mobile phases are particularly useful for
   (a) water-soluble basic compounds
   (b) acidic drugs
   (c) water-insoluble drugs
   (d) biologics

7. Which one is NOT a modern trend in HPLC method development?
   (a) Isocratic methods for purity assays
   (b) Use of quat pump, PDA, and MS
   (c) Column and mobile phase screening using UHPLC
   (d) Development of all-strength methods for drug products with the same API

8. The three-pronged template approach is NOT particularly suitable for
   (a) early-stage drug development
   (b) environment samples
   (c) method development of complex NCEs
   (d) QC method improvements

**9.** Multisegment gradient methods are particularly suited for
   (a) dissolution methods for drug products
   (b) chiral separation
   (c) complex NCEs with many isomeric impurities
   (d) QC methods for regulatory starting materials

**10.** The proposed universal gradient method has these attributes EXCEPT:
   (a) use short columns packed with SPP
   (b) MS-compatible
   (c) modified quickly for stability-indicating applications
   (d) cannot be altered without validation

**11.** Isocratic methods are usually not stability-indicating because of
   (a) provides higher resolution
   (b) UHPLC and HPLC compatible
   (c) less QC friendly
   (d) may miss dimers and have poor retention of early eluting compounds

**12.** Method development tasks are usually performed by the experienced chemists because
   (a) they are more energetic
   (b) they know more about GMP regulations
   (c) they make better judgments to develop methods that are fit for purpose
   (d) they are conservative

### 10.8.1 Bonus Quiz

1. Describe in your own words the reasons why RPC is most often used for stability-indicating assays.
2. Describe the difference between RPC and HILIC. Why are HILIC methods more difficult to develop?
3. Describe in your own words the five steps in the traditional method development approach and the three-pronged template approach.

## 10.9 REFERENCES


1. Snyder, L.R., Kirkland, J.J., and Glajch, J.L. (1997). *Practical HPLC Method Development*, 2e. New York: Wiley-Interscience.
2. Rasmussen, H.T., Li, W., Redlich, D., and Jimidar, M.I. (2005). HPLC method development. In: *Handbook of Pharmaceutical Analysis by HPLC* (ed. S. Ahuja and M.W. Dong), 145–190. Amsterdam, Netherlands: Elsevier. Chapter 6.
3. Kazakevich, Y.V. and LoBrutto, R. (eds.) (2007). *HPLC for Pharmaceutical Scientists*. Hoboken, NJ: Wiley. (Chapters 11, 13, 14).
4. Dong, M.W. (2006). *Modern HPLC for Practicing Scientists*, 193–220. Hoboken, NJ: Wiley. Chapter 8.
5. Dong, M.W. and Zhang, K. (2014). *Trends Anal. Chem.* 63: 21.
6. Dong, M.W. (2015). *LCGC North Am.* 33 (11): 764.
7. Dong, M.W. (2013). *LCGC North Am.* 31 (6): 472.
8. Dong, M.W. (2014). *LCGC North Am.* 32 (8): 552.
9. Wong, M., Murphy, B., Pease, J.H., and Dong, M.W. (2015). *LCGC North Am.* 33 (6): 402.

10. Fekete, S., Guillarme, D., and Dong, M.W. (2014). *LCGC North Am.* 32 (6): 420.
11. ICH Harmonized Tripartite Guideline (2003). Stability Testing of New Drug Substances and Products, Q1A (R2); Validation of Analytical Procedures, Q2 (R1), 2015; Impurities in New Drug Substances, Q3A(R2), 2006; Impurities in New Drug Products, Q3B(R2), 2006; Specifications: Test Procedures and Acceptance Criteria for New Drug Substances and New Drug Products, Q6A, 1999, International Conference on Harmonization, Geneva, Switzerland.
12. Fekete, S., Fekete, J., Molnár, I., and Ganzler, K. (2009). *J. Chromatogr. A* 1216: 7816.
13. Snyder, L.R. (2012). *LCGC North Am.* 25: 437.
14. Hewitt, E.F., Lukulay, P., and Galushko, S. (2006). *J. Chromatogr. A* 1107: 79.
15. Xiang, R. and Horváth, C. (2002). *J. Am. Chem. Soc.* 124 (48): 14504.
16. Verseput, R.P. and Turpin, J.A. (2015). *Chromatogr. Today* .
17. Dong, M.W., Miller, G., and Paul, R. (2003). *J. Chromatogr.* 987: 283.
18. Dong, M.W., Guillarme, D., Fekete, S. et al. (2014). *LCGC North Am.* 32 (11): 868.
19. Dong, M.W. (2013). *LCGC North Am.* 31 (8): 612.
20. Dong, M.W., Zhao, E.X., Yazzie, D.T. et al. (2012). *Amer. Pharm. Rev.* 15 (6): 10.
21. Dong, M.W. (2016). *LCGC North Am.* 34 (6): 408.
22. Fountain, K.J., Hewitson, H.B., Iraneta, P.C., and Morrison, D. (2010). Waters Corporation, Milford, Massachusetts, *720003720EN*.

# 11

# REGULATIONS, HPLC SYSTEM QUALIFICATION, METHOD VALIDATION, AND TRANSFER

## 11.1 INTRODUCTION

### 11.1.1 Scope

High-performance liquid chromatography (HPLC) is widely used for the analysis of pharmaceuticals, food products, and environmental samples to safeguard public standards in public health and product quality. This chapter discusses the regulatory aspects relating to HPLC analysis, focusing on regulations, public standards, HPLC system qualification, instrument calibration, system suitability testing (SST), and method validation and transfer. Case studies in drug development are used to illustrate the current regulatory practices and processes in the industry. This chapter offers a brief overview of the regulatory environment allow the practicing scientist to gain insights on this topic of the pharmaceutical industry. The reader is referred to guidance documents from the various regulatory agencies for the actual requirements.

### 11.1.2 Glossary and Abbreviations

A listing of the common terms or abbreviation (acronyms) relating to regulatory aspects of HPLC can be found in Section 9.1.2.

## 11.2 REGULATORY ENVIRONMENT IN THE PHARMACEUTICAL INDUSTRY

Ensuring the safety and efficacy of drug products is the ultimate goal of regulatory compliance. While the goal is straightforward, a cross-functional effort is required, which may

*HPLC and UHPLC for Practicing Scientists*, Second Edition. Michael W. Dong.

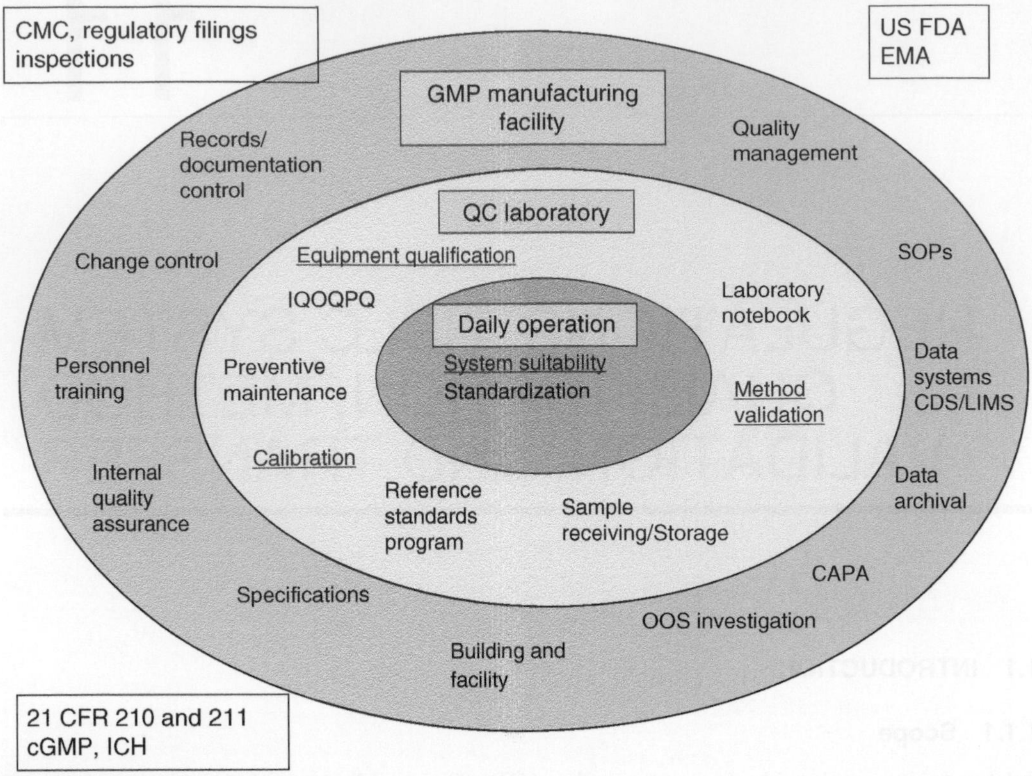


**Figure 11.1.** A schematic diagram depicting the various quality systems and processes used to ensure HPLC data reliability and regulatory compliance under a GMP environment. The three concentric ovals represent the GMP facility, HPLC QC laboratory, and its daily operation. External health authorities and regulations are shown outside the ovals.

be complex to achieve [1, 2]. To help the reader to familiarize with systems and processes of a Good Manufacturing Practice (GMP) Facility, a schematic diagram of the GMP facility, the quality control (QC) Laboratory, and its daily operation is shown in Figure 11.1. Outside the three concentric ovals are the regulatory authorities and external regulations. Compliance with the regulations is ensured by an internal quality system, which includes a quality unit (quality assurance or QA), a set of standard operating procedures (SOPs), documents and data control systems, product specifications, personnel training programs, and QC laboratories.

This chapter focuses on the role the laboratory plays in meeting the compliance goals of the company. Testing performed by the QC lab determines whether a product is within the appropriate specifications for public use. All analytical equipment used for critical quality assessments must be qualified and calibrated for performance verification. Analytical methods for quality assessments must be validated. Reference standards are used to calibrate the system before daily use. Sources of the reference standards are internal process chemistry labs, custom synthesis companies, and compendial (USP, Ph. Eur.) or chemical suppliers. Batches of active pharmaceutical ingredients (APIs) and drug products samples are received, stored, and tested by QC to generate Certificates of Analysis (COAs).

All pertinent laboratory data must be recorded in laboratory notebooks (paper-based or electronic). Results including electronic data are reviewed and archived according to GMP regulations. Analytical equipment is checked before use to determine if it is operating as intended before testing begins. This type of check can take various forms, depending on the type of instrument that is being used. An instrument, such as an analytical balance is checked daily (or before use) with calibrated weights to determine that the balance is operating accurately for the range in which it will be used. More sophisticated instruments, such as HPLCs, are standardized and checked for system suitability for the intended application before use. In most cases, analytical instruments are routinely qualified to ensure that the overall operation meets established requirements. Details of regulations, equipment qualification, and method validation/transfer are available in textbooks, regulatory guidance, and journal articles [1–11]. GMP regulations must be followed in the production of commercial products (also for clinical trial materials, CTMs) with increasing quality/regulatory oversights in late-phase drug development.

To ensure HPLC data reliability and integrity in all regulatory testing, a three-step process is generally adopted (located in the two inner circles in Figure 11.1). The first step is the *initial HPLC system qualification*, followed by *periodic calibration*. The second step is analytical *method validation*, which verifies the performance of the entire analytical procedure. The final step is *system suitability testing* (SST), which verifies the holistic functionality of the entire analytical system (instrument, CDS, column, and mobile phase) before beginning any actual testing. These steps are discussed further in later sections.

### 11.2.1 Regulations

Regulations pertaining to the pharmaceutical development and production are covered by numerous books [2, 3, 7]. A brief synopsis is included here followed by discussions on the roles of the U.S. FDA and USP.

***11.2.1.1 Good Manufacturing Practice (GMP)*** The regulations describing a quality system governing the manufacture and testing of finished pharmaceutical products and APIs – commonly referred to as GMP are found in the United States Codes of Federal Regulations – 21 CFR 210 and 211 [5]. Medical devices are covered by 21 CFR 808, 812, 820, and food regulations are covered by 21 CFR 110.

21 CFR 210 serves as an introduction to the GMP regulations, providing an overview of what is covered and including general information and definitions. 21 CFR Part 211 has the following sections: A. General Provisions; B. Organization and Personnel; C. Buildings and Facilities; D. Equipment; E. Control of Components and Drug Product Containers and Closures; F. Production and Process Controls; G. Packaging and Labeling Control; H. Holding and Distribution; I. Laboratory Controls; J. Records and Reports; K. Returned and Salvaged Drug Products [5]. 21 CFR 600 and 610 are GMP regulations for biologics. The GMP regulations are also discussed in the ICH Guidelines on Good Manufacturing Practice Guide for Active Pharmaceutical Ingredients (Q7) and Pharmaceutical Quality Systems (Q10) [9].

***11.2.1.2 International Council for Harmonization (ICH) Guidelines*** The International Council for Harmonization of Technical Requirements for Pharmaceuticals for Human Use (ICH) was formed to standardize global regulations regarding the production of drug products and substances. This organization brings together regulatory authorities of Europe, Japan, Canada, Brazil, China, and the United States, as well as pharmaceutical experts from industry to discuss and harmonize regulations governing different geographic regions and the scientific and technical aspects required for product registration. It publishes extensive volumes of ICH Guidelines categorized into four groups: Quality (Q), Safety (S),

Efficacy (E), and Multidisciplinary (M). While these guidelines are not legal requirements, they are widely recognized and followed closely by pharmaceutical companies in marketing applications and late-stage development. The quality sections are: Q1A–Q1F, Stability; Q2, Analytical Validation; Q3A–Q3D, Impurities; Q4–Q4B, Pharmacopeias; Q5A–Q5E, Quality of Biotechnological Products; Q6A–Q6B, Specifications; Q7, GMP; Q8, Pharmaceutical Development; Q9, Quality Risk management; Q10, Pharmaceutical Quality system; Q11, Development and Manufacture of Drug Substances; Q12, Life Cycle Management [9].

During early development (Phase 1 and 2a), the stringent ICH guidelines can serve as a reference but are not intended to be applied directly. Here, the International Consortium on Innovation and Quality in Pharmaceutical Development (IQ Consortium, iqconsortium .org), published a series of position papers, presentations, and guidelines, and organizing scientific conferences, offering useful guidance on drug product manufacturing, specifications, validation, and stability for early-phase development (https://iqconsortium.org) [11]. IQ is a technically focused organization comprising of pharmaceutical and biotechnology companies with a mission of advancing science and technology to augment the capability of member, regulators, and the broader R&D community. Some IQ guidelines are shown in case studies in the chapter.

***11.2.1.3 Good Laboratory Practice (GLP)*** While GMP regulations are followed by production facilities, Good Laboratory Practice (GLP) regulations (described in 21 CFR 58) [5, 7] describe a system of management controls to ensure consistency and reliability of results in nonclinical studies. They are followed by laboratories, animal care facilities, and Contract Research Organizations (CROs) conducting pivotal animal toxicology evaluations or bioanalytical clinical studies [1]. GLP uses distinctively different terminologies and languages from GMP regulations as it focuses on shorter-term projects and standalone studies. Terms such as study director, protocol, test and control articles, and in-life inspection are common terms in GLP regulations and practices.

### 11.2.2 The Role of the United States Food and Drug Administration (U.S. FDA)

The United States Food and Drug Administration (U.S. FDA) is responsible for the protection of public health by ensuring the safety, efficacy, and security of human and veterinary drugs, biological products, and medical devices and by ensuring the safety of the nation's food supply, cosmetics, and products that emit radiation. FDA also has responsibilities for regulating tobacco products. It is responsible for the enforcement of regulations (21 CFR for GLP, GMP, and Good Clinical Practice). It monitors companies on compliance and provides advice and guidance on new drug development. The FDA reviews and approves marketing applications (NDA, clinical trial protocols, BLA, and IND). It inspects manufacturing sites and audits data and quality systems, issues 483's warning letters (official document to the sponsor describing the violations), and consent decrees for those found to be noncompliant and has the power to shut down manufacturing facilities. The FDA monitors drug safety (e.g. MedWatch) and can request product recalls. The counterpart to the U.S. FDA is the European Medicines Agency (EMA) for the European Union, the Pharmaceutical and Food Safety Bureau in Japan, Health Canada, and the China Food and Drug Administration (CFDA).

### 11.2.3 The United States Pharmacopeia (USP)

The U.S. Pharmacopeial Convention (USP) is a scientific nonprofit organization that sets standards for the identity, strength, quality, and purity of medicines, food ingredients, and dietary supplements manufactured, distributed, and consumed worldwide. USP's drug standards are enforceable in the United States by FDA, and these standards are used in more than 140 countries. The Convention publishes the United States Pharmacopeia and The National Formulary (USP–NF) [10], a book of public pharmacopeial standards for chemical and biological drug substances, dosage forms, compounded preparations, excipients, medical devices, and dietary supplements. The USP contains monographs for drug substances and dosage forms, which includes the name of the ingredient or preparation, packaging, storage, labeling requirements, and specifications. The specification consists of a series of tests, procedures, and acceptance criteria. Most of these tests require the use of official USP Reference Standards. USP standards and monographs are particularly important for the generic drug industry, and common USP testing procedures are often quoted in specifications and COAs for new drugs, dietary supplements and herbal medicines (see Table 9.6). Note that similar information pertaining to the European Union, the European Pharmacopeia (Ph. Eur.) is available from www.edqm.eu.

The USP also provides general information regarding the preparation of common volumetric solutions, analytical techniques, requirements for use, qualification of analytical instrumentation, and recommendations for calculations, which allows QC labs to operate consistently.

## 11.3 HPLC SYSTEM QUALIFICATION

Equipment qualification is a formal process that provides documented evidence that an instrument is fit for its intended use. The entire process depicted schematically in Figure 11.2, consists of four parts: design qualification (DQ), installation qualification (IQ), operational qualification (OQ), and performance qualification (PQ) (also referred to as Performance Verification or PV). This topic is also described in USP<1058> of analytical instrument qualification (AIQ). After initial qualification, the system is kept in a "qualified state" using a periodic calibration program. Also, its performance for the specific application is verified at the time of use using SST. This section describes the procedures and documentation used for HPLC equipment qualification. Detailed discussions of this topic can be found in many published books and articles elsewhere [7, 8]. The regulations and qualification of computerized data systems and networks with their associated software are described elsewhere [12, 13].

### 11.3.1 Design Qualification (DQ)

Design Qualification (DQ) describes the user requirements and defines the functional and operational specifications of the instrument, which are used for testing in the OQ phase. Table 11.1 shows an example of a DQ for an HPLC system used for method development.

### 11.3.2 Installation Qualification (IQ)

Installation Qualification (IQ) verifies that the instrument is received as designed and meets manufacturers' specifications upon installation in the user's environment. Table 11.2 lists the steps recommended for IQ before and during the installation. IQ should include the analysis of a test sample to verify the correct installation of all modules including electrical, fluidics, and data connections.

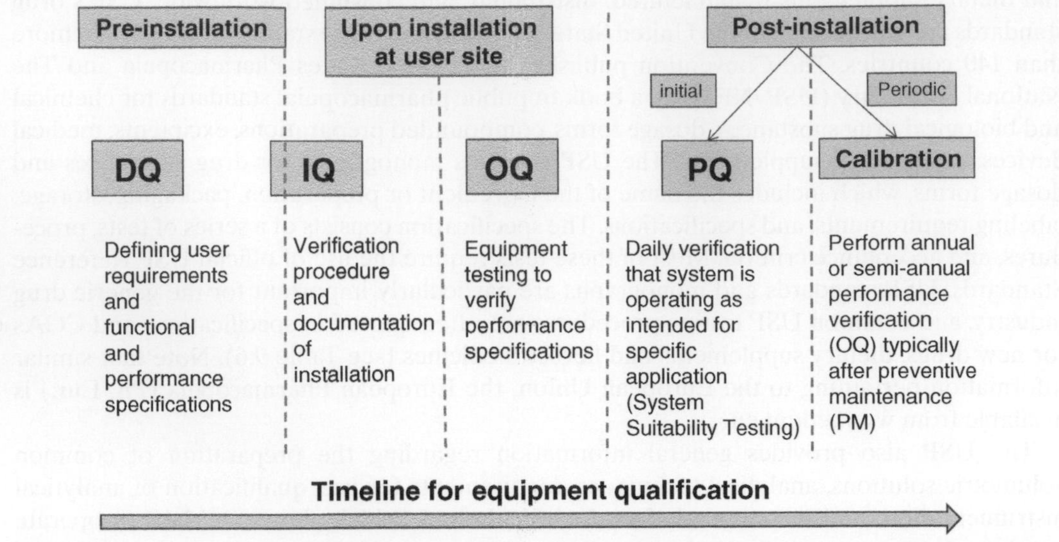


**Figure 11.2.** A schematic diagram illustrating the timeline and documents in the various stages in Equipment Qualification including those for HPLC systems.

**Table 11.1. An Example of a Design Qualification (DQ) for an HPLC System for Method Development**

| Design elements | Examples |
|---|---|
| **Intended use** | Analysis of drug products and substances |
| User requirement specification | Automated analysis of up to 100 samples per day |
| | Limit of quantitation <0.03% |
| | Automated confirmation of peak purity and identity with diode-array detection |
| | Compatible to 2.1-mm i.d. HPLC columns |
| **Functional specifications** | |
| Pump | Quaternary gradient pump with on-line degassing with a flow range of 0.01–5 ml/min |
| Detector | UV/Vis diode array, 190–600 nm with 1-nm resolution |
| Autosampler | >100 sample vials, 0.5–200 μl injection |
| Column oven | 5–80 °C, Peltier |
| Computer | System control and data acquisition by computer workstation or network with remote access |
| **Operational specifications** | • Pump precision of retention time <0.5% RSD<br>• Composition accuracy <1% absolute<br>• Detector noise, <±1 × $10^{-5}$ AU<br>• Autosampler peak area precision <0.5% RSD<br>• System dwell volume <1 ml<br>• Instrumental bandwidth <30 μl (4$\sigma$)<br>• Column temperature control precision <±1 °C |
| **User instruction** | Operation manual |
| **Qualification** | Vendor must provide procedures and services for IQ and OQ |
| **Maintenance** | Vendor must provide procedures and services for maintenance |
| | System must have built-in diagnostic functions |
| **Training** | Vendor must provide familiarization and training |

**Table 11.2. Steps Recommended for Installation Qualification (IQ)**

Before installation
- Obtain manufacturer's recommendations for installation site requirement
- Check site for fulfillment of these requirements (space, electricity, utilities, environmental conditions, and storage space for manuals, software, and log books)

During installation
- Compare equipment as received with a purchase order (including software, accessories)
- Check documentation for completeness (manual, installation qualification operation qualification [IQOQ], certifications)
- Install hardware and software (by service specialists from the manufacturer)
- Power up instrument to perform start-up diagnostic tests
- Run a test sample to verify installation
- Prepare and signoff installation report, which includes the names and serial numbers of instrument and components, software and firmware versions, actual locations of instrument and manuals, and qualification documentation

### 11.3.3 Operational Qualification (OQ)

Operational qualification (OQ) verifies that the instrument functions according to its manufacturer's specifications. Testing procedures and acceptance criteria in OQ are often similar to those in system calibration shown in Table 11.3 [14]. Both hardware and software under normal operating conditions are verified. In most laboratories, OQ testing is performed by a service specialist from the manufacturer (often together with installation qualification, IQ/OQ), while subsequent performance qualification (PQ) and calibration may be performed by an in-house metrologist or the user. Operational qualification is performed using a protocol, which may be generated by the manufacturer or by a laboratory user.

### 11.3.4 Performance Qualification (PQ)

Performance qualification (PQ) is the process of demonstrating that an instrument can consistently perform an intended application within some predefined acceptance criteria by the user. In practice, PQ testing is often interpreted as synonymous with SST conducted with specified columns, mobile phases, and test compounds. PQ is performed during initial system qualification or after the system is relocated or modified, whereas SST is conducted on a daily basis or before the instrument is used for regulatory analysis. PQ is often combined into an IQ/OQ protocol during initial installation. SST is discussed in Section 11.3.7.

### 11.3.5 System Qualification Documentation

Upon completion of equipment qualification, a set of the documentation should be available such as the DQ document, the executed IQ/OQ/PQ protocols, and a final report. These protocols and documentation usually require QA reviews, approvals, and sign-offs. Final records are kept as proofs of equipment qualification.

### 11.3.6 System Calibration

"System calibration" refers to the periodic operational qualification of the HPLC, typically every 6–12 months in most regulated laboratories. This calibration procedure is often coordinated with annual preventative maintenance (PM) program and is performed immediately

**Table 11.3. A Summary of an HPLC Calibration Test Procedures and Acceptance Criteria**

| HPLC module | Typical test | Procedure (suggested) | Acceptance criteria (suggested) |
|---|---|---|---|
| **Detector** UV/Vis or PDA | Wavelength accuracy | Measure $\lambda_{max}$ or maximum absorbance of anthracene solution (1 μg/ml). | 251 ± 3 nm<br>340 ± 3 nm |
| **Pump** | Flow accuracy | Run pump at 0.3 and 1.5 ml/min (65% acetonitrile/water) and collect 5 ml from detector into a volumetric flask. Measure time. | < ±5% |
| | Flow precision | Determine retention time RSD of six 10-μl injection of ethylparaben (same as in autosampler test) | RSD<br>< ±0.5% |
| | Compositional accuracy | Test all solvent lines at 2 ml/min with 0.1% acetone/water, step gradients at 0%, 10%, 50%, 90%, and 100%. Measure peak heights of respective step relative to 100% step. | ±1% absolute |
| **Autosampler** | Precision | Determine the peak area RSD of six 10-μl injection of ethylparaben (20 μg/ml) | RSD<br>< ±0.5% |
| | Linearity | Determine coefficient of linear correlation of injection of 5-, 10-, 40-, and 80-μl of ethylparaben solution | $R > 0.999$ |
| | Carryover | Determine carryover of peak area from injecting 80-μl of mobile phase following 80-μl injection of ethylparaben | <0.5% (<0.1% for some applications) |
| | Sampling accuracy | Determine gravimetrically the average volume of water withdrawn from a tared vial filled with water after six 50-μl injections | 50 ± 2 μl |
| **Column oven** | Temperature accuracy | Check actual column oven temperature with validated thermal probe | 30 ± 2 °C<br>50 ± 2 °C |
| **Additional tests recommended during Operation Qualification** | | | |
| **Detector** PDA or UV/Vis | Baseline noise and drift | ASTM method E19.09 | $<5 \times 10^{-5}$ AU $<2 \times 10^{-3}$ AU/hr |
| **System** Required only for systems or <2.1-mm i.d. columns | Dwell volume | Perform linear gradient with 0.1% acetone/water in 10 minutes at 1 ml/min without column and extrapolate the intersection of baseline with the gradient profile | <1 ml |
| | Bandwidth (Dispersion) | Measure the 4 $\sigma$ bandwidth of a 1-μl injection of a 0.1% caffeine solution without the column | <40 μl |

Source: Data extracted from Ref. [14].

after PM. A calibration sticker is placed on the instrument to indicate its calibration status and readiness for GMP work. The reader is referred to the principles and procedures of HPLC calibration published elsewhere [14]. A summary of HPLC system calibration procedures and acceptance criteria from a pharmaceutical laboratory designed to be completed by a user in one day is listed in Table 11.3 [14]. It is recommended that the user is responsible for performing the calibration procedures since this promotes increased familiarity and understanding of equipment operation.

### 11.3.7 System Suitability Testing (SST)

SST is used to verify the performance of the analysis system to ensure its adequacy for the intended application on a daily basis [15]. According to the latest USP and ICH guidelines, SST must be performed before and throughout all regulated assays. Primary SST parameters are resolution ($R_s$), repeatability (relative standard deviation [RSD] of peak response and retention time), column efficiency/plate number ($N$), sensitivity, and tailing factor ($T_f$). Table 11.4 summarizes guidelines in setting SST limits from the U.S. FDA's Center for Drug Evaluation and Research (CDER) and those proposed by Hsu and Chien [16], which include recommendations on biologics and trace components. SST acceptance limits should represent the minimum acceptable system performance levels rather than the optimal levels. A practical approach is to set SST limits (or acceptance criteria) based on the three-sigma rule gathered from historical performance data of the method [15].

An example chromatogram of an SST for a stability-indicating method is shown in Figure 11.3. System Suitability Sample (SSS, containing the API and key analytes) are analyzed before testing.

A typical SST initial injection sequence for a stability-indicating assay is listed as follows:

1. *Blank (diluent)*: to verify no interference from the diluent or system contaminants.
2. *SSS or retention marker solution*: to verify that the system is capable of a resolution of key components.
3. *Sensitivity solution*: to ensure that the system is capable of achieving a limit of quantitation (LOQ) of 0.05% for drug substance (DS) methods and 0.1% for drug product (DP) methods.
4. *Two injections of Calibration Standard A*: prepare a reference standard solution at 100% of the API concentration for system calibration.

**Table 11.4. Comparison of SST Criteria According to FDA's CDER and Hsu and Chien**

| SST limits | CDER guidelines | Hsu and Chien recommendation |
|---|---|---|
| Repeatability (RSD) of peak response | ≤1.0% for five replicates | ≤1.5% general<br>5–15% for trace<br><5% for biologics |
| Resolution ($R$) | >2.0 general | >2.0 general<br>>1.5 quantitation |
| Tailing factor ($T$) | ≤2.0 | <1.5 2.0 |
| Plate count ($N$) | >2000 | NA |
| Capacity factor ($k$) | >2 | 2–8 |

NA = not available.
Source: Table with data extracted from Ref. [16].

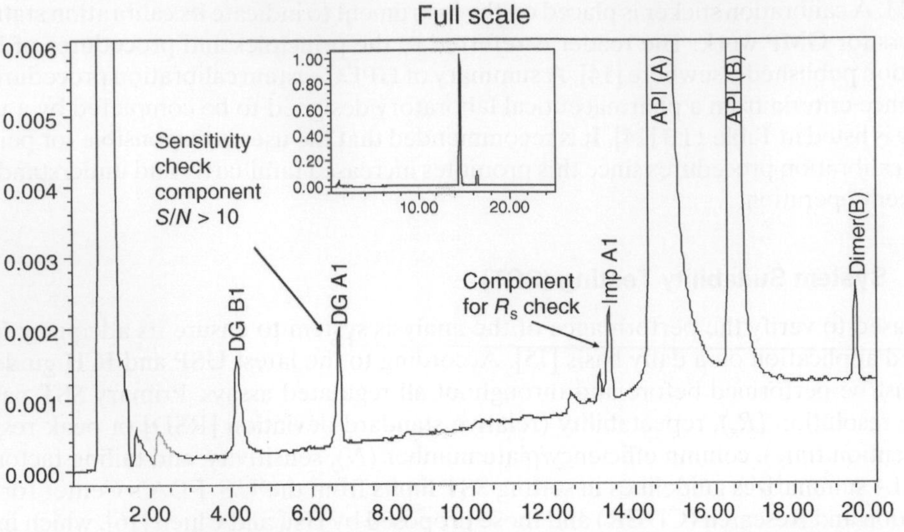


**Figure 11.3.** HPLC chromatogram of a system suitability solution (SSS) for a stability-indicating method for a drug product containing two APIs. This SSS contains both APIs and several key degradants and impurities at their expected concentrations (as retention time markers). Impurity A1 is used as a resolution check. One of the components DGA1 present at 0.10% level also serves as a sensitivity check and must meet the acceptance criterion of having an $S/N > 10$.

5. *Two injections of Calibration Standard B*: prepare a duplicate reference standard solution at 100% of the API concentration to verify that preparation of the standard solution. The response factors of reference solutions A and B must be within ±2.0% of each other.
6. *Three additional injections of Calibration Standard B*: to ensure that system precision of peak area is generally required to be < ±0.73% RSD for five repetitive injections of Standard B (according to USP<621>).
7. Typically the calibration standard is interspersed and injected every 10 assay or 12 dissolution samples to verify that system precision is maintained during the sample sequence.

If initial SST fails, the analyst should stop the sequence immediately, diagnose the problem, make necessary adjustments or repairs, and reperform SST. Analysis of actual samples should commence only after passing all SST limits, not only the failed criteria. Most SST failures are traced to problems from the autosampler, pump, column, or mobile phase (see Chapter 8 on troubleshooting guides). If one of the interspersed SSS injections fails, data from all samples after the last passing SSS become invalid and must be repeated.

## 11.4 METHOD VALIDATION

Method validation is the process of ensuring that a test procedure is accurate, reproducible, and sensitive within the specified analyte range for the intended application. Although validation is required by law for all regulatory methods, the actual implementation is somewhat open to interpretation and may differ between organizations. This section describes data elements required for method validation (listed in Figure 11.4) from ICH guidelines Q2(R1) [9]

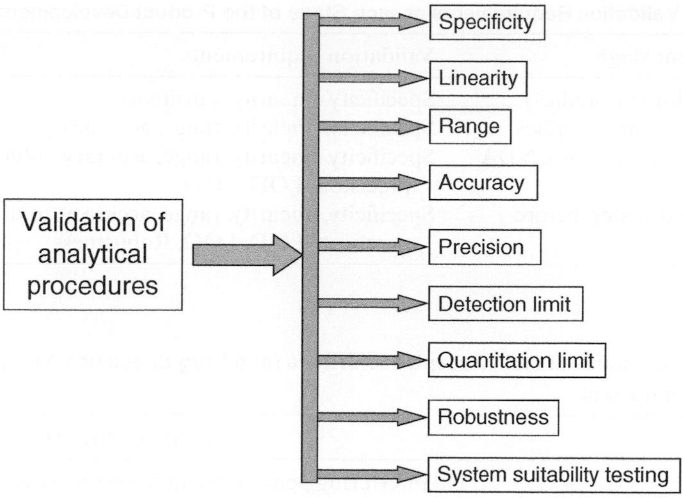


**Figure 11.4.** A diagram listing of validation requirements of analytical procedures from USP<1225> and ICH Q2 (R1).

**Table 11.5. Data Elements Required for Validation of Analytical Procedures**

| Analytical performance characteristics | Assay category I | Assay category II quantitative | Assay category II limit test | Performance assay category III | Category IV identification |
|---|---|---|---|---|---|
| Accuracy | Yes | Yes | [a] | [a] | No |
| Precision | Yes | Yes | No | Yes | No |
| Specificity | Yes | Yes | Yes | [a] | Yes |
| Detection limit | No | No | Yes | [a] | No |
| Quantitation limit | No | Yes | No | [a] | No |
| Linearity | Yes | Yes | No | [a] | No |
| Range | Yes | Yes | [a] | [a] | No |

[a]May be required, depending on nature of the specific test.
Source: Table adapted from the USP <1225>.

and USP <1225> [10]. The reader is referred to books and references for a detailed discussion on the subject [17–19]. Tables 11.5–11.7 summarize the validation requirements of different method types and at various product development cycles. Method validation data summaries from two case studies from early- and late-stage methods are shown here to illustrate the actual data.

### 11.4.1 Data Required for Method Validation

***11.4.1.1 Specificity*** Specificity is the ability of a method to discriminate between the key analyte(s) and other interfering components in the sample. Since UV/Vis detector would detect any compounds with absorbance at a set wavelength, coeluting peaks would result in inaccurate results. Specificity of the HPLC method is demonstrated by the physical separation of the analytes from other potential components such as impurities, degradants, or excipients. Also, stressed samples under forced degradation experiments (acid, base, heat,

**Table 11.6. Typical Validation Requirements at each Stage of the Product Development Cycle**

| Product development stage | Validation requirements |
|---|---|
| Preclinical (before human studies) | Specificity, linearity, sensitivity |
| Initiation of phase 1 clinical studies | Specificity, linearity, range, accuracy, precision, LOD, LOQ |
| Initiation of registration batches, NDA submission | Specificity, linearity, range, accuracy, solution stability, precision, LOD, LOQ |
| Completion of tech transfer, before product launch | Specificity, linearity, range, accuracy, solution stability, precision, LOD, LOQ, Robustness |

**Table 11.7. Validation Parameters and Acceptance Criteria for a Drug Substance Assay/impurities HPLC Method in Early Development**

| Parameters | Acceptance criteria |
|---|---|
| Specificity | No interfering peaks present in the blank in the regions of interest |
| Linearity[a] | $R^2 > 0.995$ over a range of LOQ (0.05%) to 120% |
| Precision[a] | RSD ≤2.0% ($n = 5$) for main compound, 10–20% for impurities |
| Recovery/accuracy of main compound | 98.0–102.0% at 80%, 100%, and 120% of nominal concentration ($n = 3$ at each level) |
| Recovery/accuracy of impurities[a] | 80–120% at LOQ to specification limit (or higher than spec limit) |
| Sensitivity – LOQ | Mean signal-to-noise ratio ≥ 10 ($n = 3$) |
| Range | The range of demonstrated linearity, recovery, and precision |
| Solution stability[a] | The API assay changes ≤2%. No new impurities are greater than the reporting threshold. Impurities at the reporting threshold change ≤30%; impurities at levels between the reporting threshold and the specification limit change ≤20%; and impurities at or above the specification limit change ≤15%. |

[a] Acceptance criteria recommended by the IQ Consortium Working Group.

moisture, light, and oxidation) are used to challenge the method to demonstrate method specificity. A method specificity study includes a demonstration of the noninterference of contaminants and reagents by running a procedural blank and a placebo extract for a drug product method (Figure 11.5). A placebo is a mock drug product that contains the exact formulation components minus the API. In addition, specificity is demonstrated by a peak purity assessment using a photodiode array (PDA) or a mass spectrometer (MS) (Figure 11.6) and by comparing the results of the sample to those obtained by a second well-characterized technique (e.g. infrared, qNMR, gas chromatography [GC], supercritical fluid chromatography [SFC], and ion chromatography [IC]). In practice, this secondary technique tends to be another "orthogonal" reversed-phase chromatography (RPC) method using a different mobile phase or column with different selectivity (Section 2.7) [20].

***11.4.1.2 Linearity and Range*** The **linearity** of a method is its ability to obtain test results that are directly proportional to the sample concentration or amount over a given range. For HPLC methods, the relationship between detector response (peak area) and sample

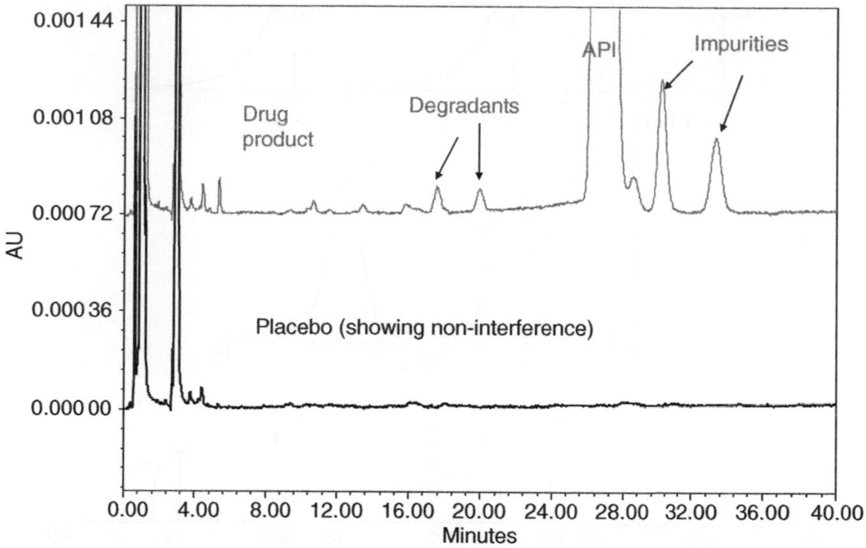


**Figure 11.5.** Example chromatograms demonstrating the noninterference of placebo in a stability-indicating assay of a drug product. The upper chromatogram shows the separation of key analytes (API and various impurities and degradants) in an extract of the drug product. The lower chromatogram shows a similar extract of the placebo showing the absence of these key analytes in the placebo extract.

**Table 11.8. Linearity Ranges and Acceptance Criteria for Various Pharmaceutical Methods**

| Test | Linearity | |
|---|---|---|
| | Levels and ranges | Acceptance criteria |
| Assay and content uniformity | 5 levels, 80–120%<br>70–130% (CU) | Correlation coefficient ($R$)<br>$R \geq 0.999$ (% y-intercept NMT 2.0%) |
| Dissolution | 5–8 levels, 10–120% or ±20% of the specified range | $R \geq 0.99$ (% y-intercept NMT 5.0%) |
| Impurities | 5 levels, 50–120% of specification | $R \geq 0.98$ |
| Cleaning surface validation | 5 levels, LOQ to 20 times LOQ | $R \geq 0.98$ |
| Bioanalytical | 6–8 levels covering the dynamic range | $R \geq 0.98$ |

NMT, not more than.

concentration (or amount) is used to make this determination. Table 11.8 summarizes the typical linearity levels and ranges as well as the acceptance criteria for various pharmaceutical methods. Figure 11.7 shows the results of a linearity study of an assay method of a drug product at levels of 50%, 75%, 100%, 125%, and 150% of label claim (a wider range of 50–150% was used to show a method bias scenario). Note that although data show acceptable linear correlation coefficient ($R > 0.999$), the percentage of y-intercept was found to be biased (%y intercept ~7%) and did not pass typical assay linearity criterion (e.g. a %y-intercept of <2%). The likely cause of this bias may be UV detector saturation, interaction of the analyte with excipients, sample preparation issue, or other factors to be investigated.

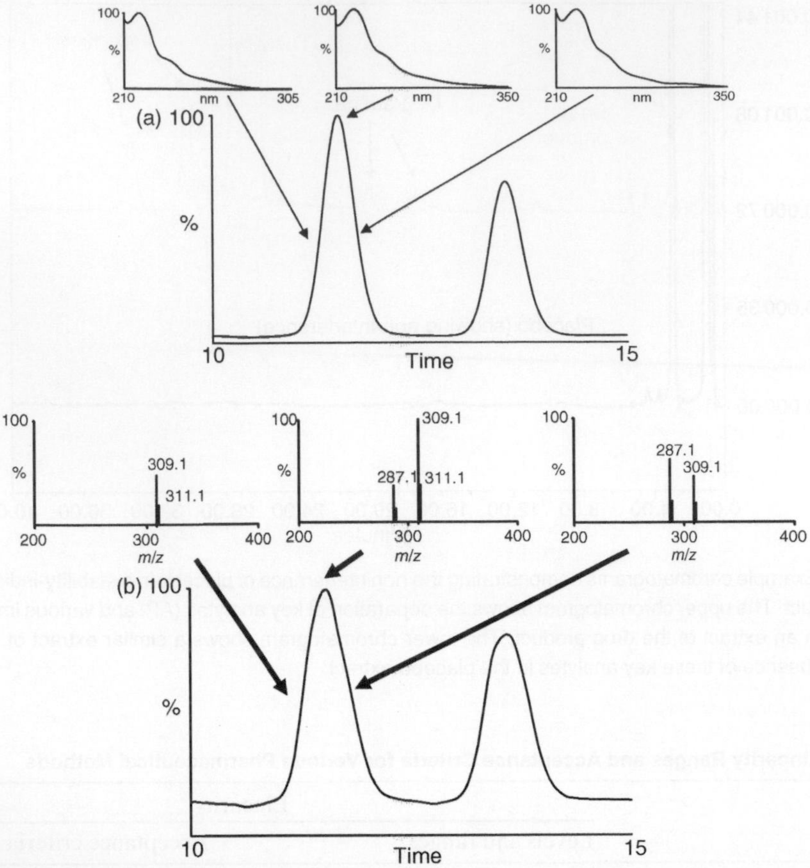


**Figure 11.6.** Diagram illustrating peak purity assessment using PDA and MS. (a) Shows that the UV spectra of the leading, apex, and trailing parts of the first peak look fairly similar visually, one might conclude erroneously that the first peak contains only one single component. (b) Shows that the first peak contains two components with parent M + 1 ions at 309 and 287, respectively. These two components might have similar UV spectra, making UV spectroscopy an insensitive assessment tool for peak purity evaluation. Source: Courtesy of Waters Corporation.

The **range** of an analytical method is the interval between the upper and lower analytical concentration of a sample that has been demonstrated to show acceptable levels of accuracy, precision, and linearity.

***11.4.1.3 Accuracy*** The **accuracy** of an analytical procedure is the closeness of the rest results obtained by that procedure to the true value. Accuracy studies are usually evaluated by determining the recovery of spiked analytes into the matrix of the sample (a placebo) or by comparison of the result to a reference standard of known purity or those obtained from another well-characterized procedure. If a placebo is not available, the technique of standard addition or sample spiking is used.

***11.4.1.4 Precision: Repeatability, Reproducibility*** Method **precision** is a measure of the ability of the method to generate reproducible results (i.e. precision is a measure of the degree of scattering of results). The precision of a method is evaluated for repeatability, intermediate precision, and reproducibility.

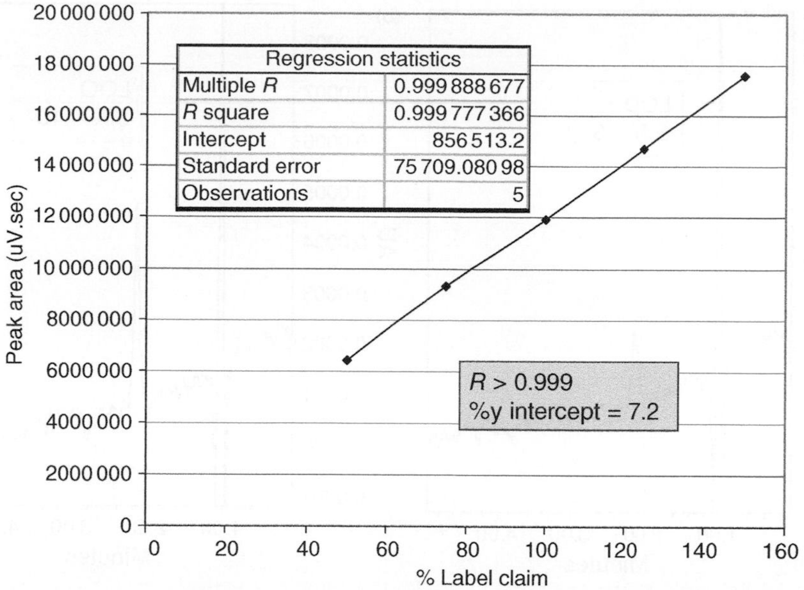


**Figure 11.7.** Linearity data of a potency assay method of a drug product method showing the peak area responses at 50%, 75%, 100%, 125%, and 150% of the API concentrations. Regression analysis of the data shows a good coefficient of linear correlation ($R > 0.999$), but a method bias since the %y intercept is about 7.2%.

**Repeatability** is a measure of the ability of the method to generate similar results for multiple preparations of the same sample by one analyst using the same instrument in short time duration (e.g. on the same day). For instance, method repeatability for pharmaceutical assays may be measured by making six sample determinations at 100% concentration or by preparing three samples at 80%, 100%, and 120% of concentration levels each.

**Intermediate precision**, synonymous with the term "ruggedness," is a measure of the variability of method results where samples are tested and compared to those using different analysts, different equipment, and on different days. This study is a measure of the intralaboratory variability and is a measure of the precision that can be expected within a laboratory. Intermediate precision is strongly affected by the design of the protocol (e.g. how many sources of variance are included).

**Reproducibility** is the precision obtained when samples are prepared and compared between different testing sites. Method reproducibility is often assessed during collaborative studies at the time of technology or method transfer (e.g. from a research facility to quality control laboratory of a manufacturing plant).

***11.4.1.5 Sensitivity: Detection Limit and Quantitation Limit*** **Detection Limit or Limit of detection (LOD)** is the smallest amount or concentration of analyte that can be detected. There are several ways for the calculation of LOD, as discussed in the ICH Q2(R1) guidelines on method validation [9]. The simplest way to calculate LOD is to determine the amount (or concentration) of an analyte that yields a peak height with a signal-to-noise ratio ($S/N$) of 3 (Figure 11.8a).

**Quantitation limit or limit of quantitation (LOQ)** is the lowest level that an analyte can be quantitated with some degree of certainty (e.g. with a precision of ±5%). The simplest way of calculating LOQ is to determine the amount (or concentration) of an analyte that

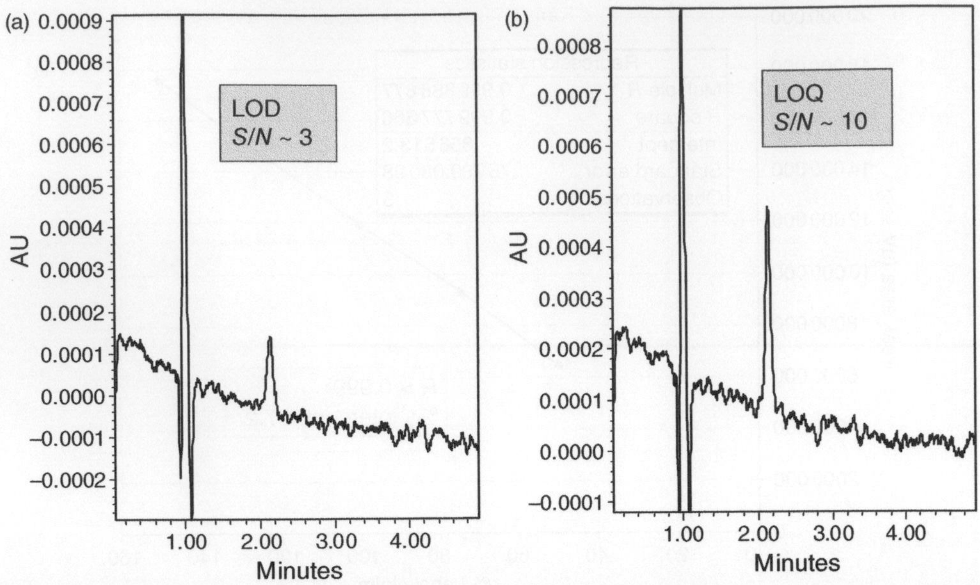


**Figure 11.8.** (a) Chromatograms showing analyte peaks at the limit of detection (LOD, *S/N* = 3) and (b) limit of quantitation (LOQ, *S/N* = 10).

yields a peak with a signal-to-noise ratio of 10 (Figure 11.8b). Thus, LOQ is roughly equal to three times of LOD. As noted in Table 11.5, LOQ or LOD determination is required only for high-sensitivity methods involving trace components such as impurity methods or cleaning validation methods.

***11.4.1.6 Robustness*** **Robustness** is a measure of the performance of a method when small, deliberate changes are made to the specified method parameters. Robustness validation is to identify critical parameters for the successful implementation of the method. Robustness is partially evaluated during method development when conditions are optimized to improve resolution and other method performance criteria (e.g. specificity, peak shape, sensitivity). Robustness validation is a formalized evaluation of the written method by varying some of the operating parameters within a reasonable range (Table 11.9). These factors can be evaluated one factor at a time or systematically by the use of Design of Experiments (DoEs) software packages (e.g. Plackett–Burman Designs).

### 11.4.2 Case Studies and Summary Data on Method Validation

Table 11.10 shows a summary of validation results for a Phase 0 stability-indicating assay method for a drug substance shown in Figure 10.7. The development process of this particular method is described in Section 10.4.5.1.

The key analytes of this assay are the API, an impurity eluting at ~6.4 minutes (impurity 1) that has been identified as an isomer of the API and the immediate synthetic precursor eluting at ~7.6 minutes.

Method specificity was demonstrated by resolving all impurities from each other and the API with a focus on two impurities (impurity 1 and the precursor). The evaluation of the peak purity of the API by PDA peak purity and MS indicated that the API is spectrally

**Table 11.9. Typical Parameters and Range for Method Robustness Evaluation**

Column consistency
- Three columns packed by bonded phases from three different silica lots

Mobile phase
- pH (±0.1–0.2 units)
- Buffer concentration (±5–10 mM)
- Percentage organic modifier (±1–2% MPB)

Sample
- Injection volume or sample concentration
- Solvent strength for the final solution

Column temperature (±5 °C)

Detector wavelength (±3 nm)

Gradient
- Dwell volume
- Gradient time ($t_G$, ±2–5 min)

**Table 11.10. Validation Results of a Stability-indicating Assay Method of an NCE**

- Specificity
  - o Resolves API from precursor, isomer, and other impurities
  - o API peak purity demonstrated by MS and PDA
- Linearity
  - o API: 5–150% (8 levels, $R = 1.000$, y% intercept = 1.24%)
  - o Precursor: 0.05–2% (5 levels, $R = 1.000$)
- Accuracy (recovery of spiked, triplicate)
  - o API: 101.86% (70%), 100.36% (100%), 99.71% (130%)
  - o Precursor: 104.5% (1%), 105.2% (0.4%), 109.8% (LOQ)
- Precision RSD ($n = 6$)
  - o 0.55% (API at 100%) and 4.5% (precursor at LOQ)
- LOQ = 0.05%, LOD = 0.02%
- Solution stability
  - o Standard solution stability: >10 days (refrigerated)
  - o Mobile phase stable for >14 days
- Robustness
  - o Demonstrated toward perturbations ($T$, $\lambda$, $F$, $t_G$, %TFA) and found to be robust to these variations
  - o Three different lots of columns showing no significant resolution difference.

pure. Note that PDA peak purity evaluation is not a particularly effective technique for peak purity assessment since impurities of the API often have similar UV spectra, though the data is easily obtainable from PDA and automated peak purity algorithms in chromatography data system (CDS). The MS spectra of the up-slope, apex, and down-slope of the API peak showed that no new $m/z$ ion >1% level was observed.

Method linearity, accuracy, and precision studies were performed, and validation data are listed in Table 11.10 showing acceptable performance for the API and the precursor. A robustness study toward perturbations of five parameters ($T$, $\lambda$, $F$, $t_G$, %TFA [trifluoroacetic acid] in MPA) was designed using a DoE software (i.e. Design Expert). Eight experimental runs were performed, and the results were evaluated toward their effects on six system suitability responses (area precision, retention time precision, peak tailing, resolution of the API and impurity 1, and signal-to-noise ratio of the precursor peak [*Pre*]). Results showed that the method is robust to these variations. Note that robustness validation is typically only performed for late-stage methods.

Table 11.11 shows an example of method validation summary data of a late-stage stability-indicating ultra-high-pressure liquid chromatography (UHPLC) method for an antiviral oral drug product with four APIs for human immunodeficiency virus (HIV) indications described in Figure 5.6 and Section 5.3.2 [21].

## 11.5 METHOD TRANSFER

The need for method transfer may arise when drug substance and drug product manufacturing and testing is outsourced to a contract manufacturing organization (CMO) or a different laboratory within the same company [18]. If a method has not been validated and is not anticipated to be used by the originating lab, full method validation by the receiving lab is preferred. Method transfer is not required for compendial methods, although the laboratories performing the compendial test must verify that the method is suitable for use for the specific samples (see USP<1224>). A method transfer protocol is written by the originating laboratory and approved by QA at both the originating and receiving laboratories. After the method transfer laboratory work is completed, a method transfer report is written by the receiving unit and reviewed and approved by QC and QA at the receiving and originating companies. Table 11.12 gives an example of method transfer parameters and acceptance criteria for late development [22]. The method transfer requirements for early development are more flexible than those during late development. For example, it is acceptable to use fewer batches and samples, and wider ranges of acceptance criteria can be justified. Table 11.13 shows the acceptance criteria of a Method Transfer Protocol of an early-phase stability-indicating method for new drug substance.

## 11.6 REGULATORY FILINGS

Pharmaceutical products cannot be marketed without the approval of the New Drug Application (NDA) in the United States or Marketing Authorization Application (MAA) in Europe, and clinical trials cannot proceed without the filing of the Investigational New Drug Application (IND) or Clinical Trial Application (CTA) to the appropriate health authority [1]. Data generated by analytical chemists contribute to a large portion of the Chemistry, Manufacturing, and Control (CMC) sections of IND/CTA and NDA/MAA filings.

The FDA has published many guidance documents for the preparation of INDs [11]. Certain CMC sections often receive scrutiny from health authorities, such as specifications, control of genotoxic impurities, stability data, and shelf-life. Failure to satisfy regulatory inquiries can result in a hold on clinical trials. The analytical or QC chemist should work closely with the technical team and Regulatory Affairs to establish a well-defined QC strategy, including well-justified specifications and reliable methods that are appropriate for the phase of development. Before any changes in specifications and methods are made, the impact on regulatory filings should be assessed.

## 11.7 COST-EFFECTIVE REGULATORY COMPLIANCE STRATEGIES

No arguments can be made against the need for strict regulatory compliance for production facilities for public consumption. However, the interpretation of the regulations is often

**Table 11.11. Summary of Method Validation Data of a Late-stage Drug Product UHPLC Method**

| Validation element | Acceptance criteria | Results | | | |
|---|---|---|---|---|---|
| | | API 1 | API 2 | API 3 | API 4 |
| **Accuracy** | | | | | |
| *Assay levels* | 98.0–102.0% | 101.3% | 99.9% | 100.4% | 100.1% |
| *Low level* | 70–130% | 86% | 100% | 105% | 108% |
| **Precision** | | | | | |
| *Repeatability at assay levels* | RSD ≤ 2.0% | 0.5% | 0.3% | 0.2% | 0.2% |
| *Repeatability at low level* | RSD ≤ 10% | 0% | 5% | 4% | 2% |
| *Intermediate precision* | ≤3% difference | ≤1.6% | ≤2.5% | ≤2.0% | ≤2.0% |
| **Detection limit (DL)** | ≤0.05% | 0.03% | 0.03% | 0.03% | 0.03% |
| **Quantitation limit (QL)** | ≤0.15% (RSD ≤ 10%) | 0.05% (4%) | 0.05% (7%) | 0.05% (4%) | 0.05% (2%) |
| **Linearity** | $r \geq 0.999$ linear from QL – 120% | $r = 1.000$ (0.05–151%) | $r = 1.000$ (0.05–152%) | $r = 1.000$ (0.05–152%) | $r = 1.000$ (0.05–151%) |
| **Range** | QL – 120% | 0.05–151% | 0.05–152% | 0.05–152% | 0.05–151% |
| **Robustness** | | | | | |
| *Chromatographic conditions* | Passes system suitability criteria | Passed system suitability criteria | | | |
| *Adsorption on sample solution filters* | Filtrate is within 98.5–101.5% of unfiltered std | 99.0–99.8% | 99.4–100.2% | 99.5–100.2% | 99.5–100.2% |
| *Sample preparation* | ≤2% difference | ≤0.6% | ≤0.7% | ≤1.6% | ≤1.5% |
| **Specificity** | No significant interference from sample filters and excipients | No significant interference from sample filters and excipients | | | |
| | Resolution ≥ 1.0 | Resolution ≥ 1.2 | | | |
| | No significant interfering components | No significant interference | | | |
| **Solution stability** | | **Ambient** | | **2-8 °C** | |
| *Standard* | Report established expiry | 3 days | | 14 days | |
| *Stock and working sample* | | 3 days | | 14 days | |
| *Sensitivity standard* | | — | | 22 days | |
| *System suitability standard* | | — | | 1 month | |

Source: Data acapted with permission from Ref. [21].

**Table 11.12. Typical Method Transfer Parameters and Acceptance Criteria**

| Type of method | Number of analysts | # of Lots or units | Acceptance criteria | Notes |
|---|---|---|---|---|
| Assay | 2 | 3 lots in triplicate | A two one-sided *t*-test with intersite differences of ≤2% at 95% confidence level (CL) | Each analyst should use different instrumentation and columns and independently prepare all solutions. All applicable system suitability criteria must be met. |
| Content uniformity | 2 | 1 lot | A direct comparison of the means, ±3% and variability of the results, (%RSD), i.e. a two one-sided *t*-test with intersite differences of 3% at 95% CL. | If the method for content uniformity is equivalent to the assay method, then a separate amount is not required. |
| Impurities, degradation products | 2 | 3 lots of duplicate (triplicate if done together with the assay) | For high levels, a two one-sided *t*-test with intersite differences of 10% at 95% CL. For low levels, criteria are based on the absolute differences of the means, ±25%. | Notes for assay apply here. The LOQ should be confirmed in the receiving lab, and chromatograms should be compared to the impurity profile. All samples should be similar with respect to age, homogeneity, packaging, and storage. If samples do not contain impurities above the reporting limit, then spiked samples are recommended. |
| Dissolution | N/A | 6 units for immediate release, 12 units for extended | Meet dissolution specs in both labs, and the two profiles should be comparable or based on the absolute difference of the means, ±5% | A statistical comparison of the profiles (e.g. F2) or the data at the Q time point (s) similar to that for the assay may be performed. |
| ID | N/A | 1 unit | Chromatography: confirm retention time. Spectral identification and chemical testing can also be used, assuming operators are sufficiently trained and the instrumentation can provide equivalent results. | |
| Cleaning validation | N/A | 2 spiked samples, one above, one below spec | Spiked levels should not deviate from the spec by an amount 3× the validated SD of the method, or 10% of the spec, whichever is greater. | Essentially a limit test. Low and high samples to confirm both positive and negative outcomes are required. |

Source: Table adapted from Ref. [22].

**Table 11.13. Acceptance Criteria in an Early-phase Method Transfer Protocol**

| Parameters | Acceptance criteria |
|---|---|
| System suitability | Meets method requirements |
| Assay (% w/w) | Average API assay determined by the recipient and originating laboratory is within ±2% (w/w) for each sample on a solvent-free, anhydrous free base basis.<br>Method precision from four replicate preparations has RSD ≤2.0% |
| Impurity profile (% area) | Average individual unspecified and specified impurity content above LOQ determined by the recipient and originating laboratory is as follows for each sample:<br>• Within ±50% (relative) for impurities present at <0.1%<br>• Within ±40% (relative) for impurities present at ≥0.1% and <0.5%<br>• Within ±20% (relative) for impurities present at ≥0.5% |

subjective and inexact, and the amount of work and documentation required by many organizations has escalated significantly in recent years. Maintaining regulatory compliance has become costly, bureaucratic, and often excessive and inefficient. A case in point is in equipment qualification reports from one organization, which contains hundreds of pages of documentation with numerous signatories. While such time-consuming qualification may be warranted for equipment employing new technologies, they are less meaningful for an off-the-shelf instrument such as an HPLC system. A delicate balance of laboratory productivity and compliance may be achieved by some research facilities adopting a two-tier or risk-based approach, allowing less quality compliance requirements in early analytical development laboratories and full GMP compliance for QC and late-stage development.

Ultimately, regulatory flexibility in manufacturing processes would be achieved by a deeper understanding of the science behind the physicochemical properties of the new chemical entities (NCEs) and the material and process parameters by the judicious applications of Quality by Design (as described in ICH Q8).

### 11.7.1 Regulatory Compliance in Other Industries

In this chapter, the focus is on HPLC testing in small-molecule drug development or production facilities, which require compliance with stringent GMP regulations. QC assessment using HPLC for recombinant biologics are covered briefly in Chapter 12. The regulatory pathways for these more complex biologics do mirror those for small-molecule drugs although the regulatory bodies and marketing applications have different names in the United States (Center for Biologics Evaluation and Research [CBER] vs. Center of Drug Evaluation and Research (CDER), Biologic License Applications (BLAs) vs. NDA). Some of the HPLC applications and regulations pertaining to food analysis and environmental testing are described in Chapter 13.

A generics drug company must file an Abbreviated NDA (ANDA) to demonstrate product bioequivalence to the original drug product to waive the need for clinical trials. Alternatively, a 505(b) NDA can be filed for a drug product with an approved API in a new formulation [23]. The dietary supplement manufacturers must follow GMP regulations, although there are no rules or requirements for clinical trials [24].

## 11.8 SUMMARY AND CONCLUSIONS

This chapter discusses the regulatory aspects of HPLC testing with a focus on procedures and regulatory requirements in the pharmaceutical industry for system qualification, calibration, method validation, method transfer, and SST. An example of a GMP pharmaceutical production facility is used to illustrate the various processes and systems to ensure regulatory compliance and the accuracy/integrity of HPLC data for critical assays. The roles of the US FDA, EMA, UP, ICH, and IQ consortium are briefly described. Examples of system qualification, calibration, SST, method validation, and transfer are included to illustrate these processes.

## 11.9 QUIZZES

**1.** Which regulation is considered the most important in the manufacturing of drug products?
(a) GLP
(b) ICH
(c) GMP
(d) USP

**2.** Which document would contain information most similar to an HPLC calibration procedure?
(a) OQ
(b) DQ
(c) IQ
(d) PQ

**3.** Which three acceptance criteria are typically required in SST?
(a) Robustness, precision, accuracy
(b) Linearity, accuracy, specificity
(c) Requirements from FDA, ICH and USP
(d) $R_s$, sensitivity, precision

**4.** Which method validation requirement is generally not performed officially in early-stage drug development?
(a) Linearity
(b) LOQ, LOD
(c) Robustness
(d) Precision

**5.** Peak purity by MS can typically detect an impurity coeluting with the API down to this level.
(a) 1%
(b) 0.05%
(c) 5%
(d) 10%

**6.** According to ICH guidelines, a drug product method for stability studies must detect a degradation product down to 0.1% level. Which S/N ratio is related to LOQ?
(a) 10%
(b) 3

(c) 10
(d) 100

7. The GLP regulations can be found in which document?
   (a) 21 CFR 210
   (b) 21 CFR 58
   (c) 21 CFR 10
   (d) ICH Q2 (R1)

8. Determination of LOD is required for this type of test.
   (a) Potency assay
   (b) Dissolution testing
   (c) Performance testing
   (d) Limit test

9. An accuracy evaluation for a potency test is typically conducted at which three levels?
   (a) LOQ, specification, 20% above spec
   (b) 10%, 50%, 100%
   (c) 50%, 100%, 150%
   (d) 80%, 100%, 120%

10. Which regulatory guidelines are most appropriate for early drug development of a small-molecule NCE?
   (a) GLP
   (b) ICH
   (c) IQ Consortium
   (d) USP

### 11.9.1 Bonus Quiz

1. Explain in your own words the difference between GLP and GMP regulations.
2. Explain in your own words the difference between USP and ICH.

## 11.10 REFERENCES


1. Hill, R.G. and Rang, H.P. (eds.) (2012). *Drug Discovery and Development: Technology in Transition*, 2e. Edinburgh, Scotland: Churchill Livingston, Elsevier.
2. Miller, J.M. and Crowther, J.B. (2000). *Analytical Chemistry in a GMP Environment: A Practical Guide*. New York: Jossey-Bass.
3. Sarker, D.K. (2008). *Quality Systems and Controls for Pharmaceuticals*. Hoboken, NJ: Wiley.
4. (2016). *Code of Federal Regulations, Title 21, Parts 210 and 211*, Current Good Manufacturing Practice for Finished Pharmaceuticals. Washington, D.C.: Government Publishing Office.
5. (2011). *Code of Federal Regulations, Title 21, Parts 58*, Good Laboratory Practice for Nonclinical Laboratory Studies. Washington, D.C.: Government Publishing Office.
6. (2006). *FDA Guidance for Industry: Quality Systems Approach to Pharmaceutical CGMP Regulations*. Rockville, MD: U. S. Food and Drug Administration.
7. Huber, L. (2002). *Good Laboratory Practice and Current Good Manufacturing Practice, for Analytical Laboratories*. Walbronn, Germany: Agilent Technologies.
8. Reuter, W.M. (2005). *Handbook of Pharmaceutical Analysis by HPLC* (ed. S. Ahuja and M.W. Dong). Amsterdam, the Netherlands: Elsevier. Chapter 12.

9. International Conference on Harmonization (ICH) Q2 (R1), *Validation of Analytical Procedures: Methodology*, Geneva, Switzerland (November 1996), updated 2015. Published in the Federal Register, vol. 62, no. 96, May 19, 1997, pp. 27463–27467.
10. (2013). *Validation of Compendial Methods*. Rockville, MD: United States Pharmacopeial Convention *USP 37/NF 32, General Chapter <1225>*.
11. Kou, D., Wigman, L., Yehl, P., and Dong, M.W. (2015). *LCGC North Am.* 33 (12): 900–909.
12. McDowall, R. (2017). *Validation of Chromatography Data Systems: Ensuring Data Integrity, Meeting Business and Regulatory Requirements*, 2e. Cambridge, United Kingdom: Royal Society of Chemistry.
13. (2003). *Code of Federal Regulations, Title 21, Parts 11*, Electronic Records; Electronic Signatures. Washington, D.C.: Government Publishing Office.
14. Dong, M.W. (2005). *Handbook of Pharmaceutical Analysis by HPLC* (ed. S. Ahuja and M.W. Dong). Amsterdam, the Netherlands: Elsevier. Chapter 11.
15. Dong, M.W., Paul, R., and Gershanov, L. (2001). *Today's Chemist at Work* 10 (9): 38.
16. Hsu, H. and Chien, C.S. (1994). *J. Food Drug Anal.* 2 (3): 161.
17. Lister, A. (2005). *Handbook of Pharmaceutical Analysis by HPLC* (ed. S. Ahuja and M.W. Dong). Amsterdam, the Netherlands: Elsevier. Chapter 7.
18. Kazakevich, Y.V. and LoBrutto, R. (eds.) (2007). *HPLC for Pharmaceutical Scientists*. Hoboken, NJ: Wiley. Chapters 9 and 16.
19. (2015). *FDA guidance for Industry, Analytical Procedures and Method Validation for Drugs and Biologics*. Rockville, MD: U. S. FDA.
20. Rasmussen, H.T., Li, W., Redlich, D., and Jimidar, M.I. (2005). *Handbook of Pharmaceutical Analysis by HPLC* (ed. S. Ahuja and M.W. Dong). Amsterdam, the Netherlands: Elsevier. Chapter 6.
21. Dong, M.W., Guillarme, D., Fekete, S. et al. (2014). *LC GC North Am.* 32 (11): 868.
22. Krull, I. and Swartz, M.E. (2006). *LCGC North Am.* 24 (11): 1204.
23. (2017). *Draft Guidance for Industry: Determining Whether to Submit an ANDA or a 505(b) 2*. Silver Spring, MD: U.S. FDA, CDER.
24. (2010). *Guidance for Industry: Current Good Manufacturing Practice in Manufacturing, Packaging, Labeling, or Holding Operations for Dietary Supplements; Small Entity Compliance Guide*. College Park, MD: U. S. FDA.

# 12

# HPLC AND UHPLC FOR BIOPHARMACEUTICAL ANALYSIS

**JENNIFER REA and TAYLOR ZHANG**

## 12.1 INTRODUCTION

Modern biological drug development stems mostly from using recombinant DNA technology in living microorganisms such as *Escherichia coli* or Chinese hamster ovary (CHO) cells to produce therapeutic agents [1]. Coincidentally, modern high-performance liquid chromatography (HPLC) emerged around the same time as these biotechnological advancements, and each field promotes and compliments the development of the other. The complexity of biological products calls for the development of sophisticated analytical tools, including improvements in HPLC and mass spectrometry (MS), which noticeably contributes to the understanding of biologics and facilitates their development.

Biologics differ in many ways from synthesized small-molecule pharmaceuticals. Biologics can be comprised of proteins, peptides, or nucleic acids and include vaccines, blood components, living cells, or extracts of living organisms. Biologics are typically made through cell culture fermentation and bioprocessing, which result in products that are complex and heterogeneous due to the molecule size and nature of cell-based manufacturing processes. On the other hand, small molecules are produced using well-defined and synthetic chemical processes and are highly purified and homogeneous. The molecular weight of biologics can be in the tens or hundreds of kDa, while the molecular weight of small-molecule drugs is typically less than 1 kDa.

Among the biologics, recombinant monoclonal antibodies (mAbs) are now the standard therapy for oncology and immunology indications. In 2016, 5 out the top 10 best-selling drug products (based on revenue) were mAbs [2]. mAbs are commonly developed as therapeutics due to their specificity to a chosen biological target. Antibody–drug conjugates (ADCs)

*HPLC and UHPLC for Practicing Scientists*, Second Edition. Michael W. Dong.

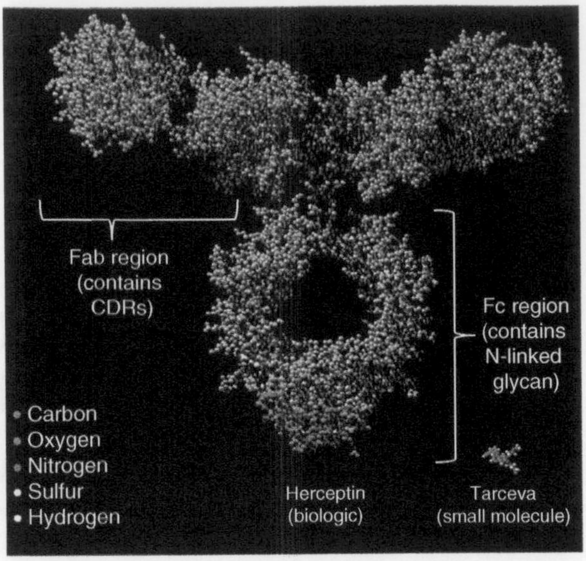


**Figure 12.1.** Structure of a monoclonal antibody therapeutic (Herceptin®) compared to a small-molecule therapeutic (Tarceva). Fab region and Fc region of the monoclonal antibody (mAb) are denoted. CDRs refer to complementary determining regions. Source: Reproduced with permission of Rea et al. 2012 [3].

leverage this specificity by having small-molecule drugs attached to the mAbs, such that when the mAb binds to a specific target cell, the entire complex is internalized and subsequently releases its small-molecule drug payload into the target cells. This approach reduces the chance of off-target side effects, since the potent small molecule drug is only released into target cells and not systemically. The maturity of the mAb market has allowed for the development of nonstandard mAb formats (e.g. fusion proteins, bispecifics, etc.) and biosimilars, which are molecules that are developed to perform similarly to the innovator mAb and are intended to be marketed after the innovator mAb's patent protection expires.

Antibody therapeutics are complex biomolecules [3] (Figure 12.1) and possess critical quality attributes (CQAs), which can be analyzed by different modalities of chromatographic and electrophoretic separations. The introduction of ultra-high-pressure liquid chromatography (UHPLC) in the last decade has shifted the paradigm from HPLC- to UHPLC-based analysis, which offers higher resolution and throughput. UHPLC has been adopted into quality control (QC) laboratories, and characterization procedures of these complex biologic molecules often employ UHPLC in conjunction with UV or MS detection.

Product quality consistency of mAbs is ensured when each drug batch conforms to product specifications determined by the manufacturer [4]. Table 12.1 shows an example of drug substance specifications for an antibody product. Note that chromatographic methods are routinely employed for product quality testing, and the testing results are reported on a Certificate of Analysis (COA) for each batch. In addition to lot release testing, chromatographic methods are sometimes employed during the manufacturing process to ensure consistent manufacturing at various unit operations. In regulatory filings submitted to health authorities and regulators around the world, such as the U.S. FDA, chromatographic methods are used in conjunction with MS to further characterize the product beyond what is tested in-process during manufacturing and for lot release and stability testing [5].

After several decades of thorough studies and detailed characterizations, mAbs still challenge analytical chemists and biochemists to make greater efforts in elucidating their structures and functions [6]. Here we summarize the different modes of chromatographic

**Table 12.1. Mock Specifications and Batch Release Results of an Antibody Product**

| Test | Purpose | Specification | Batch #1 | Batch #2 | Batch #3 |
|---|---|---|---|---|---|
| Protein concentration (mg/ml) | Quantity | 45–55 | 50 | 49 | 50 |
| Percent purity by SEC | Purity (size) | ≥98% | 99 | 99 | 99 |
| Percent purity by IEC | Purity (charge) | ≥95% | 97 | 98 | 97 |
| Percent purity by CE-SDS | Purity (size) | ≥95% | 98 | 99 | 98 |
| Peptide mapping | Identity | Positive identity | Positive identity | Positive identity | Positive identity |
| Cell-based assay | Potency | 80–120% | 101 | 95 | 98 |
| Host cell proteins (ppm) | Impurities | ≤30 | 1 | 1 | 1 |
| Endotoxin (EU/ml) | Impurities | ≤3 | 1 | 1 | 1 |
| Bioburden (CFU/ml) | Impurities | ≤10 CFU/10 ml | 0 | 0 | 0 |
| Osmolality (mOsm/kg) | General | 300–400 | 367 | 380 | 340 |
| pH | General | 6.0–7.0 | 6.5 | 6.4 | 6.6 |
| Appearance | General | Colorless, slightly opaque, free from particulates | Colorless, slightly opaque, free from particulates | Coloreless, slightly opaque, free from particulates | Colorless, slightly opaque, free from particulates |

separations of protein biologics, especially for mAbs, and include general method conditions for practicing scientists and analysts. Readers are referred to books, journal articles, and Section 13.6.1 for additional details on protein separations [7–14].

## 12.2 SIZE-EXCLUSION CHROMATOGRAPHY (SEC)

### 12.2.1 SEC Introduction

Size-exclusion chromatography (SEC) is a commonly used analytical technique for protein therapeutics, in part because it can resolve high-molecular-weight forms (HMW) of proteins. Size variants are product-specific impurities that are monitored from the earliest stages of clinical development, as aggregates may pose a safety risk to the patient with immunogenic responses. Aggregates and low-molecular-weight forms (LMW) may also affect the potency and pharmacokinetics (PKs) of the drug product. HMW are generally dimers and larger non-covalently aggregated forms; these aggregates can be formed under a variety of conditions including increased temperature and pressure, varied solvent conditions such as pH, and agitational stress [15]. LMW are generally protein fragments that can be formed under a variety of conditions including thermal stress and oxidative stress [16–18]. Size variants are CQAs and are monitored in drug substance and drug product per regulatory requirements [19] with SEC being the primary method used in QC to measure HMW for lot release and stability testing while LMW is often quantitated by an orthogonal technique such as denaturing capillary electrophoresis–sodium dodecyl sulfate (CE-SDS) [20]. Another technique that is orthogonal to SEC is analytical ultracentrifugation (AUC), which can serve to verify the accuracy of an SEC method [21]. Both SEC and AUC are performed under native conditions, which have advantages over denaturing conditions in that the aggregates are kept in their native conformation. Therefore, SEC can also be used to measure the amount of dissociable aggregates; by injecting the sample neat and comparing against a diluted sample, one can measure the difference in HMW levels and attribute the difference to aggregates that can dissociate during sample dilution.

### 12.2.2 SEC Theory and Fundamentals

SEC is typically used to separate HMW from the main peak, which primarily contains the monomeric protein species. This separation is achieved due to the differences in hydrodynamic radii of the size variants and monomer. SEC columns are generally packed with porous particles with hydrophilic diol ligands that serve as the stationary phase. The pores of the stationary phase particles restrict access of molecules based on their Stokes radius. Larger proteins are excluded from the pores, thus eluting first (see Section 1.2.4). Subsequent proteins elute in order of decreasing size. Varying pore size may modulate the resolution of HMW and LMW, as separation curves shift with different SEC pore sizes (Figure 12.2) [22]. In addition to pore size, the resolution of an SEC method can be modulated by changing various chromatographic conditions, including particle size, mobile phase composition, flow rate, column ID and length, injection volume, and protein load. It should be noted that secondary interactions, such as hydrophobic and charge interactions, between the SEC stationary phase and protein analytes may be observed. These secondary interactions sometimes result in increased retention time and peak tailing, which may be mitigated by mobile phase additives. Increased ionic strength may reduce charge interactions, and the addition of organic modifiers may reduce hydrophobic interactions between the protein analytes and the stationary phase [23].

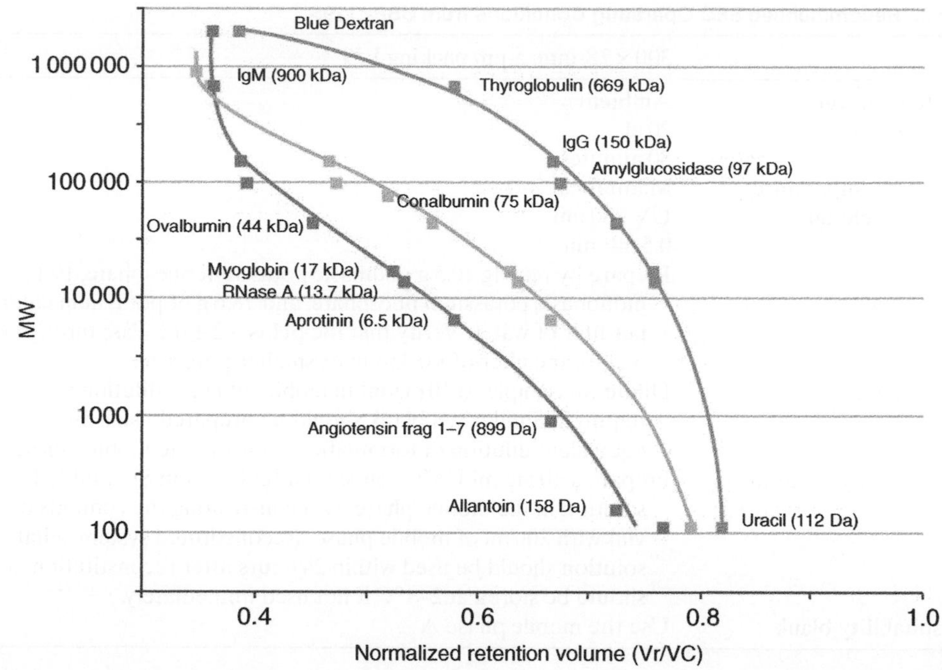


**Figure 12.2.** Separation curves by SEC. Varying column resin pore size results in different separation curves, with larger pore sizes having a shallower retention curve at higher molecular weights. Leftmost curve = 125 Å pore size, 1.7 μm particle size. Middle curve = 200 Å pore size, 1.7 μm particle size. Rightmost curve = 450 Å pore size, 2.5 μm particle size. Source: Reproduced with permission of Waters Corporation [22].

Traditionally, SEC was performed on HPLC instruments. However, advances in instrumentation and the development of SEC resins with smaller particles (i.e. ≤2 μm) enable the increasing use of high-performance size-exclusion chromatography (HP-SEC) with UHPLC columns and systems (see Figure 13.29). HP-SEC separation was initially regarded as challenging, as the appropriate pore volume for size-based separations of proteins may compromise the physical strength of porous particles at higher pressures. More recently, SEC packings with sufficient mechanical strengths to withstand high backpressures and higher flow rates have been introduced for UHPLC separations. HP-SEC methods have been shown to achieve better resolution with shorter run times than conventional SEC methods. The effect of extra-column dispersion can be problematic for isocratic SEC separations; thus, low-dispersion UHPLC systems are necessary to achieve the full benefits of the resolution and throughput gains by HP-SEC columns. Sample preparation should also be carefully considered for SEC applications; certain diluents, as well as vortexing or excess agitation of the sample, can cause protein aggregation [24]. In these cases, the sample may be diluted in its formulation buffer and gently mixed before injection onto an SEC column to improve sample stability and avoid artificially introducing aggregates into the sample.

### 12.2.3 SEC Method Conditions

The USP has recently published General Chapter USP<129>, "Analytical Procedures for Recombinant Therapeutic Monoclonal Antibodies," which contains the conditions for HP-SEC analysis of mAbs. The HP-SEC conditions in USP<129> are outlined in Table 12.2. It should be noted that USP<129> also contains conditions for other methods,

**Table 12.2. Recommended SEC Operating Conditions from USP<129>**

| Column | 300 × 7.8-mm; 5-µm packing L59 |
|---|---|
| Column temperature | Ambient |
| Injection volume | 20 µl |
| Runtime | 30 minutes |
| Autosampler temperature | Maintain at 2–8 °C |
| Detection wavelength | UV 280 nm |
| Flow rate | 0.5 ml/min |
| Mobile phase composition | Prepare by mixing 10.5 g of dibasic potassium phosphate, 19.1 g of monobasic potassium phosphate, and 18.6 g of potassium chloride per liter of water. Verify that the pH is 6.2 ± 0.1. Pass through a membrane filter of ≤0.45-µm or smaller pore size. |
| Sample solution | Dilute the sample to 10 mg/ml in mobile phase if dilution is required. Similarly, a blank should be prepared using an equivalent dilution of formulation buffer in the mobile phase. |
| System suitability solution | Prepare a 10 mg/ml USP Monoclonal IgG System Suitability RS solution in the mobile phase by reconstituting the contents of one vial with 200 ml of mobile phase. Reconstituted system suitability solution should be used within 24 hours after reconstitution and should be stored at 2–8 °C if not used immediately. |
| System suitability blank | Use the mobile phase A |

including CE-SDS, but does not contain conditions for UHPLC analysis. Accompanying the USP general chapter is a recently developed IgG (immunoglobulin G) system suitability reference standard (RS). The USP RS is available for purchase through the USP and is used in the physicochemical assays described in General Chapter <129> for system suitability assessments. For SEC, the USP<129> monoclonal IgG system suitability RS solution must meet the following system suitability requirements: (i) the RS injections must bracket the injections of sample solutions, and the RS chromatograms must be consistent with the typical chromatogram provided in the USP certificate, and (ii) the area percent of the HMW species, the main peak, and the LMW species of the RS must meet the acceptance criteria published in USP<129>. Note that NIST (National Institute of Standards and Technology) also released a mAb standard available for purchase, which can be used as a product-independent reference standard.

SEC methods can vary beyond what is published in USP<129>. For example, MS can be used in conjunction with SEC, provided that the SEC mobile phase is MS-friendly, that is, contains volatile buffer components. SEC mobile phase modifiers can be used to improve resolution; for example, an organic solvent can be added to SEC mobile phase to mitigate unwanted hydrophobic interactions between the analyte and the stationary phase [25]. In addition, alternative wavelengths such as 214 nm can boost signal-to-noise (S/N) as opposed to 280 nm [26].

Column manufacturers, such as Waters Corporation and Tosoh Bioscience, have released UHPLC columns for commercial use. Each of those manufacturers has released a variety of application notes that an analyst can use when planning and executing experiments. SEC methods that utilize UHPLC tend to have shorter run time compared to HPLC methods, often with improved peak resolution. Operating conditions for HP-SEC are similar to the SEC conditions in Table 12.1, with lower flow rates to accommodate the increased backpressure that results from the smaller particle sizes in UHPLC columns. Also, protein load and injection volume may be reduced for UHPLC methods compared to HPLC methods, due to the decreased column diameters typically found in UHPLC columns.

### 12.2.4 SEC Applications

Size variants are a key concern of health authorities in the development and manufacturing of protein therapeutics. SEC is often used for routine analysis during batch release and stability testing to monitor these size variants. Figure 12.3 and Figure 12.4 show typical chromatograms of a mAb analyzed by SEC or HP-SEC, respectively. Note that HP-SEC (Figure 12.4) demonstrates improved resolution and shorter run time when compared to traditional SEC (Figure 12.3). In particular, the Fab-Fc form (also called "Fab/c") elutes after the main peak shown in Figure 12.4 (HP-SEC) but is not resolved in Figure 12.3. In addition to routine testing, SEC is often used to demonstrate comparability during manufacturing site transfers or process changes throughout clinical development. SEC can be used to characterize size variants; for example, fractions can be directly collected after the SEC column and tested to determine their masses and their potency. SEC can also be coupled to a light scattering detector or a mass spectrometer to obtain an online approximation of size variant masses; these techniques are referred to as SEC-MALS [27] (size-exclusion chromatography–multi-angle light scattering) or SEC-MS (size-exclusion chromatography–mass spectrometry) [28]. Characterization results by SEC are typically reported in filings to health authorities and regulatory agencies.

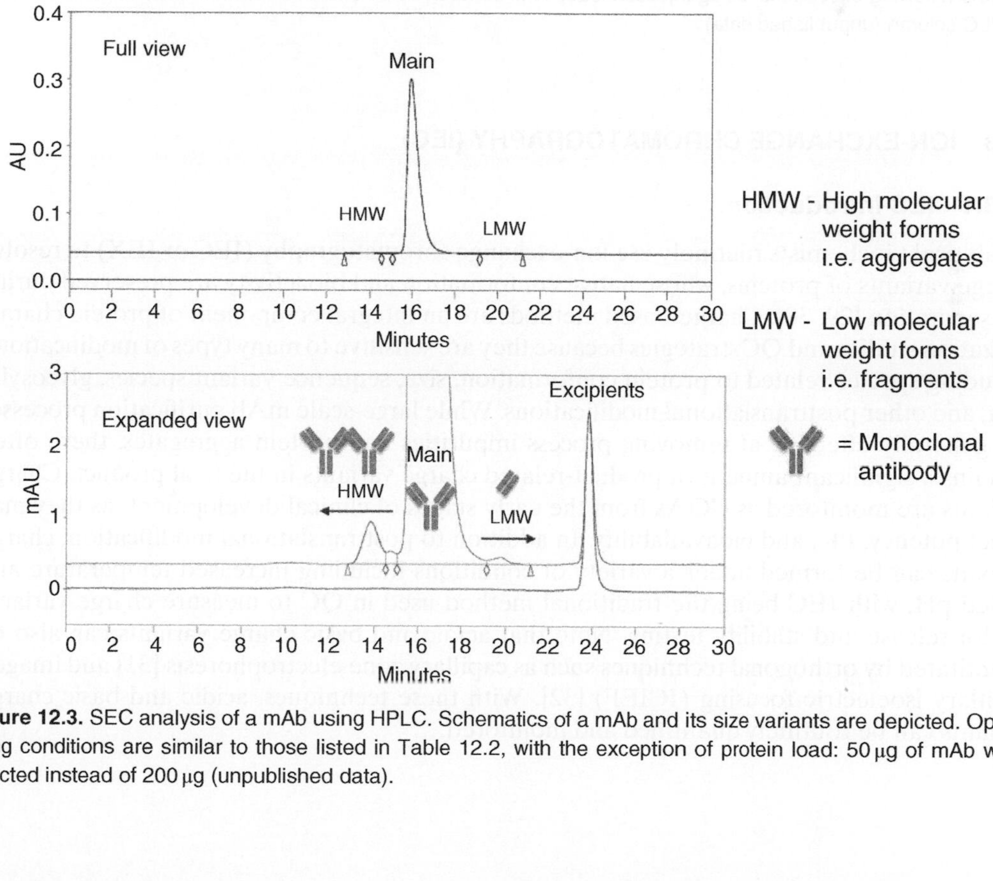


**Figure 12.3.** SEC analysis of a mAb using HPLC. Schematics of a mAb and its size variants are depicted. Operating conditions are similar to those listed in Table 12.2, with the exception of protein load: 50 μg of mAb was injected instead of 200 μg (unpublished data).

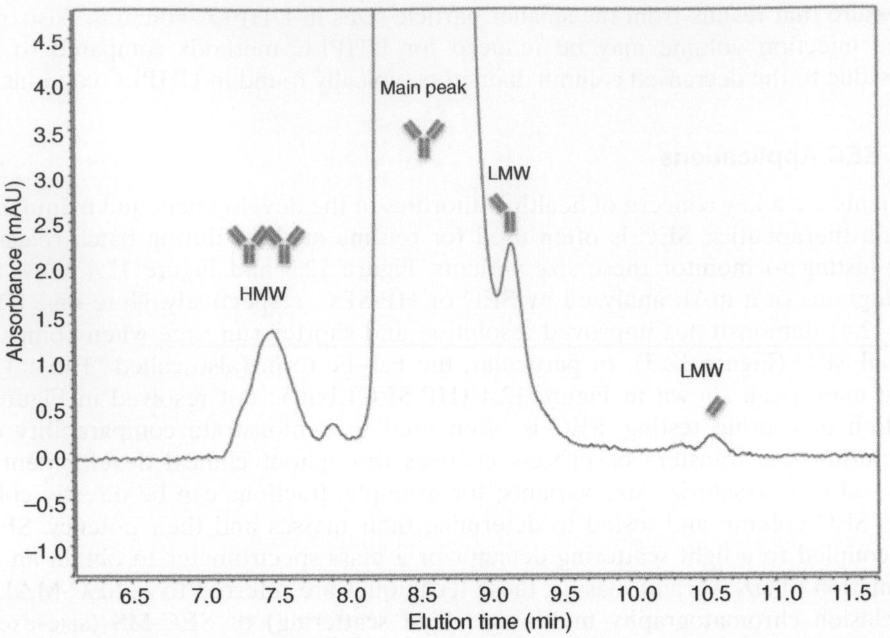


**Figure 12.4.** High-performance size-exclusion chromatography (HP-SEC) analysis of a mAb using UHPLC. Schematics of a mAb and its size variants are depicted. Operating conditions are similar to those in Table 12.2, with the following exceptions: 50 μg injection load, 10 μl injection volume, 0.3 ml/min flow rate, commercial SEC UHPLC column (unpublished data).

## 12.3 ION-EXCHANGE CHROMATOGRAPHY (IEC)

### 12.3.1 IEC Introduction

Analytical biochemists routinely use ion-exchange chromatography (IEC or IEX) to resolve charge variants of proteins, whose native conformation and bioactivity are preserved during the separation [29, 30]. Charge-based methods are an integral component of protein characterization studies and QC strategies because they are sensitive to many types of modifications, including species related to protein conformation, size, sequence variant species, glycosylation, and other posttranslational modifications. While large-scale mAb purification processes are generally effective at removing process impurities and protein aggregates, there often remains a significant amount of product-related charge variants in the final product. Charge variants are monitored as CQAs from the early stages of clinical development, as they may affect potency, PK, and bioavailability. In addition to posttranslational modification, charge variants can be formed under a variety of conditions including increased temperature and varied pH, with IEC being the traditional method used in QC to measure charge variants for lot release and stability testing. Note that acidic and basic charge variants can also be quantitated by orthogonal techniques such as capillary zone electrophoresis [31] and imaged capillary isoelectric focusing (ICIEF) [32]. With these techniques, acidic and basic charge variants can be routinely quantified and monitored.

### 12.3.2 IEC Theory and Fundamentals

IEC separates proteins based on differences in the surface charge of the molecules, with separation being dictated by the protein interaction with the stationary phase (see Section 1.2.3). While cation-exchange chromatography (CEX) has been the primary chromatographic mode for mAb charge variants analysis, anion-exchange chromatography (AEX) has been utilized for molecules with relatively low pI's. Proteins are eluted from the column using a salt gradient or pH gradient; the increasing salt in the mobile phase disrupts the charge interaction of the proteins with the stationary phase, whereas the pH gradient changes the charge of the protein, thus reducing the charge interactions with the column. Method parameters such as column type, mobile phase pH and salt concentration, and gradient slope may be optimized for each protein.

Charge variants that are separated from the main peak typically fall into two categories: acidic and basic species. Acidic species are defined as the charge variants that elute earlier than the main peak during CEX analysis or after the main peak during AEX analysis. The formation of acidic species can be caused by protein modifications including deamidation of Asn residues and glycation of Lys residues. Basic species are defined as the charge variants that elute later than the main peak during CEX or earlier than the main peak during AEX analyses. Modifications that result in the generation of basic species include incomplete removal of C-terminal Lys, conversion of the N-terminal glutamine to pyroGlu, and proline amidation.

### 12.3.3 IEC Method Conditions

IEC is widely used for profiling the charge heterogeneity of proteins [33, 34]. There are many published IEC methods in the literature, but these methods are often product-specific and may not be suitable for other molecules. In addition to published salt- and pH-gradient IEC methods, there are commercially available multiproduct IEC kits that utilize a pH gradient, which comes with low and high pH mobile phases (for example, pH 5.6 and pH 10.2, respectively) and a column designed for protein separations. IEC kits have been shown to effectively separate charge variants in different mAbs whose pI's fall within the range of the pH gradient [35]. Another column manufacturer has published application notes using a quaternary HPLC system and method development software to streamline the method development process by altering salt and pH gradients using the solvents employed by the quaternary pump [36].

Regarding instrument setup for analytical IEC methods, it is recommended that HPLC systems with inert flow paths (see Section 4.1.2) be used to prevent metal leachates from the HPLC system that foul the IEC column [37]. If noninert HPLC systems are used for salt-gradient IEC, it is recommended to perform regular passivation, particularly for steel flow paths. Online pH monitoring of column effluent can also be a useful tool, particularly for pH-gradient IEC method development [38]. It should be noted that although IEC column particle sizes have decreased over the years, the large resolution gains are typically seen with UHPLC using other modalities (e.g. SEC, RPC) have not yet been fully realized with IEC. Therefore, IEC methods used in the biopharmaceutical industry typically still use HPLC columns rather than UHPLC columns.

Regarding sample preparation, note that because carboxypeptidase B (CpB) can specifically remove C-terminal basic amino acid residues, comparison of the chromatograms of

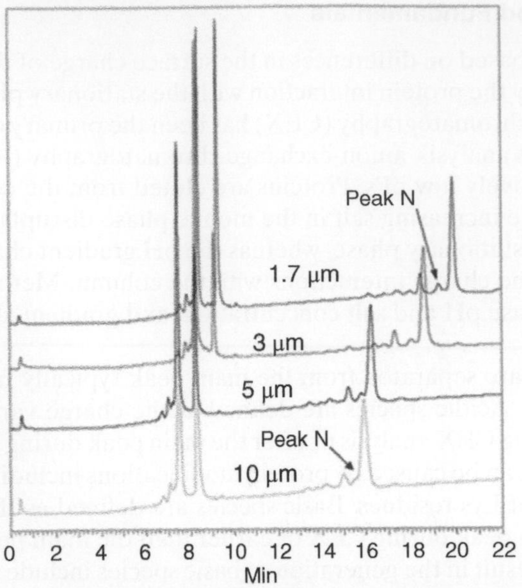


**Figure 12.5.** Overlaid IEC chromatograms of proteins using weak cation-exchange columns with particle size ranging from 1.7 to 10 μm. Column dimensions were 4.6 × 50 mm. The sample was a mixture of ribonuclease A (first major peak), cytochrome C (second major peak), and lysozyme (third major peak). Peak N is unidentified but is denoted to demonstrate improved resolution with decreasing particle size. Source: Reproduced with permission of Sepax Technologies.

antibodies before and after CpB digestion can clearly demonstrate the relative contribution of C-terminal Lys to the formation of basic species [39]. This step is sometimes added to the sample preparation protocol to simplify the chromatographic profiling. Finally, similarly to SEC, it has been shown that decreasing stationary phase particle size can improve resolution for IEC (Figure 12.5). Thus, column selection is another important factor in IEC method development.

### 12.3.4 IEC Applications

Protein charge variants can affect product quality. Thus, a charge variant method is often used for routine analysis during lot release, stability testing, and to demonstrate comparability of different lots. Unlike SEC, which is primarily used to measure aggregates, IEC can be used to characterize many types of protein modifications [40]. Fractions can be collected from IEC separations and further characterized to determine the identity and the potency of acidic and basic variants. These characterization results can be reported in filings to the health authorities during clinical development and for product licensure.

## 12.4 AFFINITY CHROMATOGRAPHY

### 12.4.1 Affinity Chromatography Introduction

Affinity chromatography is used both on a large preparative scale to purify therapeutic mAbs from cell culture fluid and on an analytical scale to determine antibody concentration, purify samples for further analysis, or for functional characterization. Affinity chromatography columns typically employ ligands immobilized on a solid support for the capture and subsequent elution of the retained antibodies. There are many commercially available affinity chromatography columns on the market, each with different ligands, immobilization chemistries, solid support materials, column hardware, and dimensions. Each type of affinity column may be targeted for specific applications, including purification of proteins during manufacturing, high-throughput small-scale preparation of protein samples in cell-culture fluid or biological matrices, and the characterization of binding attributes of the protein.

### 12.4.2 Affinity Chromatography Theory and Fundamentals

Affinity chromatography columns are packed with a solid support with immobilized ligands, such as Protein A, that bind to the protein of interest. Protein A binds to the Fc portion of antibodies between the CH2 and CH3 domains, permitting common use as a multiproduct or platform column for antibody purification and analysis [41]. Other affinity ligands are also available, such as Protein G and Protein L, which have different binding properties compared to Protein A.

To manufacture affinity columns, recombinant ligands are covalently attached to the hydrophilic surface of a polymeric resin, such as spherical agarose. The functionalized resin is then packed into the column body. The hydrophilic nature of the backbone minimizes nonspecific binding and therefore enables accurate quantification of the antibody concentration or titer. Since titer is often measured in harvested cell-culture fluid (HCCF) or other biological matrices, it is important that Protein A affinity columns are tolerant to a wide variety of operational conditions required for different samples. In addition, the nonporous particle produces a highly efficient column at high flow rates, enabling high-throughput applications with the ability to automate sample preparation.

To measure titer using a Protein A affinity column, samples are applied to a column equilibrated with mobile phase buffers at ~pH 7.5. To elute the target protein from the affinity column, a low pH mobile phase is utilized (~pH 3–4), which typically produces in a single peak in the chromatogram for titer analysis (Figure 12.6). A standard curve is prepared from standards injected onto the column, and the titer of the sample can be calculated based on the results of the standard curve. It should be noted that the standards are prepared from the same protein type as the sample to ensure calibration accuracy. This practice is particularly important for quantifying concentration in low titer samples, such as early cell-culture samples for bioprocess development and monitoring.

Affinity chromatography for functional characterization, such as Fc binding to predict pharmacokinetic properties, has also been developed to characterize protein therapeutics [42]. In this application, target ligands are covalently attached to chromatographic media (e.g. biotin–streptavidin conjugation chemistry). Proteins of interest are then applied to the column under various conditions to study the binding interactions of the protein and target ligand.

### 12.4.3 Affinity Chromatography Method Conditions

For bench-top protein purification applications, affinity chromatography can be performed using gravity flow in hand-packed columns or with bench-top centrifuges for faster flow rates. Bench-top protein purification kits that utilize various affinity columns can be purchased through a variety of vendors. For HPLC applications specifically in titer analysis, Protein A affinity columns for mAbs are commercially available. Typically, columns are kept at

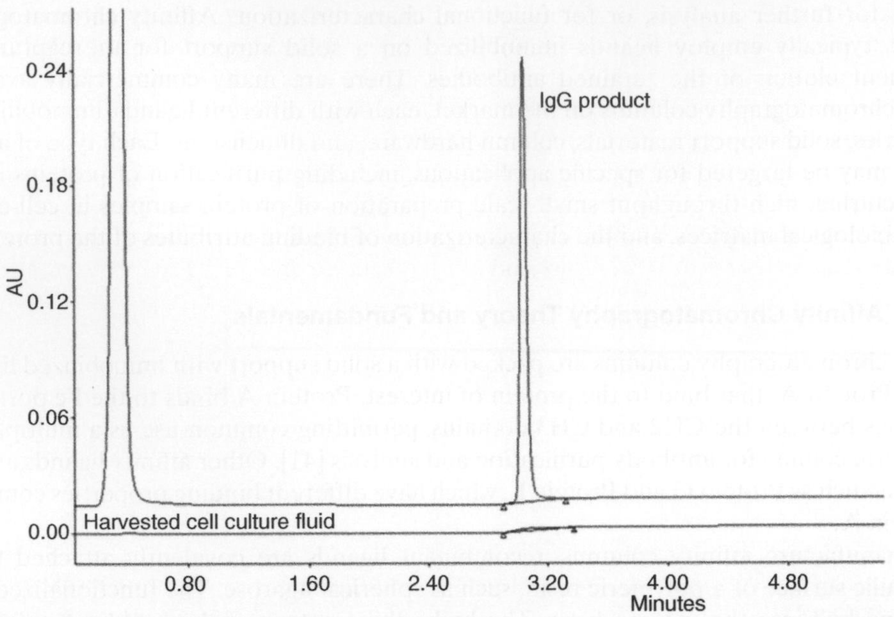


**Figure 12.6.** Affinity chromatography profile for titer determination of a mAb in a harvested cell culture fluid. Blank injection is shown below the product profile. Operating conditions are as follows: phosphate/sodium chloride mobile phase A, pH 7.4, acetic acid/glycine mobile phase B, pH 3.0, 25 μl injection volume, 5 ml/min flow rate, 4.6 × 100 mm Protein A affinity column (unpublished data).

temperatures ranging from ambient to 35 °C, equilibrated with a mobile phase buffer at ~pH 7.5, with analysis times as short as two minutes with relatively fast flow rates (e.g. 3–5 ml/min). A low pH mobile phase (~pH 3–4) is typically used to elute the target protein from the Protein A affinity column. If fractions are collected from an affinity chromatography separation, neutralization buffer should be added to the eluent immediately after elution to prevent protein degradation due to the low pH. Alternatively, a neutralizing solution can be added to the collection vessel or fractionation vials prior to fractionation, such that the eluent fractions are neutralized immediately upon collection. For affinity chromatography columns used for functional characterization, elution conditions depend on the nature of the protein–ligand interaction. In some cases, an increasing pH gradient, rather than a decreasing pH gradient, is employed to elute the protein from the affinity chromatography column.

### 12.4.4 Affinity Chromatography Applications

Protein A affinity chromatography has been used as the large-scale mAb capture step in the majority of health authority submissions. Protein A affinity purification removes the majority of host cell proteins (HCPs), which are present in the supernatant of mammalian cell culture. Analytical-scale affinity chromatography is heavily used for the measurement of protein concentration (i.e. titer) in HCCF, which determines yield and the appropriate amount of HCCF to load onto a purification-scale affinity column during manufacturing. Thus, for titer analysis, HPLC is commonly used both in development laboratories and at QC testing sites. In high-throughput laboratories, affinity chromatography can be used as a sample clean-up step prior to further chromatographic analysis. For example, Protein A resin can be packed into pipette tips or small prepacked minicolumns for automation applications, where robotic liquid handling systems can take up sample solutions and elute the mAbs from the Protein A tips by aspirating a low-pH solution and collecting the sample in 96-well plates. The purified samples are then pH-neutralized by a liquid handling robot or by adding neutralization buffer to the collection plate prior to elution [43].

Affinity chromatography for functional characterization of protein therapeutics is also routinely implemented [42]. In this application, proteins modified through amino acid residue changes or posttranslational modifications are applied to the affinity chromatography column to determine changes in retention time and binding characteristics. This approach has been used to determine which protein domains affect ligand binding and also correlate affinity to results obtained from *in vivo* pharmacokinetic studies.

## 12.5 HYDROPHILIC INTERACTION LIQUID CHROMATOGRAPHY (HILIC)

### 12.5.1 HILIC Introduction

Many protein-based biopharmaceuticals are glycosylated proteins. Protein glycosylation has implications for a variety of biological functions, including cell–cell signaling, protein stability, pharmacokinetics, half-life *in vivo*, and affinity to a target molecule. The glycan profile is particularly relevant for some therapeutic proteins because it can impact efficacy, pharmacokinetics, and safety of the product [44, 45]. Thus, glycan profiles of biopharmaceuticals are often monitored, both on the protein level and by quantitation of individual glycan structures. For glycan analysis, several approaches can be utilized, including analysis of intact glycoproteins by intact/reduced MS, glycopeptides after enzymatic digestion by LC/MS, or released glycans by CE or HPLC analysis. For HPLC glycan analysis, the glycans released by chemical or enzymatic methods often are labeled with UV- or fluorescence-active reagents, followed by HILIC analysis [46].

### 12.5.2 HILIC Theory and Fundamentals

Glycosylation, or the enzymatic addition of polysaccharides to proteins, is a common post-translational modification of therapeutic proteins and antibodies. Protein glycosylation can be either O-linked on serine and threonine residues or N-linked on asparagine residues, most commonly in the Fc region of the mAb [44]. Glycans vary in the number of galactose and sialic acids as well as in the presence or absence of fucose or bisecting *N*-acetylglucosamine. Often these different glycans influence the mechanisms of action of a therapeutic protein or antibody. Figure 12.7 shows common N-linked glycans that are found on IgG antibodies [47].

Mammalian cell lines are the most common antibody production cell lines because of their innate ability to produce near human and largely nonimmunogenic antibodies. Growth and culture conditions can affect the glycan profile of therapeutic IgG antibodies. Therefore, process factors that are important for correct and consistent glycosylation must be identified and controlled during manufacturing.

Often the enzyme PNGase F is used to release N-linked glycans. The released glycans are then labeled with UV- or fluorescence active reagents, followed by appropriate HPLC or UHPLC separation. The most common fluorescent labeling reagent used today is 2-aminobenzamide or 2-AB, which is rapidly becoming a standard method for glycoprofiling. UHPLC columns for glycan analysis are based on HILIC [48], in which the strongly retained polar analytes are eluted with an increasingly aqueous gradient.

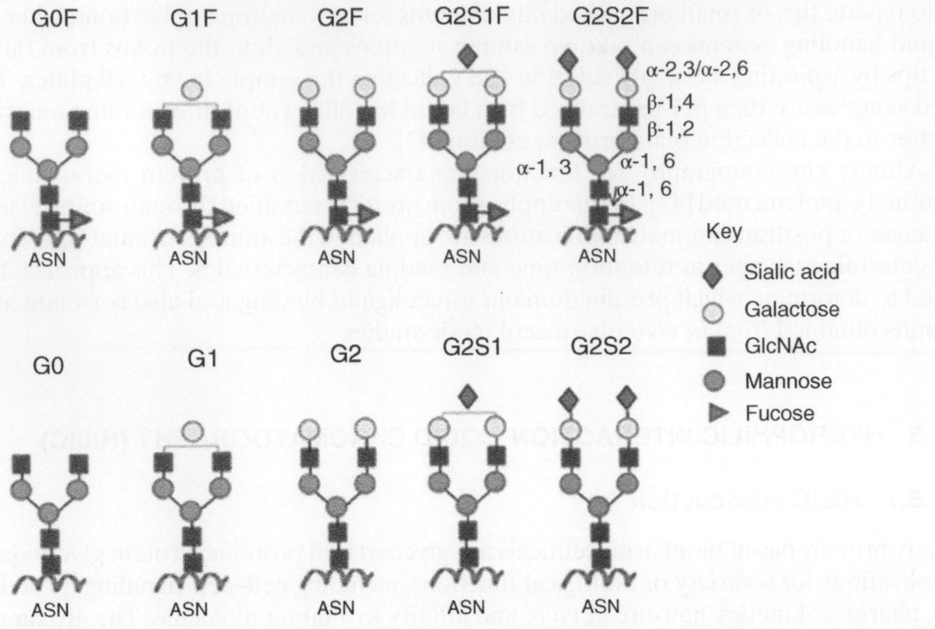


**Figure 12.7.** Major N-linked glycoforms of mAb therapeutics. Source: Liu 2015 [47]. Reprinted with permission of Elsevier.

### 12.5.3 HILIC Method Conditions

There are commercially available UHPLC columns for glycan analysis that are optimized for use with UHPLC systems with fluorescence detection, separating the released glycans of biopharmaceuticals as their 2-aminobenzamide derivatives, including high mannose, complex, hybrid, and sialylated glycans (Figure 12.8) [50]. Several manufacturers have released application notes that an analyst can use when planning and executing glycan experiments, which include the glycan release and labeling procedure, as well as the chromatographic method [49, 51]. HILIC glycan methods that utilize UHPLC tend to be shorter in runtime compared to HPLC methods, often with improved peak resolution.

For glycoprotein and glycopeptide analysis, commercial columns are available that are designed specifically for glycoprotein profiling and glycopeptide mapping, in which the intact glycoproteins and glycopeptides can be analyzed in addition to released glycans [51, 52]. Both HILIC and reversed-phase chromatography (RPC) can be used for glycoprotein and glycopeptide analysis and can be particularly informative when used in conjunction with mass spectrometry. The MS data can be interrogated to identify the different peaks in the glycoprotein and glycopeptide separations (Figure 12.9) [51]. Note that sialylated glycoforms can be well separated using ion-exchange chromatography instead of HILIC or RPC, due to the changing net charge accompanying different sialic acid occupancy [53].

### 12.5.4 HILIC Applications

Glycan analysis and identification can be completed in several ways. In the biopharmaceutical industry, orthogonal methods are often employed to evaluate glycan structure, as it is a CQA for many mAb therapeutics. First, glycosylation sites and levels of occupancy can be determined by peptide mapping and mass spectrometry of both native and deglycosylated

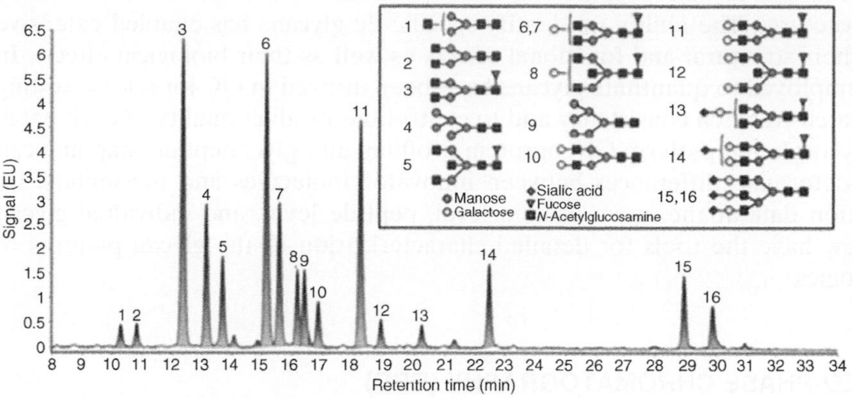


| Peak | 2-AB labeled glycan |
|---|---|
| 1 | G0-GN |
| 2 | G0 |
| 3 | G0F |
| 4 | Man5 |
| 5 | G0FN |
| 6 | G1F |
| 7 | G1F |
| 8 | G1FN |
| 9 | Man6 |
| 10 | G2 |
| 11 | G2F |
| 12 | G2FN |
| 13 | G1FS1 |
| 14 | G2FS1 |
| 15 | A3 |
| 16 | A3 |

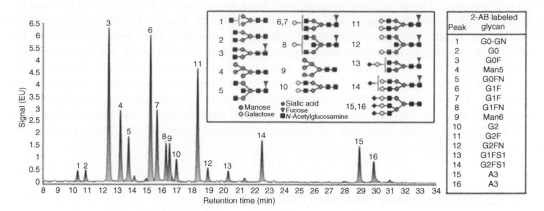

**Figure 12.8.** HILIC analysis of released glycans. 3 pmol of the sample in 2.5 μl was injected onto an ACQUITY UPLC GST Amide BEH Glycan column, 1.7 μm particle size, 2.1 × 150 mm column dimensions. Source: Lauber and Koza 2015 [49]. Reproduced with permission of Waters Corporation.

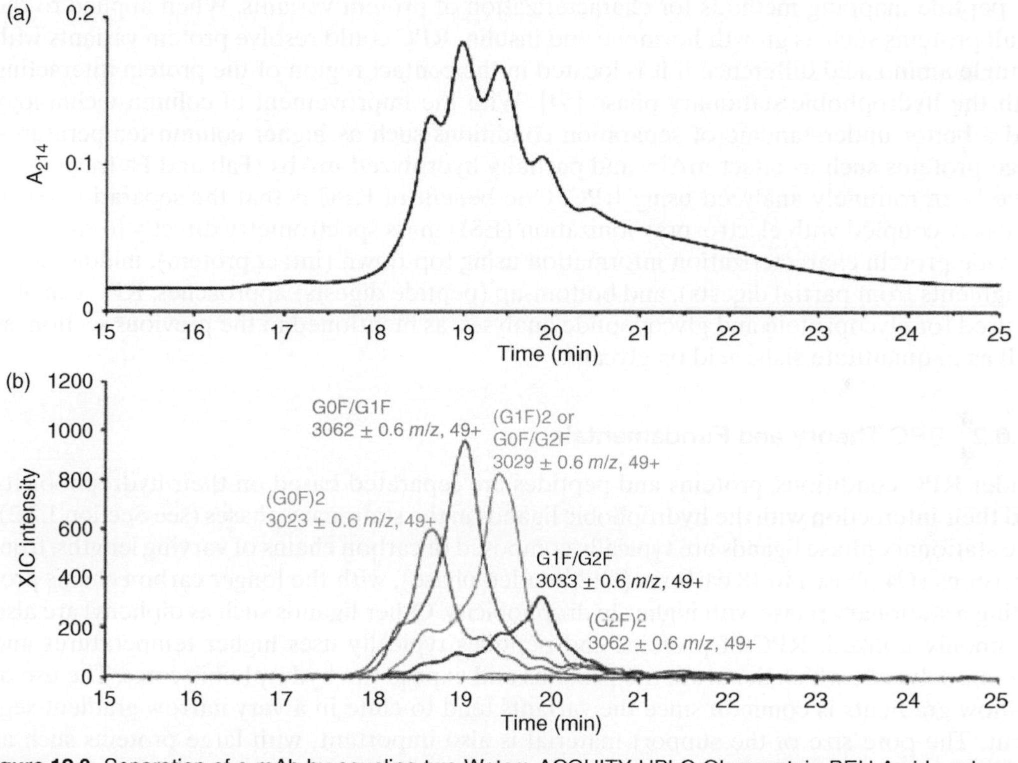


**Figure 12.9.** Separation of a mAb by coupling two Waters ACQUITY UPLC Glycoprotein BEH Amide columns. A UV chromatogram and extracted ion chromatograms for each of the major heterogeneous glycoforms are displayed. The column used was an ACQUITY UPLC Glycoprotein BEH Amide column, 1.7 μm particle size, 2.1 × 150 mm column dimensions. Mobile phase A was 0.1% v/v TFA in water, and mobile phase B was 0.1% v/v TFA in acetonitrile. The flow rate was set to 0.2 ml/min, and the column temperature was set to 30 °C. Column pressure at the retention time of the mAb was approximately 7000 psi. Source: Lauber et al. 2015 [51]. Reproduced with permission of Waters Corporation.

antibody product. HILIC glycan mapping and mass spectrometry are then used to identify specific glycan structures. The ability to identify specific Fc glycans has enabled extensive investigation of their structural and functional effects as well as their biological effects. In addition, assays employed to quantitate glycans have been utilized in QC lot release testing to demonstrate batch-to-batch consistency and to control the product quality of each batch with regard to glycosylation pattern. Glycoprotein profiling and glycopeptide mapping can also be used to determine differences between innovator molecules and biosimilars. By having glycosylation data at the intact protein level, peptide level, and individual glycan level, analysts now have the tools for detailed characterization of the glycan patterns of glycosylated biologics.

## 12.6 REVERSED-PHASE CHROMATOGRAPHY (RPC)

### 12.6.1 RPC Introduction

RPC is a key method type for separation of proteins and peptides based on their hydrophobicity under denaturing conditions. RPC offers relatively high-resolution separation, especially for peptide mapping methods for characterization of protein variants. When applied to the small proteins such as growth hormone and insulin, RPC could resolve protein variants with a single amino acid difference if it is located in the contact region of the protein interacting with the hydrophobic stationary phase [54]. With the improvement of column technology and a better understanding of separation conditions such as higher column temperatures, large proteins such as intact mAbs and partially hydrolyzed mAbs (Fab and Fc fragments) have been routinely analyzed using RPC. One benefit of RPC is that the separation could be easily coupled with electrospray ionization (ESI) mass spectrometry directly to routinely provide protein characterization information using top-down (intact protein), middle-down (fragments from partial digests), and bottom-up (peptide digests) approaches. RPC can also be used for glycoprotein and glycopeptide analyses, as mentioned in the previous section, as well as to quantitate sialic acid on glycans.

### 12.6.2 RPC Theory and Fundamentals

Under RPC conditions, proteins and peptides are separated based on their hydrophobicity and their interaction with the hydrophobic ligands in the stationary phases (see Section 1.2.2). The stationary phase ligands are typically composed of carbon chains of varying lengths, from 4 carbons (C4 phase) to 18 carbons (C18 bonded phase), with the longer carbon chains providing a stationary phase with higher hydrophobicity. Other ligands such as diphenyl are also commonly utilized. RPC of proteins and peptides typically uses higher temperatures and organic solvents, which denature the proteins and expose the hydrophobic cores. The use of shallow gradients is common since the variants tend to elute in a very narrow gradient segment. The pore size of the support material is also important, with large proteins such as antibodies using 300 Å pore materials, while small proteins or peptides are better resolved in columns packed with smaller pore supports (e.g. 120 Å). The phenomenon of "entangled diffusion" of large proteins in small-pore support materials leading to poor mass transfer and band broadening is discussed in Section 13.6.1.

Many protein variants can be separated by RPC, including those related to oxidation, sequence variants, misassembled molecules (e.g. bispecific antibodies and homodimer variants), cysteine linkage related variants, and variably conjugated ADCs [55]. The peptides generated after enzymatic digestion of proteins are separated based on posttranslational modifications, such as oxidation and deamidation, and sequence variants. Overall, the high resolution of RPC provides greater details about the proteins compared to lower-resolution techniques such as SEC.

### 12.6.3 RPC Method Conditions

For peptides, the RPC methods have been long established using C8 or C18 columns and trifluoroacetic acid (TFA) or formic acid as a mobile phase additive to water/acetonitrile mobile phases. For antibodies and very large proteins, a C4, C8, or diphenyl column and mobile phases containing water/acetonitrile (or isopropyl alcohol [IPA] for hydrophobic proteins) with ~0.1% TFA as modifier are often employed. Higher column temperatures (up to 110 °C) are often used to reduce carryover, while the column temperature could be lower for small protein and peptide separation. Depending on the molecules, IPA and *n*-butanol could also be added to improve the recovery of the proteins, especially antibodies. To reduce the ionization suppression in LC/MS applications while maintaining excellent peak shapes, other volatile acids to replace TFA (e.g. 3,3,3-trifluoropropionic acid, 2,2-difluoroacetic acid) have been proposed [56].

In addition to commonly used acidic mobile phases, alkaline buffers have been used to separate small proteins and peptides due to the unique selectivity at high pHs. However, columns packed with bonded phases to hybrid particles designed to tolerate high-pH mobile phases, such as bridged ethylene hybrid (BEH) columns or other similar columns, must be used (see Section 3.5.2).

Recently, superficially porous particles (SPPs) have been developed for RPC applications. These particles consist of a solid core, which is impervious to the solutes, and a porous outer layer, which provides shorter diffusion pathways in the particle, thus yielding significantly higher column efficiency (20–40%) vs. those packed with totally porous particles (see Section 3.5.7). SPP columns can have lower backpressures than other particle types – a reason for the popularity of small-particle SPP (<3 μm) [57].

### 12.6.4 RPC Applications

The protein variants separated by RPC are commonly related to oxidation and other post-translational modifications, as well as sequence variants. In addition, RPC has been used to separate the disulfide variants of IgG2 [58]. Because of the tendency of proteins to denature under RPC conditions, the wild-type antibody and antibody with free thiols can be well separated. Recently, RP-UHPLC of such free thiol variants has been shown with improved resolution (Figure 12.10) [59]. RP-UHPLC has also been developed as a middle-down approach for Fc oxidation analysis, which involves enzymatically cleaving antibodies at the hinge region to produce smaller protein fragments [60, 61].

RP-UHPLC is routinely applied to peptide mapping analysis, often in conjunction with MS, as shown in Figure 12.11 [62]. Posttranslational modifications such as oxidation and deamidation can be monitored using peptide mapping, as well as mapping disulfide linkages. Peptide mapping methods are also commonly used in QC testing (e.g. identity confirmation), since the peptide map of a protein has higher specificity than other identification methods such as capillary zone electrophoresis. Peptide maps are also used to compare innovators' drugs to biosimilars. Sequence variants analysis is routinely performed using peptide mapping RP-UHPLC to detect the single amino acid modification that may occur during the manufacturing process. RPC with MS can be used to obtain molecular weight information on peaks of interest for studying proteins by bottom-up proteomics [63] and for characterizing protein therapeutics [64].

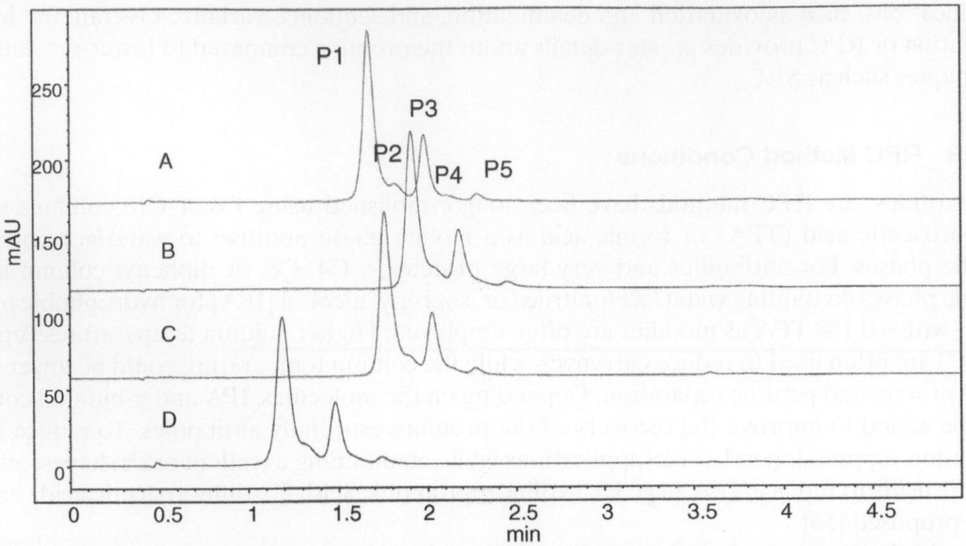


**Figure 12.10.** Different RP-UHPLC columns reveal the free thiol variants for an IgG1 mAb (mAb1). HPLC columns: Zorbax StableBond diphenyl (A), Zorbax StableBond C3 (B), Zorbax StableBond C8 (C), and ACQUITY® BEH-300 C4 (D). The diphenyl bonded column shows the best resolution for separating mAb1 free thiol variants. Running conditions: 0.6 ml/min at 75 °C using absorbance at 280 nm for detection. The mobile phases consisted of 0.08–0.1% trifluoroacetic acid (TFA) in water (mobile phase A) and 0.1% TFA in acetonitrile (mobile phase B). The major peaks (P1, P3, and P5) are most likely due to unpaired Cys 22 and Cys 96 in the heavy chain variable domain. The minor peaks (P2 and P4) are likely cysteine related. Source: Wei et al. 2017 [58]. Reprinted with permission of Elsevier.

## 12.7 HYDROPHOBIC INTERACTION CHROMATOGRAPHY (HIC)

### 12.7.1 HIC Introduction

Hydrophobic Interaction Chromatography (HIC) separates proteins based on their hydrophobicity under native (nondenaturing) conditions. HIC has been historically used in protein purification because of its unique selectivity and its ability to separate proteins while retaining the protein's biological activities. In addition, HIC is routinely used for separating ADCs with different drug-to-antibody ratios (DARs), thus allowing for the calculation of average DAR, which is a CQA for ADCs [65], and fractionation of ADC species with different DARs for further characterization. Because of the use of high salt in the mobile phase, it is challenging to couple HIC with MS directly, although some conditions have been reported to enable HIC-MS for proteins [66].

### 12.7.2 HIC Theory and Fundamentals

Under HIC conditions, proteins are separated based on their hydrophobicity and retain their native structures for further structure-function studies. The high salt concentration at the beginning of the mobile phase gradient drives the proteins to partition into the hydrophobic

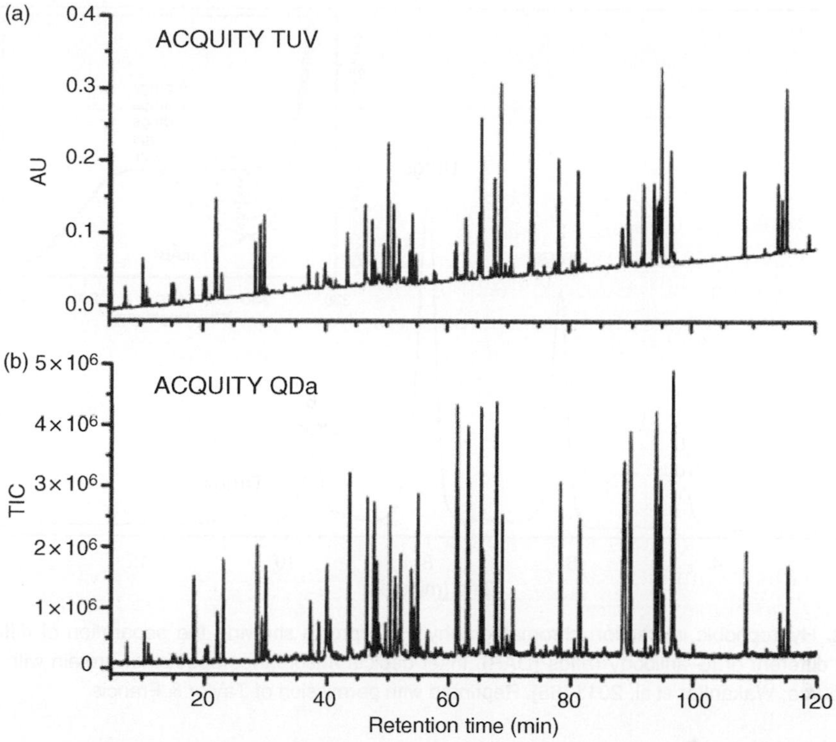


**Figure 12.11.** Peptide map of a mAb generated by reverse-phase liquid chromatography coupled to MS. Panel (a) shows the UV trace, while panel (b) shows the total ion chromatogram from the mass spectrometer. Data was obtained on an ACQUITY UPLC H-Class system coupled to an ACQUITY QDa mass spectrometer. The column was an ACQUITY UPLC Peptide CSH 130A C18 column with 1.7 μm particles, 2.1 × 100 mm column dimensions. The column temperature was 65 °C, injection volume was 8 μl. The flow rate was 0.2 ml/min. The mobile phases consisted of 0.1% TFA in water (mobile phase A) and 0.1% TFA in ACN (mobile phase B). Source: Birdsall and McCarthy 2013 [62]. Reproduced with permission of Waters Corporation.

stationary phase from the mobile phase. Decreasing the salt concentration allows the proteins to preferentially partition to the mobile phase and elute based on increasing hydrophobicity. It is a common methodology for protein purification because it offers orthogonal selectivity compared with IEC.

### 12.7.3 HIC Method Conditions

A variety of mAb variants can be separated by HIC [67]. For analytical HIC for proteins, a salt gradient (from high salt to low salt) is typically applied, and separation is at room temperature to preserve the native structure of proteins. The stationary phases are largely alkyl-based C4 or C3 ligands, although often at lower ligands density compared with the bonded phases for RPC. Depending on the molecules and ligands, mobile phase additives such as ammonium sulfate and ammonium acetate could be evaluated to achieve the separation.

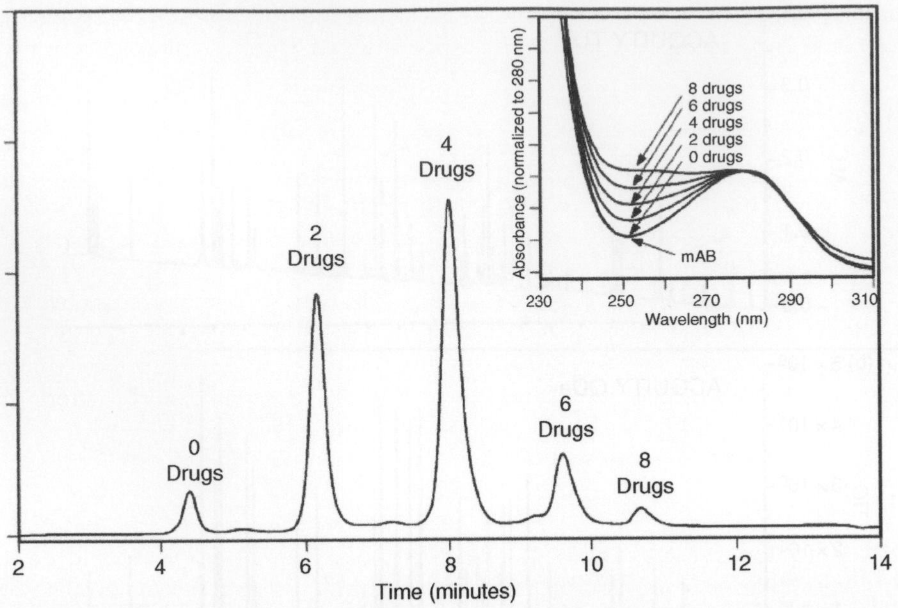


**Figure 12.12.** Hydrophobic interaction chromatography (HIC) profile showing the separation of different ADC variants with different drug–antibody ratios (DAR). Inset depicts absorbance spectra of protein with increasing drug load. Source: Wakankar et al. 2011 [68]. Reprinted with permission of Taylor & Francis.

### 12.7.4 HIC Applications

ADC DAR determination is a key application of HIC in biotherapeutics [68]. As more small-molecule drugs (toxins that are typically hydrophobic) are attached to the antibodies, the variants with different ADC drug loads could be resolved depending on the number of drug molecules that are attached on the antibodies (Figure 12.12) [68].

Additionally, HIC has been used to separate protein variants. For example, if aspartic acid residues in the antibodies formed isoaspartic isomers, the resulting variants could be resolved using HIC [69]. Oxidation in the complementary determining region (CDR) of antibodies can be determined at the intact protein level using HIC [70]. Also, HIC has been used to separate the free thiol mAb variants and the free thiol Fab variants [71]. These species have been enriched to study their functional activities, which demonstrated an advantage of HIC as a separation mode that preserves the activities of proteins for further analysis. Finally, HIC has been used to separate mAb aggregates as an orthogonal method to SEC [72].

## 12.8 MIXED-MODE CHROMATOGRAPHY (MMC)

### 12.8.1 MMC Introduction

Since proteins, particularly antibodies, are complex molecules with many functional groups, they can exhibit multiple characteristics, for example, low pI, high hydrophobicity, and so on. The chromatographic mode of separation can be purposely modulated and tuned to achieve the desired separation by taking into consideration these multiple physicochemical properties of the protein. In large-scale separation, mixed-mode chromatography (MMC) is a common purification technique. For analytical separation, MMC could be employed to resolve variants that are not easily separated by other means, especially under native conditions for proteins.

### 12.8.2 MMC Theory and Fundamentals

MMC, as its name implies, involves multiple separation mechanisms to resolve the protein variants in a single column. By relying on the different selectivities, subtle improvements in separation, which could be difficult in other modes of chromatography, could be accomplished. A common MMC application uses the large pore size of the SEC columns and residual hydrophobic interactions with the stationary phase to separate proteins by both size and hydrophobicity. Although packings designed for SEC separation typically have reduced hydrophobicity, often there is some residual hydrophobicity, which could be exploited to achieve unique selectivity [73]. Other MMC columns utilize combinations of different interactions, such as charge interactions or specific ligand binding [74].

### 12.8.3 MMC Method Conditions

Since MMC conditions will vary based on the modes of chromatography employed, MMC method conditions cannot be generalized, as they can be for other types of chromatography (SEC or RPC, for example). For mixed-mode separation using SEC columns, the separation relies on high salt concentrations of the mobile phase and SEC columns with a residual hydrophobic nature [75]. The mobile phase salts could include sodium chloride and sodium sulfate, used at up to 2.4 M in concentration in neutral buffers; isocratic elution conditions are typically used to resolve the variants. The challenge of developing a method to resolve oxidation variants from the HMWS species, for example, may involve careful selection of salt type and concentrations to achieve the unique selectivity needed. Method conditions for other MMC applications may be found in the appropriate reference.

### 12.8.4 MMC Applications

One application using MMC is the separation of oxidation variants of a monoclonal antibody [75]. In this application, an SEC method was purposely designed to separate the different oxidation degradants by varying the salt concentration of the mobile phase. As oxidation could occur in one or two arms of the heavy chains, multiple species could then be resolved (Figure 12.13) [75]. The advantage of the mixed-mode separation compared with the typical peptide mapping or middle-down approach is the lack of tedious sample preparation. In addition, the oxidation variants can be recovered for biological activity assessment because of the native MMC conditions.

MMC has been utilized for separation of 2AA-labeled *N*-linked glycans [76]. Amide-HILIC columns are particularly useful for the separation of 2-aminobenzamide-(2AB) labeled *N*-linked glycans released from mAbs, in which the majority of glycans harbor no charge. However, amide-HILIC columns do not provide adequate separations where glycans that harbor two or more charge states are present (e.g. neutral and sialylated glycans) because glycan isoforms with different charge states coelute in the separation envelope. A weak anion exchange HILIC mixed-mode column that further improves separations by resolving glycans into different charge groups was developed and separates glycans within each charged group based on isomerization and size. This substantially increases the resolution of complex N-linked glycan structures and helps differentiate isomeric structures not resolved by other approaches.

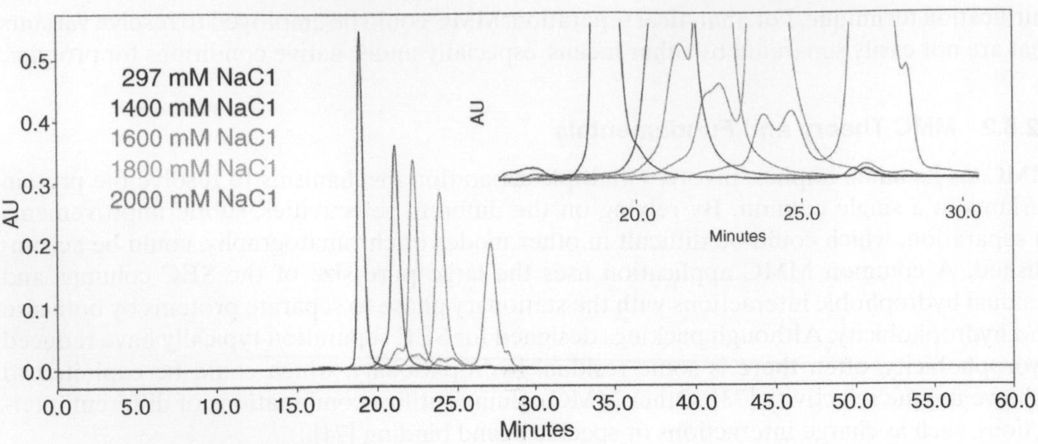


**Figure 12.13.** Profiles of mAb oxidation variants using increasing mobile phase salt concentration on a SEC column. The mixed-mode HPLC method condition: Zenix SEC-300MK column (7.8 × 300 nm, 5 μm) at a flow rate of 0.5 ml/min. Elution was monitored by UV at 280 nm. The mobile phase was phosphate buffered saline (PBS, pH 7.4) containing different concentrations of added NaCl, ranging from 297 to 2400 mM. Source: Pavon et al. 2016 [75]. Reprinted with permission of Elsevier.

## 12.9 MULTIDIMENSIONAL LIQUID CHROMATOGRAPHY

### 12.9.1 Multidimensional LC Introduction

Multidimensional LC, in which two or more orthogonal chromatography columns are used in tandem either in comprehensive (all peaks) or heart-cutting (selected peaks) mode, is emerging as a powerful characterization tool to characterize biotherapeutics [77, 78]. Analysts can take advantage of the multiple separation mechanisms involved in multidimensional chromatography to achieve improved separation or to identify and monitor variants that could not be monitored using a single chromatography mode.

### 12.9.2 Multidimensional LC Theory and Fundamentals

In multidimensional liquid chromatography, two or more independent liquid phase separation modes are applied sequentially to a sample. This is in contrast to MMC, in which several mechanisms of separation are innate in a single column. In multidimensional liquid chromatography, eluent from the first chromatographic mode (i.e. first dimension) can be applied to the second dimension. If only portions of the first dimension are applied to the second dimension (i.e., certain peaks of interest), this is referred to as "heart-cutting" chromatography. If the entirety of the first dimension eluent is applied to the second dimension, this is referred to as "comprehensive 2D-LC" (two-dimensional liquid chromatography). Furthermore, for heart-cutting chromatography, peak fractions of interest could be applied to RPC "traps," which serve as desalting sample loops for the second dimension [79]. Alternatively, fractions from peaks of interest can be automatically collected offline in a fraction collector. The fraction collector may contain sample vials with neutralizing buffer, for example, to neutralize and dilute the collected fractions before injection into the second dimension.

### 12.9.3 Multidimensional LC Method Conditions

Several instrument manufacturers produce commercially available 2D-LC systems and have released application notes for 2D-LC biopharmaceutical characterization applications [80, 81]. Instruments achieve heart-cutting by either trapping peaks of interest in RPC trapping columns, collecting fractionated peaks and reinjecting, or storing the fractions in multiple loops on the instrument for direct injection into the second dimension.

### 12.9.4 Multidimensional LC Applications

Multidimensional liquid chromatography has been used to characterize biotherapeutics more efficiently than offline manual fraction collecting method. One powerful and relatively simple use of 2D-LC is to use it to fractionate and desalt samples online for RPC/MS. In this application, portions of the eluent from the first dimension, that is, peaks of interest, are "trapped" in small RPC cartridge columns in the second dimension for MS analysis [82]. Thus, masses of peaks of interest are obtained in a single workflow. Other advantages of this application are that it can be used to purify samples in biological matrices prior to MS analysis and also include repetitive injections for sensitivity enhancements of low-abundance peaks. Figure 12.14 shows one peak in an IEX profile being heart-cut and analyzed by RPC [83]. Repeating the heart-cuts in subsequent samples prior to RPC analysis enables a linear increase in signal and improves sensitivity. Alternatively, affinity immunodepletion columns can be used in the first dimension to selectively remove higher abundant proteins to make it easier to identify other proteins in the sample in the second dimension [84].

One important attribute of ADCs is the residual content of unconjugated small-molecule drug present from either incomplete conjugation or degradation of the ADC.

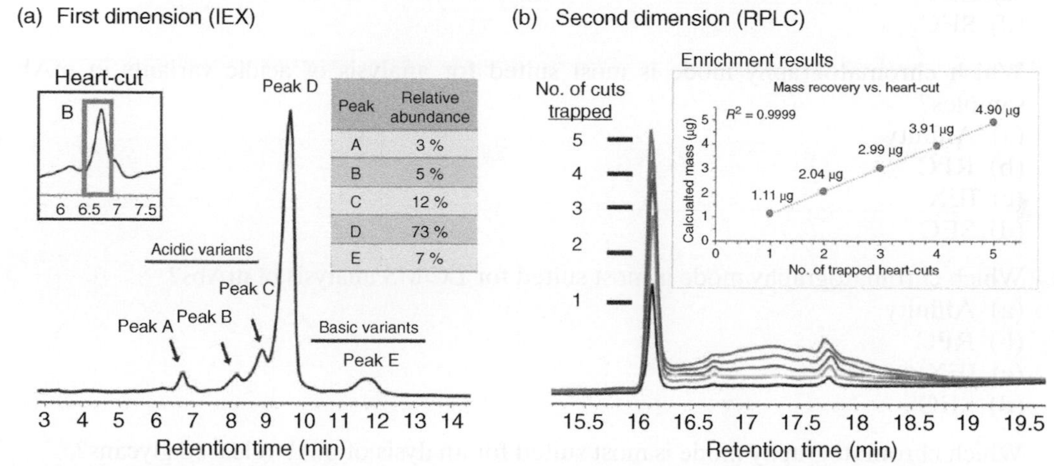


**Figure 12.14.** Multidimensional liquid chromatography using IEX as the first dimension and RPC as the second dimension. (a) mAb sample analyzed by IEX, with a heart-cut performed from 6.4 to 6.9 minutes as shown in the inset. (b) Heart-cuts were eluted off the second dimension column using a reversed-phase gradient. Overlays of chromatographic traces with enrichment of one to five heart-cuts are shown with their corresponding recovered mass plotted in the inset graph. Source: Birdsall et al. 2015 [83]. Reproduced with permission of Waters Corporation.

A two-dimensional heart-cutting method coupling SEC with RPC has been used to analyze the free linker-drug species in ADC samples [85].

Multidimensional LC applications can also include tandem column chromatography, in which two columns of different column chemistries are connected in series [86]. Tandem column chromatography is different than MMC in that multiple columns are used, as opposed to MMC where a single column is used that exhibits multiple separation mechanisms in one column. Tandem LC methods can be challenging to develop, but has the advantage of being highly modular and customizable, and does not require a multidimensional chromatography system (i.e. multiple pumps and/or switching valves).

## 12.10 SUMMARY

The rapid development of biotherapeutics has accompanied the improvements in HPLC resolution and throughput in the past few decades. The different HPLC modes can resolve different protein modifications, critical to the understanding of the quality attributes of the biopharmaceutical products. The holistic applications of HPLC are becoming indispensable methodologies in the development of well-characterized biologicals as effective therapeutics.

## 12.11 QUIZZES

**1.** Which chromatography mode is most suited for analysis of aggregates in mAb samples?
(a) Affinity
(b) RPC
(c) IEX
(d) SEC

**2.** Which chromatography mode is most suited for analysis of acidic variants in mAb samples?
(a) Affinity
(b) RPC
(c) IEX
(d) SEC

**3.** Which chromatography mode is most suited for LC/MS analysis of mAbs?
(a) Affinity
(b) RPC
(c) IEX
(d) SEC

**4.** Which chromatography mode is most suited for analysis of mAb released glycans?
(a) HILIC
(b) HIC
(c) IEX
(d) SEC

**5.** Which chromatography mode is most suited for drug antibody ratio analysis for ADCs?
(a) HIC
(b) RPC
(c) IEX
(d) SEC

**6.** Which analytical technique is routinely used for protein characterization?
(a) DSC
(b) FTIR
(c) GC
(d) HPLC/MS

**7.** Which statement is NOT generally true for SEC of mAb?
(a) Most components elute within the column void volume
(b) mAb fragments elute later than the main mAb
(c) High amounts of organic solvent (>50%) are added to the mobile phase to reduce hydrophobic interactions
(d) Dimer elutes before the main component

**8.** Peptide mapping is NOT generally used for which of these analyses of mAb?
(a) *De novo* sequencing
(b) Identification
(c) Analysis of aggregation
(d) Analysis of degradation products

**9.** Large-scale affinity chromatography is mostly used for:
(a) Analysis of tryptic digests
(b) Harvesting mAb from fermentation broth
(c) Analysis of aggregation
(d) Detailed analysis of degradation products

**10.** Which format of 2D-LC is most likely NOT compatible?
(a) IEX/SEC
(b) IEX/RPC
(c) HIC/IEX
(d) SEC/RPC

## 12.12 REFERENCES


1. Reichert, J.M., Rosensweig, C.J., Faden, L.B., and Dewitz, M.C. (2005). Monoclonal antibody successes in the clinic. *Nat. Biotech.* 23 (9): 1073.
2. Philippidis, A. (2017). The top 15 best-selling drugs of 2016. *Genet. Eng. Biotechnol. News*.
3. Rea, J.C., Wang, Y.J., Moreno, G.T. et al. (2012). *Monoclonal Antibody Development and Physiochemical Characterization by High-Performance Ion Exchange Chromatography*. Innovations in Biotechnology (ed. E.C. Agbo), 439–464. InTech.
4. Rathore, A. (2010). Setting specifications for a biotech therapeutic product in the quality by design paradigm. *Biopharm Int.* 23 (1).
5. Geigert, J. (ed.) (2016). *The Challenge of CMC Regulatory Compliance for Biopharmaceuticals and Other Biologics*, 2e. New York: Springer.
6. Beck, A., Wagner-Rousset, E., Ayoub, D. et al. (2012). Characterization of therapeutic antibodies and related products. *Anal. Chem.* 85: 715.
7. Gooding, K.M. and Regnier, F.E. (2002). *HPLC of Biological Macro-Molecules, Revised and Expanded*. Boca Raton, FL: CRC Press.
8. Rodriguez-Diaz, R., Wehr, T., and Tuck, S. (2005). *Analytical Techniques for Biopharmaceutical Development*. Boca Raton, FL: CRC Press.
9. Hearn, M. (1991). *HPLC of Proteins, Peptides and Polynucleotides*. Weinheim, Germany: VCH.
10. Kazakevich, Y.V. and LoBrutto, R. (eds.) (2007). *HPLC for Pharmaceutical Scientists*. Hoboken, NJ: Wiley.

11. Guillarme, D. and Veuthey, J.-L. (eds.) (2012). *UHPLC in Life Sciences*. Cambridge, United Kingdom: RSC.
12. Zhang, T., Quan, C., and Dong, M.W. (2014). HPLC for characterization and quality control of therapeutic monoclonal antibodies. *LCGC North Am.* 32 (10): 796.
13. Fekete, S., Veuthey, J.-L., and Guillarme, D. (2017). Achievable separation performance and analysis time in current liquid chromatographic practice for monoclonal antibody separations. *J. Pharm. Biomed. Anal.* 141 (50): 59.
14. Ossipow, V. and Fischer, N. (eds.) (2014). *Monoclonal Antibodies: Methods and Protocols*, 2e. New York: Humana Press.
15. Hong, P., Koza, S., and Bouvier, E.S.P. (2012). Size-exclusion chromatography for the analysis of protein biotherapeutics and their aggregates. *J. Liq. Chromatogr. Rel. Technol.* 35: 2923.
16. Cordoba, A.J., Shyong, B.J., Breen, D., and Harris, R.J. (2005). Non-enzymatic hinge region fragmentation of antibodies in solution. *J. Chromatogr. B* 818: 115.
17. Vlasak, J. and Ionescu, R. (2011). Fragmentation of monoclonal antibodies. *mAbs* 3: 253.
18. Ravuluri, S., Bansal, R., Chhabra, N., and Rathore, A.S. (2018). Kinetics and characterization of non-enzymatic fragmentation of monoclonal antibody therapeutics. *Pharm. Res.* 35 (142).
19. Schnerman, M.A., Sunday, B.R., Kozlowski, S. et al. (2004). CMC strategy forum report: analysis and structure characterization of monoclonal antibodies. *Bioprocess Int.* 2: 42.
20. Kubota, K., Kobayashi, N., Yabuta, M. et al. (2017). Identification and characterization of a thermally cleaved fragment of monoclonal antibody-A detected by sodium dodecyl sulfate-capillary gel electrophoresis. *J. Pharm. Biomed. Anal.* 140: 98.
21. Gandhi, A.V., Pothecary, M.R., Bain, D.L., and Carpenter, J.F. (2017). Some lessons learned from a comparison between sedimentation velocity analytical ultracentrifugation and size exclusion chromatography to characterize and quantify protein aggregates. *J. Pharm. Sci.* 106: 2178.
22. Waters Corporation (2012). *ACQUITY UPLC SEC Columns*, Waters Application Note 720003398EN. Milford, MA: Waters Corporation.
23. Arakawa, T., Ejima, D., Li, T., and Philo, J.S. (2010). The critical role of mobile phase composition in size exclusion chromatography of protein pharmaceuticals. *J. Pharm. Sci.* 99: 1674.
24. Cromwell, M.E.M., Hilario, E., and Jacobson, F. (2006). Protein aggregation and bioprocessing. *AAPS J.* 8: E572.
25. Chen, T., Chen, Y., Stella, C. et al. (2016). Antibody-drug conjugate characterization by chromatographic and electrophoretic techniques. *J. Chromatogr. B* 1032: 39.
26. Rea, J.C., Moreno, G.T., Vampola, L. et al. (2011). Capillary size exclusion chromatography with picogram sensitivity for analysis of monoclonal antibodies purified from harvested cell culture fluid. *J. Chromatogr. A* 1219: 140.
27. Boyd, D., Ebrahimi, A., Ronan, S. et al. (2018). Isolation and characterization of a monoclonal antibody containing an extra heavy-light chain Fab arm. *mAbs* 10: 346.
28. Haberger, M., Leiss, M., Heidenreich, A.K. et al. (2016). Rapid characterization of biotherapeutic proteins by size-exclusion chromatography coupled to native mass spectrometry. *mAbs* 8: 331.
29. Khawli, L.A., Goswami, S., Hutchinson, R. et al. (2010). Charge variants in IgG1: isolation, characterization, in vitro binding properties and pharmacokinetics in rats. *mAbs* 2: 613.
30. Miao, S., Xie, P., Zou, M. et al. (2017). Identification of multiple sources of the acidic charge variants in an IgG1 monoclonal antibody. *Appl. Microbiol. Biotechnol.* 101: 5627.
31. Goyon, A., Francois, Y.N., Colas, O. et al. (2018). High-resolution separation of monoclonal antibodies mixtures and their charge variants by an alternative and generic CZE method. *Electrophoresis* 0: 1. (online ahead of print).
32. Wu, G., Yu, C., Wang, W., and Wang, L. (2018). Interlaboratory method validation of icIEF methodology for analysis of monoclonal antibodies. *Electrophoresis* 39: 2091–2098.
33. Vlasak, J. and Ionescu, R. (2008). Heterogeneity of monoclonal antibodies revealed by charge-sensitive methods. *Curr. Pharm. Biotech.* 9: 468.
34. Weitzhandler, M., Farnan, D., Rohrer, J.S., and Avdalovic, N. (2001). Protein variant separations using cation exchange chromatography on grafted, polymeric stationary phases. *Proteomics* 1: 179.

35. Lin, S., Baek, J., and Pohl, C. (2014). *A Fast and Robust Linear pH Gradient Separation Platform for Monoclonal Antibody (mab) Charge Variant Analysis* Thermo Fisher Scientific Application Note 20946. Sunnyvale, CA: Thermo Fisher Scientific.
36. Birdsall, R., Wheat, T., and Chen, W. (2013). *Developing Robust and Efficient IEX Methods for Charge Variant Analysis of Biotherapeutics Using ACQUITY UPLC H-Class System and AutoBlend Plus*, Waters® Application Note 720004847EN. Milford, MA: Waters Corporation.
37. Rao, S. and Pohl, C. (2011). Reversible interference of $Fe^{3+}$ with monoclonal antibody analysis in cation exchange columns. *Anal. Biochem.* 409: 293.
38. Zhang, L., Patapoff, T., Farnan, D., and Zhang, B. (2013). Improving pH gradient cation-exchange chromatography of monoclonal antibodies by controlling ionic strength. *J. Chromatogr. A* 1272: 56.
39. Seo, N., Polozova, A., Zhang, M. et al. (2018). Analytical and functional similarity of Amgen biosimilar ABP 215 to bevacizumab. *mAbs* 10: 678.
40. Schmid, I., Bonnington, L., Gerl, M. et al. (2018). Assessment of susceptible chemical modification sites of trastuzumab and endogenous human immunoglobulins at physiological conditions. *Comm. Biol.* 1: 28.
41. Choe, W., Durgannavar, T.A., and Chung, S.J. (2016). Fc-binding ligands of immunoglobulin G: an overview of high affinity proteins and peptides. *Materials* 9: 994.
42. Schlothauer, T., Rueger, P., Stracke, J.O. et al. (2013). Analytical FcRn affinity chromatography for functional characterization of monoclonal antibodies. *mAbs* 5 (4): 576.
43. Schmidt, P.M., Abdo, M., Butcher, R.E. et al. (2016). A robust robotic high-throughput antibody purification platform. *J. Chromatogr. A* 1455: 9.
44. Reusch, D. and Tejada, M.L. (2015). Fc glycans of therapeutic antibodies as critical quality attributes. *Glycobiology* 25: 1325.
45. Zheng, K., Bantog, C., and Bayer, R. (2011). The impact of glycosylation on monoclonal antibody conformation and stability. *mAbs* 3: 568.
46. Zhang, L., Luo, S., and Zhang, B. (2016). Glycan analysis of therapeutic glycoproteins. *mAbs* 8 (2): 205.
47. Liu, L. (2015). Antibody glycosylation and its impact on the pharmacokinetics and pharmacodynamics of monoclonal antibodies and Fc-fusion proteins. *J. Pharm. Sci.* 104: 1866.
48. Spencer, D., Freeke, J., and Barattini, V. (2013). *Analysis of Human IgG Glycans on a Solid Core Amide HILIC Stationary Phase*, Thermo Scientific™ Application Note 20703. United Kingdom: Thermo Fisher Scientific.
49. Lauber, M.A. and Koza, S.M. (2015). *Measuring the Glycan Occupancy of Intact mAbs Using HILIC and Detection by Intrinsic Fluorescence*, Waters Application Note 720005435EN. Milford, MA: Waters Corporation.
50. Lauber, M.A., Koza, S.M., and Fountain, K.J. (2013). *Optimization of GlycoWorks® HILIC SPE for the Quantitative and Robust Recovery of N-linked Glycans from mAb-Type Samples*, Waters Application Note 720004717EN. Milford, MA: Waters Corporation.
51. Lauber, M.A., McCall, S.A., Alden, B.A. et al. (2015). *Developing High-Resolution HILIC Separations of Intact Glycosylated Proteins Using a Wide-Pore Amide-Bonded Stationary Phase*, Waters® Application Note 720005380EN. Milford, MA: Waters Corporation.
52. D'Atri, V., Dumont, E., Vandenheede, I. et al. (2017). Hydrophilic interaction chromatography for the characterization of therapeutic monoclonal antibodies at protein, peptide, and glycan levels. *LCGC* 30: 424.
53. Zhenyu, D., Qun, X., Liang, L., and Rohrer, J. (2016). *Separation of Intact Monoclonal Antibody Sialylation Isoforms by pH Gradient Ion-Exchange Chromatography*, Thermo Scientific™ Application Note 1092. Sunnyvale, CA: Thermo Fisher Scientific.
54. Josic, D. and Kovac, S. (2010). Reversed-phase high performance liquid chromatography of proteins. *Curr. Protoc. Protein Sci.*, Chapter 8, Unit 8.7 .
55. Liu, L. and Coffey, A. (2017). *Determination of Drug-to-Antibody Distribution in Cysteine-Linked ADCs*, Agilent Application Note 5991-7192EN. Wilmington, DE: Agilent Technologies.
56. Boyes, B.E. (2017). *Structure of a Monoclonal Antibody*. Chinese American Chromatography Association (CACA) webinar, Advanced Materials Technology.

57. Kirkland, J.J., Schuster, S.A., Johnson, W.L., and Boyes, B.E. (2013). Fused-core particle technology in high-performance liquid chromatography: an overview. *J. Pharm. Anal.* 3 (5): 303.
58. Wei, B., Zhang, B., Boyes, B., and Zhang, Y.T. (2017). Reversed-phase chromatography with large pore superficially porous particles for high throughput immunoglobulin G2 disulfide isoform separation. *J. Chromatogr. A* 1526: 104.
59. Liu, H., Jeong, J., Kao, Y.-H., and Zhang, Y.T. (2015). Characterization of free thiol variants of an IgG1 by reversed phase ultra high pressure liquid chromatography coupled with mass spectrometry. *J. Pharm. Biomed. Anal.* 109: 142.
60. Zhang, T., Zhang, J., Hewitt, D. et al. (2012). Identification and characterization of buried unpaired cysteines in a recombinant monoclonal IgG1 antibody. *Anal. Chem.* 84: 7112.
61. Zhang, B., Jeong, J., Burgess, B. et al. (2016). Development of a rapid RP-UHPLC–MS method for analysis of modifications in therapeutic monoclonal antibodies. *J. Chromatogr. B* 1032: 172.
62. Birdsall, R.E. and McCarthy, S.M. (2013). *Adding Mass Detection to Routine Peptide-Level Biotherapeutic Analyses with the ACQUITY QDa Detector*, Waters® Application Note 720004568EN. Milford, MA: Waters Corporation.
63. Yates, J.R., Ruse, C.I., and Nakorchevsky, A. (2009). Proteomics by mass spectrometry: approaches, advances, and applications. *Annu. Rev. Biomed. Eng.* 11: 49.
64. Yu, X.C., Joe, K., Zhang, Y. et al. (2011). Accurate determination of succinimide degradation products using high fidelity trypsin digestion peptide map analysis. *Anal. Chem.* 83 (15): 5912.
65. Wagh, A., Song, H., Zeng, M. et al. (2018). Challenges and new frontiers in analytical characterization of antibody-drug conjugates. *mAbs* 10: 222.
66. Chen, B., Peng, Y., Valeja, S.G. et al. (2016). Online hydrophobic interaction chromatography-mass spectrometry for top-down proteomics. *Anal. Chem.* 88 (3): 1885.
67. Haverick, M., Mengisen, S., Shameem, M., and Ambrogelly, A. (2014). Separation of mAbs molecular variants by analytical hydrophobic interaction chromatography HPLC: overview and applications. *mAbs* 6: 852.
68. Wakankar, A., Chen, Y., Gokarn, Y., and Jacobson, F.S. (2011). analytical methods for physicochemical characterization of antibody drug conjugates. *mAbs* 3: 161.
69. Cacia, J., Keck, R., Presta, L.G., and Frenz, J. (1996). Isomerization of an aspartic acid residue in the complementarity-determining regions of a recombinant antibody to human IgE: identification and effect on binding affinity. *Biochemistry* 35: 1897.
70. Boyd, D., Kaschak, T., and Yan, B. (2011). HIC resolution of an IgG1 with an oxidized Trp in a complementarity determining region. *J. Chromatogr. B* 879: 955.
71. Chaderjian, W.B., Chin, E.T., Harris, R.J., and Etcheverry, T.M. (2005). Effect of copper sulfate on performance of a serum-free CHO cell culture process and the level of free thiol in the recombinant antibody expressed. *Biotechnol. Progr.* 21: 550.
72. Baek, J., Lin, S., and Liu, X. (2016). *HIC as a Complementary, Confirmatory Tool to SEC for the Analysis of mAb Aggregates*, Thermo Fisher Scientific Application Note 21206. Sunnyvale, CA: Thermo Fisher Scientific.
73. Wong, C., Strachan-Mills, C., and Burman, S. (2012). Facile method of quantification for oxidized tryptophan degradants of monoclonal antibody by mixed mode ultra performance liquid chromatography. *J. Chromatogr. A* 1270: 153.
74. Zhang, K. and Liu, X. (2016). Mixed-mode chromatography in pharmaceutical and biopharmaceutical applications. *J. Pharm. Biomed. Anal.* 130: 19.
75. Pavon, A., Li, X., Chico, S. et al. (2016). Analysis of monoclonal antibody oxidation by simple mixed mode chromatography. *J. Chromatogr. A* 1431: 154.
76. Aich, U., Saba, J., Liu, X. et al. (2013). *Separation of 2AA-Labeled N-linked Glycans from Glycoproteins on a High Resolution Mixed-Mode Column*, Thermo Fisher Scientific Application Note 20910. Sunnyvale, CA: Thermo Fisher Scientific.
77. Doneanu, C., Rainville, P., and Plumb, R. (2012). *Advantages of Online Two-Dimensional Chromatography for MRM Quantification of Therapeutic Monoclonal Antibodies in Serum*, Waters Application Note, 720004510EN. Milford, MA: Waters Corporation.

78. Stoll, D.R., Harmes, D.C., and Danforth, J. (2015). Direct identification of rituximab main isoforms and subunit analysis by online selective comprehensive two-dimensional liquid chromatography-mass spectrometry. *Anal. Chem.* 87: 8307.
79. Alvarez, M., Tremintin, G., Wang, J. et al. (2011). On-line characterization of monoclonal antibody variants by liquid chromatography-mass spectrometry operating in a two-dimensional format. *Anal. Biochem.* 419: 17.
80. McCarthy, S.M., Wheat, T.E., Yu, Y.Q., and Mazzeo, J.R. (2012). *Two-Dimensional Liquid Chromatography for Quantification and MS Analysis of Monoclonal Antibodies in Complex Samples*, Waters® Application Note 720004304EN. Milford, MA: Waters Corporation.
81. Schneider, S. (2016). *Online 2D-LC Characterization of Monoclonal Antibodies Using Protein A and Weak Cation Exchange Chromatography*, Agilent Application Note 5991-6848EN. Waldbronn, Germany: Agilent Technologies.
82. Gendeh, G., Decrop, W., and Swart, R. (2011). *Development of an Automated Method for Monoclonal Antibody Purification and Analysis*, Thermo Fisher Scientific Application Note LPN 2945-01. Sunnyvale, CA: Thermo Fisher Scientific.
83. Birdsall, R., Ivleva, V., McCarthy, S.M., and Chen, W. (2015). *Characterization of Biotherapeutics: ACQUITY UPLC H-Class Bio with 2D Part 3 of 3: Online Enrichment of Low Abundance Species*, Waters® Application Note 720005327EN. Milford, MA: Waters Corporation.
84. Mrozinski, P., Zolotarjova, N., and Chen, H. (2008). *Human Serum and Plasma Protein Depletion – Novel High-Capacity Affinity Column for the Removal of the "Top 14" Abundant Proteins*, Agilent Application Note 5989-7839EN. Wilmington, DE: Agilent Technologies.
85. Li, Y., Stella, C., Zheng, L. et al. (2016). investigation of low recovery in the free drug assay for antibody drug conjugates by size exclusion – reversed phase two dimensional-liquid chromatography. *J. Chromatogr. B* 1032: 112.
86. Koza, S., Lauber, M., and Fountain, K.J. (2013). *The Analysis of Multimeric Monoclonal Antibody Aggregates by Size-Exclusion UPLC*, Waters® Application Note 720004713EN. Milford, MA: Waters Corporation.

# 13

# HPLC APPLICATIONS IN FOOD, ENVIRONMENTAL, CHEMICAL, AND LIFE SCIENCES ANALYSIS

## 13.1 INTRODUCTION

### 13.1.1 Scope

This chapter provides a cursory overview of the use of high-performance liquid chromatography/ultra-high-pressure liquid chromatography (HPLC/UHPLC) in food, environmental, chemical (polymers, ion chromatography), and life sciences applications (proteins, proteomics, glycomics). The advantages and trends of HPLC applications are summarized with illustrative examples and detailed separation conditions with UV or mass spectrometry (MS) detection. Key references are cited for further information. Note that HPLC applications for pharmaceutical analysis and recombinant biotherapeutics are discussed in Chapters 9 and 12, respectively.

## 13.2 FOOD APPLICATIONS

HPLC is widely used in food analysis, particularly in product research, quality control (QC), nutritional labeling, and multiresidue testing of contaminants and pesticides. HPLC is a cost-effective technique, ideally suited for the testing of labile components in complex matrices. Tables 13.1 and 13.2 summarize the common analytes, advantages, trends, and chromatographic modes, and detector options in food analysis by HPLC. Analysis of natural components (sugars, fats, proteins, amino acids, and organic acids), food additives (preservatives, colors, flavors, and sweeteners), and contaminants are described. For further details on specific applications, the reader is referred to books [1–4] and methodologies from the Association of Official Analytical Chemists (AOAC International) [5] or the American

*HPLC and UHPLC for Practicing Scientists*, Second Edition. Michael W. Dong.

**Table 13.1. Summary of Food Applications by HPLC**

| | |
|---|---|
| Key applications | Characterization and quantitation of natural components, additives, and contaminants |
| HPLC advantages | Offers high resolution, multicomponent, and quantitative analysis<br>Provides optimal sensitivity and specificity, using diverse detectors and columns; Amenable to labile analytes |
| Trends | Increasing use of UHPLC, SFC, ELSD, or CAD (for nonchromophoric analytes), and LC/MS (SQ-MS, MS/MS, or TOF), for trace multicomponent residue analysis |

**Table 13.2. Summary of HPLC Modes and Detection in Food Analysis**

| Food analysis | HPLC mode | Detection options |
|---|---|---|
| Natural food components | | |
| Carbohydrates | Mixed-mode, RPC, IEC | RI, PAD, ELSD, MS |
| Lipids, triglycerides, and cholesterols | NARP, NPC, SFC | RI, UV, ELSD |
| Fatty acids and organic acids | IEC, RPC (also GC) | UV, RI, MS |
| Proteins, peptides | IEC, RPC, SEC | UV,MS |
| Amino acids | IEC, RPC | UV/Vis,[a] FL[a] |
| Food additives | | |
| Acidulants, sweeteners, flavors, | RPC (also GC) | UV, ELSD, RI |
| Antioxidants and preservatives | RPC | UV |
| Colors and dyes | RPC | UV/Vis, MS |
| Vitamins | RPC, HILIC | UV, FL |
| Contaminants | | |
| Mycotoxins (aflatoxins) | RPC | FL, UV, MS |
| Pesticide and drug residues | RPC | UV, FL, MS |
| PAHs and nitrosamines | RPC | UV, FL, MS |

HILIC = hydrophilic interaction chromatography, NARP = nonaqueous reversed-phase, PAD = pulsed amperometric detector, ELSD (evaporative light scattering detector or charged aerosol detection), MS = SQ, MS/MS, or hybrids.

[a]Precolumn or postcolumn derivatization required.

Society of Testing and Materials (ASTMs). The increase in regulations in food labeling and public safety is fueling the growth in food testing in recent years [6].

### 13.2.1 Natural Food Components

***13.2.1.1 Sugars*** Common carbohydrates are monosaccharides (glucose, fructose), disaccharides (sucrose, maltose, and lactose), trisaccharides (raffinose), and polysaccharides (starch). HPLC offers a direct, quantitative methodology for simple sugars, which requires a specialty cationic resin-based column and a refractive index (RI), evaporative light scattering detector (ELSD), or charged aerosol detector (CAD) [2, 7]. UV detection at low wavelengths (195–210 nm) can be used but is more prone to interferences [7]. Resin columns with various counterions ($Ca^{++}$, $Pb^{++}$, $H^+$, $Ag^+$, etc.) are used for different sample types, with calcium as the common form and water as the mobile phase at 60–85 °C [7] (see Figure 13.1). The basis of separation is a mixed-mode mechanism, which includes size-exclusion, ion-exclusion, ligand exchange, and hydrophobic interaction with the polystyrene support. This reliable assay

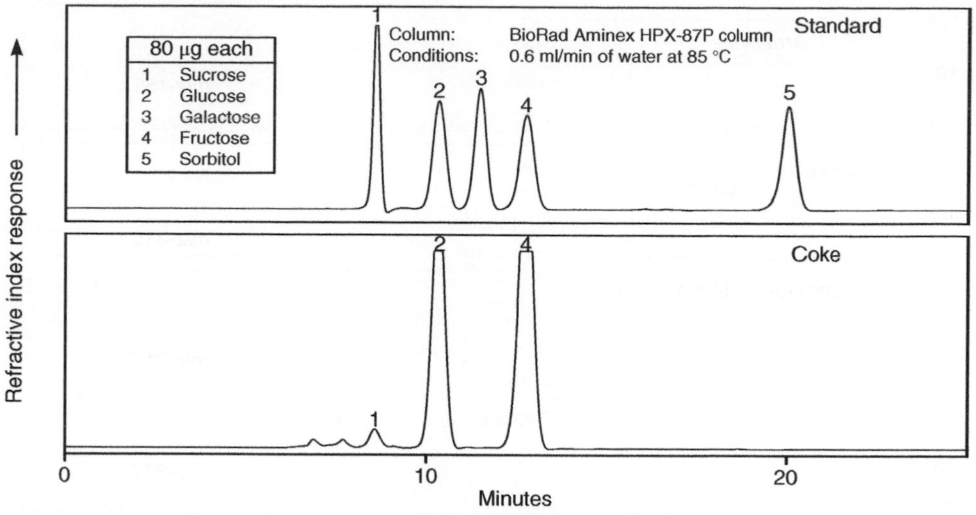


**Figure 13.1.** HPLC analysis of carbohydrates (simple sugars) in a soft drink sample using a resin column and refractive index detection. This is a common method for QC of sugar contents in beverages. HPLC conditions: Column – BioRad Aminex HPX-87C (9 μm, 300 × 7.8 mm i.d.); mobile phase – water; flow rate – 0.6 ml/min at 85 °C; detection – refractive index; sample – 10 μl of degassed beverage filtered through 0.45 μm membrane filter. Source: Courtesy of PerkinElmer, Inc.

has excellent performance on precision (retention time of 0.1% and peak area of 0.7% RSD [relative standard deviation]), sensitivity (10–20 ng), linearity (0.050–800 μg with $r > 0.9999$), and good column lifetime [7]. This method is preferred for QC of beverages, juices, syrups, cereals, and yogurts. An amino column with acetonitrile/water (80/20) (AOAC method) can also be used; however, sensitivity is lower and column lifetime is limited [2].

A sensitive approach utilizing anion-exchange with a high pH and with pulsed amperometric detection was developed in the 1990s for bioscience research applications. Currently, these assays are performed by liquid chromatography-mass spectrometry (LC/MS). [2] An example of LC/MS analysis of sugar and sugar alcohols, using hydrophilic interaction chromatography (HILIC) separation with a compact single-quadrupole single quadrupole-mass spectrometry (SQ/MS), is shown in Figure 13.2 [8]. LC/MS with selected ion recording (SIR) mode has excellent sensitivity, linearity, and separation power suited for the profiling of sugars and sugar alcohols in beverages, juices, liquors, beers, and wines. As MS detection becomes more popular and affordable, LC/MS-based methods in QC and product development applications will be more prevalent.

***13.2.1.2 Fats, Oils, and Triglycerides*** Fats and oils are triesters of glycerol with fatty acids (triglycerides). Fatty acids are commonly analyzed as their methyl esters by gas chromatography (GC), after sample transesterification. Free acids and triglycerides can also be analyzed by HPLC with RI, low UV detection, CAD, ELSD, or MS detection [2, 9]. Phospholipids (lecithin, cephalin, or phosphotidyl-inositol) from soybeans are used as food additives. Common sterols are cholesterols and phytosterols (sitosterols and stigmasterol). These sterols can readily be separated on silica columns with UV at 210 nm. Isomers of tocopherols and tocotrienols in foodstuff and cereal products can also be separated by normal-phase chromatography (NPC) with sensitive fluorescence (FL) detection (Figure 13.3) [10].

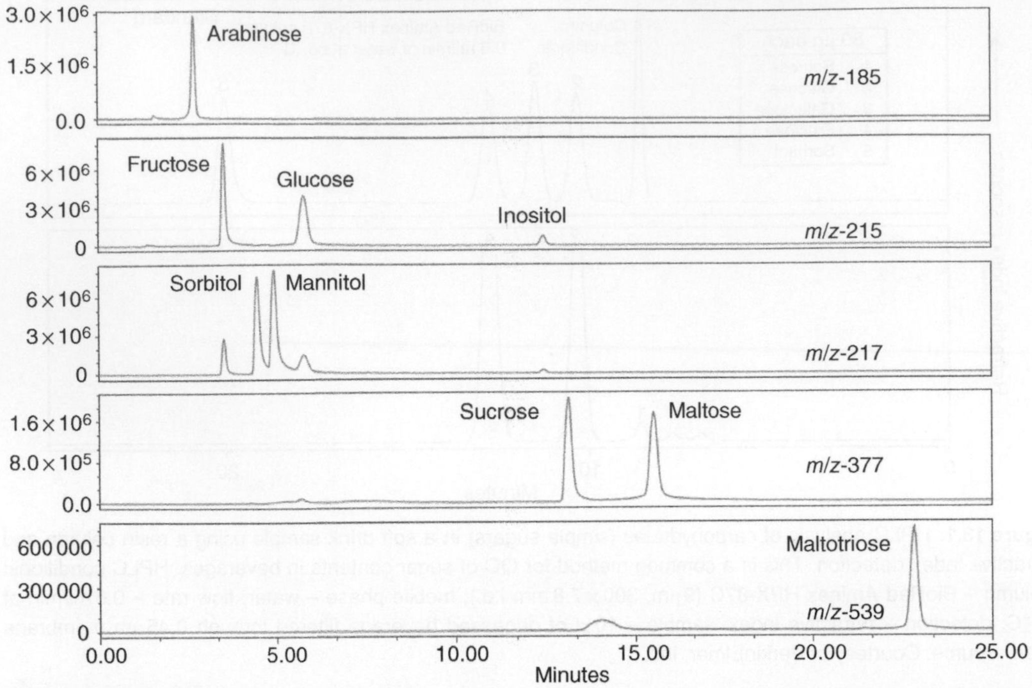


**Figure 13.2.** Chromatograms of HILIC/MS analysis of nine saccharide standards with selected ion recording (SIR) of the $[M+Cl]^-$ ions for the analysis of mono- and disaccharides and alditols in beverage profiling and quantitation. HPLC conditions: Column – XBridge XP Amide, 2.5 μm, 150×3.0 mm; MPA – 90/5/5 (ACN, IPA, Water with 500 ppb of guanidine HCl and 0.05% dimethylamine [DEA]); MPB – 80/20 (ACN, Water with 500 ppb of guanidine HCl and 0.05% DEA); flow rate – 0.8 ml/min at 85 °C; gradient – 0%B 4.5 minutes initial; 0–100%B in 13.5 minutes; 100%B for 5 minutes; 100–0%B in 0.1 minutes; 0%B for 15 minutes; MS – QDa with electrospray ionization (ESI), quantitation with SIR. Source: Benvenuti et al. 2017 [8]. Reproduced with permission of Waters Corporation.

HPLC analysis of triglycerides is simple and quantitative. It can be used for QC and product testing for adulteration. Edible oils can be injected directly as a 5% solution in acetone using nonaqueous reversed-phase chromatography (NARP) using refractive index detection under isocratic conditions or with CAD, ELSD, and MS detection with higher resolution under gradient conditions (Figure 13.4) [9].

#### *13.2.1.3 Free Fatty Acids and Organic Acids*

Free fatty acids can be readily analyzed by reversed-phase chromatography (RPC) with an acidified mobile phase and detection at 210 nm [2]. However, precolumn derivatization of free fatty acids to form derivatives such as *p*-bromophenacyl esters, 2-nitrophenyl hydrazides, or anthrylmethylesters can improve the chromatographic performance and detection sensitivity by UV or fluorescence detection [2]. Nevertheless, GC of their methyl esters remains the more common assay method. Organic acids occur naturally in foodstuffs as a result of biochemical processes, hydrolysis, or bacterial growth. Common organic acids are as follows:

- *Unsubstituted*: Formic, acetic, propionic, butyric
- *Substituted*: Glycolic, lactic, pyruvic, glyoxylic
- *Di- or tricarboxylic*: Oxalic, succinic, fumaric, maleic, malic, citric acid

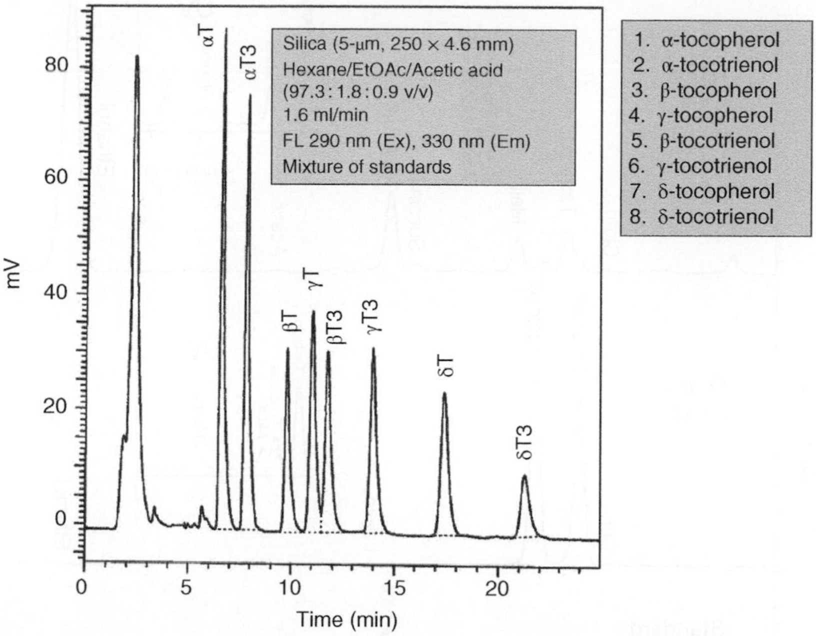


**Figure 13.3.** HPLC analysis of tocopherols and tocotrienols using NPC with fluorescence detection. Source: Panfili et al. 2003 [10]. Copyright 2003. Reprinted with permission of American Chemical Society.

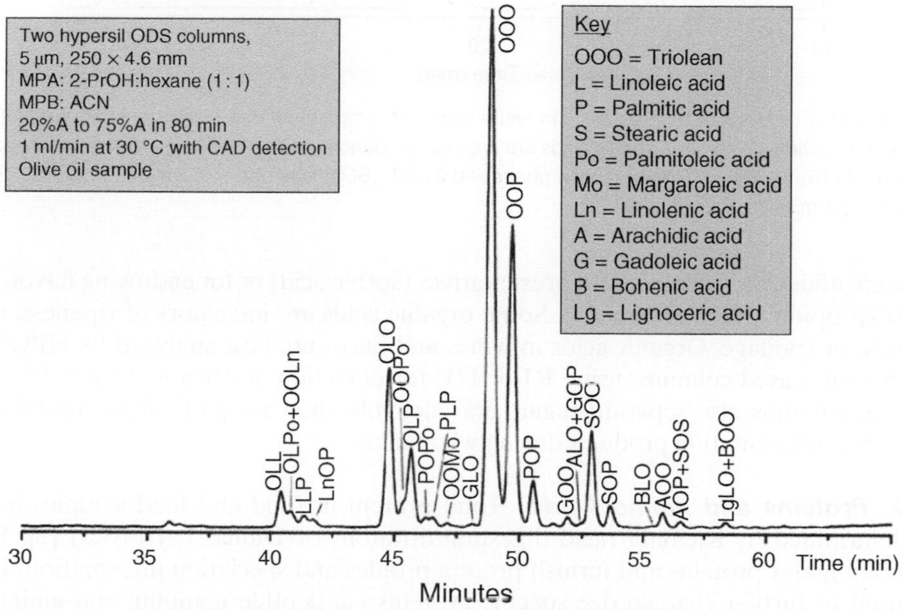


**Figure 13.4.** HPLC analysis of triglycerides in olive oil using nonaqueous reversed-phase chromatography (NARP) with charged aerosol detection (CAD) with peak identification by MS. Source: Lisa et al. 2007 [9]. Reprinted with permission of Elsevier.

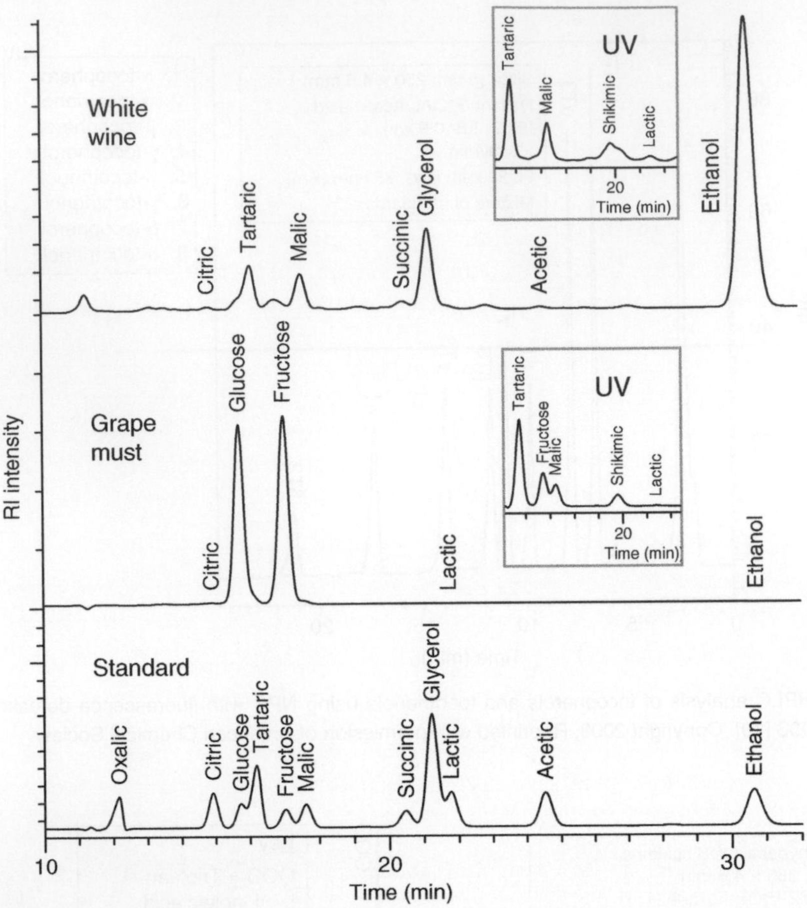


**Figure 13.5.** HPLC analysis of organic acids in white wine and grape must use a resin column with refractive index and UV detection at 210 nm. Organic acids are important organoleptic flavor components. HPLC conditions: two polypore H (10 µm, 220 × 4.6 mm); mobile phase – 0.01 N $H_2SO_4$; flow rate – 0.2 ml/min at 60 °C. Source: Reprinted with permission of Dong 1998 [11].

These are added as stabilizers and preservatives (sorbic acid) or for endowing flavor, taste, or aroma (propionic acid, citric acid). Some organic acids are indicators of ripeness, bacterial activity, or spoilage. Organic acids in wines and juices are best analyzed by HPLC with hydrogen resin-based columns, using RI or UV for detection at 210 nm (Figure 13.5) [11]. Since these columns also separate sugars and alcohols, they are particularly useful in the monitoring of fermentation products during winemaking.

***13.2.1.4 Proteins and Amino Acids*** Total protein in food and feed samples is commonly determined by Kjeldahl (acid digestion/titration) or Dumas (pyrolysis) [2]. HPLC can separate major proteins and furnish protein profiles and speciation information. HPLC can be used to further characterize specific proteins via peptide mapping and amino acid sequence analysis. HPLC modes used for protein include ion-exchange chromatography (IEC), size-exclusion chromatography (SEC), RPC, and affinity chromatography with typical UV detection at 215 or 280 nm (detection of peptide bonds or aromatic amino acids). Details on protein separations are discussed in Section 13.6 on life sciences. Amino acid analysis is useful for assessment of the nutritional value of food and feed [1, 2]. The essential amino acids are methionine, cysteine, lysine, threonine, valine, isoleucine, leucine, phenylalanine, tyrosine, and tryptophan. Glutamic acid is a commercial flavor enhancer. Since amino acids are nonchromophoric (possess low UV absorbance), either precolumn or postcolumn derivatization is typically used:

*Postcolumn derivatization*: IEC is used to separate free amino acids followed by ninhydrin or *o*-phthaldehyde (OPA) postcolumn derivatization to form chromophoric derivatives. This technique is used in dedicated amino acid analyzers, preferred for QC testing. An example of the ion-exchange chromatogram with postcolumn derivatization is shown in Figure 1.7.

*Precolumn derivatization*: RPC analysis of phenylisothiocyanate (PITC), OPA, 9-fluorenylmethyl chloroformate (FMOC), or 6-aminoquinolyl-*N*-hydroxysuccinimidyl carbamate (Waters AccQ.Tag) with UV or fluorescence detection is most common. This is the preferred methodology for life science research because of its higher sensitivity. An example of precolumn derivatization of amino acids is shown in Figure 13.32.

### 13.2.2 Food Additives

The analysis of food additives is important for nutritional labeling and QC [1, 2, 6]. Common additives are as follows:

- Acidulants
- *Sweeteners*: Aspartame, saccharin, cyclamate, sucralose (Figure 13.6)
- Colors and food dyes (Figure 13.7)
- *Vitamins*: Water-soluble and fat-soluble vitamins (Figures 13.3, 13.8–13.10)
- Antioxidants and preservatives (Figures 13.10 and 13.11)
- *Flavor compounds*: Bitter, pungent, aromatic compounds, as well as flavor enhancers (Figures 13.5 and 13.12)

Figure 13.6 shows UHPLC/MS SIR chromatograms for 12 sweeteners using both positive and negative ionization under 10 minutes. This method can be used for quantitative assays of natural and artificial sweeteners in beverages, puddings, and other food samples with minimal sample clean-up [12]. Figure 13.7 shows the HPLC analysis of various food colors using UV detection at 290 nm. Food dyes can be detected with high specificity at their respective $\lambda_{max}$ in the visible region or at 290 nm as a universal monitoring wavelength for all dyes.

Water-soluble vitamins were traditionally analyzed using isocratic RPC/UV with an ion-pairing reagent [18]. Here, the HPLC analysis is straightforward, while the quantitative extraction of all the vitamins from multivitamin tablets or foodstuff proves to be more challenging. Figure 13.8 shows a fast UHPLC gradient separation of 10 water-soluble vitamins with detection at multiple wavelengths of 205, 220, and 266 nm in 1.6 minutes. The multiwavelength detection using a photodiode array (PDA) detector allows for the sensitive analysis of vitamins with low chromophoric activities such as calcium pantothenate, biotin, and cyanocobalamin (B12) [13]. Figure 13.9 shows a high-resolution UHPLC separation of 15 carotenoids in 20 minutes, including several difficult-to-resolve isomers and related compounds [14]. High-resolution UHPLC offers an effective solution for the analysis of complex samples vs. conventional HPLC. Figure 13.10 shows a HILIC separation of its ingredients and additives in an energy drink sample with UV detection of chromophoric

| | |
|---|---|
| Column: | Waters CORTECS® T3, 3 × 100 mm, 2.7 μm |
| Column temp.: | 40 °C |
| Injection volume: | 2 μl |
| Flow rate: | 0.55 ml/min |
| Mobile phase A: | Water with 0.1% formic acid |
| Mobile phase B: | Acetonitrile with 0.1% formic acid |

| | Time (min) | Flow rate (ml/min) | %A | %B |
|---|---|---|---|---|
| 1. | Initial | 0.55 | 90 | 10 |
| 2. | 2.0 | 0.55 | 90 | 10 |
| 3. | 8.0 | 0.55 | 40 | 60 |
| 4. | 8.5 | 0.55 | 0 | 100 |
| 5. | 9.5 | 0.55 | 0 | 100 |
| 6. | 9.6 | 0.55 | 90 | 10 |

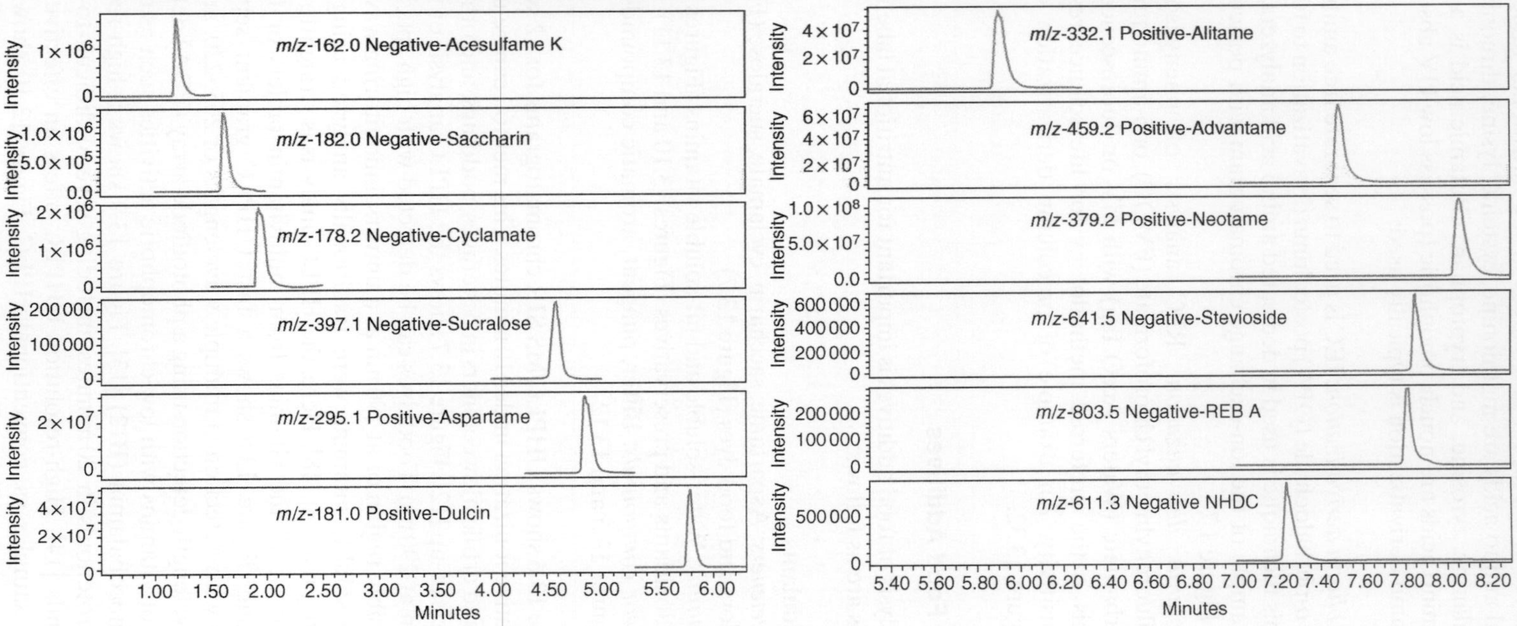

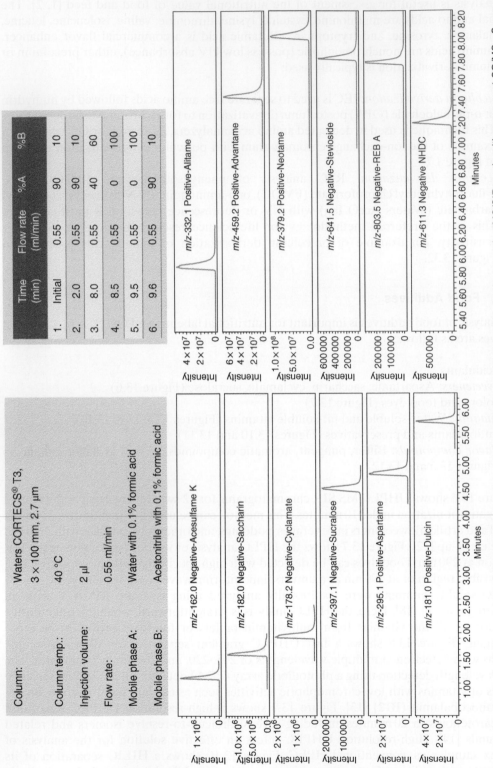

**Figure 13.6.** Chromatograms of UHPLC/MS analysis of 12 natural and artificial sweeteners under 7 minutes using MS/SIR mode with a compact SQ/MS. Source: Benvenuti et al. 2017 [12]. Reproduced with permission of Waters Corporation.

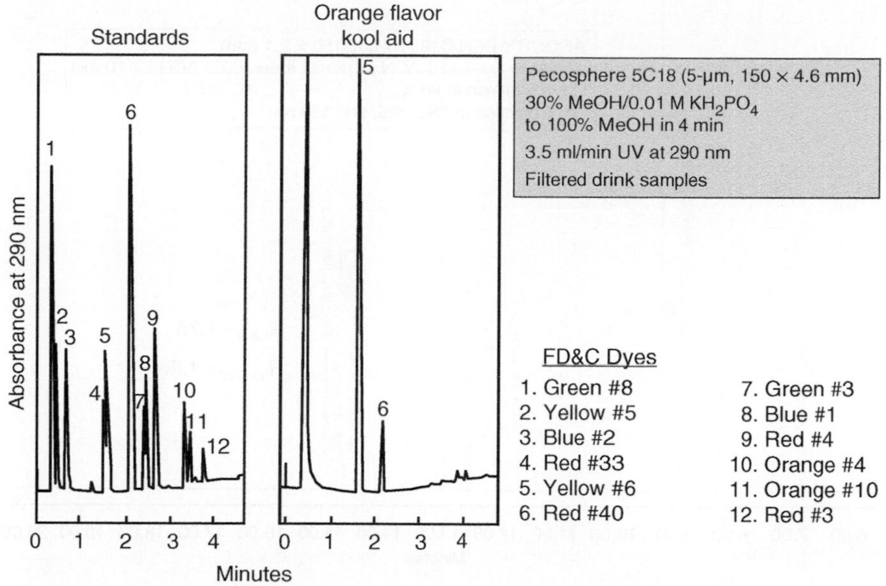


**Figure 13.7.** Fast HPLC analysis of food dyes in a beverage powder sample (Kool-Aid) using RPC/UV at 290 nm, which can detect most colored dyes. Source: Courtesy of PerkinElmer, Inc.

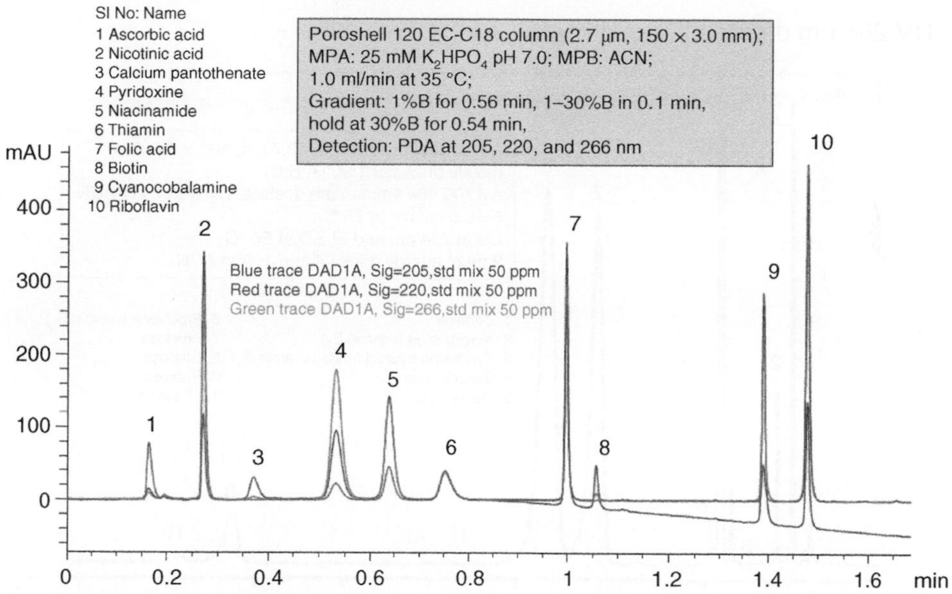


**Figure 13.8.** Fast UHPLC analysis of water-soluble vitamins with multiple wavelengths PDA detection at 205, 220, and 266 nm for sensitive assay of all vitamins including those with low chromophoric activities. An SPP column was used for enhanced column efficiency. Source: Joseph 2011 [13]. Reproduced with permission of Agilent Technologies.

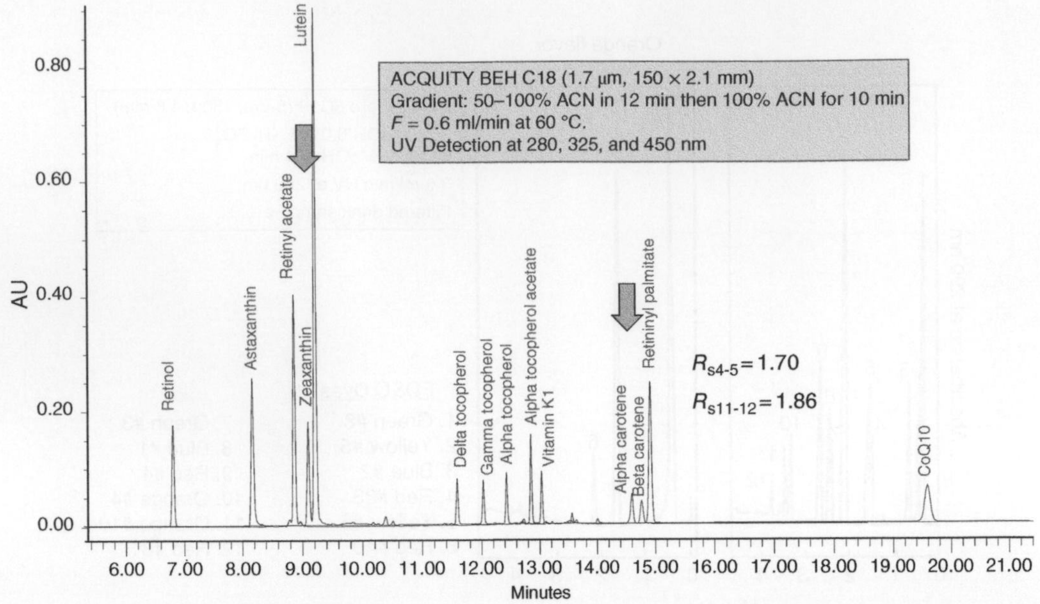


**Figure 13.9.** Chromatogram showing high-resolution UHPLC analysis of 14 carotenoids including difficult-to-resolve isomers and related compounds. UV detection: 280 nm for $\alpha$-tocopherol, $\delta$-tocopherol, $\gamma$-tocopherol, $\alpha$-tocopherol acetate, vitamin K1, and CoQ10; 325 nm for retinol, retinyl acetate, and retinol palmitate, and 450 nm for astaxanthin, zeaxanthin, lutein, $\alpha$-carotene, and $\beta$-carotene. Source: Reprinted with permission of Dong 2017 [14].

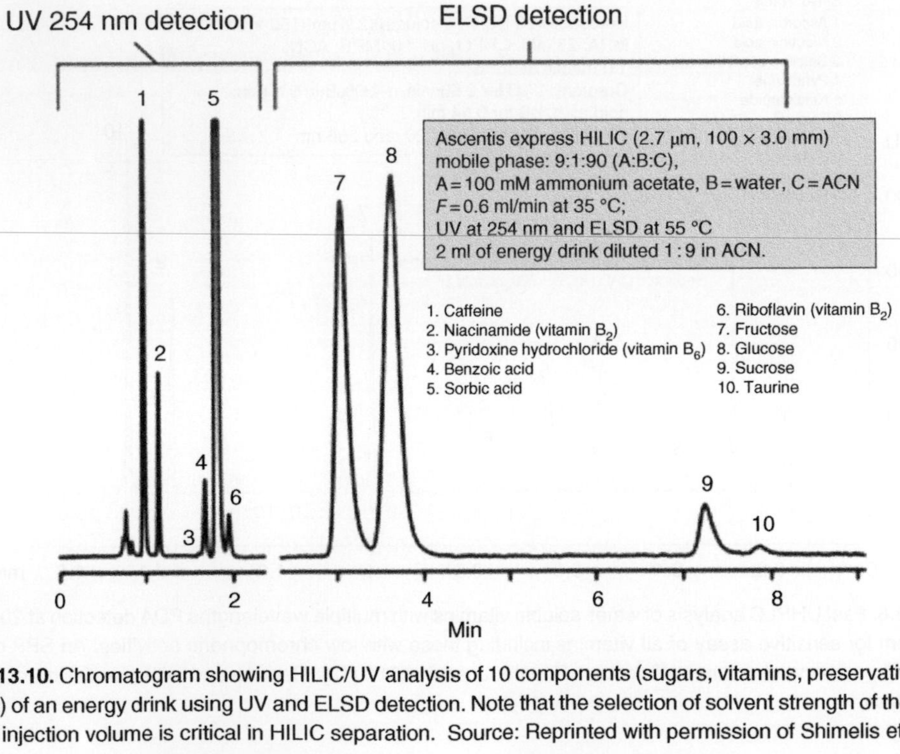


**Figure 13.10.** Chromatogram showing HILIC/UV analysis of 10 components (sugars, vitamins, preservatives, and caffeine) of an energy drink using UV and ELSD detection. Note that the selection of solvent strength of the diluent and the injection volume is critical in HILIC separation. Source: Reprinted with permission of Shimelis et al. [15].

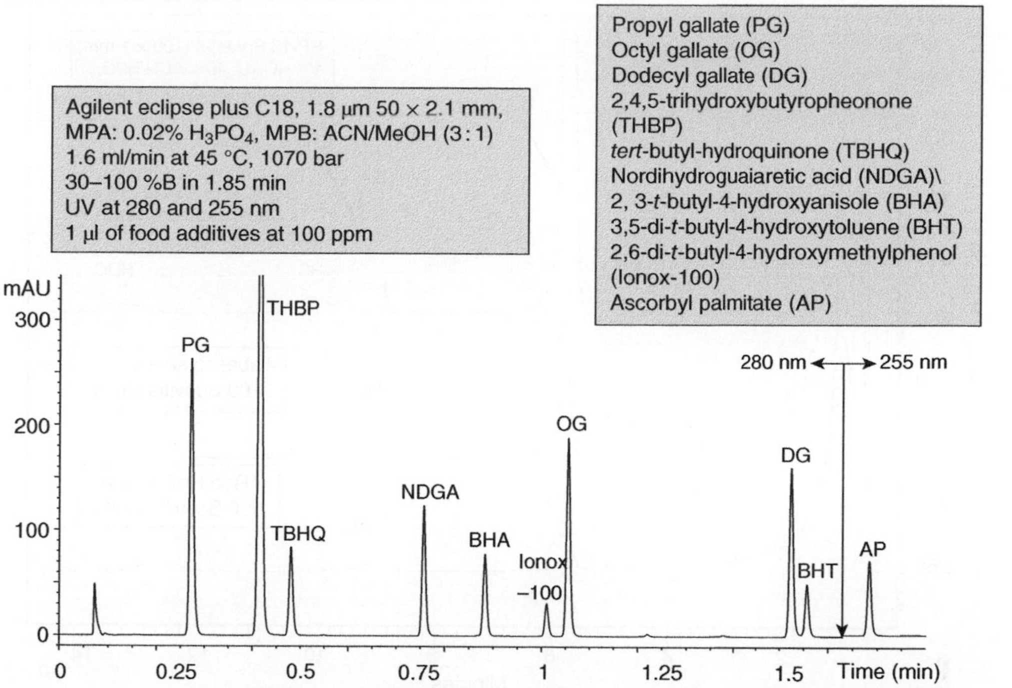


**Figure 13.11.** Overlaid chromatograms of UHPLC/UV analysis of 10 antioxidants for food preservatives for vegetable oils under high-throughput screening conditions with UV detection at 280 and 256 nm. Source: Vanhoenacker et al. 2009 [16]. Reproduced with permission of Agilent Technologies.

components (caffeine, vitamins, and preservatives) augmented by ELSD detection of sugars and taurine [15]. Figure 13.11 shows a UHPLC separation of 10 antioxidants for vegetable oils at a 100-ppm level under high-throughput screening (HTS) conditions in two minutes with UV detection at 280 and 256 nm [16].

***13.2.2.1 Flavors: A Case Study on HPLC Analysis of Capsaicins*** GC is used mostly in the analysis of fragrances and flavors due to their volatility. One notable exception is the analysis of capsaicins, the components responsible for the "heat" in hot pepper [17]. The degree of the pungency of hot peppers differs significantly due to the varying concentrations of capsaicins in different species (Table 13.3). The traditional way to evaluate the "hotness" in peppers or hot sauces is through a taste panel, measured by the "Scoville" scale, through a series of comparative evaluations of diluted samples. The taste panel method is subjective, costly, and time-consuming. Capsaicins can be analyzed directly (without derivatization) by HPLC. The method is precise and rapid (20 minutes) and requires only a drop of filtered hot sauce. Figure 13.12 shows the HPLC separation of individual capsaicins in the calibration standard and two hot sauces. Best of all, the Scoville units of each sample can be automatically reported by summing up the percentage of each capsaicin via a custom calculation in the chromatography data system (CDS) [17].

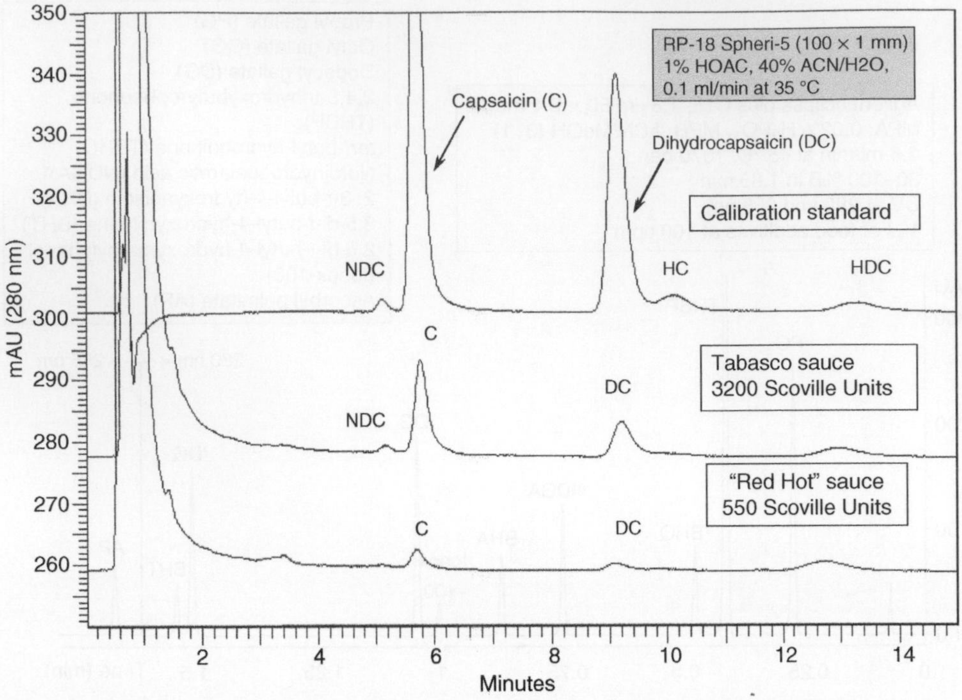


**Figure 13.12.** Microbore HPLC analysis of capsaicins, the principal components responsible for the "heat" in hot pepper in two samples of hot sauces by RPC/UV. Note that the Scoville scale of the hot sauce can be calculated automatically by the CDS. Source: Dong 2000 [17]. Copyright 2000. Reprinted with permission of American Chemical Society.

**Table 13.3. The Pungency Levels of Various Peppers**

| Pepper | *Pungency* (in Scoville units) |
|---|---|
| Bell pepper | *0* |
| Anaheim | *250–1400* |
| Jalapeno | *3000* |
| Hungarian yellow | *4000* |
| Japanese chili | *20 000–30 000* |
| Tabasco | *30 000–50 000* |
| Cayenne | *50 000–100 000* |
| Indian birdeye | *100 000–125 000* |
| Japanese kumataka | *125 000–150 000* |
| Habanero | *300 000* |

Source: Reprinted with permission of Ref. [17].

### 13.2.3 Contaminants

The analysis of food contaminants (toxic or biologically active residues) is essential for public health and food safety reasons. Examples are mycotoxins (including aflatoxins) and pesticides or drug residues. Sample preparation is elaborate and may involve deproteinization, solvent extraction, and clean-up via solid-phase extraction (SPE). The use of HPLC coupled with MS/MS or high-resolution mass spectrometry (HRMS) (TOF, time of flight) has simplified most of the sample preparation procedures, due to the inherent specificity of the MS detector [4, 19, 20].

***13.2.3.1 Mycotoxins*** A mycotoxin is a toxic secondary metabolite produced by organisms of the fungus kingdom and is capable of causing disease and death in both humans and other animals. Details on the analysis of mycotoxins are found elsewhere [1, 2]. Examples of mycotoxins are as follows:

- *Aflatoxins*: Toxic metabolites from molds (Figure 13.13)
  — B1, B2, G1, G2 (in grains and nuts) or M1, M2 (in dairy products)
- Ochratoxins A, B, and C fungal metabolites found in maize, wheat, or oats
- *Zearalenone*: Found in maize
- *Citrinine*: From molds found in grains
- *Trichothecenes*: Fungal poisons found in grains
- *Patulin*: Found in apple juice

Figure 13.13 shows the analysis of common aflatoxins in peanut butter samples using RPC with fluorescence detection, taking advantage of an innovative postcolumn electrochemical derivatization to increase the sensitivity for regulatory testing [21].

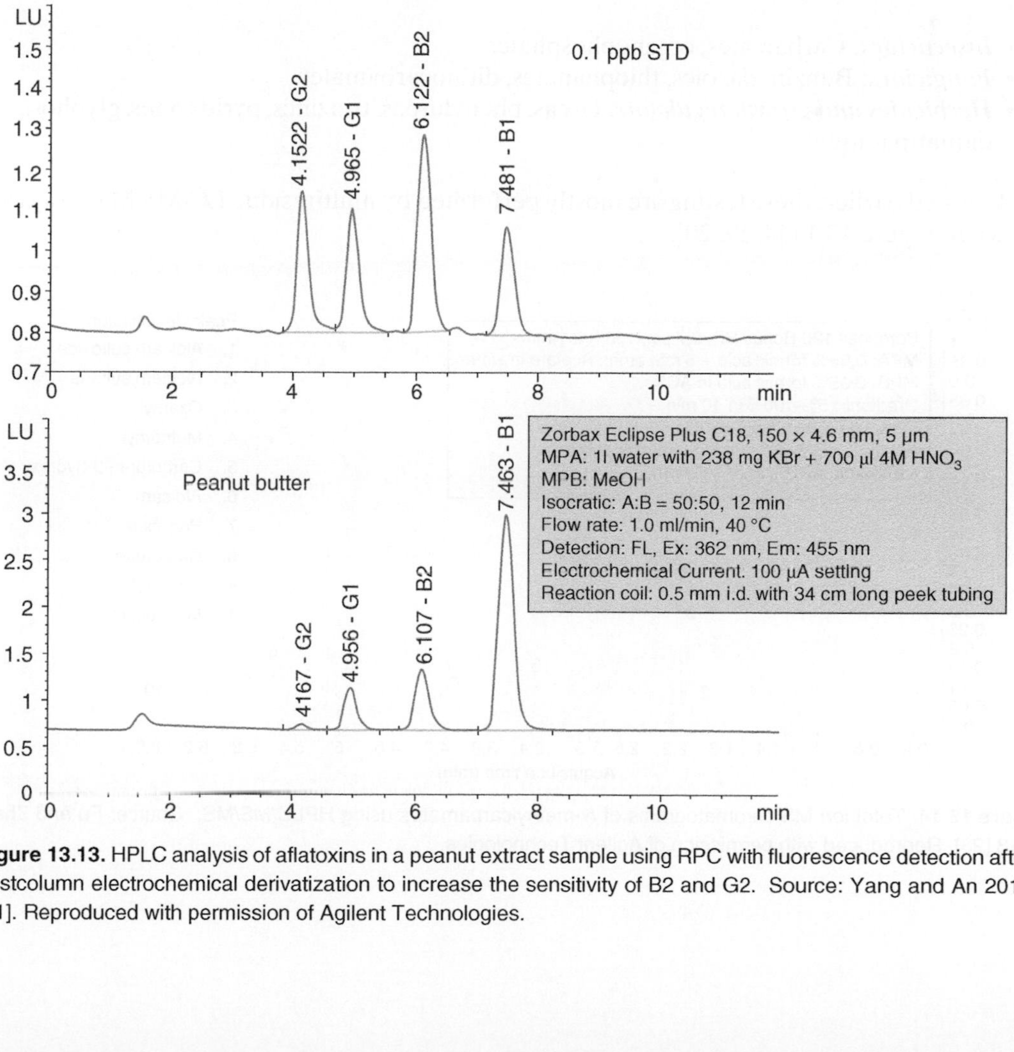


**Figure 13.13.** HPLC analysis of aflatoxins in a peanut extract sample using RPC with fluorescence detection after postcolumn electrochemical derivatization to increase the sensitivity of B2 and G2. Source: Yang and An 2011 [21]. Reproduced with permission of Agilent Technologies.

***13.2.3.2 Antimicrobial Additives*** Common antimicrobials and their HPLC detection wavelength are listed as follows. Details can be found elsewhere [2].

- Tetracycline antibiotics (UV 355 nm)
- Beta-lactam antibiotics (UV 210 nm)
- Polyether antibiotics (FL)
- Aminoglycoside antibiotics (FL)
- Macrolide antibiotics (UV 231 nm)
- Nitrofurans (UV 362 nm)
- Sulfonamides (UV 275 nm)
- Quinoxaline 1,4-dioxin (UV 350 nm)

***13.2.3.3 Pesticide Residues*** Pesticide residue analysis often requires elaborate sample preparation procedures and analyte enrichment, which might involve extraction (homogenization, solvent extraction), clean-up (solvent partitioning, gel permeation chromatography [GPC], SPE), and analyte derivatization [19, 20]. There appears to be increasing acceptance of QuEChERS (Quick, Easy, Cheap, Effective, Rugged, and Safe) sample preparation procedures based on homogenization and SPE for pesticide residues in foodstuffs [22]. Common pesticide residues found in food crops, vegetables, or fruit are as follows:

- *Insecticides*: Carbamates, organophosphates
- *Fungicides*: Benzimidazoles, thiophanates, dithiocarbamates
- *Herbicides and growth regulators*: Ureas, phenylureas, triazines, pyridazines, glyphosate, diquat/paraquat

As noted earlier, these testing are mostly performed by multiresidue LC/MS/MS assays as shown in Figure 13.14 [4, 19, 20].

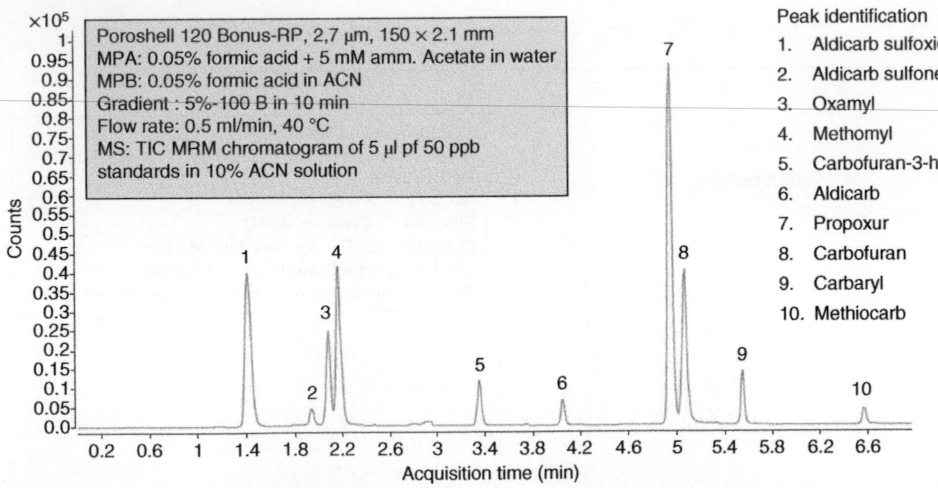


**Figure 13.14.** Total ion MS chromatograms of *N*-methylcarbamates using HPLC/MS/MS. Source: Fu and Zhai 2013 [23]. Reproduced with permission of Agilent Technologies.

## 13.3 ENVIRONMENTAL APPLICATIONS

Environmental testing is performed by environmental testing labs, water authorities, utilities, government and state health labs, and agrochemical or pesticide manufacturers [24, 25]. The United States Environmental Protection Agency (U.S. EPA) is the regulatory agency responsible for protecting the environment and public health in response to regulations such as the Clean Water Act, Clean Air Act, *Toxic Substance Control Act (TSCA) and Resource Conservation and Recovery Act (RCRA) passed by the U.S. Congress* [25]. Hundreds of official test methods for the various environmental monitoring programs (drinking water, waste water, toxic substances and pesticides, solid wastes, and air samples) are available, with many of them using HPLC [25]. Table 13.4 summarizes the common sample types, analytes, and the HPLC advantages in environmental testing. Brief descriptions and examples of ion chromatography (IC) applications can be found in Section 13.5.

### 13.3.1 Listing of U.S. EPA Test Methods Using HPLC

Many U.S. EPA methods utilize HPLC. These are 500-Series methods for drinking water, 600-Series methods for waste water, 8000-Series methods for solid wastes (SW-846), and toxic organics (TO) methods for TO in ambient air samples [25]. An updated list can be found on the U.S. EPA website (www.epa.gov), and an index published in 2003 [26]. Selected examples are listed in Table 13.5. Note the emergence of HPLC/MS/MS as the platform methodology for multiple-contaminant assays with reduced sample preparation.

### 13.3.2 Pesticides Analysis

Pesticide analysis methods developed by manufacturers, in support of registration petitions to the regulatory agencies, are usually company proprietary. These are mostly HPLC methods with UV or MS detection. These methods are used to generate assay, stability, residue, and metabolism data. In the United States, regulations require pesticide manufacturers to notify the EPA using a Pre-Manufacture Notice (PMN), describing the structure, impurities, by-products, environmental fate, and toxicology data. Pesticide determinations are performed on drinking water, waste water, and solid waste using EPA test methods. Residue analysis of foodstuffs is performed using ASTM, AOAC, or National Pesticides

**Table 13.4. Summary of Environmental Applications of HPLC**

| | |
|---|---|
| Samples | Drinking water, wastewater, solid waste, toxic organics, and air samples. |
| Key applications | Pesticides, herbicides, and plant growth regulators (carbamates, glyphosate, diquat/paraquat, pyrethrins, organophosphorus compounds, |
| | Industrial pollutants (dyes, surfactants, amines, PAHs, PCBs, dioxins, phenols, aldehydes, explosives, pharmaceutics) |
| HPLC advantages | Amenable to thermally labile, nonvolatile, and polar components |
| | Rapid and automated analysis of complex samples |
| | Precise and quantitative (with spectral confirmation) |
| | Quantitative recovery (good sample preparation technique) |
| | Sensitive and selective detection (UV, FL, MS, MS/MS) |
| | Can tolerate injections of large volumes of aqueous samples (0.5–1 ml) |

**Table 13.5. US EPA Test Methods Using HPLC**

| US EPA method | Analyte | Detection |
|---|---|---|
| 531.2/8318 | Carbamates | FL, postcolumn derivatization |
| 537 | Organic contaminants in drinking water | MS/MS |
| 547 | Glyphosate | FL, postcolumn derivatization |
| 549.2 | Diquat/Paraquat | UV |
| 550.1/610/6440B/ 8310/TO13 | PAHs | UV/FL |
| 1694 | Pharmaceuticals and personal care products | MS/MS |
| 8315A | Carbonyl compounds | UV, precolumn derivatization |
| 8321B | Nonvolatiles | MS, MS/MS, or UV |
| 8330B | Explosives (nitroaromatics, nitramines, nitrate esters) | UV |
| TO-11 | Formaldehyde | UV, precolumn derivatization |

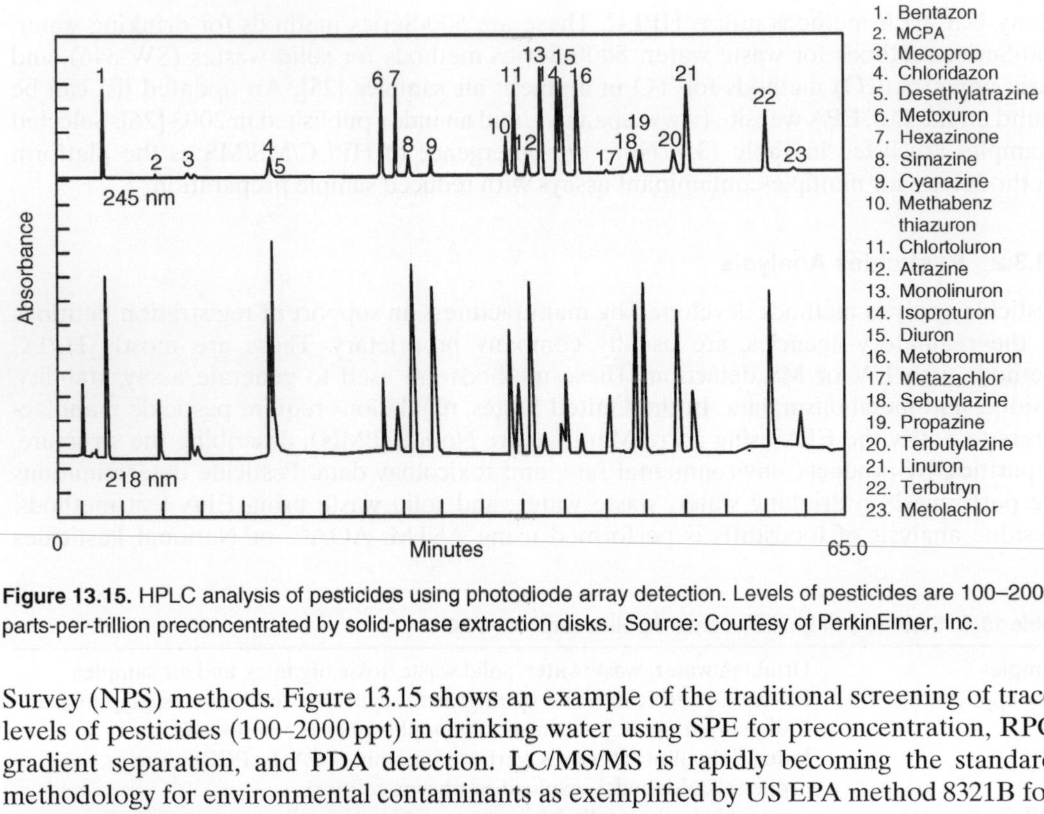


**Figure 13.15.** HPLC analysis of pesticides using photodiode array detection. Levels of pesticides are 100–2000 parts-per-trillion preconcentrated by solid-phase extraction disks. Source: Courtesy of PerkinElmer, Inc.

Survey (NPS) methods. Figure 13.15 shows an example of the traditional screening of trace levels of pesticides (100–2000 ppt) in drinking water using SPE for preconcentration, RPC gradient separation, and PDA detection. LC/MS/MS is rapidly becoming the standard methodology for environmental contaminants as exemplified by US EPA method 8321B for nonvolatiles including many pesticides and herbicides.

***13.3.2.1 Carbamates and Glyphosate*** Carbamates are important insecticides for food crop applications. Carbamate testing (U.S. EPA methods 531.2 and 8318) is performed using postcolumn derivatization with OPA for fluorescence detection [27]. Figure 13.14 shows an example of a carbamate analysis using UHPLC/MS/MS for foodstuff or other environmental samples [23]. Glyphosate is a common herbicide that is monitored in drinking water using U.S. EPA method 547 with a postcolumn reaction system similar to EPA method 531.2.

### 13.3.3 Polynuclear Aromatic Hydrocarbons (PAH)

Polynuclear aromatic hydrocarbons (PAHs) are formed by pyrosynthesis during the combustion of organic matter and have widespread occurrence in the environment. PAHs are found as trace pollutants in soil, air particulate matter, water, tobacco tar, coal tar, used engine oil, and foodstuffs such as barbequed meat. Many PAHs are carcinogenic in experimental animals (e.g. benzo(a)pyrene) and are implicated as the causative agents in human lung cancer for cigarette smokers. Analytical techniques for the separation, identification, and quantitation of PAHs include HPLC and GC. Advantages of HPLC are the ability to resolve isomeric PAHs, and the selective and sensitive quantitation by UV, fluorescence or MS detection. U.S. EPA methods for PAHs are 550.1, 610, and 8310.

#### *13.3.3.1 Case Study: Quick Turnaround Analysis of PAHs in Multimedia Samples*

This case study describes a quick turnaround HPLC method for the analysis of 16 priority pollutant PAHs in multimedia samples such as contaminated soil, air particulate matter, used engine oil, and water samples [28]. The use of rapid extraction, direct injection of sample extracts, and improved HPLC separation coupled with programmed fluorescence and UV detection reduce the total assay time from one to two days for the traditional EPA test methods to about one hour. Method sensitivity was 20 ppb for soil/sediment, low μg/1000 $M^3$ for air particulate matter, and low ppt levels for water samples [28]. Example chromatograms of the selective detection of PAHs in marine sediment and an air particulate matter extract are shown in Figures 13.16 and 13.17. Advantages, limitations, and method validation data are described in the reference [28].

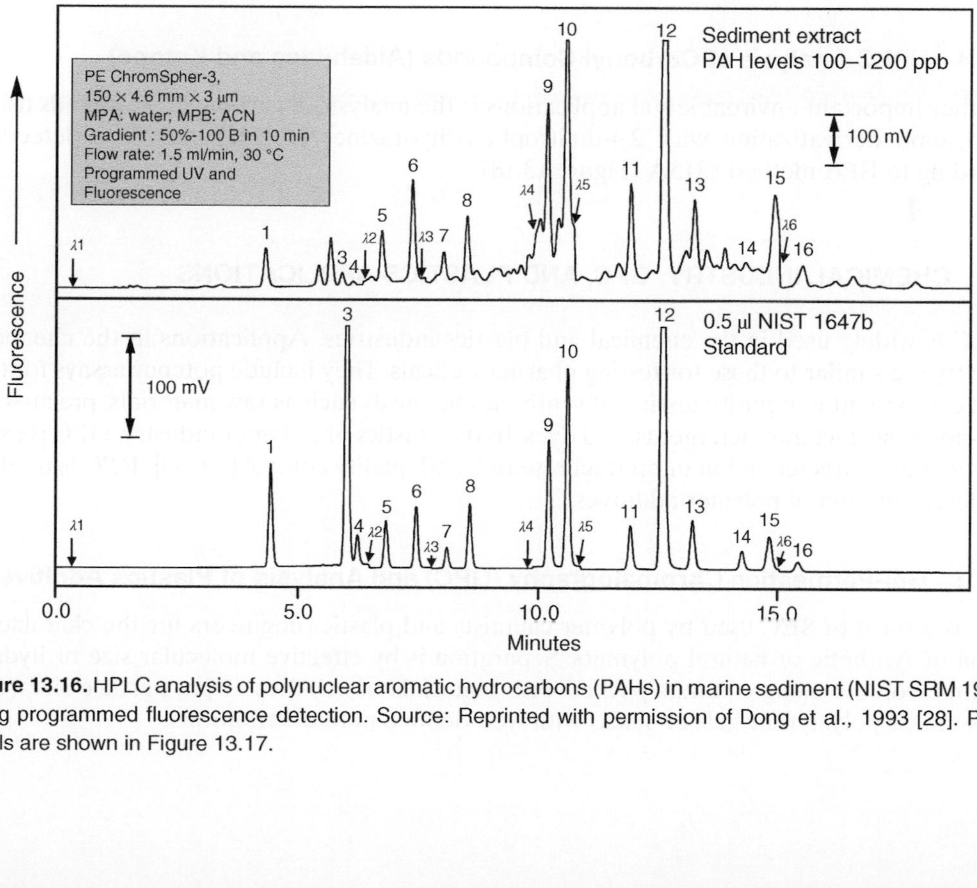


**Figure 13.16.** HPLC analysis of polynuclear aromatic hydrocarbons (PAHs) in marine sediment (NIST SRM 1941) using programmed fluorescence detection. Source: Reprinted with permission of Dong et al., 1993 [28]. Peak labels are shown in Figure 13.17.

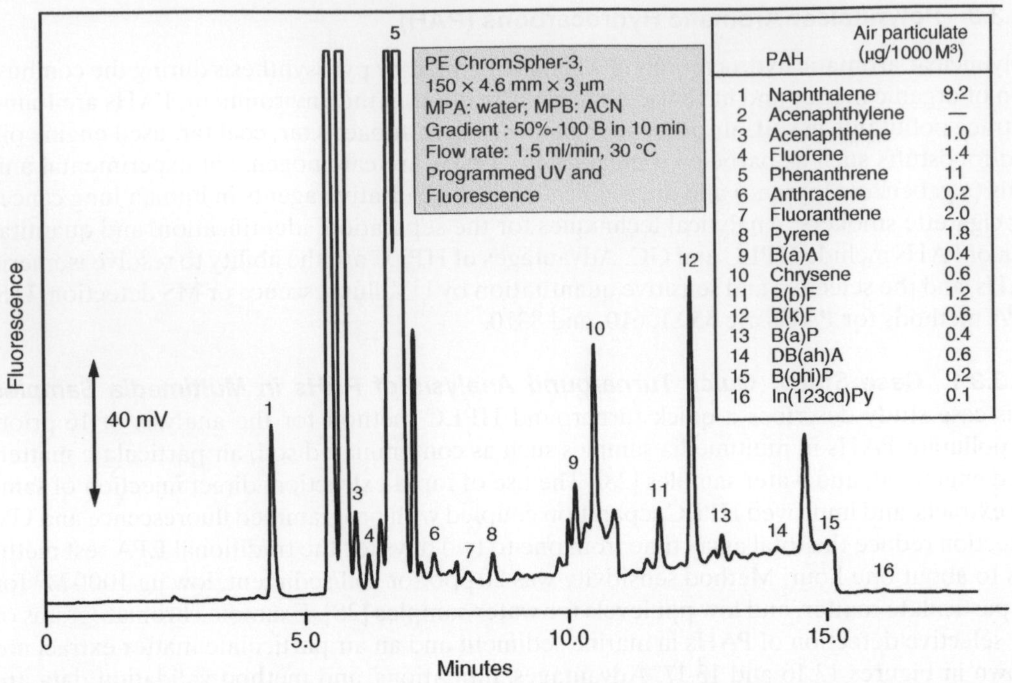


**Figure 13.17.** HPLC analysis of PAHs in an air particulate extract sample collected at a site near Long Island Expressway in New York, using programmed fluorescence detection. Source: Reprinted with permission of Dong et al. 1993 [28].

### 13.3.4 HPLC Analysis of Carbonyl Compounds (Aldehydes and Ketone)

Another important environmental applications is the analysis of carbonyl compounds using pre-column derivatization with 2,4-dinitrophenylhydrazine (DNPH) with UV detection according to EPA method 8315A (Figure 13.18).

## 13.4 CHEMICAL INDUSTRY, GPC, AND PLASTICS APPLICATIONS

HPLC is widely used in the chemical and plastics industries. Applications in the chemical industry are similar to those for testing pharmaceuticals. They include potency assays for the main component and purity testing of synthetic chemicals such as raw materials, precursors, monomers, surfactants, detergents, and dyes. In the plastics or polymer industry, GPC is used for polymer characterization in product research and quality control [29, 30]. RPC is used in the determination of polymer additives.

### 13.4.1 Gel-Permeation Chromatography (GPC) and Analysis of Plastics Additives

GPC is a form of SEC used by polymer chemists and plastics engineers for the characterization of synthetic or natural polymers. Separation is by effective molecular size or hydrodynamic volume using columns packed with materials of sub-3 to 10 μm particle size (e.g. cross-linked polystyrene gels or silica) with well-defined pore distributions.

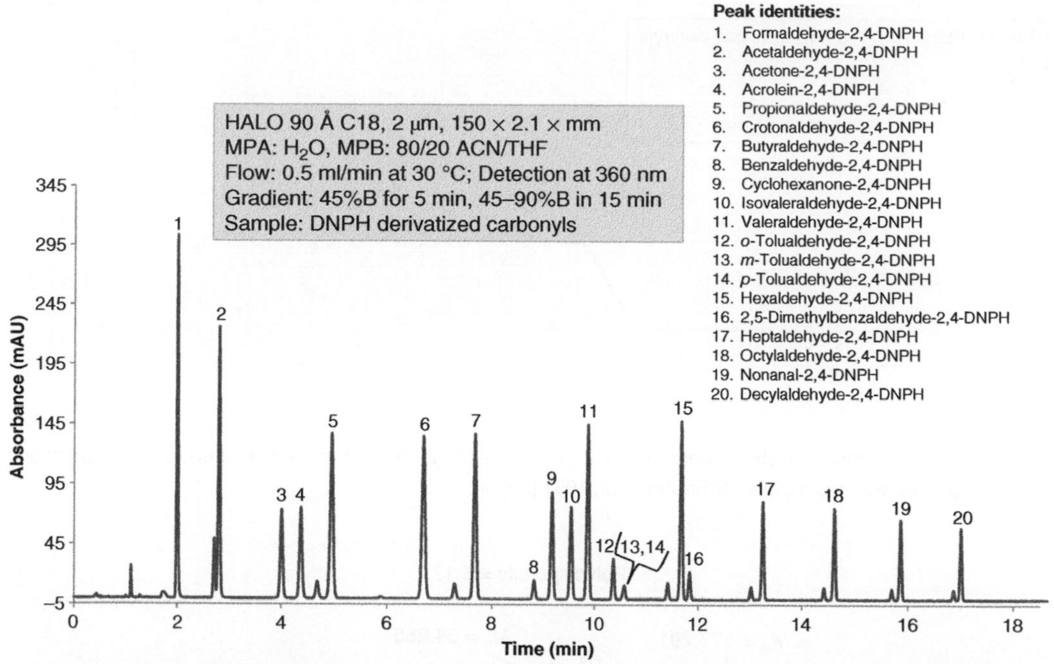


**Figure 13.18.** HPLC analysis of carbonyl compounds (aldehydes and ketone) using precolumn derivatization with 2,4-dinitrophenylhydrazine (DNPH) with UV detection according to EPA method 8315A. The method is also used to monitor industrial or air pollution samples (EPA TO-11). Source: Reproduced with permission of Advanced Materials Technology (AMT).

A typical GPC system uses an organic solvent as the mobile phase (e.g. tetrahydrofuran [THF] or toluene) delivered by a precise pump and a refractive index detector [30]. In GPC, large macromolecules are excluded from most intraparticular pores and emerge first. Small molecules penetrate more pores and elute the latest. Medium-size molecules permeate a fraction of the pores and elute with intermediate retention times according to their molecular weights (sizes). GPC is best used for high-molecular-weight components (molecular weight >1 000 Da). All solutes should elute before the sample diluent or any total permeation marker. Any adsorptive interaction of the solute with the packing must be suppressed by mobile phase additives. GPC has short and predictable retention times but with moderate resolution and limited peak capacity. Separating analytes with size differences of <5–10% is difficult in GPC.

GPC is often used in sample profiling studies shown in Figure 13.19 for use in production QC of plastics materials. Here, "good and bad" batches of alkyd resins (adhesive) were profiled [30]. The high-molecular-weight components in the "bad" batch were causing processing problems and deficiency in product quality. Molecular weight calibration may not be required for this type of profiling or screening application.

For characterization of molecular weight averages and distribution (MWD), the column set must first be calibrated using a set of molecular weight standards (such as polystyrene or polyethylene glycol, see Figure 1.8). These molecular weight averages and MWD are important because they control the thermal, physical, mechanical, and processing properties of the polymer (e.g. number-average molecular weight ($M_n$) – tensile strength, hardness;

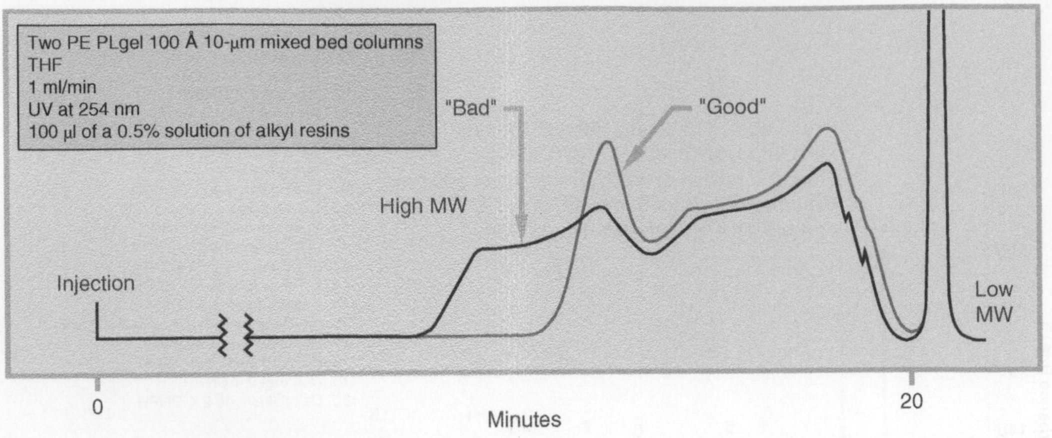


**Figure 13.19.** Gel-permeation chromatography (GPC) analysis of "good and bad" alkyl resins using UV detection. Source: Reprinted with permission of Reuter et al. 1991 [30].

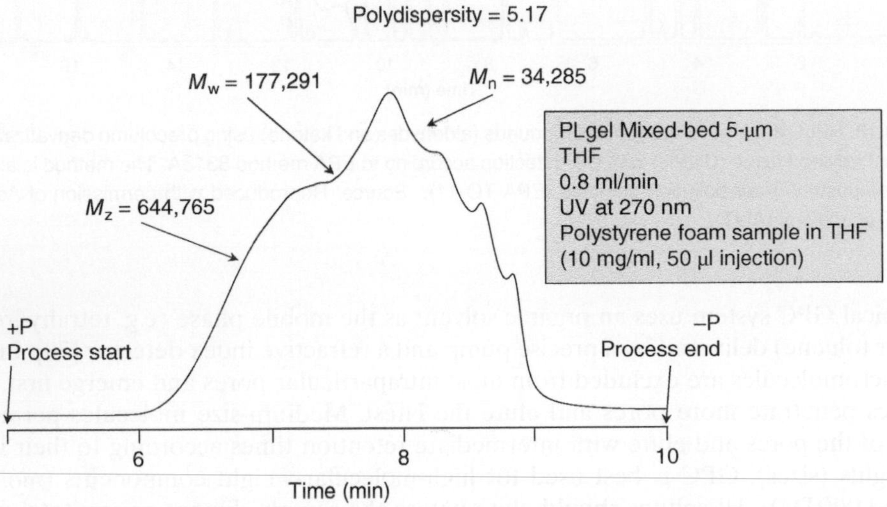


**Figure 13.20.** GPC analysis of a polystyrene foam sample using UV detection. Annotations are calculated molecular weight averages and polydispersity.

weight-average molecular weight ($M_w$) – brittleness, flow properties; z-average molecular weight ($M_z$) – flexibility, stiffness; viscosity-average molecular weight ($M_v$) – extrudability, molding properties). Readers are referred elsewhere for details for these calculations [29, 30]. Specialized GPC software coupled to the CDS is required for these calculations, using a set of molecular weight calibration standards with GPC columns of different pore sizes [30]. Mixed-bed columns are useful for screening samples of unknown molecular weights or of wide MWD. Figure 13.20 shows a sample of polystyrene foam (from a white coffee cup) with annotations of various molecular weight averages and polydispersity. If an appropriate set of calibration standard is not available, other techniques such as universal calibration,

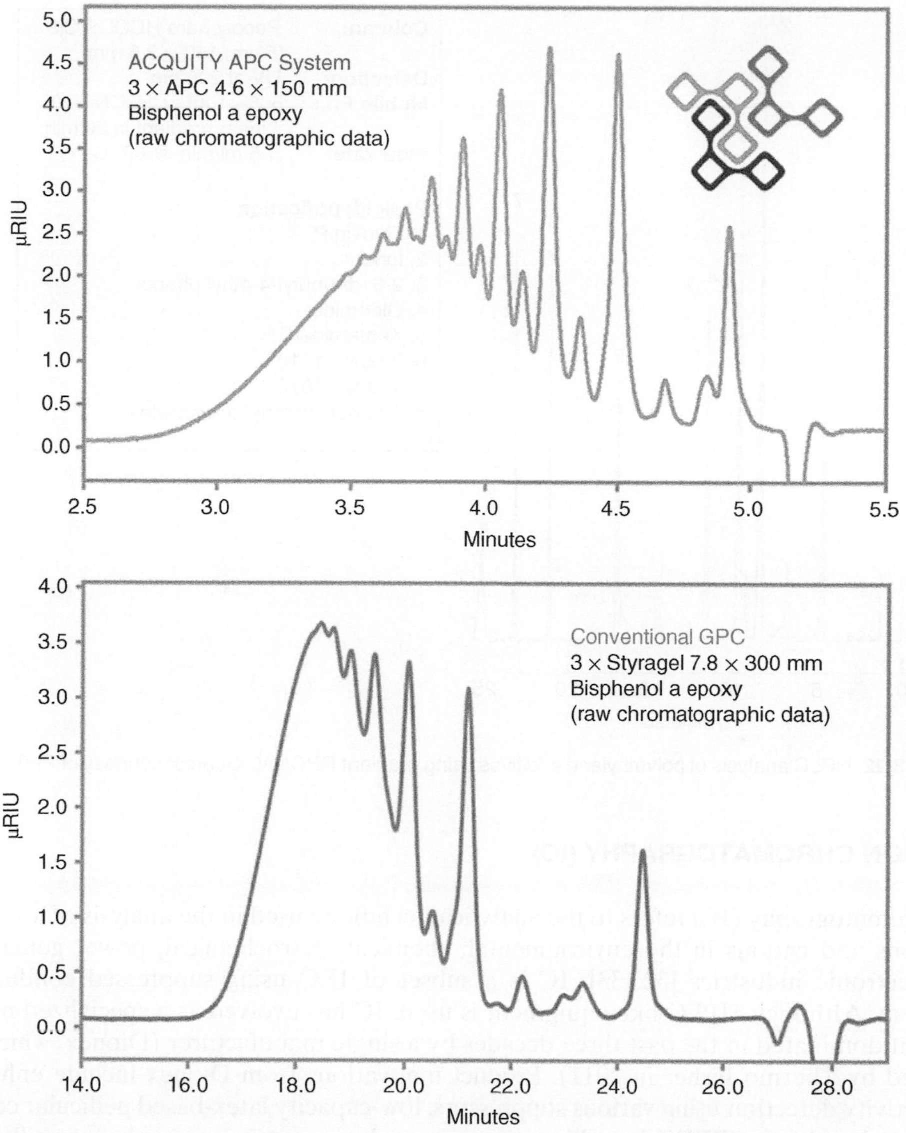


**Figure 13.21.** Comparative GPC chromatogram of a Bisphenol A Epoxy sample using RI detection using a conventional and an acquity advanced polymer chromatography (APC) system. GPC conditions in the bottom chromatogram: three ACQUITY APC XT columns 150 × 4.6 mm, 125 and 45 Å in series; mobile phase – THF at 40 °C; RI detection. Source: Reproduced with permission of Waters Corporation [31].

multiple light scattering detector, or MS can be used [29]. Note that UHPLC technology is impacting GPC analysis as exemplified by the introduction of a high-performance GPC system capable of faster analysis as shown in Figure 13.21 [31].

Figure 13.22 shows an example of the gradient analysis of polyethylene additives using RPC and UV detection.

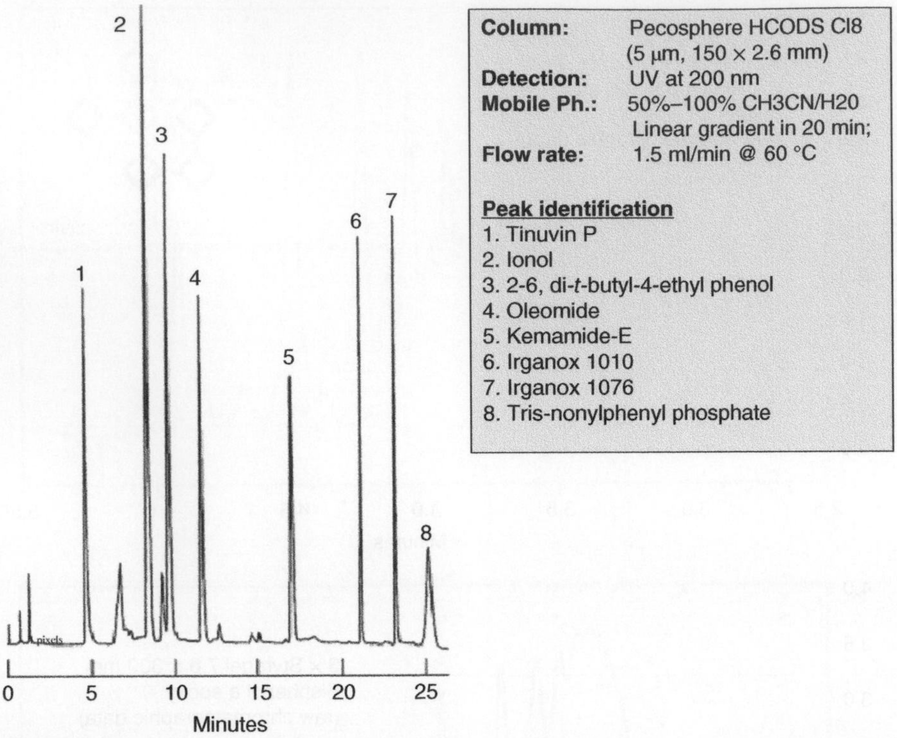
Column: Pecosphere HCODS Cl8 (5 μm, 150 × 2.6 mm)
Detection: UV at 200 nm
Mobile Ph.: 50%–100% CH3CN/H20 Linear gradient in 20 min;
Flow rate: 1.5 ml/min @ 60 °C
Peak identification
1. Tinuvin P
2. Ionol
3. 2-6, di-t-butyl-4-ethyl phenol
4. Oleomide
5. Kemamide-E
6. Irganox 1010
7. Irganox 1076
8. Tris-nonylphenyl phosphate
1
2
3
4
5
6
7
8
0
5
10
15
20
25
Minutes

**Figure 13.22.** HPLC analysis of polyethylene additives using gradient RPC/UV. Source: courtesy of PerkinElmer, Inc.

## 13.5 ION CHROMATOGRAPHY (IC)

Ion chromatography (IC) refers to the analytical technique used in the analysis of low levels of anions and cations in the environmental, chemical, petrochemical, power generation, and electronic industries [32, 33]. IC is a subset of IEC using suppressed conductivity detection. Although HPLC-like equipment is used, IC has evolved as a specialized market segment dominated in the past three decades by a single manufacturer (Dionex, which was acquired by Thermo Fisher in 2011). Product innovations from Dionex include enhanced conductivity detection using various suppressors, low-capacity latex-based pellicular column packing, metal-free (PEEK-based) equipment, and reagent-free microbore/capillary IC systems. Other manufacturers include Metrohm and Shimadzu. Figures 13.23 and 13.24 show examples of IC analysis of anions and cations showing the excellent sensitivity and selectivity of this technique [33]. IC is used routinely for environmental testing (e.g. EPA Method 300) and pharmaceutical analysis (described in USP<1065> and used in many USP monographs) [34].

## 13.6 LIFE SCIENCES APPLICATIONS

HPLC and HPLC/MS life science applications focus on the separation, quantitation, and purification of biomolecules such as proteins, peptides, amino acids, lipids, nucleic acids, nucleotides, and glycans and in bioscience research areas such as proteomics, metabolomics, glycomics, and clinical diagnostics.

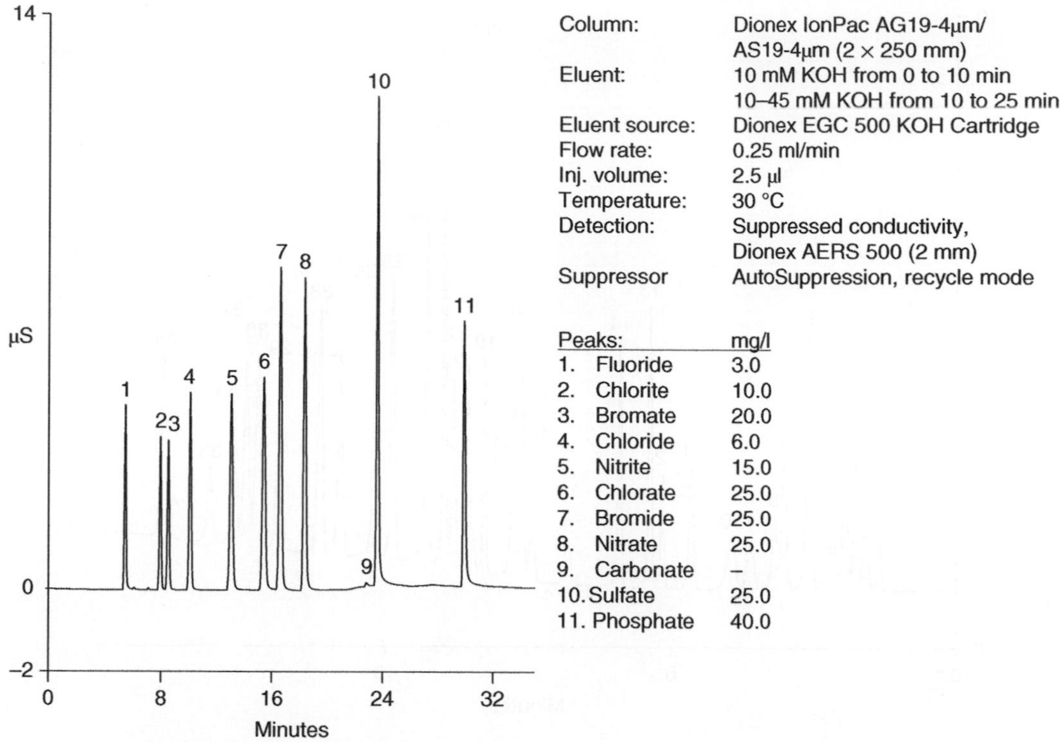
14
μS
0
−2
0
8
16
24
32
Minutes
1
2
3
4
5
6
7
8
9
10
11
Column: Dionex IonPac AG19-4μm/ AS19-4μm (2 × 250 mm)
Eluent: 10 mM KOH from 0 to 10 min
10–45 mM KOH from 10 to 25 min
Eluent source: Dionex EGC 500 KOH Cartridge
Flow rate: 0.25 ml/min
Inj. volume: 2.5 μl
Temperature: 30 °C
Detection: Suppressed conductivity, Dionex AERS 500 (2 mm)
Suppressor AutoSuppression, recycle mode
Peaks: mg/l
1. Fluoride 3.0
2. Chlorite 10.0
3. Bromate 20.0
4. Chloride 6.0
5. Nitrite 15.0
6. Chlorate 25.0
7. Bromide 25.0
8. Nitrate 25.0
9. Carbonate –
10. Sulfate 25.0
11. Phosphate 40.0

**Figure 13.23.** IC Separation of the common anions on IonPac AS19-4μm/AS-19-4μm (250 × 2.0 mm) column. Source: Courtesy of Thermo Fisher Scientific.

### 13.6.1 Proteins, Peptides, and Amino Acids

Chromatography of proteins is performed by IEC, RPC, and SEC, as well as affinity and hydrophobic interaction chromatography (HIC). Historically, the chromatography of large proteins is problematic using RPC with silica-based materials for the following reasons [35–39].

- Hydrophobic and ionic characteristics of proteins, leading to their adsorption in most HPLC packings with acidic silanols
- Possible conformational changes and denaturing of proteins during the RPC process
- Slow diffusivities and sorption kinetics (large Van Deemter C term)
- “Restricted” diffusion in small-pore supports leads to band broadening

As mentioned in Chapter 3, protein separations require specialized columns packed with wide-pore polymer supports or silica materials with low silanol activity. Figure 13.25 shows an example of an RPC gradient separation of a protein mixture using a column packed with Vydac C4-bonded phase on a wide-pore (300 Å) silica support. Vydac was a first-generation high purity silica material developed in the 1980s and found to be amenable to RPC of proteins. Vydac (manufactured by the Separation Group) was acquired by W. R. Grace in 2000. Gradient elution using 0.1% trifluoroacetic acid (TFA) and acetonitrile is commonly used with UV detection at 215 or 280 nm. The C4-bonded phase is preferred over the more hydrophobic C18 for protein separations. Polymer support materials such

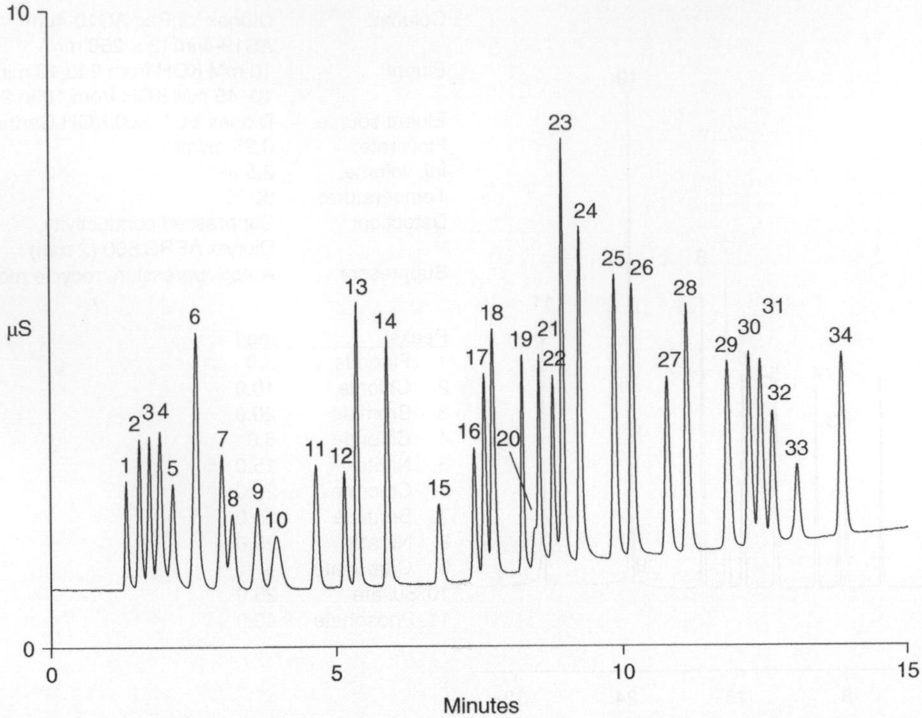


**Figure 13.24.** Fast gradient separation of inorganic anions and organic acids on an IonPac AS11 column. Column: 4 × 250 mm IonPac AS11. Eluent: 0.5 mM NaOH for 2 minutes, followed by a linear gradient from 0.5 to 5 mM NaOH in 3 minutes and a second linear gradient from 5 to 38.25 mM NaOH for the last 10 minutes. Flow rate: 2.0 ml/min. Injection volume: 10 μl. Detection: Suppressed conductivity utilizing the anion self-generating suppressor (4 mm), recycle mode. Ions: (1) isopropylmethylphosphonate (5 mg/l); (2) quinate (5 mg/l); (3) fluoride (1 mg/l); (4) acetate (5 mg/l); (5) propionate (5 mg/l); (6) formate (5 mg/l); (7) methylsulfonate (5 mg/l); (8) pyruvate (5 mg/l); (9) chlorite (5 mg/l); (10) valerate (5 mg/l); (11) monochloroacetate (5 mg/l); (12) bromate (5 mg/l); (13) chloride (2 mg/l); (14) nitrite (5 mg/l); (15) trifluoroacetate (5 mg/l); (16) bromide (3 mg/l); (17) nitrate (3 mg/l); (18) chlorate (3 mg/l); (19) selenite (5 mg/l); (20) carbonate (5 mg/l); (21) malonate (5 mg/l); (22) maleate (5 mg/l); (23) sulfate (5 mg/l); (24) oxalate (5 mg/l); (25) ketomalonate (10 mg/l); (26) tungstate (10 mg/l); (27) phthalate (10 mg/l); (28) phosphate (10 mg/l); (29) chromate (10 mg/l); (30) citrate (10 mg/l); (31) tricarballylate (10 mg/l); (32) isocitrate (10 mg/l); (33) cis-aconitate and (34) trans-aconitate(10 mg/l combined). Source: Pohl et al. 2005 [33]. Copyright 2005. Reprinted with permission of Elsevier.

as cross-linked polystyrene-divinylbenzene, polyethers, and polymethacrylates have been used successfully in bioseparations [39]. Their strength and performance have improved though they still lag behind silica in efficiency. Major advantages are the wider pH range (1–14) and the absence of active silanol groups. Figure 13.26 shows an RPC separation of proteins using a polymer column illustrating an excellent peak shape performance of these columns.

Note that small-pore packings are problematic for large biomolecules, which can become entangled in the pores leading to slower mass transfer and additional band broadening. Figure 13.27 shows comparative chromatograms of a protein mixture on columns packed with 100 and 300 Å polymeric materials, respectively [39]. Large proteins (peaks 5 and 6)

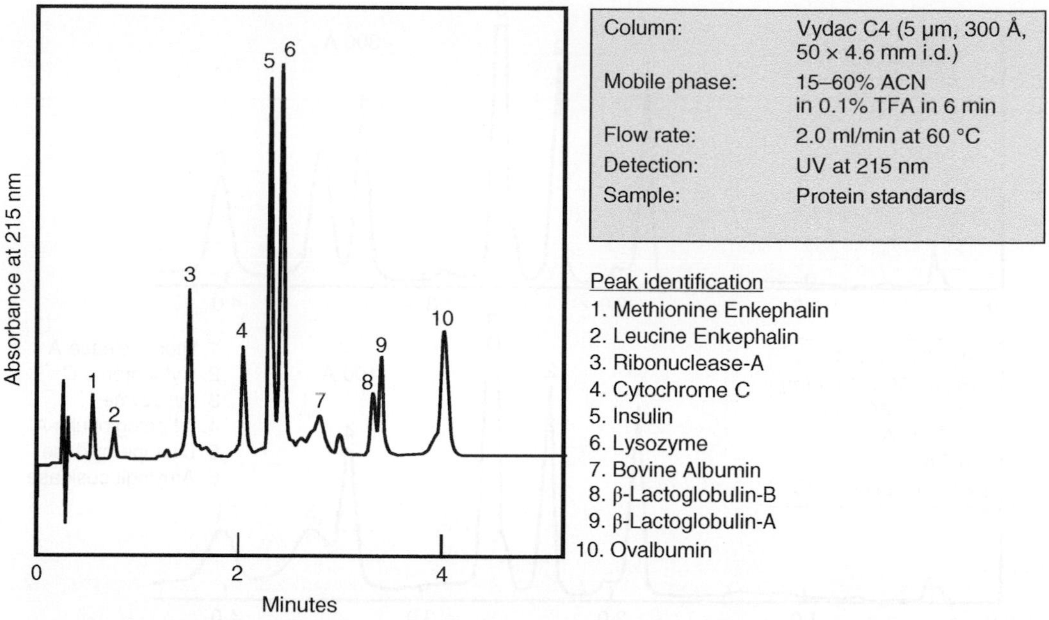


**Figure 13.25.** HPLC analysis of a protein mixture using RPC/UV with a short Vydac C4 column using a high-purity silica support material.

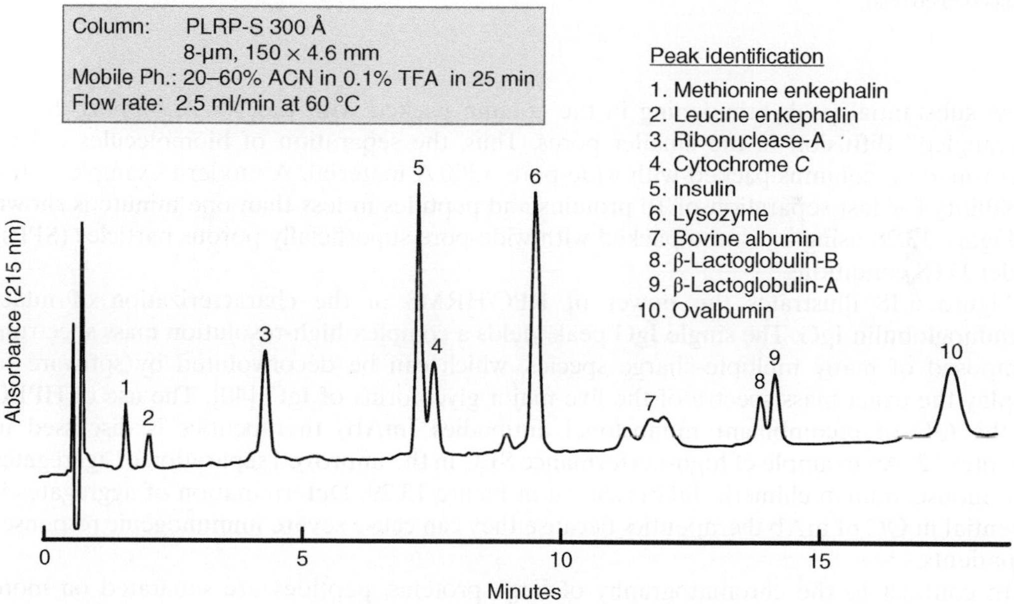


**Figure 13.26.** RPC/UV chromatogram of a protein mixture using a column packed with polymeric materials

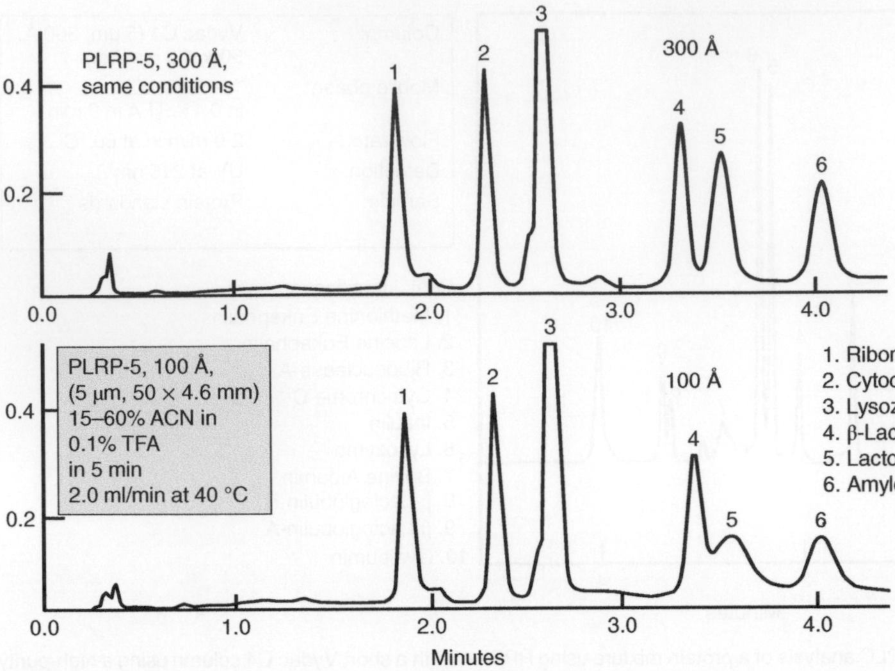


**Figure 13.27.** Comparative chromatograms illustrating the effect of pore size on protein separation. Two larger proteins (peaks 5 and 6) show substantial peak broadening due to entangled diffusion within the smaller pores of the second column.

show substantial peak broadening in the column packed with the 100 Å support due to "entangled" diffusion in the smaller pores. Thus, the separation of biomolecules is best performed on columns packed with wide-pore >300 Å material. A modern example of the capability for fast separation of 10 proteins and peptides in less than one minute is shown in Figure 13.28 using a column packed with wide-pore superficially porous particles (SPPs) under HTS conditions.

Figure 6.18 illustrates the power of RPC/HRMS in the characterization of intact immunoglobulin IgG. The single IgG peak yields a complex high-resolution mass spectrum composed of many multiple-charge species, which can be deconvoluted by software to display the exact mass spectra of the five major glycoforms of IgG [40]. The use of HPLC in the QC of recombinant monoclonal antibodies (mAb) therapeutics is discussed in Chapter 12. An example of high-performance SEC in the improved separation of aggregates of a mouse–human chimeric IgG is shown in Figure 13.29. Determination of aggregates is essential in QC of mAb therapeutics because they can cause severe immunogenic responses in patients.

In contrast to the chromatography of large proteins, peptides are separated on more conventional small-pore C18-bonded phases [38, 39]. Peptide mapping is a common protein characterization technique in which a protein is cleaved by an enzyme to yield peptide fragments, which are then separated by HPLC to yield a peptide map [38] (see Figure 12.9). Before the 1990s, these peptide fragments would be collected for Edman sequencing to determine the amino acid sequence of the protein. This tedious process is now superseded by the use of LC/MS to yield sequence information directly during peptide mapping. A hybrid

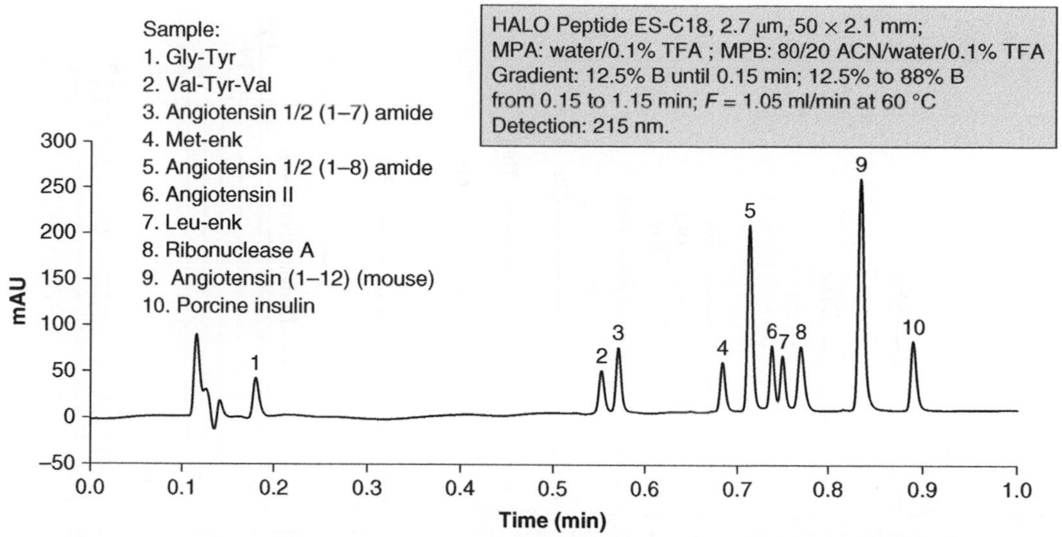


**Figure 13.28.** Fast separation of peptides and small proteins using a short column packed with wide-pore SPP under high-throughput screening conditions. Source: Reproduced with permission of Advanced Materials Technology (AMT).

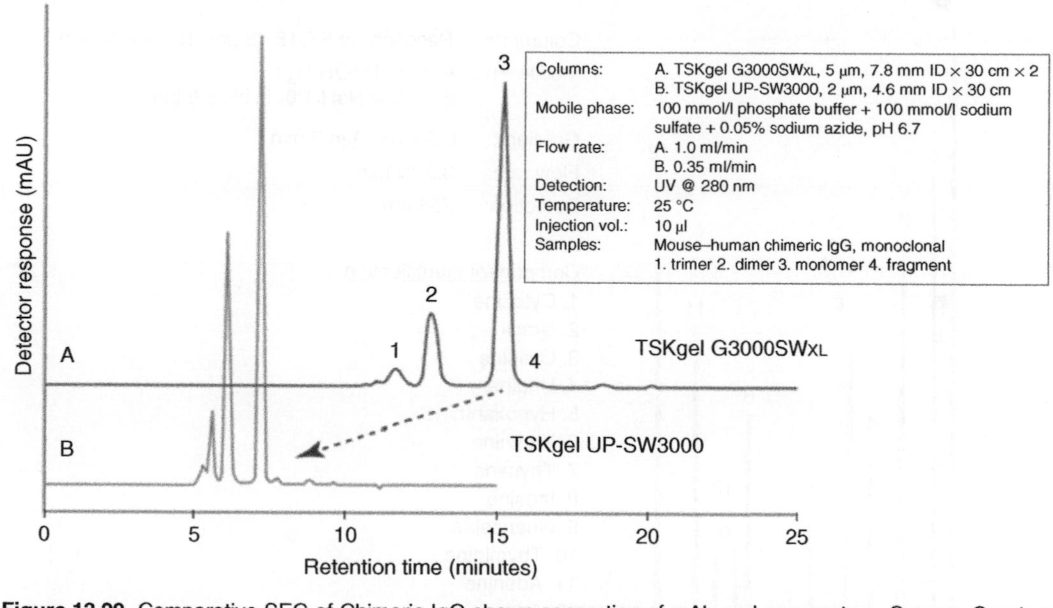


**Figure 13.29.** Comparative SEC of Chimeric IgG shown separation of mAb and aggregates. Source: Courtesy of TOSOH Bioscience.

quadrupole-time-of-flight (Q-TOF) MS or an OrbiTrap is highly suited for this *de novo* sequencing technique.

Amino acid analysis has been discussed in the food applications section of this chapter. In life science applications where sample size can be very limited, the high-sensitivity precolumn derivatization technique is preferred as shown in Figure 13.30 [41].

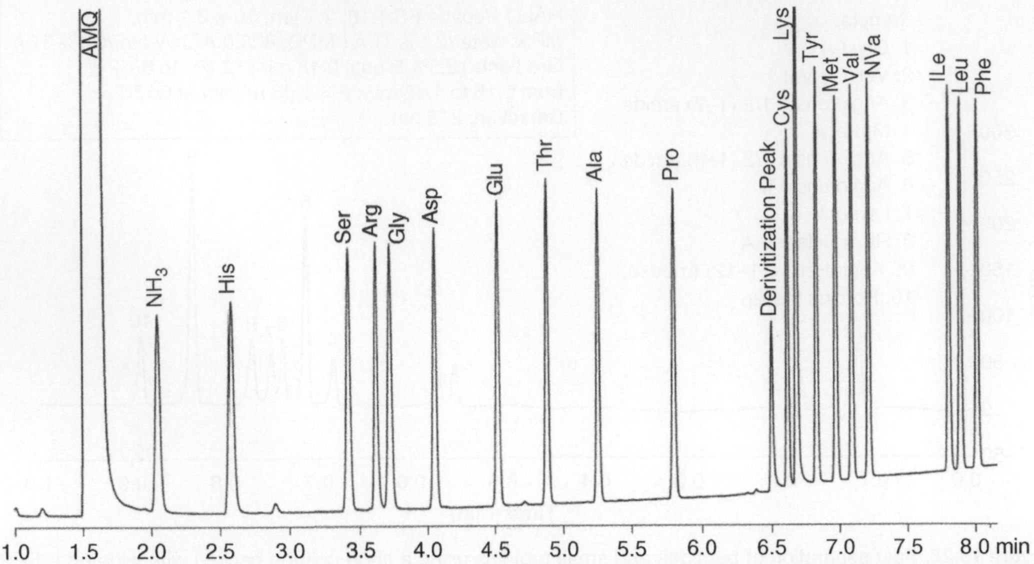


**Figure 13.30.** UHPLC analysis of precolumn derivatized amino acids (10 picomoles) using Waters AccQ.Tag reagents with RPC/UV. Analysis performed on Waters AccQ.Tag Ultra 1.7 μm, 100 × 2.1 mm column and eluted with AccQ.Tag Ultra Eluent A and B. Source: Reproduced with permission of Waters Corporation [41].

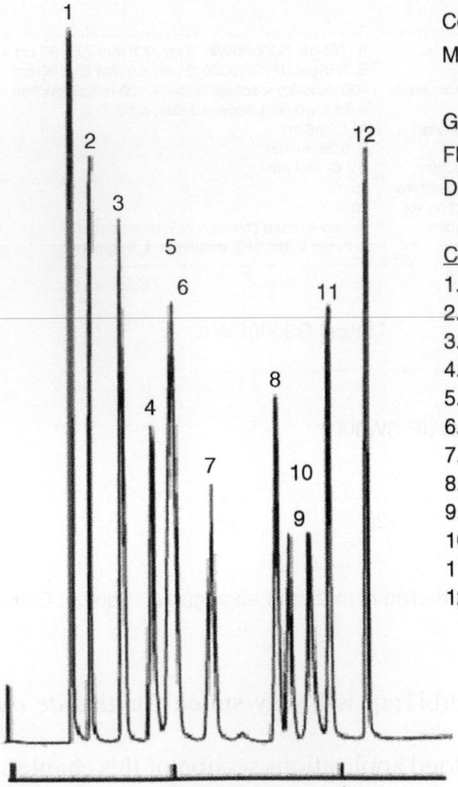

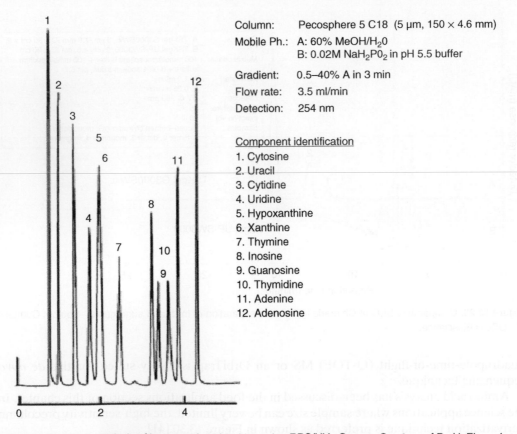

Column: Pecosphere 5 C18 (5 μm, 150 × 4.6 mm)
Mobile Ph.: A: 60% MeOH/$H_2O$
B: 0.02M $NaH_2PO_2$ in pH 5.5 buffer
Gradient: 0.5–40% A in 3 min
Flow rate: 3.5 ml/min
Detection: 254 nm

Component identification
1. Cytosine
2. Uracil
3. Cytidine
4. Uridine
5. Hypoxanthine
6. Xanthine
7. Thymine
8. Inosine
9. Guanosine
10. Thymidine
11. Adenine
12. Adenosine

**Figure 13.31.** HPLC analysis of bases and nucleosides using RPC/UV. Source: Courtesy of PerkinElmer, Inc.

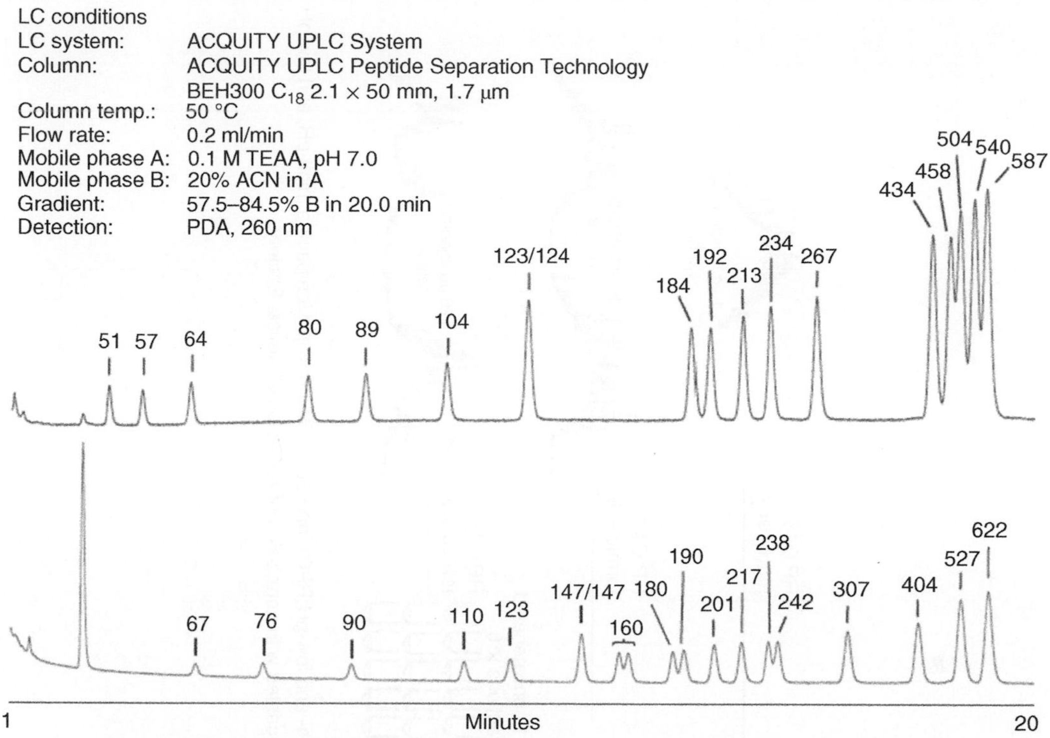


**Figure 13.32.** HPLC analysis of duplex nucleic acids (pBR322-DNA HAE-III digest) using gradient RPC/UV detection at 260 nm. Source: McCarthy and Gilar 2015 [42]. Reproduced with permission of Waters Corporation.

### 13.6.2 Bases, Nucleosides, Nucleotides, Oligonucleotides, and Nucleic Acids

HPLC analysis of bases and small nucleotides is best accomplished by RPC or IEC, as exemplified by the example in Figure 13.31 [42]. Oligonucleotides, DNA restriction fragments, and polymerase chain reaction (PCR) products are analyzed by gel electrophoresis, capillary electrophoresis, chip-based devices, ion-pair RPC [42], or HPLC using IEC [43] (as shown in Figure 13.32). HPLC is useful for quantitation and purification of oligonucleotides and PCR products [43].

### 13.6.3 Bioscience Research in Proteomics, Metabolomics, Glycomics and Clinical Diagnostics

The emergence of HPLC/MS as a preeminent investigative tool has opened up many "OMICS" research areas such as proteomics, metabolomics, and glycomics with strong impacts on the discovery of biomarkers for disease diagnostics and new drug development.

Proteomics is the study of the entire set of proteins encoded by a genome [44–46]. As proteomics samples are too complex for a single HPLC separation, 2D-LC operating in the off-line or online mode is often performed. The traditional approach is to use IEC (strong cationic, SCX) to fractionate the complex sample, followed by RPC-MS to characterize each fraction off-line superseded by automated 2D-LC/MS workflow as exemplified in Figure 13.33 [47].

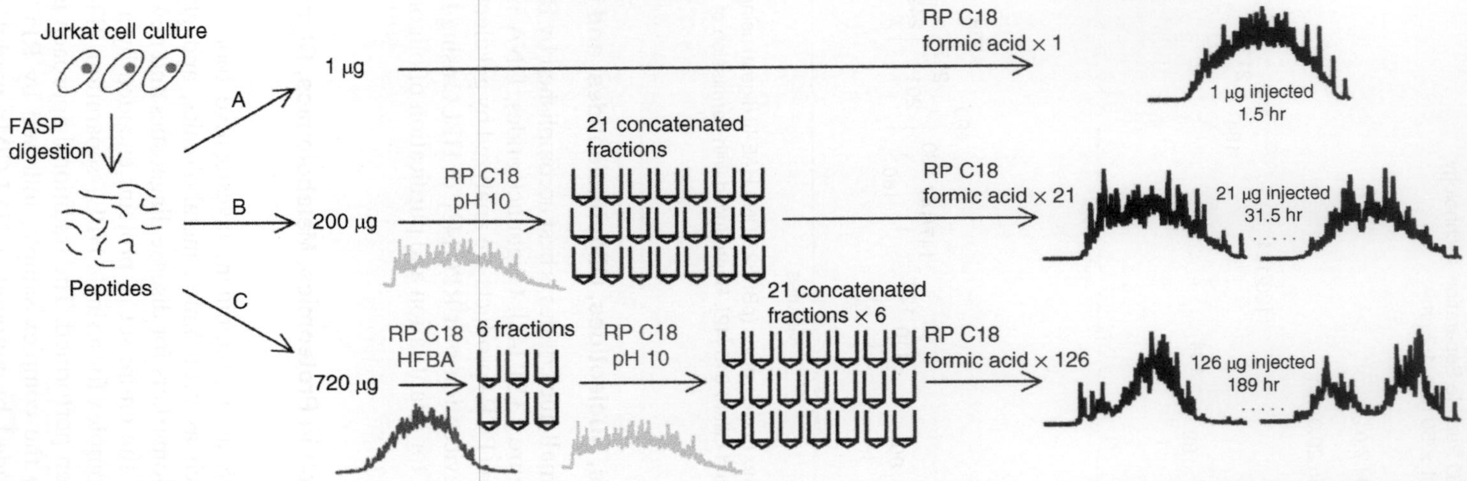


**Figure 13.33.** Overview of separation setup and workflow used in 1D (90 min), 2D (high- and low-pH RPC), and 3D LC/MS (low-pH ion pairing, high-pH, and low-pH formic acid) of proteomic samples. Source: Spicer et al. 2016 [47]. Copyright 2016. Reprinted with permission of American Chemical Society.

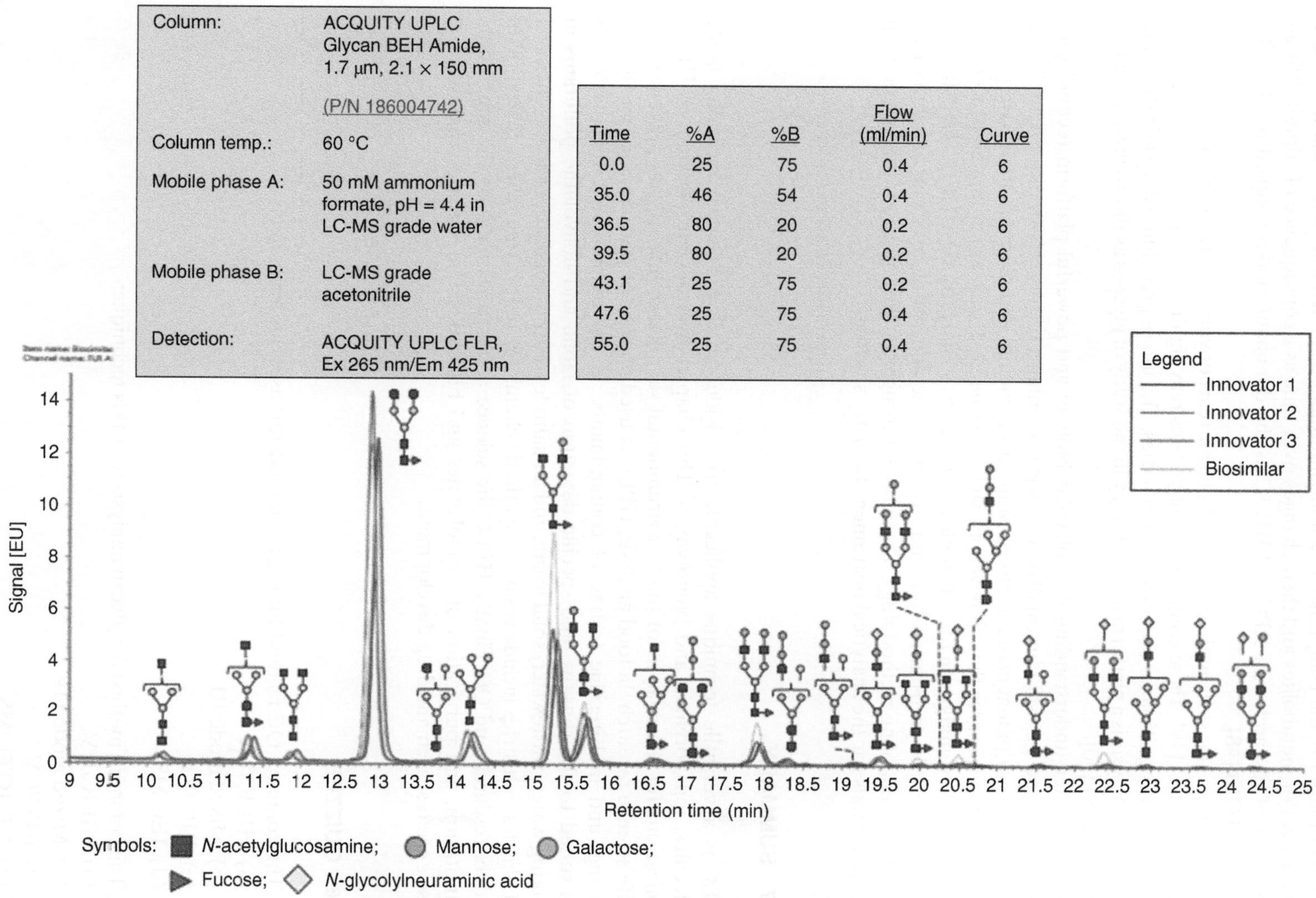


**Figure 13.34.** Glycan profiles using HILIC of innovator vs. biosimilar Infliximab (Remicade). Source: Alley and Yu 2016 [50]. Reproduced with permission of Waters Corporation.

Metabolomics is the comprehensive and simultaneous systematic determination of the complete set of metabolites and their changes over time as a consequence of stimuli. It is an active area of research with RPC and HILIC as the dominant modes coupled to MS (TQ, QTOF, or TOF) [48].

Glycomics is the systematic study of all glycan (carbohydrate) structures of a given cell type or organism [49]. These complex carbohydrates play significant roles in structural diversity and cell recognition for monoclonal antibodies. The use of precolumn derivatization and HILIC/MS analysis is a platform methodology in the study of glycans as shown in the example in Figure 13.34 [50].

With a better understanding of molecular biology and powerful platform technologies such as PCR and LC/MS, clinical diagnostics is becoming a major opportunity for analytical chemists and equipment manufacturers. Many instrumental and pharmaceutical companies are investing in this area though new approaches are needed, such as automated procedures to isolate the key analytes in these complex matrices [51–53]. Here, 2D-UHPLC-HRMS with automated sample preparation frontends to isolate the appropriate isoforms as biomarkers could be the enabling methodology. These breakthroughs in clinical diagnostics would reap huge potentials for the analytical instrument Industry shortly.

## 13.7 SUMMARY

HPLC is a versatile technique applicable to a wide range of analytes, including labile molecules, ions, organic, and biopolymers. This chapter provides an overview of HPLC applications for the analysis of food, environmental samples, chemicals, and polymers and in life science research. In food analysis, HPLC is used in product research, QC, nutritional labeling, and multiresidue testing of contaminants. In environmental testing, HPLC is well-suited to the sensitive and specific detection of labile and nonvolatile pollutants in drinking water, wastewater, solid waste, and air samples. HPLC applications in the chemical and plastics industries include assays of synthetic chemicals and polymer characterization in product research and QC. Finally, HPLC life science applications focus on the separation, quantitation, and purification of biomolecules and biomarkers for disease diagnostics in bioscience research and drug development.

## 13.8 QUIZZES

1. The common QC method for sugars in foodstuff uses this technique:
   (a) HILIC/UV
   (b) Mixed mode/RI
   (c) RPC/RI
   (d) IEC/MS

2. The common method for glycan analysis uses this technique:
   (a) HILIC/UV
   (b) Mixed mode/UV
   (c) RPC/UV
   (d) HILIC/FL/MS

**3.** The common method for detecting adulteration of edible oil uses this technique:
(a) HILIC/MS
(b) Mixed mode/RI
(c) RPC/CAD
(d) NARP/RI

**4.** The common method for testing organic acids in wine DOES NOT use this technique:
(a) RPC/CAD
(b) Mixed mode/RI
(c) RPC/RI/UV
(d) IEC/MS

**5.** Trace analysis of carbamates in foodstuff and environmental sample DOES NOT use this detection technique:
(a) RPC/MS
(b) RI
(c) Post-column FL
(d) HILIC/MS

**6.** PAHs are NOT generally analyzed by HPLC with detection by this technique:
(a) Programmed FL
(b) UV
(c) Postcolumn FL
(d) MS

**7.** In GPC, the primary separation mode is based on differentiation by
(a) hydrogen bonding
(b) hydrophobicity
(c) molecular weight
(d) molecular size

**8.** Ion chromatography is a subset of
(a) IEC
(b) SEC
(c) RPC
(d) mixed mode

**9.** Separation of large proteins requires silica or polymer support of this pore size:
(a) 5000 Å
(b) 100 Å
(c) 60 Å
(d) 300 Å

**10.** A critical quality attribute of aggregates in mAb therapeutics, which can illicit immune response of the patient require an assay using this chromatographic mode:
(a) RPC
(b) IEC
(c) SEC
(d) SFC

### 13.8.1 Bonus Questions

In your own words, describe the challenges in the bioseparations of proteins using column packed with silica support.

What are the most prominent applications of 2D-LC?

## 13.9 REFERENCES


1. Nollet, L.M.L. (ed.) (2012). *Food Analysis by HPLC*, 3e. Boca Raton, FL: CRC Press.
2. Nollet, L. (ed.) (1992). *Food Analysis by HPLC*, 2e. New York: Marcel Dekker.
3. Grasfeld-Husgen, A. and Schuster, R. (1996). *HPLC for Food Analysis: A Primer*. Waldbronn, Germany: Agilent Technologies, 2011 Publication number 12–5965–5124E.
4. Naushad, M., Khan, M.R., and Alothman, Z.A. (2014). *Ultra Performance Liquid Chromatography Mass Spectrometry: Evaluation and Applications in Food Analysis*. Boca Raton, FL: CRC Press.
5. (2016). *Official Methods of Analysis of the Association of Official Analytical Chemists (AOAC)*, 20e. Maryland: Rockville.
6. (2016). Food Labeling Regulations: Final Rule. *Federal Register* 81 (103): 33742, 21 CFR Part 101, U.S. Food and Drug Administration, College Park, Maryland.
7. Dong, M.W. (1993). *HPLC System for Carbohydrate Analysis, Carbohydrate Analysis Cookbook*. Norwalk, CT: Perkin-Elmer.
8. Benvenuti, M.E., Cleland, G., and Burgess, J. (2017). Waters Application Note, 720005609EN, Milford, Massachusetts.
9. Lisa, M., Lynen, R., Holcapek, M., and Sandra, P. (2007). *J. Chromatogr. A* 1176: 135.
10. Panfili, G., Fratianni, A., and Irano, M. (2003). *J. Agri. Food Chem.* 51: 3940.
11. Dong, M.W. (1998). *LCGC* 16 (12): 1092.
12. Benvenuti, M.E., Cleland, G., and Burgess, J. (2017). Waters Application Note, Milford, Massachusetts, 720005017EN.
13. Joseph, S. (2011). Agilent Technologies, Bangalore, India, 5990-7950EN.
14. Dong, M.W. and Guillarme, D. (2017). *LCGC North Amer.* 35 (8): 486–495.
15. Shimelis, O., Cramer, H., and Buchanan, M.D. US Reporter Volume 30.3, Supelco, Bellefonte, Pennsylvania.
16. Vanhoenacker, G., David, F., and Sandra, P. (2009). Agilent Application Note, Waldbronn, Germany, 5990-4694EN.
17. Dong, M.W. (2000). *Today's Chemist at Work* 9 (5): 17.
18. Dong, M.W. and Pace, J.L. (1996). *LCGC* 14 (9): 794.
19. Tuzumski, T. and Sherma, J. (eds.) (2015). *HPLC in Pesticide Residue Analysis*. Boca Raton, FL: CRC Press.
20. Romero-González, R. and Frenich, A.G. (eds.) (2017). *Applications in High Resolution Mass Spectrometry: Food Safety and Pesticide Residue*. Oxford, United Kingdom: Elsevier.
21. Yang, X. and An, R. (2011). Agilent Technologies Application Note, Shanghai, China, 5990-9125EN.
22. Payá, P., Anastassiades, M., Mack, D. et al. (2007). *Anal. Bioanal. Chem.* 389 (6): 1697.
23. Fu, R.J. and Zhai, A. (2013). Agilent Technologies Application Note, 5991-2124EN, Shanghai, China.
24. Patnaik, P. (2017). *Handbook of Environmental Analysis: Chemical Pollutants in Air, Water, Soil, and Solid Wastes*, 3e. Boca Raton, FL: CRC Press.
25. Grosser, Z.A., Ryan, J.F., and Dong, M.W. (1993). *J. Chromatogr.* 642 (1–2): 75.
26. Nelson, P. (2003). *Index to EPA Test Methods*. Boston, MA: U.S. EPA.
27. Dong, M.W., Pickering, M.V., Mattina, M.J., and Pylypiw, H.M. (1992). *LCGC* 10 (6): 442.
28. Dong, M.W., Duggan, J.X., and Stefanou, S. (1993). *LCGC* 11 (11): 802.
29. Mori, S. and Barth, H.G. (1999). *Size Exclusion Chromatography*. Berlin, Germany: Springer-Verlag.

30. Reuter, W.M., Dong, M.W., and McConville, J. (1991). *Amer. Lab.* 23 (5): 45.
31. (2016). ACQUITY Advanced Polymer Chromatography (APC) Applications Notebook, Waters Corporation, 720004649EN, Milford, Massachusetts.
32. Fritz, J.S. and Gjerde, D.T. (2009). *Ion Chromatography*, 4e. Weinheim, Germany: VCH.
33. Pohl, C. (2005). *Handbook of Pharmaceutical Analysis by HPLC* (ed. S. Ahuja and M.W. Dong). Amsterdam, the Netherlands: Elsevier, Chapter 8.
34. Bhattacharyya, L. (April 2011). Ion chromatography in USP-NP. *Amer. Pharm. Rev.* .
35. Harrison, R.G., Todd, P.W., Rudge, S.R., and Petrides, D.P. (2015). *Bioseparations Science and Engineering*, 2e. Oxford, United Kingdom: Oxford University Press.
36. Zhang, T., Quan, C., and Dong, M.W. (2014). *LCGC North Am.* 32 (10): 796.
37. (2004). *Waters Life Sciences Chemistry Solutions Brochure*, 720000831EN, Waters Corporation, Milford, Massachusetts
38. Dong, M.W. (1992). *Advances in Chromatography*, vol. 32 (ed. P. Brown), 21–51. New York: Marcel Dekker.
39. Dong, M.W., Gant, J.R., and Larsen, B. (1989). *Biochromatography* 4 (1): 19.
40. Gudihal, R. and Babu, S. (2014). Agilent Application Note, 5991-4266 EN, Bangalore, India.
41. (2014). AccQ Tag Ultra Derivatization Kit, 715001331EN rev C, Waters Corporation, Milford, Massachusetts.
42. McCarthy, S.A. and Gilar, M. (2015). Waters Application Note, 7200002741EN, Waters Corporation, Milford, Massachusetts.
43. Katz, E. and Dong, M.W. (1990). *BioTechniques* 8 (5): 546.
44. Issaq, H.J. and Veenstra, T.D. (eds.) (2013). *Proteomic and Metabolomic Approaches to Biomarker Discovery*. Amsterdam, the Netherlands: Academic Press/Elsevier.
45. Lovric, J. (2011). *Introducing Proteomics: From Concepts to Sample Separation, Mass Spectrometry and Data Analysis*. Oxford, United Kingdom: Wiley-Blackwell.
46. Matthiesen, R. (ed.) (2013). *Mass Spectrometry Data Analysis in Proteomics (Methods in Molecular Biology)*, 2e. New York, NY: Humana Press.
47. Spicer, V., Ezzati, P., Neustaeter, H. et al. (2016). *Anal. Chem.* 88: 2847.
48. Putri, S.P. and Fukusaki, E. (eds.) (2015). *Mass Spectrometry-Based Metabolomics: A Practical Guide*. Boca Raton, FL: CRC Press.
49. Packer, N.H. and Karlsson, N.G. (eds.) (2009). *Glycomics: Methods and Protocols*. New York, NY: Humana Press.
50. Alley, W.R. and Yu, Y.Q (2016). Waters Application Note, 720005753EN, Milford, MA.
51. Nicoll, D., Lu, C.M., and McPhee, S.J. (2017). *Guide to Diagnostic Tests*, 7e. New York, NY: McGraw-Hill Education.
52. Sannes, L. (2011). Commercializing Biomarkers in Therapeutic and Diagnostic Applications – Overview. Insight Pharma Report 2011.
53. Regnier, F.E. (2012). *LCGC North Amer.* 30 (8): 622.

# Appendix

# KEYS TO QUIZZES

Chapter 1: 1b, 2d, 3d, 4a, 5b.

Chapter 2: 1b, 2b, 3a, 4c, 5d, 6b, 7d, 8b, 9d, 10a.

Chapter 3: 1b, 2c, 3a, 4b, 5a, 6c, 7d, 8c, 9a, 10a.

Chapter 4: 1c, 3b, 3c, 4a, 5a, 6d, 7d, 8a, 9d, 10c, 11b, 12a, 13c, 14a, 15a.

Chapter 5: 1b, 2c, 3d, 4a, 5c, 6a, 7d, 8c, 9d, 10b, 11d, 12c, 13d, 14a, 15c.

Chapter 6: 1b, 2c, 3a, 4d, 5b, 6a, 7d, 8d, 9b, 10c.

Chapter 7: 1c, 2a, 3d, 4b, 5a.

Chapter 8: 1d, 2b, 3c, 4a, 5c, 6b, 7a, 8d, 9d, 10d.

Chapter 9: 1b, 2c, 3d, 4b, 5b, 6c, 7b, 8d, 9b, 10c.

Chapter 10: 1c, 2d, 3a, 4b, 5a, 6a, 7a, 8d, 9c, 10d, 11d, 12c.

Chapter 11: 1c, 2a, 3d, 4c, 5a, 6c, 7b, 8d, 9d, 10c.

Chapter 12: 1d, 2c, 3b, 4a, 5a, 6d, 7c, 8c, 9b, 10c.

Chapter 13: 1b, 2d, 3d, 4d, 5b, 6c, 7d, 8a, 9d, 10c.

*HPLC and UHPLC for Practicing Scientists*, Second Edition. Michael W. Dong.

# INDEX



*HPLC and UHPLC for Practicing Scientists*, Second Edition. Michael W. Dong.